Advances in Intelligent Systems and Computing

Volume 368

About this Series

The series "Advances in Intelligent Systems and Computing" contains publications on theory, applications, and design methods of Intelligent Systems and Intelligent Computing. Virtually all disciplines such as engineering, natural sciences, computer and information science, ICT, economics, business, e-commerce, environment, healthcare, life science are covered. The list of topics spans all the areas of modern intelligent systems and computing.

The publications within "Advances in Intelligent Systems and Computing" are primarily textbooks and proceedings of important conferences, symposia and congresses. They cover significant recent developments in the field, both of a foundational and applicable character. An important characteristic feature of the series is the short publication time and world-wide distribution. This permits a rapid and broad dissemination of research results.

More information about this series at http://www.springer.com/series/11156

Álvaro Herrero · Javier Sedano
Bruno Baruque · Héctor Quintián
Emilio Corchado
Editors

10th International Conference on Soft Computing Models in Industrial and Environmental Applications

Editors
Álvaro Herrero
Department of Civil Engineering
University of Burgos
Burgos
Spain

Javier Sedano
Technological Institute of Castilla y León
Burgos
Spain

Bruno Baruque
Department of Civil Engineering
University of Burgos
Burgos
Spain

Héctor Quintián
University of Salamanca
Salamanca
Spain

Emilio Corchado
University of Salamanca
Salamanca
Spain

ISSN 2194-5357 ISSN 2194-5365 (electronic)
Advances in Intelligent Systems and Computing
ISBN 978-3-319-19718-0 ISBN 978-3-319-19719-7 (eBook)
DOI 10.1007/978-3-319-19719-7

Library of Congress Control Number: 2015940738

Springer Cham Heidelberg New York Dordrecht London

Printed on acid-free paper

Springer International Publishing AG Switzerland is part of Springer Science+Business Media
(www.springer.com)

Preface

This volume of Advances in Intelligent and Soft Computing contains accepted papers presented at the *10th International Conference on Soft Computing Models in Industrial and Environmental Applications* (SOCO 2015), held in the beautiful and historic city of Burgos (Spain), in June 2015.

Soft computing represents a collection or set of computational techniques in machine learning, computer science, and some engineering disciplines, which investigate, simulate, and analyze very complex issues and phenomena. This conference is mainly focused on its industrial and environmental applications.

After a through peer-review process, the SOCO 2015 International Program Committee selected 41 papers, written by authors from 15 different countries. These papers are published in present conference proceedings, achieving an acceptance rate of 40 %.

The selection of papers was extremely rigorous in order to maintain the high quality of the conference and we would like to thank the members of the International Program Committees for their hard work during the review process. This is a crucial issue for creation of a high standard conference, and the SOCO conference would not exist without their help.

SOCO'15 enjoyed outstanding keynote speeches by distinguished guest speakers: Prof. Senén Barro—University of Santiago de Compostela (Spain), and Prof. Hans J. Briegel—University of Innsbruck (Austria).

For this SOCO'15 edition, as a follow-up of the conference, we anticipate further publication of selected papers in a special issue of the following prestigious journals: Neurocomputing (published by Elsevier), Soft Computing: A Fusion of Foundations, Methodologies and Applications (published by Springer), Journal of Multiple-Valued Logic and Soft Computing (published by Old City Publishing), and Journal of Applied Logic (published by Elsevier).

Particular thanks go as well to the conference main sponsors, IEEE—Spain Section, IEEE Systems, Man and Cybernetics—Spanish Chapter, and The International Federation for Computational Logic, who jointly contributed in an active and constructive manner to the success of this initiative. We want also to extend our warm gratitude to all the Special Sessions chairs for their continuing support to the SOCO Series of conferences.

We would like to thank all the Special Session organizers, contributing authors, as well as the members of the Program Committees and the Local Organizing Committee for their hard and highly valuable work. Their work has helped to contribute to the success of the SOCO 2015 event. We are proud of successfully organizing the *tenth edition* of this remarkable and appealing conference.

June 2015

Álvaro Herrero
Javier Sedano
Bruno Baruque
Héctor Quintián
Emilio Corchado

SOCO 2015

Organization

General Chair

Emilio Corchado—University of Salamanca (Spain)

Honorary Chairs

Alfonso Murillo—Rector of the University of Burgos (Spain)
José Mª Vela—Director of the Technological Centre ITCL (Spain)

Local Chair

Álvaro Herrero—University of Burgos (Burgos)

International Advisory Committee

Aditya Ghose—University of Wollongong (Australia)
Ajith Abraham—Machine Intelligence Research Labs—MIR Labs (Europe)
Amparo Alonso Betanzos—University of Coruña and President of the Spanish Association for Artificial Intelligence (AEPIA) (Spain)
Amy Neustein—Linguistic Technology Systems (USA)
Ashraf Saad—Armstrong Atlantic State University (USA)
Henri Pierreval—LIMOS UMR CNRS 6158 IFMA (France)
Isidro Laso-Ballesteros—European Commission Scientific Officer (Europe)
Jaydip Sen—Innovation Lab, Tata Consultancy Services Ltd (India)
Jon G. Hall—The Open University (UK)

Michael Gabbay—Kings College London (UK)
Paulo Novais—Universidade do Minho (Portugal)
Rajkumar Roy—The EPSRC Centre for Innovative Manufacturing in Through-life Engineering Services (UK)
Saeid Nahavandi—Deakin University (Australia)

Program Committee

Emilio Corchado—University of Salamanca (Spain) (PC Co-chair)
Álvaro Herrero—University of Burgos (Spain) (PC Co-chair)
Bruno Baruque—University of Burgos (Spain) (PC Co-chair)
Javier Sedano—Technological Institute of Castilla y León (Spain) (PC Co-Chair)
Aboul Ella Hassanien—Cairo University (Egypt)
Abraham Duarte—Universidad Rey Juan Carlos (Spain)
Ajith Abraham—Machine Intelligence Research Labs (USA)
Alberto Freitas—University of Porto (Portugal)
Alfredo Cuzzocrea—ICAR-CNR and University of Calabria (Italy)
Alicia Troncoso—Universidad Pablo de Olavide (Spain)
Amy Neustein—Linguistic Technology Systems (USA)
Ana Gil—University of Salamanca (Spain)
Andre Carvalho—University of São Paulo (Brazil)
Ángel Arroyo—University of Burgos (Spain)
Anna Bartkowiak—University of Wroclaw (Poland)
Anna Burduk—University of Technology (Poland)
Antonio Bahamonde—Universidad de Oviedo at Gijón (Spain)
Antonio Peregrin—University of Huelva (Spain)
Ashraf Saad—Armstrong Atlantic State University (USA)
Aureli Soria-Frisch—Starlab Barcelona, S.L. (Spain)
Benoit Otjacques—Centre de Recherche Public (France)
Bogdan Gabrys—Bournemouth University (UK)
Camelia Chira—Technical University Cluj-Napoca (Romania)
Carlos Laorden—DeustoTech, University of Deusto (Spain)
Carlos Pereira—ISEC (Spain)
Cesar Analide—University of Minho (Portugal)
Cesar Hervas—Full Professor (Spain)
Cui Zhihua—Complex System and Computational Intelligence Laboratory (China)
Damian Krenczyk—Silesian University of Technology (Poland)
Daniela Perdukova—Technical University of Kosice (Czech Republic)
Daniela Zaharie West—University of Timisoara (Romania)
David Griol—Universidad Carlos III de Madrid (Spain)
Dilip Pratihar—Department of Mechanical Engineering (Indian)
Dimitris Mourtzis—University of Patras (Greece)

Dragan Simic—University of Novi Sad (Serbia)
Edward Chlebus—Wroclaw University of Technology (Poland)
Eleni Mangina—University College Dublin (Ireland)
Enrique Onieva—Deusto Institute of Technology (Spain)
Esteban García-Cuesta—iSOCO (Spain)
Eva Volna—University of Ostrava (Czech Republic)
Fanny Klett—German Workforce ADL Partnership Laboratory (Germany)
Florentino Fdez-Riverola—University of Vigo (Spain)
Francesco Marcelloni—University of Pisa (Italy)
Francesco Moscato—Second University of Naples (Italy)
Francisco Herrera—University of Granada (Spain)
Francisco Martínez-Álvarez—Universidad Pablo de Olavide (Spain)
George Georgoulas—TEI of Epiruw (Greece)
Georgios Ch. Sirakoulis—Democritus University of Thrace (Greece)
Gerald Schaefer—Loughborough University (UK)
Giuseppe Cicotti—National Research Council of Italy (Italy)
Héctor Quintián—University of Salamanca (Spain)
Henri Pierreval—LIMOS-IFMA (France)
Horia-Nicolai Teodorescu—"Gheorghe Asachi" Technical University of Iasi (Romania)
Humberto Bustince—UPNA (Italy)
Igor Santos—University of Deusto (Spain)
Jaroslava Žilková—KEM, FEI TU of Košice (Czech Republic)
Javier Nieves—University of Deusto (Spain)
Jose Luis Calvo-Rolle—University of A Coruña (Spain)
Jiri Pospichal—Slovak University of Technology (Slovakia)
Jose Gamez—University of Castilla-La Mancha (Spain)
Jose Alfredo Ferreira Costa UFRN—Universidade Federal do Rio Grande do Norte (Brazil)
José Luis Casteleiro—Roca University of Coruña (Spain)
Jose Manuel Molina—Universidad Carlos III de Madrid (Spain)
José Manuel Benítez—University of Granada (Spain)
José Ramón Villar—University of Oviedo (Spain)
José Riquelme—University of Sevilla (Spain)
José Valente De Oliveira—Universidade do Algarve (Portugal)
Josef Tvrdik—University of Ostrava (Czech Republic)
Jose-Maria Peña—Universidad Politécnica de Madrid (Spain)
Jouni Lampinen—University of Vaasa (Finland)
Juan Gomez Romero—Universidad de Granada (Spain)
Juan Álvaro Muñoz Naranjo—University of Almería (Spain)
Juan Manuel Corchado—University of Salamanca (Spain)
Krzysztof Kalinowski—Silesian University of Technology (Poland)
Lenka Lhotska—Czech Technical University in Prague (Czech Republic)
Leocadio G. Casado—University of Almeria (Spain)
Leticia Curiel—University of Burgos (Spain)

Luís Nunes—Instituto Universitário de Lisboa (ISCTE-IUL) (Portugal)
Luis Paulo Reis—University of Minho (Portugal)
M. Chadli—University of Picardie Jules Verne (Italy)
Mª Belén Vaquerizo—University of Burgos (Spain)
Maciej Grzenda—Warsaw University of Technology (Poland)
Manuel Graña—University of Basque Country (Spain)
Manuel Mejía Lavalle—CENIDET (Spain)
Marcin Paprzycki—IBS PAN and WSM (Poland)
Marco Mora—Catholic University of Maule (China)
María Martínez Ballesteros—University of Seville (Spain)
María N. Moreno García—University of Salamanca (Spain)
Mario Koeppen—Kyushu Institute of Technology (Japan)
Marius Balas Aurel Vlaicu—University of Arad (Romania)
Martin Macas—Czech Technical University in Prague (Czech Republic)
Martin Stepnicka IRAFM—University of Ostrava (Czech Republic)
Mehmet Emin Aydin—University of Bedfordshire (UK)
Michael Vrahatis—University of Patras (Greece)
Michal Wozniak—Wroclaw University of Technology (Poland)
Miroslav Burša—Czech Technical University in Prague (Czech Republic)
Mitiche Lahcene—University of Djelfa (India)
Mohamed Mostafa—Arab Academy for Science, Technology & Maritime Transport (Egypt)
Noelia Sánchez-Maroño—University of A Coruña (Spain)
Oliviu Matei North—University of Baia Mare (Romania)
Oscar Fontenla-Romero—University of A Coruña (Spain)
Paulo Novais—University of Minho (Portugal)
Pedro Caballero-Lozano—CARTIF Technology Centre (Spain)
Pedro Antonio Gutierrez—University of Córdoba (Spain)
Petrica Claudiu—Pop North University of Baia Mare (Romania)
Petro Gopych—Universal Power Systems USA-Ukraine LLC (Ukraine)
Przemyslaw Korytkowski—West Pomeranian University of Technology in Szczecin (Poland)
Ramón Ferreiro García—University of Coruña (Spain)
Raquel Redondo—University of Burgos (Spain)
Richard Duro—Universidad de Coruna (Spain)
Robert Burduk—Wroclaw University of Technology (Poland)
Roman Senkerik—TBU in Zlin (Czech Republic)
Rosa Basagoiti—Mondragon University (Spain)
Rosario Girardi—Federal Universty of Maranhão (Brazil)
Rui Sousa—University of Minho (Portugal)
Sebastián Ventura—Universidad de Cordoba (Spain)
Stefano Pizzuti—Italian National Agency for New Technologies, Energy and Sustainable Economic Development (Italy)
Sung-Bae Cho—Yonsei University (South Korea)
Tzung-Pei Hong National—University of Kaohsiung (Taiwan)

Urko Zurutuza—Mondragon University (Spain)
Vaclav Snasel—University of Ostrava (Czech Republic)
Veronica Tricio—University of Burgos (Spain)
Vivian F. López—University of Salamanca (Spain)
Wei-Chiang Hong—Oriental Institute of Technology (China)
Wilfried Elmenreich—Alpen-Adria Universität Klagenfurt (Germany)
Witold Pedrycz—University of Alberta (Canada)
Yin Hujun—University of Manchester (UK)
Zhihua Cui Taiyuan—University of Science and Technology (China)
Zita Vale GECAD—ISEP/IPP (Portugal)
Zuzana Oplatkova—Tomas Bata University in Zlin (Czech Republic)

Special Sessions

Soft Computing Methods in Bioinformatics

Camelia Chira—Technical University of Cluj-Napoca (Romania) (session chair)
Emilio Corchado—University of Salamanca (Spain) (session chair)
Javier Sedano—Technological Institute of Castilla y León (Spain) (session chair)
Jose Ramon Villar—University of Oviedo (Spain) (session chair)
Anca Andreica—Babes-Bolyai University (Romania)
Camelia Pintea—Technical University of Cluj-Napoca (Romania)
Carmen Vidaurre—TU-Berlin (Germany)
David Peña—ITCL (Spain)
Dominik Olszewski—Warsaw University of Technology (Poland)
Gerardo M. Méndez—Instituto Tecnologico de Nuevo León (Mexico)
Ioana Zelina—North University of Baia Mare (Romania)
Irene Diaz—University of Oviedo (Spain)
Jerónimo Gonzalez—University of Burgos (Spain)
Jose Luis Calvo-Rolle—University of A Coruña (Spain)
José María Trejo—HUBU (Spain)
Madalina Drugan—Vrije Universiteit Brussel (Romania)
Mohammad Reza—Bonyadi Shahid beheshti University (Indian)
Petrica Claudiu Pop—North University of Baia Mare (Hungry)
Ramiro Varela—University of Oviedo (Spain)
Stefan Oniga—Universitatea de Nord din Baia Mare (Romania)
Viorica Rozina—Technical University of Cluj-Napoca (Romania)

Optimization, Modelling and Control Systems (OMCS)

José Luis Calvo—Universitiy of Coruña (Spain) (session chair)
Eloy Irigoyen—University of Deusto (Spain) (session chair)
Emilio Corchado—University of Salamanca (Spain) (session chair)
Matilde Santos—Complutense University of Madrid (Spain) (session chair)
Aldo Cipriano—Catholic University of Chile (Chile)
Alejo Nevado—University of Bristol (UK)
Alex Gammerman—Royal Holloway University of London (UK)
Amaury Lendasse—Seamans Center for the Engineering Arts and Sciences (Belgium)
Andrés José Piñón Pazos—University of A Coruña (Spain)
Benigno Antonio Rodríguez Gómez—University of A Coruña (Spain)
Bruno Apolloni—University of Milan (Italy)
Carlos Lamela—Universidad Centroccidental Lisandro Alvarado
Cecilia Zanni-Merk—INSA de Strasbourg (France)
Emilio Del Moral—Sao Paulo University (Brazil)
Guang Bin Huang—University of Wollongong (Australia)
Héctor Alaiz Moretón—University of Leon (Spain)
Hong Yue Zhang—School of Automation Science and Electrical Engineering (China)
José Luis Casteleiro-Roca—University of Coruña (Spain)
Joshué Pérez—Centre for Automation and Robotics (Spain)
Juan Albino Méndez Pérez—University of La Laguna (Spain)
Kazami Nakamatsu—School of Human Science and Environment (Japan)
Luis Alfonso Fernández-Serantes—FH Joanneum—University of Applied Sciences (Austria)
FH Joanneum—University of Applied Sciences (Austria)
Marco Mora—Catholic University of Maule (Chile)
María Tomás-Rodríguez—University of Shefiield (UK)
María Del Carmen Meizoso López—University of A Coruña (Spain)
Oscar Fontenla-Romero—University of A Coruña (Spain)
Roberto Muñoz-Soto—University of Valparaiso (Chile)
Rossi Setchi—Cardiff University (UK)

Soft Computing Approaches for Knowledge Extraction

Cristóbal J. Carmona—University of Burgos (Spain) (session chair)
Julián Luengo—University of Burgos (Spain) (session chair)
Alberto Fernandez—University of Jaen (Spain)
Antonio Jesús Rivera Rivas—University of Jaen (Spain)
Francisco Charte Ojeda—Universidad de Jaén (Spain)
María Dolores Pérez Godoy—University of Jaen (Spain)

Maria Jose Del Jesus—Universidad de Jaén (Spain)
Pedro Gonzalez—Universidad de Jaén (Spain)
Salvador García López—University of Granada (Spain)

Soft Computing Methods in Manufacturing and Management Systems

Anna Burduk—Wroclaw University of Technology (Poland) (session chair)
Edward Chlebus—Wroclaw University of Technology (Poland) (session chair)
Bożena Skołud—Silesian University of Technology (Poland) (session chair)
Krzysztof Kalinowski—Silesian University of Technology (Poland) (session chair)
Piotr Michalski—Silesian University of Technology (Poland) (session chair)
Persona Alessandro—University of Padova (Italy) (session chair)
Franjo Jović—Josip Juraj Strossmayer University of Osijek (Croatia) (session chair)
Przemyslaw Korytkowski—West Pomeranian University of Technology in Szczecin (Poland) (session chair)
Joze Balic—University of Maribor (Slovenia) (session chair)
Damian Krenczyk—Silesian University of Technology (Poland) (session chair)
Arkadiusz Kowalski Wrocław—University of Technology (Poland)
Jaroslaw Chrobot Wrocław—University of Technology (Poland)
Marcin Zemczak—University of Bielsko-Biala (Poland)
Mieczyslaw Jagodzinski—Silesian University of Technology (Poland)
Paul-Eric Dossou—ICAM (France)
Sebastian Saniuk—University of Zielona Gora (Poland)

Organising Committee

Álvaro Herrero—University of Burgos (Chair)
Bruno Baruque—University of Burgos (Co-chair)
Javier Sedano—Technological Institute of Castilla y León (Co-chair)
Emilio Corchado—University of Salamanca (Spain)
Ángel Arroyo—University of Burgos (Spain)
Raquel Redondo—University of Burgos (Spain)
Leticia Curiel—University of Burgos (Spain)
Belkis Díaz—University of Burgos (Spain)
Belén Vaquerizo—University of Burgos (Spain)
Pedro Burgos—University of Burgos (Spain)
Juán Carlos Pérez—University of Burgos (Spain)
Héctor Quintian—University of Salamanca (Spain)
José Luis Calvo—University of La Coruña (Spain)
José Luis casteleiro—University of La Coruña (Spain)
Amelia García—Technological Institute of Castilla y León (Spain)
Mónica Cámara—Technological Institute of Castilla y León (Spain)
Silvia González—Technological Institute of Castilla y León (Spain)

Contents

Applications

User Modeling Optimization for the Conversational Human-Machine Interfaces . . . 3
David Griol and José Manuel Molina

A View Based Approach for Enhancing Web Services Availability . . . 15
Hela Limam and Jalel Akaichi

Using Magentix2 in Smart-Home Environments . . . 27
S. Valero, E. del Val, J. Alemany and V. Botti

Analyzing Accelerometer Data for Epilepsy Episode Recognition . . . 39
José R. Villar, Manuel Menéndez, Javier Sedano, Enrique de la Cal and Víctor M. González

A Preliminary Cooperative Genetic Fuzzy Proposal for Epilepsy Identification Using Wearable Devices . . . 49
E.A. de la Cal, J.R. Villar, P.M. Vergara, J. Sedano and A. Herrero

Analysis of Knowledge Management in Industrial Sectors by Means of Neural Models . . . 65
Álvaro Herrero, Emilio Corchado, Lourdes Sáiz-Bárcena and Miguel A. Manzanedo

Classification and Clustering Methods

Towards a Soft Evaluation and Refinement of Tagging in Digital Humanities . . . 79
Gonzalo A. Aranda-Corral, Joaquín Borrego Díaz and Juan Galán Páez

Neural Networks and Linear Predictive Coding Coefficients Used for European Starling Detection in Vineyards 91
Petr Dolezel and Martin Mariska

Stroke-Based Intelligent Word Recognition Using a Formal Language 101
David Álvarez, Ramón Fernández and Lidia Sánchez

Link Similarity Reveals Multiscale Link Communities 111
Guishen Wang and Lan Huang

A Comparison of Clustering Techniques for Meteorological Analysis 117
Ángel Arroyo, Verónica Tricio, Emilio Corchado and Álvaro Herrero

Evolutionary Computation and Optimization

Evolutionary Battery Scheduling Optimization Under Variable Electricity Prices in Micro-Grids with Renewable Generation 133
R. Mallol-Poyato, S. Salcedo-Sanz, S. Jiménez-Fernández and P. Díaz-Villar

Reliability-Based Design Optimization of Structures Combining Genetic Algorithms and Finite Element Reliability Analysis 143
Luis Celorrio

Improved Energy-Aware Stochastic Scheduling Based on Evolutionary Algorithms via Copula-Based Modeling of Task Dependences 153
Zorana Banković and Pedro López-García

Relational Crossover in Evolutionary Ontologies 165
Oliviu Matei, Diana Contraş and Honoriu Vălean

Intelligent Systems

A Movement Control System Based on Qualitative Reasoning 179
Przemysław Wałęga and Emilio Muñoz-Velasco

A Multi-agent Framework for the Analysis of Users Behavior over Time in On-Line Social Networks 191
E. del Val, C. Martínez and V. Botti

An Emotional-Based Hybrid Application for Human-Agent Societies 203
J.A. Rincon, V. Julian and C. Carrascosa

Modeling and Analysis of Agent-Based Labor Market 215
Jae-Min Yu and Sung-Bae Cho

SOCO15-SS01 - Soft Computing Methods in Bioinformatics

Medical Edge Detection Combining Fuzzy Mathematical Morphology with Interval-Valued Relations 229
Agustina Bouchet, Pelayo Quirós, Pedro Alonso, Irene Díaz and Susana Montes

Shape-Output Gene Clustering for Time Series Microarrays 241
Camelia Chira, Javier Sedano, José R. Villar, Monica Camara and Carlos Prieto

Heuristics for Apnea Episodes Recognition 251
Silvia González, José Ramón Villar, Javier Sedano, Joaquín Terán, María Luz Alonso Álvarez and Jerónimo González

Energy-Efficient Sound Environment Classifier for Hearing Aids Based on Multi-objective Simulated Annealing Programming . . . 261
Alberto Cocaña-Fernández, Luciano Sánchez, José Ranilla, Roberto Gil-Pita and David Ayllón

SOCO15-SS02 - Optimization, Modeling and Control Systems (OMCS)

Modeling the Electromyogram (EMG) of Patients Undergoing Anesthesia During Surgery 273
José Luis Casteleiro-Roca, Juan Albino Méndez Pérez, Andrés José Piñón-Pazos, José Luis Calvo-Rolle and Emilio Corchado

Real Time Parallel Robot Direct Kinematic Problem Computation Using Neural Networks 285
Asier Zubizarreta, Mikel Larrea, Eloy Irigoyen and Itziar Cabanes

Reinforcement Learning in Single Robot Hose Transport Task: A Physical Proof of Concept 297
Jose Manuel Lopez-Guede, Julián Estévez and Manuel Graña

Bridging Classical Control with Nature Inspired Computation Through PID Robust Design 307
P.B. de Moura Oliveira, Hélio Freire, E.J. Solteiro Pires and J. Boaventura Cunha

A First Intelligent Approach to Path Following Algorithm for Quadcopters 317
Pablo García Auñón, Matilde Santos Peñas and Jesus Manuel de la Cruz García

Adaptive Neural Control-Oriented Models of Unmanned Aerial Vehicles 329
J. Enrique Sierra and Matilde Santos

SOCO15-SS03 - Soft Computing Approaches for Knowledge Extraction

Data Mining for Predicting Traffic Congestion and Its Application to Spanish Data 341
E. Florido, O. Castaño, A. Troncoso and F. Martínez-Álvarez

Summarizing Information by Means of Causal Sentences Through Causal Questions 353
C. Puente, A. Sobrino, E. Garrido and J.A. Olivas

Discovering the Dialog Rules by Means of a Soft Computing Approach 365
David Griol and José Manuel Molina

Plant Identification: Two Dimensional-Based Vs. One Dimensional-Based Feature Extraction Methods 375
Tarek Gaber, Alaa Tharwat, Vaclav Snasel and Aboul Ella Hassanien

A First Approach in the Class Noise Filtering Approaches for Fuzzy Subgroup Discovery 387
C.J. Carmona and J. Luengo

SOCO15-SS04 - Soft Computing Methods in Manufacturing and Management Systems

Computation of Mechanical Properties of Parts Manufactured by Fused Deposition Modeling Using Finite Element Method 403
Filip Górski, Wiesław Kuczko, Radosław Wichniarek and Adam Hamrol

New Method for Assessment of Raters Agreement Based on Fuzzy Similarity 415
Magdalena Diering, Krzysztof Dyczkowski and Adam Hamrol

The Role of Artificial Neural Network Models in Ensuring the Stability of Systems 427
Anna Burduk

Coordination in the Supply Chain 439
Pawel Pawlewski

ERP, APS and Simulation Systems Integration to Support Production Planning and Scheduling 451
Damian Krenczyk and Mieczyslaw Jagodzinski

Multi-agent System to Support Decision-Making Process in Ecodesign 463
Ewa Dostatni, Jacek Diakun, Damian Grajewski, Radosław Wichniarek and Anna Karwasz

Preparatory Stages of the Production Scheduling of Complex and Multivariant Products Structures 475
Krzysztof Kalinowski and Marcin Zemczak

Author Index 485

Applications

User Modeling Optimization for the Conversational Human-Machine Interfaces

David Griol and José Manuel Molina

Abstract In this paper, we test the applicability of a statistical user modeling technique to develop a simulated user agent that generates dialogs which are similar to real human-machine spoken interactions. The proposed user simulation technique decides the next response of the agent taking into account the previous user turns, the last system answer and the objective of the dialog. In our contribution, we present the results of the comparison between a corpus acquired from real interactions of users with a conversational agent and a corpus acquired by means of the proposed user simulation technique. To do so, we describe the practical application of our proposal for a conversational agent providing tourist information in natural language, employing a comprehensive set of measures for its evaluation.

Keywords Agent simulation · User modeling · Conversational agents · Human machine interfaces · Spoken interaction

1 Introduction

Multi-agent systems (MAs) are designed as a collection of interacting autonomous agents, each having their own capacities and goals that are situated to a common environment. This way, the development of a MAs offers the capability of simulating autonomous agents and the interaction among them. Agent-based simulation (ABS) is a relatively recent modeling technique widely used to model complex systems with applications in many industrial and environmental disciplines. Detailed studies can be found in [1, 2].

D. Griol (✉) · J.M. Molina
Computer Science Department, Carlos III University of Madrid,
Avda. de la Universidad 30, 28911 Leganés, Spain
e-mail: david.griol@uc3m.es

J.M. Molina
e-mail: josemanuel.molina@uc3m.es

Á. Herrero et al. (eds.), *10th International Conference on Soft Computing Models in Industrial and Environmental Applications*, Advances in Intelligent Systems and Computing 368, DOI 10.1007/978-3-319-19719-7_1

Considering the growing interest of adapting agent-based approaches to modeling and simulation, it is not surprising the number of software frameworks specifically aimed at supporting the realization of agent-based simulation systems. A first category of these platforms provides general purpose frameworks in which agents mainly represent passive abstractions interacting in an overall simulation process (e.g., NetLogo [3]). A second category of platforms are based on general purpose programming languages providing very similar support tools (e.g., Repast [4]). A third category of platforms represents an attempt to provide a higher level linguistic support, trying to reduce the distance between agent-based models and their implementations (e.g., SimSesam [5]).

Research in techniques for user modeling has a long history within the fields of language processing and conversational interfaces. The design practices of conventional commercial dialog systems are currently well established in industry. In these practices, voice user interface (VUI) experts [6] handcraft a detailed dialog plan based on their knowledge about the specific task and the business rules (e.g., to verify the user's identity before providing certain information). In addition, designers commonly define the precise wording for the system prompts according to the dialog state and context, and also the expected types of user's utterances for each turn. As described in [7, 8], this approach is well-documented [9] and has been used to develop hundreds of successful commercial dialog systems.

As described in [10], there are three main categories of elements of the spoken dialog interaction where the use of massive amounts of data generated by industrial systems can potentially improve automation rate, and ultimately, the penetration and acceptance of speech interfaces in the wider consumer market. They are task-independent behaviors (e.g., error correction and confirmation behavior), task-specific behaviors (e.g., logic associated with certain customer-care practices), and task-interface behaviors (e.g., prompt selection). However, these three categories have in common today the lack of robust guiding principles which are validated by empirical evidence. Thus, rule-based approaches are not suitable in order to exploit the full potential of the available data. Statistical methodologies can be used to tackle this problem while still allowing designers to have enough control to ensure VUI completeness.

The main purpose of a simulated user related to these task-independent behaviors is to improve the usability of a conversational interface through the generation of corpus of interactions between the system and simulated users [11], reducing time and effort required for collecting large samples of interactions with real users. Moreover, each time changes are made to the system it is necessary to collect more data in order to evaluate the changes. Thus the availability of large corpus of simulated data should contribute positively to the development of the conversational agent.

The construction of user models based on statistical methods has provided interesting and well-founded results in recent years and is currently a growing research area. A thorough literature review on the application of how data mining techniques to user modeling for system personalization can be found in [12]. It is possible to classify the different approaches with regard to the level of abstraction at which they model dialog. This can be at either the acoustic level, the word level or the intention-level. The latter is a particularly useful representation of human-computer interaction [12]. Modeling interaction on the intention-level avoids the need to reproduce the enormous variety of human language on the level of speech signals or word sequences.

In this paper, we propose a statistical approach to acquire a labeled dialog corpus from the interaction of a user agent simulator and a conversational agent. In our methodology, the new user response is selected using the probability distribution provided by a neural network, which models the user's intention-level after each one of the system responses. By means of the interaction of the conversational agent and the user simulator, an initial dialog corpus can be extended by increasing its variability and detecting dialog situations in which a commercial conversational agent does not provide an appropriate answer.

Different studies have been carried out to compare corpus acquired by means of different techniques and to define the most suitable measures to carry out this evaluation [13, 14]. In this work, we have applied our proposal for user simulation to acquire a corpus for the *Enjoy Your City* information system, which provides user-adapted tourist information. This corpus has been compared with a corpus recorded from real users interactions with the same system

We use a set of measures to carry out the evaluation of the acquired corpus based on prior work in the dialog literature. The results of this comparison show that the simulated corpus obtained is very similar to the corpus recorded from real user interactions in terms of number of turns, confirmations and dialog acts distribution among other evaluation measures.

2 Our Proposal to Develop a User Agent Simulator

The user simulator that we present in this paper replaces the user agent by simulating the user intention level, that is, the simulator provides concepts and attributes that represent the intention of the user utterance. Therefore, the user simulator carries out the functions of the Automatic Speech Recognizer (ASR) and the Natural Language Understanding (NLU) modules of a spoken conversational agent. The user responses are generated taking into account the information provided by the simulator throughout the history of the dialog, the last system turn, and the objective(s) predefined for the dialog.

In order to control the interaction, our user simulator uses the representation the dialogs as a sequence of pairs (A_i, U_i), where A_i is the output of the conversational agent (the system response) at time i, expressed in terms of dialog acts; and U_i is the semantic representation of the user turn (the result of the understanding process of the user input) at time i. This way, each dialog is represented by $(A_1, U_1), \dots, (A_i, U_i), \dots, (A_n, U_n)$, where A_1 is the greeting turn of the system (the first turn of the dialog), and U_n is the last user turn. We refer to a pair (A_i, U_i) as S_i, the state of the dialog sequence at time i.

The lexical, syntactic and semantic information associated to the speaker u's ith turn (U_i) is usually represented by:

- the words uttered;
- part of speech tags, also called word classes or lexical categories. Common linguistic categories include noun, adjective, and verb, among others;
- predicate-argument structures, used by SLU modules in various contexts to represent relations within a sentence structure. They are usually represented as triples (subject-verb-object).
- named entities: sequences of words that refer to a unique identifier. This identifier may be a proper name (e.g., organization, person or location names), a time identifier (e.g., dates, time expressions or durations), or quantities and numerical expressions (e.g., monetary values, percentages or phone numbers).

In this framework, we consider that, at time i, the objective of the user simulator is to find an appropriate user answer U_i. This selection is a local process for each time i and takes into account the sequence of dialog states that precede time i, the system answer at time i, and the objective of the dialog $\mathcal{O}$. If the most probable user answer U_i is selected at each time i, the selection is made using the maximization:

$$\hat{U}_i = \operatorname*{argmax}_{U_i \in \mathcal{U}} P(U_i | S_1, \dots, S_{i-1}, A_i, \mathcal{O})$$

where set $\mathcal{U}$ contains all the possible user answers.

As the number of possible sequences of states is very large, we establish a partition in this space (i.e., in the history of the dialog preceding time i). This data structure, that we call *User Register* (*UR*), contains the information provided by the user throughout the previous history of the dialog. After applying the above considerations and establishing the equivalence relations in the histories of the dialogs, the selection of the best U_i is given by:

$$\hat{U}_i = \operatorname*{argmax}_{U_i \in \mathcal{U}} P(U_i | UR_{i-1}, A_i, \mathcal{O})$$

As in our work on dialog management [15], we propose the use of a multilayer perceptron (MLP) to make the assignation of a user turn. The values of the output layer can be viewed as the a posteriori probability of selecting the different user answers defined for the simulator given the current situation of the dialog. The choice of the most probable user answer of this probability distribution leads to the previous equation. In this case, the user simulator will always generate the same answer for the same situation of the dialog. Since we want to provide the user simulator with a richer variability of behaviors, we base our choice on the probability distribution supplied by the MLP on all the feasible user answers.

One of the main problems which must be considered during the interaction with a conversational agent is the propagation of errors through the different modules in the system. In our proposal, the user simulator provides the conversational agent with the frame representation associated to the user input together with its confidence scores. To do this, an error simulation agent has been implemented to include semantic errors in the generation of dialogs.

For the study presented in this paper, we have improved this agent using a model for introducing errors based on the method presented in [16]. The generation of confidence scores is carried out separately from the model employed for error generation. This model is represented as a communication channel by means of a generative probabilistic model $P(c, a_u|\tilde{a}_u)$, where a_u is the true incoming user dialog act $\tilde{a}_u$ is the recognized hypothesis, and c is the confidence score associated with this hypothesis.

The probability $P(\tilde{a}_u|a_u)$ is obtained by Maximum-Likelihood using the initial labeled corpus acquired with real users and considers the recognized sequence of words w_u and the actual sequence uttered by the user $\tilde{w}_u$. This probability is decomposed into a component that generates a word-level utterance from a given user dialog act, a model that simulates ASR confusions (learned from the reference transcriptions and the ASR outputs), and a component that models the semantic decoding process.

$$P(\tilde{a}_u|a_u) = \sum_{\tilde{w}_u} P(a_u|\tilde{w}_u) \sum_{w_u} P(\tilde{w}_u|w_u)P(w_u|a_u)$$

Confidence score generation is carried out by approximating $P(c|\tilde{a}_u, a_u)$ assuming that there are two distributions for c. These two distributions are handcrafted, generating confidence scores for correct and incorrect hypotheses by sampling from the distributions found in the training data corresponding to our initial corpus.

$$P(c|a_w, \tilde{a}_u) = \begin{cases} P_{corr}(c) & if \ \ \tilde{a}_u = a_u \\ P_{incorr}(c) & if \ \ \tilde{a}_u \neq a_u \end{cases}$$

A maximum number of user turns per dialog is defined for acquiring a corpus using our user simulator. A user request for closing the dialog is selected once the system has provided the information defined in the objective(s) of the dialog. The simulated dialogs are automatically considered unsuccessful when the following conditions take place: (i) The dialog exceeds a maximum number of user turns; (ii) The response selected by the conversational agent corresponds with a query not required by the user simulator; (iii) The conversational agent provides an error warning because the user simulator has not provided the mandatory information needed to carry out a specific query; (iv) The conversational agent provides a error warning when the selected response involves the use of a data not provided by the user simulator.

3 Case Application: The *Enjoy Your City* System

We have applied our proposal to develop and evaluate the adaptive system *Enjoy Your City*, which provides tourist information in natural language in Spanish. The information provided by the system includes places of interest, weather forecast, hotel booking, restaurants and bars, shopping, street guide and "how to get there" functionalities, cultural activities (cinema, theater, music, exhibitions, literature and science), sport activities, festivities, and public transportation. The system has been developed using an hybrid dialog management approach that combines the VoiceXML standard with user modeling and statistical estimation of optimized dialog strategies [17]. The information offered to the user is extracted from different web pages and several databases are also used to store this information and automatically update the data that is included in the application.

We have defined ten concepts to represent the different queries that the user can perform (*Places-Interest*, *Weather-Forecast*, *Hotel-Booking*, *Restaurants-Bars*, *Shopping*, *Street-how-toget*, *Cultural*, *Sport*, *Festivities*, and *Public-Transport*). Three task-independent concepts have also been defined for the task (*Acceptance*, *Negation*, and *Not-Understood*). A total of 115 system actions (dialog acts) were defined taking into account the information that is required by the system to provide the requested information.

The *UR* defined for the task is a sequence of 129 fields:

- The 10 concepts defined for the dialog act representation;
- The total of 115 possible attributes for the concepts;
- The 3 task-independent concepts that users can provide (*Acceptance*, *Negation*, and *Not-Understood*).

A set of 100 dialogs was acquired with the *Enjoy Your City* system by means of its interaction with 20 recruited users. These users followed a set of scenarios that specifies a set of objectives that must be fulfilled by the user at the end of the dialog and are designed to include the complete set of functionalities previously described for the system. The acquisition process resulted in a spontaneous Spanish speech dialog corpus with 60 different speakers. The total number of user turns was 422. The recorded material has a duration of 150 min. Figure 1 shows an example of a dialog extracted from this corpus. Turns with *S* refer to system turns, and turns with *U* refer to user turns.

S1: Welcome to Enjoy Your City Granada. How can I help you?
U1: I want to know about cultural activities for today.
S2: There are more than 10 activities. To hear the complete list say list. To hear about a specific kind of activity say cinema, theater, music, exhibition, literature or science.
U2: Science.
S3: There is one activity: conference about transgenic food in Palacio de la Madraza at 20:00. Do you want to know about any other activities?
U3: Yes, literature.
S4: There are two activities: conference on Scandinavian black novels in Biblioteca de Andalucía at 19:00 and conference on the Gold Century in Faculty of Philosophy at 19:00. Do you want to know about any other activities?
U4: No, thank you.
S5: Have a nice day!

Fig. 1 An example of a dialog acquired with the interaction of the *Enjoy Your City* system with recruited users

The semantics selected for the conversational agent is represented through the 115 possible system responses defined for the task. The selection of the possible user responses for the user simulator was carried out using the semantics defined for the user in the NLU module. A dialog corpus of 3,000 successful dialogs was acquired using the user agent simulator following the same set of scenarios defined for the acquisition with real users.

4 Evaluation of the Proposed Methodology

We have defined a set of measures to carry out the evaluation of the acquired corpus based on prior work in the dialog literature. A comprehensive set of quantitative evaluation measures to compare two dialog corpus is proposed in [13, 14]. These measures were adapted for use in our comparisons, based on the information available in our corpus. This set of proposed measures are divided into three types:

- High-level dialog features: These features evaluate how long the dialogs last, how much information is transmitted in individual turns, and how active the dialog participants are.
- Dialog style/cooperativeness measures: These measures try to analyze the frequency of different speech acts and study what proportion of actions is goal-directed, what part is taken up by dialog formalities, etc.
- Task success/efficiency measures: These measures study the goal achievement rates and goal completion times.

We defined six high-level dialog features for the evaluation of the dialogs: the average number of turns per dialog, the percentage of different dialogs without considering the attribute values, the number of repetitions of the most seen dialog, the number of turns of the most seen dialog, the number of turns of the shortest dialog, and the number of turns of the longest dialog. Using these measures, we tried to evaluate the success of the simulated dialogs as well as its efficiency and variability with regard to the different objectives.

For dialog style features, we define and count a set of system/user dialog acts. On the system side, we have measured the confirmation of concepts and attributes, questions to require information, and system answers generated after a database query. We have not take into account the opening and closing system turns. On the user side, we have measured the percentage of turns in which the user carries out a request to the system, provide information, confirms a concept or attribute, Yes/No answers, and other answers not included in the previous categories.

To compare the two corpus, we compute the mean value for each corpus with respect to each of the evaluation measures shown in the previous section. We then use two-tailed t-tests to compare the means across the two corpus as described in [14]. All differences reported as statistically significant have p-values less than 0.05 after Bonferroni corrections.

4.1 High-Level Dialog Features

As stated in the previous section, the first group of experiments covers the following statistical properties: (i) Dialog length, measured in the average number of turns per dialog, number of turns of the shortest dialog, number of turns of the longest dialog, and number of turns of the most seen dialog; (ii) Different dialogs in each corpus, measured in the percentage of different dialogs and the number of repetitions of the most seen dialog; (iii) Turn length, measured in the number of actions per turn; (iv) Participant activity as a ratio of system and user actions per dialog.

Table 1 shows the results of the comparison of the high-level dialog features. It can be seen that all measures have similar values in both corpus. The more significant difference is the average number of user turns. In the four types of scenarios, the dialogs acquired using the simulation technique are shorter than the dialogs acquired with real users. This can be explained because of the fact that there is a set of the dialogs acquired with real users in which the user asked for additional information not included in the definition of the corresponding scenario once the dialog objectives had been achieved.

Table 1 Results of the high-level dialog features defined for the comparison of the two corpus

	Initial corpus	Simulated corpus
Average number of user turns per dialog	7.98	6.75
Percentage of different dialogs	84.61 %	78.43 %
Number of repetitions of the most seen dialog	7	9
Number of turns of the most seen dialog	2	2
Number of turns of the shortest dialog	2	2
Number of turns of the longest dialog	16	12

4.2 Dialog Style and Cooperativeness

Tables 2 and 3 respectively show the frequency of the most dominant user and system dialog acts. Table 2 shows the results of this comparison for the system dialog acts. It can be observed that there are also only slightly differences between the values obtained for both corpus. There is a higher percentage of confirmations and questions in the corpus acquired with real users due the higher average number of turns per dialog in this corpus.

Table 3 shows the results of this comparison for the user dialog acts. The most significant difference between both corpus is the percentage of turns in which the user makes a request to the system. The percentage of these kind answers is lower in the corpus acquired with real users. This can be explained by the fact that it is less probable that simulated users provide useless information, as it is shown in the lower percentage of the users turns classified as Other answers.

Table 2 Percentages of the different types of system dialog acts in both corpus

	Initial corpus	Simulated corpus
Confirmation of concepts and attributes	13.51 %	12.23 %
Questions to require information	18.44 %	16.57 %
Answers generated after a database query	68.05 %	71.20 %

Table 3 Percentages of the different types of system dialog acts in both corpus

	Initial corpus	Simulated corpus
Request to the system	31.74 %	35.43 %
Provide information	21.72 %	20.98 %
Confirmation	10.81 %	9.34 %
Yes/No answers	33.47 %	32.77 %
Other answers	2.26 %	1.48 %

5 Conclusions

In this paper, we have described an approach for the modeling of users interacting with conversational interfaces. The proposed statistical methodology allows evaluating this kind of systems by means of the automatic generation of dialogs, which allow detecting errors in the definition of the dialog strategy of the system and improve their operation. We consider that our proposal is specially important for current commercial spoken conversational systems and dialog systems currently developed following the industrial standards. The application of our proposal to these systems allow not only detecting communication errors and possible misunderstood system responses, but also generating systems adapted to the users' specific requirements and preferences.

We have presented an application of our proposal for a practical system based on a comparison between two corpus acquired using two different techniques. First, we acquired an initial dialog corpus by means of the interaction of our dialog system with real users. Second, we have developed a statistical user simulation technique based on a classification process that takes into account the complete dialog history.

A set of measures have been used to compare both corpus. Our results show that it is feasible to obtain a realistic corpus by means of the simulation technique. This corpus has similar characteristics to a corpus acquired form interaction of real users with a spoken dialog system. Thus, the simulated dialogs can be employed for evaluation purposes, helping to detect the aspects in which the dialog strategy can be improved. As future work, we plan to study the simulated corpus in more detail to extract information which can help us to optimize the dialog strategy. We want also to validate our proposal by means of the comparison with similar related works applied to the same practical domain.

Acknowledgments This work was supported in part by Projects MINECO TEC2012-37832-C02-01, CICYT TEC2011-28626-C02-02, CAM CONTEXTS (S2009/TIC-1485).

References

1. Macal C, North M (2010) Tutorial on agent-based modelling and simulation. J Simul 4:151–162
2. Heath B, Hill R, Ciarallo F (2009) A survey of agent-based modeling practices (January 1998 to July 2008). JASSS 12(4)
3. Wilensky U, Rand W (2015) An introduction to agent-based modeling: modeling natural, social and engineered complex systems with NetLogo. MIT Press
4. North M, Collier N, Vos J (2006) Experiences creating three implementations of the repast agent modeling toolkit. ACM Trans Model Comput Simul 16:1–25
5. Klgl F, Herrler R, Oechslein C (2003) From simulated to real environments: how to use SeSam for software development. LNCS 2831:13–24
6. Barnard E, Halberstadt A, Kotelly C, Phillips M (1999) A consistent approach to designing spoken-dialog systems. In: Proceedings of the ASRU, pp 1173–1176
7. Pieraccini R, Huerta J (2005) Where do we go from here? research and commercial spoken dialog systems. In: Proceedings of the SIGdial, pp 1–10
8. Williams J (2009) The best of both worlds: unifying conventional dialog systems and pomdps. In: Proceedings of the InterSpeech, pp 1173–1176
9. Cohen M, Giangola J, Balough J (2004) Voice user interface design. Addison Wesley
10. Paek T, Pieraccini R (2008) Automating spoken dialogue management design using machine learning: an industry perspective. Speech Commun 50(8–9):716–729
11. Möller S, Englert R, Engelbrecht K, Hafner V, Jameson A, Oulasvirta A, Raake A, Reithinger N (2006) MeMo: towards automatic usability evaluation of spoken dialogue services by user error simulations. In: Proceedings of the Interspeech, pp 1786–1789
12. Schatzmann J, Weilhammer K, Stuttle M, Young S (2006) A survey of statistical user simulation techniques for reinforcement-learning of dialogue management strategies. Knowl Eng Rev 21(2):97–126
13. Schatzmann J, Georgila K, Young S (2005) Quantitative evaluation of user simulation techniques for spoken dialogue systems. In: Proceedings of the SIGdial, pp 45–54
14. Ai H, Raux A, Bohus D, Eskenazi M, Litman D (2007) Comparing spoken dialog corpora collected with recruited subjects versus real users. In: Proceedings of the SIGdial, pp 124–131

15. Griol D, Callejas Z, López-Cózar R, Riccardi G (2014) A domain-independent statistical methodology for dialog management in spoken dialog systems. Comput Speech Lang 28(3)
16. Schatzmann J, Thomson B, Weilhammer K, Ye H, Young S (2007) Agenda-based user simulation for bootstrapping a pomdp dialogue system. In: Proceedings of the HLT/NAACL, pp 149–152
17. Griol D, Molina J, Callejas Z (2012) Bringing together commercial and academic perspectives for the development of intelligent Am I interfaces. JAISE 4(3):183–207

A View Based Approach for Enhancing Web Services Availability

Hela Limam and Jalel Akaichi

Abstract With the advance of Web services technologies and the emergence of Web services into the information space, tremendous opportunities for empowering users and organizations appear in various application domains including electronic commerce, travel, intelligence information gathering and analysis, health care, digital government. Hence, Web services appear to be a solution for integrating distributed, autonomous and heterogeneous information sources. However, as Web services evolve in a dynamic environment which is the Internet many changes can occur, affect them and make them become unavailable. This presents us with the problem of substituting them, while maintaining the whole Web service functionality. In this paper, we propose an approach for solving this problem. We present several algorithms for solving several variants of this problem. Our work is illustrated with a healthcare case study.

Keywords Web services · Substitution · Schema changes · Healthcare

1 Introduction

Synchronization within highly dynamic environments such as Internet is far away from being trivial. In fact it is the trickiest environment one could image due to its unpredictable behavior. Unfortunately, there is no satisfactory solution which guaran-tees Web service availability in such a situation. It makes sense to work on a solution for the usual scenario, which, anyway, has to deal with the unavailable Web services situation due to unemployment of individual Web services caused by changes which can alter their contents. Motivation for substitution includes Web

H. Limam (✉) · J. Akaichi
BestMod Laboratory, Institut Supérieur de Gestion, Tunis, Tunisia
e-mail: Hela1.limam@laposte.net

J. Akaichi
e-mail: Jalel.akaichi@isg.rnu.tn

Á. Herrero et al. (eds.), *10th International Conference on Soft Computing Models in Industrial and Environmental Applications*, Advances in Intelligent Systems and Computing 368, DOI 10.1007/978-3-319-19719-7_2

services non-responsiveness to client requests and better arrangement with another, competitor Web service. To perform Web services substitution with less impact on the ongoing, and sometimes critical, business processes, we propose in this paper an approach based on different components and algorithms.

The solution is a mediator able to integrate information sources into Web services while addressing the substitution issue for affected Web services based on EVE framework [1].

Since EVE system proposes a prototype solution to automate view definitions re-writing [2, 3]. We proposed new algorithms for solving the problem of changes within the Web services leading to unavailability of systems and resources. Problem can be emphasized together with the growing dynamics of online services and changes of schema. Solution is based on three components including meta knowledge base, view knowledge base and web services synchronization algorithm. Approach uses service equivalents and evaluation of substitute service.

This paper is organized as follows: Sect. 2 describes the related works. Section 3 introduces the Web service model for gathering information sources. In Sect. 4, we present the Web services synchronization solution by presenting the middleware main components and our illustration by the healthcare application. In Sect. 5, we intro-duce our Web services synchronization algorithms AS2 W. Section 6 gives details of the implementation and Sect. 7 concludes our work and presents some insights for future works.

2 Related Works

In the information space, data providers are autonomous. However, they usually have control over the schemas of their information sources which raises the question of the influence of schema changes, that can render affected view definition undefined [4–6]. Different approaches for addressing this problem have been presented in the literature. Service synchronization or substitution based on the functional properties of components has been addressed by many authors [7–11].

What sets us apart from the proposed approaches is that we aim at addressing the service synchronization problem taking into account the detection of changes which can occur on information sources from which Web services are constructed. In this context, EVE project [12, 13], offers a generic framework within which a view adaptation is solved when underlying information sources change their capabilities. It neither relies on a globally fixed domain nor on ontology of permitted classes of data, both strong assumptions that are often not realistic. Instead, views are built in the traditional way over a number of base schemas and those views are adapted to base schema changes by rewriting them using information space redundancy and relaxable view queries [14]. The benefit of this approach is that no pre-defined domain is necessary, and a view can adapt to changes in the underlying data by automatically rewriting user queries, thanks to synchronization algorithms. This framework has opened up a new direction of research by identifying view synchronization as unexplored problem of current view technology in the WWW. Our approach distinguishes itself from EVE [12] by the fact that we rely on specific advanced applications on the WWW which are Web services. Another novelty of our approach is to apply our work is the health care domain.

3 Web Services Model

In today's collaborative environment, Web services appear to be a privileged mean to interconnect applications across organizations. Web services are software systems designed to support interoperable machine-to-machine interaction over a network [15]. They are modular applications with interface descriptions that can be published, located, and invoked across the Web [16].

Different formalisms are proposed in the literature for modelling Web services. In [17–19], state machine formalism is used for the description of Web services. This choice is justified by the fact that the state machine is simple especially to describe Web services conversation. The states represent phases through it passes the service while interacting. In [20, 21], Web services are modelled using state chart diagram which is a graphic representation of a state machine. The service chart diagram is based on the UML state chart diagram to specify Web services components. None of the studied formalisms can be suitable for modelling the changes that can occur on Web services.

In this section we introduce a novel approach for modelling and specifying Web services. This approach sheds the light on two types of behaviours: presentation and dynamic parts where:

- The dynamic Web service includes information sources access using views
- The static Web service part contains the presentation components

Web service presentation and dynamic parts are executed iteratively as given in Algorithm 1.

```
                    Algorithm 1
While (true)
  {       Presentation part;
          Dynamic part;
          Vi // views call
     }
```

Web services are constructed from views which are built from distributed, heterogeneous and autonomous information sources. Each information source has its own schema composed of relations and attributes. In several cases, Web service is undefined so it should be substituted by another Web service as modelled in Fig. 1.

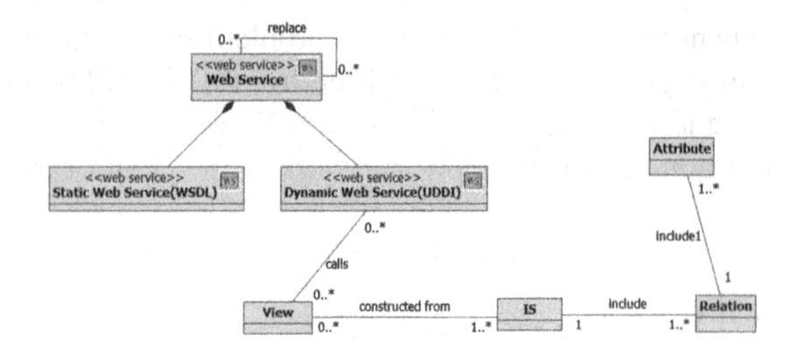

Fig. 1 Web service components relation

Table 1 Relations between web services components

Let *WS* be a Web service, $WS = \{V_1,\ldots,V_n\}$ With V_i: views called by Web service *WS*, $\lvert V_i \rvert \geq 1$, n : total number of views called by *WS*
Let *V* be a view, $V = \{IS_1,\ldots, IS_n\}$ With IS_i : information sources from which the view *V* is constructed. $\lvert IS_i \rvert \geq 1$, n : total number of information sources from which the view *V* is constructed
Let *SI* be an information source, $IS = \{R_1,\ldots,R_n\}$ With R_i : relations which belong to the information source *IS*, $\lvert R_i \rvert \geq 1$, n : total number of relations which belong to the information source *IS*
Let *R* be a relation, $R = \{A_1,\ldots,A_n\}$ With A_i : the attributes which belong to the relation *R*, $\lvert A_i \rvert \geq 1$, n : total number of attributes which belong to the relation *R*
Let *WS* be a Web service, $WS = \{WS_1,\ldots,WS_n\}$ with WS_i the Web service replacement list

The relations between the different components are formalized in Table 1

In several cases, Web services are unavailable so we need to substitute them. In our case Web services are undefined due to schema changes which may render views (dynamic part) undefined. So Web service substitution reach on substituting Web Service dynamic part by rewriting affected views.

Let WS be a Web service and Vi the set of views defined accessed by Web service dynamic part. We suppose that the view V is undefined after schema changes. The Web service WS is synchronized to the Web service WS' with V rewritten on V'$\in$ Vi according to Algorithm 2.

Algorithm 2

```
While (true)                          While (true)
{                                     {
Presentation part;                     Presentation part;
Dynamic part;                          Dynamic part';
Vi// views call                         Vi'// views call
  }                                     }
```

The substitute Web service can be equivalent (≡), superset (⊇), subset (⊆) or indifferent (≈) to the initial Web service.

- The substitute Web service is equivalent (≡) to the initial Web service, if all dynamic part views of the substitute Web service are equivalent to all dynamic part views of the initial Web service.
- The substitute Web service is a superset (⊇) of the initial Web service, if at least one of the dynamic part views of the substitute Web service is a superset of one of the dynamic part views of the initial Web service.
- The substitute Web service is a subset (⊆) of the initial Web service, if at least one of the dynamic part views of the substitute Web service is a subset of one of the dynamic part views of the initial Web service.
- The substitute Web service is indifferent (≈) of the initial Web service, if all dynamic part views of the substitute Web service are indifferent to all dynamic part views of the initial Web service.

4 Web Services Synchronization Framework

Web services are built from distributed, heterogeneous, autonomous information sources which change continuously not only contents but also their schema which may render Web services undefined. We propose therefore a synchronization process which consists of detecting schema changes and substituting affected Web services. Only the two operations attribute deletion and relation deletion affect Web services. The Web service synchronization algorithm searches possible substitution of the affected component (attribute or relation) using WSMKB constraints and WSVKB constraints. Our solution takes the form of a middleware connecting Web services to information sources as shown in Fig. 2.

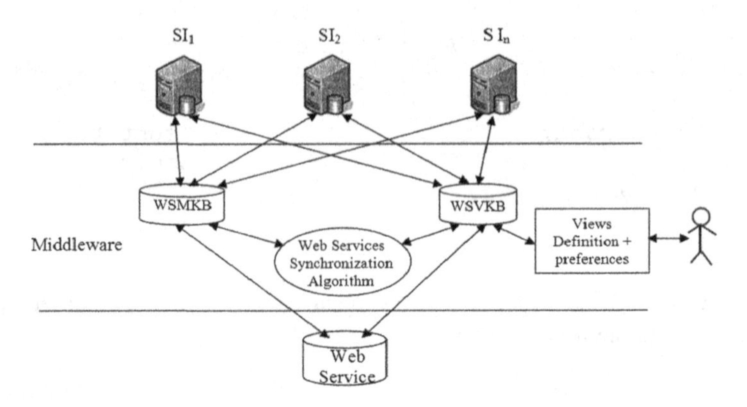

Fig. 2 The system architecture

4.1 Web Services Meta Knowledge Base (WSMKB)

Information sources joining the system must provide its structures and its contents to be stored in the WSMKB (Fig. 3). Relationships between information sources have to be added to WSMKB as substitution rules as given in Figs. 4, 5 and 6. The WSMKB constraints are represented respecting a model called MISD [3]. WSMKB can be organized as follow:

- WS (WSid, WSISidList): Web services with information sources from which they are built as given in Fig. 7.
- IS (ISid, ISRidList): information sources and their included relations. Relations (Rid, AttList): relations and their included attributes.
- The relationships between information sources or substitution constraints such as type integrity constraints, join constraints and partial/complete constraints.
- Replacement (WSid, WSreplacementList): Web services and their substitution Web services list as given in Fig. 8.

In the following, we give an example of healthcare application. Each information sources have their own schemas and contents.

S1	Patient (IdP, Name, Age, Tel, IdH) Doctor (IdD, Name, Speciality) Hospital (IdH, Name, Localization) Doctor_Hospital (IdD, IdH) Diagnostic (IdP, IdD, DateT, Result) Operation (IdP, IdD, DateO, Result)
S2	Patient (IdP, Name, Age, Tel, IdH, Med_Resp) Doctor (IdD, Name, Speciality, IdS) Hospital (IdH, Name, Localization) Doctor_Hospital (IdD, IdH) Diagnostic (IdP, IdD, DateT, Result) Operation (IdP, IdD, DateO, Result) Service (IdS, Speciality)

Fig. 3 Information sources schemas

A type integrity constraint of a relation $R(A_1,\ldots,A_n)$ states that an attribute A_i is of domain type $Type_i$. It allows verifying substitution possibility of an attribute by another while synchronizing Web services. A type integrity constraint is defined as follow: $\mathbf{TC_{R(A1,\ldots,An)} = R(A_1,\ldots,A_n) \subseteq A_1(Type_1) \times \ldots \times A_n(Type_n)}$

The type integrity constraints are expressed in the following

TC	Type integrity constraints
TC1	TCS1.Patient(IdP, Name, Age, Tel, IdH) = Patient (IdP, Name, Age, Tel, IdH)⊆ IdP(Number)×Name(String)×Age(Number)×Tel(Number)×IdH(Number)
TC2	TCS1.Doctor(IdD, Name, Speciality) = Doctor (IdD, Name, Speciality)⊆ IdD(Number) ×Name(String) ×Speciality(String)

Fig. 4 Type integrity constraints

Join constraint between two relations R1 and R2 states that attributes in R1 and R2 can be joined. It allows verifying substitution possibility of a relation by another while synchronizing Web services. Join constraint between two relations R1 and R2 is defined as follow: $\mathbf{JC_{R1,R2} = (C_1\ AND\ \ldots AND\ C_n)}$

In the following we state the list of the join constraints related to our example

JC	Join constraints
JC1	S1.Patient.Name = S2.Patient.Name
JC2	S1.Patient.Name = S3.Patient.Name

Fig. 5 Join constraints

Partial/complete constraint between two relations R1 and R2 states that the relation R1 (or a fragment of the relation R1) is a subset, a superset or equivalent to the relation R2 (or a fragment of the relation R2). Partial/complete constraint is defined as follow: $\mathbf{PC_{R1,R2} = (\pi_{Ai1,\ldots,Aik}(\sigma_{C(Aj1,\ldots,Ajp)}R1)\ \theta\ \pi_{An1,\ldots,Ank}(\sigma_{C(Am1,\ldots,Aml)}R2))}$

PC	Partial/ complete constraints
PC1	$PC_{S1.Patient,S2.Patient} = (\pi_{IdP, Name, Age, Tel}(S1.Patient) \subseteq \pi_{IdP, Name, Age, Tel}(S2.Patient))$
PC2	$PC_{S1.Doctor,S2.Doctor} = (\pi_{IdD, Name, Speciality}(S1.Doctor) \subseteq \pi_{IdD, Name, Speciality}(S2.Doctor))$
PC3	$PC_{S1.Hospital,S2.Hospital}=(\pi_{IdH, Name, Localization}(S1.Hospital) \supseteq \pi_{IdH, Name, Localization}(S2.Hospital))$

Fig. 6 Partial/complete constraints

(WS1, {S1, S2}): The Web service WS1 is construct from information sources S1 and S2
(WS2, {S1, S2}): The Web service WS2 is construct from information sources S1 and S2
(WS3, {S3}) : The Web service WS3 is construct from information sources S3

Fig. 7 Relation between Web services and information sources

(WS1, {WS2, WS3}): The Web service WS1 can be replaced by the Web service WS2 or WS3
(WS2, {WS3}): The Web service WS2 can be replaced by the Web service WS3

Fig. 8 Web services substitution constraints

4.2 Web Services View Knowledge Base WSVKB

Views definition using E-SQL and relations between Web services and its accessed views are given in Fig. 11. E-SQL [3] language allows user preferences inclusion in views definition to indicate how views can evolve after schema changes.

In an E-SQL query, each attribute, relation or condition has two evolution parameters. The dispensable parameter indicates if view components can be conserved (XD=False) or dropped (XD=True) from the substitute view. The replaceable parameter indicates if view components can be substituted (XR=True) or not (XR=False). Here X can be an attribute, a relation or a condition and the default value is False.

View extension parameters VE proposed by E-SQL states that the substitute view can be equivalent (≡), superset (⊇), subset (⊆) or indifferent (≈) to the initial view.

WSVKB contents can be organized as follow:

- VIEW (VDid, VDText): View definition using E-SQL.
- WS (WSid, VDidList): Web services and their views definition list.

The synchronization process consists on detecting schema changes (relations or attributes deletion), detecting affected Web services and applying synchronization algorithm to determine possible substitution of the affected Web services.

Suppose that Name attribute from the relation S1.Doctor is deleted, then it is substituted by Name attribute from the relation S2.Doctor since [TCS1.Doctor(IdD, Name, Speciality)=Doctor(IdD, Name, Speciality) $\subseteq$ IdD(Number) $\times$ Name (String) $\times$ Speciality(String)] and [TCS2.Doctor(IdD, Name, Speciality, IdS) = Doctor(IdD, Name, Speciality, IdS) $\subseteq$ IdD(Number) $\times$ Name(String) $\times$ Speciality (String) $\times$ IdS(Number)] and [PCS1.Doctor,S2.Doctor=(πIdD, Name, Speciality (S1.Doctor) $\subseteq$ πIdD, Name, Speciality(S2.Doctor))] and [S1.Doctor.Name = S2. Doctor.Name]. The view definition of V1 becomes:

```
CREATE VIEW V1' VE='⊇' AS
SELECT D.IdD, D2.Name (AD=false, AR=true)
FROM S1.Doctor D (RD=false, RR=true, S2.Doctor D2
(RD=false, RR=true)
 WHERE(D.Speciality= "Cardiologist") (CD=false, CR=true)
AND (D.IdD = D2.IdD);
```

5 Web Services Synchronization Algorithms

Web services are composed by presentation and dynamic parts including information sources access using views call. As previously said dynamic part includes services gathered from information sources, the latter change continuously which may render views undefined then may render Web services undefined and inaccessible. So these Web services must be substituted by other ones.

Web services synchronization consists on substituting affected Web services referring to constraints embodied into the WSMKB and into the WSVKB. So synchronization process consists of detecting change and according to this change Delete_Attribute procedure or Delete_Relation procedure will be executed as given in substitution Algorithm. Only the two operators delete attribute and delete relation are treated by our algorithm.

```
Substitution Algorithm
Step 1: Identify the type of change (relationship removal
or attribute deletion)
Step 2: Check the characteristics of the attribute or
relationship (essential or replaceable) in the view base.
Step 3: Search in the Knowledge Base the Web service
related to the attribute or relationship.
Step 4: Go to view base to verify the characteristics of
Web services (essential or replaceable).
Step 5: Find substituent of the attribute or relationship
in question.
Step 6: Do substitution
```

6 Performance Evaluation

A prototype of the proposed system has been implemented. We used AXIS 1.1 which is Java platform for creating and deploying web services applications for creating Web services. The simulation of the case study serves the purpose of determining a good trade-off between the need of speed change detection and a need of a quick replacement of affected Web services. To achieve this, we tested the substitution system with different monitoring intervals (2 s–6 s) in order to compare the curves of average response time for both detection of changes and replacement of affected Web services. The response time was almost solely impacted by the monitoring interval and therefore was growing proportionally with it (Fig. 9).

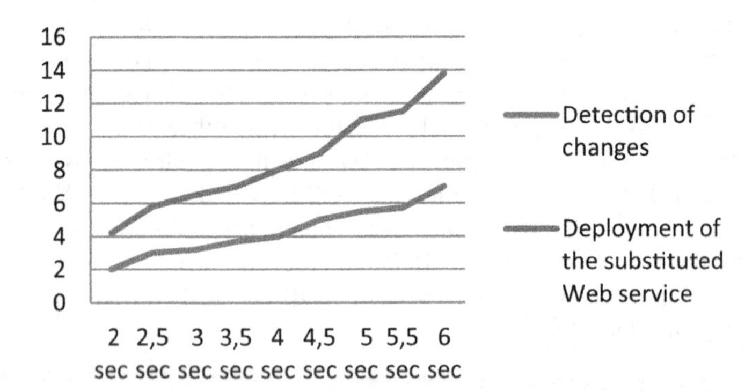

Fig. 9 Average response time

7 Conclusion

In this paper we proposed a solution to the problem of Web services synchronization caused by changes which can occur to information sources from which Web services are built and which may render Web services partially or totally inaccessible.We have presented as solution a middleware connecting Web services to information sources. The middleware is composed by a Web service Meta Knowledge Base WSMKB, a Web Service View Knowledge WSVKB and Web services synchronization algorithms. Our model proved the feasibility of marrying Web services concepts, and the EVE approach [12] which offers a solid foundation for addressing the general problem of how to maintain views in dynamic environments. Future work focus on a total synchronization of Web Services and will not be limited to the two operations attribute deletion and relation deletion which affect Web service. We also intend to develop algorithms for view maintenance of the view extent under both schema and data changes of the information sources.

References

1. Lee AJ, Nica A, Rundensteiner A (2002) The EVE approach: view synchronization in dynamic distributed environments. IEEE Trans Knowl Data Eng 14(5):931–954
2. Zhang X, Rundensteiner EA, Ding L (2001) PVM: Parallel view maintenance under concurrent data updates of distributed sources. In: Proceedings in Data Warehousing and Knowledge Discovery, pp 230–239
3. Lee AJ, Nica A, Rundensteiner EA (1997) The EVE framework: view evolution in an evolving environment, technical report WPI-CS-TR-97-4. Department of Computer Science, Worcester Polytechnic Institute
4. Blakeley JA, Larson P-E, Tompa FW (1986) Efficiently updating materialized views. In: Proceedings of SIGMOD, pp 61–71
5. Zhuge Y, Garcìa-Molina H, Wiener JL (1996) The strobe algorithms for Multi-SourceWarehouse consistency. In: International Conference on Parallel and Distributed Information Systems, pp 146–157
6. Agrawal D, El Abbadi A, Singh A, Yurek T (1997) Efficient view maintenance at data warehouses In: Proceedings of SIGMOD, pp 417–427
7. Benatallah B, Casati F, Toumani F (2006) Representing, analysing and managing web service protocols. Data Knowl Eng 58:327–357
8. Bordeaux L, Salaun G, Berardi D, Mecella M (2005) When are two web services compatible. Lect Notes Comput Sci 3324:15–28
9. Liu F, Zhang L, Shi Y, Lin L, Shi B (2005) Formal analysis of compatibility of web services via ccs. In: Proceedings of the International Conference on Next Generation Web Services Practices, IEEE Computer Society, pp 143
10. Martens A, Moser S, Gerhardt A, Funk K (2006) Analyzing compatibility of bpel processes. In: International Conference on Internet and Web Applications and Services, p 147
11. Pathak J, Basu S, Honavar V (2007) On context-specific substitutability of web services. In Proceedings of the International Conference on Web Services, IEEE Computer Society, pp 192–199
12. Rundensteiner EA, Lee AJ, Nica A (1997) On preservingviews in evolving environments. In: Proceedings of 4th International, Athens, Greece, pp 13.1–13.11

13. Koeller A, Rundensteiner EA (2000) History-Driven View Synchronization. Springer, Greenwich, pp 168–177
14. Nica A (1999) View evolution support for information integration systems over dynamic distributed information spaces. Ph.D. thesis, University of Michigan, Ann Arbor (in progress)
15. W3C (2004). Web services architecture. http://www.w3.org/TR/ws-arch/
16. Dollimore J, Kindberg T, Coulouris G (2005) Distributed systems concepts and design, 4th edn. Addison Wesley, Pearson Education
17. Benatallah B, Casati F, Toumani F (2004) Web service conversation modeling. IEEE Computer Society
18. Benatallah B, Casati F, Toumani F, Hamadi R (2003) Conceptual Modeling of Web Service Conversations. In: CAiSE. Springer
19. Ponge J (2005) Modeling and Analyzing Web Services Protocols
20. Benslimane D, Maamar Z, Ghedira C (2005) A view-based Approach for tracking composite web services. In: Proceedings of the Third European Conference on Web Services (ECOWS 2005)
21. Maamar Z, Benatallah B, Mansoor W (2003) Service chart diagrams - description and application. In Proceedings of The Alternate Tracks of The 12th International World Wide Web Conference (WWW'2003)

Using Magentix2 in Smart-Home Environments

S. Valero, E. del Val, J. Alemany and V. Botti

Abstract In this paper, we present the application of a multi-agent platform Magentix2 for the development of MAS in smart-homes. Specificallly, the use of Magentix2 (http://gti-ia.upv.es/sma/tools/magentix2/index.php) platform facilitates the management of the multiple occupancy in smart living spaces. Virtual organizations provide the possibility of defining a set of norms and roles that facilitate the regulation and control of the actions that can be carried out by the internal and external agents depending on their profile. We illustrate the applicability of our proposal with a set of scenarios.

1 Introduction

Currently, there is a number of social and economic drivers behind smart homes that do specially important the research in this area. In particular, an increasing number of old people in Europe that already live alone or prefer living independently and needs assistance services. Smart home technology can be coupled with works in other areas such as tele-health in order to improve the daily living of users. Moreover the demand for increased security, energy saving, comfort and improved quality of life in the home environment is increasingly taking more importance. Thus the value of building efficient and self-adaptive smart home systems is of primordial relevancy.

S. Valero (✉) · E. del Val · J. Alemany · V. Botti
Departamento de Sistemas Informáticos y Computación,
Universitat Politècnica de València, 46020 Valencia, Spain
e-mail: svalero@dsic.upv.es

E. del Val
e-mail: edelval@dsic.upv.es

J. Alemany
e-mail: jalemany@dsic.upv.es

V. Botti
e-mail: vbotti@dsic.upv.es

Á. Herrero et al. (eds.), *10th International Conference on Soft Computing Models in Industrial and Environmental Applications*, Advances in Intelligent Systems and Computing 368, DOI 10.1007/978-3-319-19719-7_3

The multi-agent system paradigm (MAS) is envisioned as a strong solution of challenges in the context of smart homes. MAS are one of the most representative instances among artificial intelligent systems dealing with complexity and distribution. In the context of smart homes, agents represent the entities in the environment and can be considered autonomous, adaptive, context-aware and capable of making decisions about actions (behaviors) based on their observations. They can be the software interface of the ubiquitous devices that offer their services or the software interface from the user side that demands such services. In other terms, agents are called soft-sensors in smart environments and will be designed to learn from previous experiences and to reason about their local information in order to improve their decisions and achieve their goals. In this way, a smart home (or environment) can be naturally viewed as a multi-agent system with distributed intelligence.

In this paper, we present the application of the platform Magentix2 to the context of Smart-Homes. The use of Magentix2 helps as to define a MAS with normative context that allow us to: (i) define different user profiles (roles) and living spaces (organization units); (ii) control the access to the system and to the services offered by agents that are part of the smart-home in presence of several inhabitant profiles (norms); and to trace the behavior of the inhabitants.

2 Related Work

Traditionally, smart home environments have been seen as centralized systems oriented to a single-user where home appliances are connected to the home network. Advances in ambient intelligence, ubiquitous computing, and autonomic computing and their application into smart homes have added intelligence to our housing and have facilitated the decentralization of the system information and functionality. These features provide flexibility and adaptability to the smart home system. Through the incorporation of intelligent remote control systems that supervise the home appliances and devices it has been possible to improve the quality of living in several aspects: comfort, health-care, safety, security, and energy efficiency. The research in the area of smart homes has tackled different technical issues such as heterogeneity in devices and technologies, context awareness, and security in order to facilitate the implementation of intelligent environments.

Although several technical challenges have been achieved, there are still open issues. On one hand, the major part of research on smart homes until now has focused on technical challenges required to provide ubiquitous and context-aware services considering that there is a single user in the living space [2, 14]. However, usually more than one resident occupy living spaces. The main challenges related to the multi-occupancy can be classified in three groups: activity tracking [15], profiling activities and behaviors [4], and conflicts in input preferences from multiple residents [3, 6].

There are previous works in the area of MAS that deal with some of these multiple occupancy problems. Rodriguez et al. [10] propose a MAS architecture based on virtual organizations that combines data obtained by multiple sensors in order to identify and localize the position of the inhabitants in residential home for the elderly. Sun et al. [12] propose a multi-agent design framework for smart house and home automation applications. They propose a BDI model for agent individual behavior design and a regulation policy-based method for multi-agent group behavior design. Also, a Petri-net based method is developed for system evaluation and analysis. Loseto et al. [8] propose a flexible multi-agent approach for smart-environments based on semantic resource discovery and orchestration. They also include negotiation techniques between user agents and the agents that represent the devices. However, as in the majority of approaches, the user agents are assumed that all of them play the same role. Bajo et al. [1, 9] present a previous version of the THOMAS architecture applied to a similar scenario used in this paper. Our proposal presents a set of THOMAS [5] extensions that give support to the dynamic management of virtual organizations. These extensions are implemented and provided by Magentix2 [11]. Moreover, Magentix2 includes a trace functionality in order to keep track of the activities inside organizations, which is also used in our proposal.

In this work, we propose a MAS based on organizations where different profiles for inhabitants can be defined. The proposed MAS is based on the agent platform Magentix2. Magentix2 provides concepts such as virtual organizations and norms that facilitate the management of multiple occupancy in smart-home environments. The virtual organizations provide the possibility of defining a set of norms that facilitates the regulation and control of the actions that can be carried out depending on the type of inhabitant.

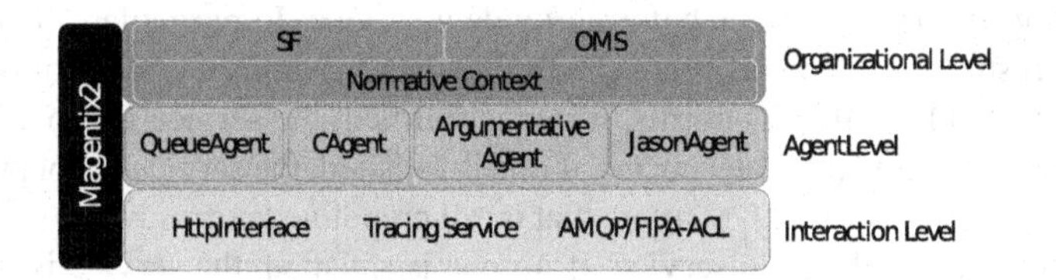

Fig. 1 Magentix 2 platform: components and services

3 Agent Platform: Magentix2

Magentix2[1] is an agent platform for open MAS which provides support at three levels (see Fig. 1):

Organizational level. It integrates the THOMAS framework [13] to provide a complete support for virtual organizations and SOA-like services. THOMAS offers a set of modular services provided by two main components, the Service Facilitator (SF) and Organization Manager Service (OMS). The SF offers a yellow/green

[1]http://www.gti-ia.upv.es/sma/tools/magentix2/index.php.

page service. The OMS is mainly responsible of the management of the organizations. Different types of virtual organizations are supported. Each organization can contain others organizations. Furthermore, it is possible to define diverse roles related to each organization. These roles are characterized by some attributes (position, accessibility and visibility), that can restrict the access to the services offered by the OMS. Also, it is possible to define a normative context to further restrict (or to permit) the access to the THOMAS services.

Interaction level. The platform supports flexible interaction protocols and conversations, indirect communication by means of a tracing service and interactions among agent organizations.

Agent level. Developers have different classes of agents available. For example, in the smart home context, it should be suitable to use two of them: CAgent (automatic creation of simultaneous conversations) and JasonAgent (BDI agents which can participate in simultaneous conversations).

3.1 Normative Context of the Organizational Level

There are some predefined norms which control the access to OMS services. Some OMS services are only available if agents which request them play a certain role inside the organization (Table 1). In this way, users can add PERMITTED or FORBIDDEN norms to relax or restrict the access control to the OMS services in an organization by means of the *registerNorm* service of the OMS. The OMS agent is responsible for verifying if a norm applies before provide a service. In particular, the order in which the restrictions and norms are checked before providing a service of the OMS is as follows: (i) PERMITTED norms, if one is fulfilled, the service is provided with no restrictions; (ii) FORBIDDEN norms, if one is satisfied, the service is not provided; and finally (iii) STRUCTURAL NORMS, that is, all predefined norms of the service are checked before providing the service. If no one is satisfied, the service is provided as usual.

Table 1 OMS service access of some of its available services taking into account the role position played by the requesting agent

	Position			
OMS services	Creator	Member	Supervisor	Subordinate
RegisterUnit	x	-	-	-
RegisterNorm	x	x	x	-
AcquiereRole	x	x	x	x (1)
AllocateRole	x	x	x	-
DeallocateRole	x	x	x	-
InformAgentRole	x	x	x	x

(1) The agent could acquire another role with posistio = creator or position = supervisor in its organization

Norm Description Norms are registered into an organizational unit with a unique name inside that organization. They should be written following the syntax of the THOMAS normative language (see [7] for a detailed explanation), which is based on AgentSpeak language. Concretely, the appearance of a norm is as follows:

$$@normName[Deontic, Target, Action, Activation, Expiration]$$

where:

- *Deontic* $\in \{f, p\}$, where f is used for forbidden norms, which restricts the access to services; and p is used for permitted norms, which relaxes the access to services.
- *Target* $=< type, id >$ where *type* $\in \{agentName, roleName, positionName\}$, whereas *id* is the associated value (for example, the position value *creator*) or the anonymous variable "_" (when any value is accepted). The field *target* allows users to determine which agents will be affected by a norm.
- *Action* is the name of the service. THOMAS only manages norms which their action corresponds to an OMS service.
- *Activation* is a well-formed formula expressed by means of first-order predicates which indicates the conditions to fulfill a norm. Users should add predicates related to data known in the THOMAS world (role details, played roles, organization structure, agent names, etc.). The *Activation* could be empty ("_"), in that case the norm is always fulfilled.
- *Expiration* is a well-formed formula expressed by means of first-order predicates which indicates when a norm expires. So, if the expiration of a norm is satisfied, the norm is not applied although the activation is fulfilled. Users should add predicates related to known data in the THOMAS world. The *Expiration* could be empty ("_"), in that case the norm never expires.

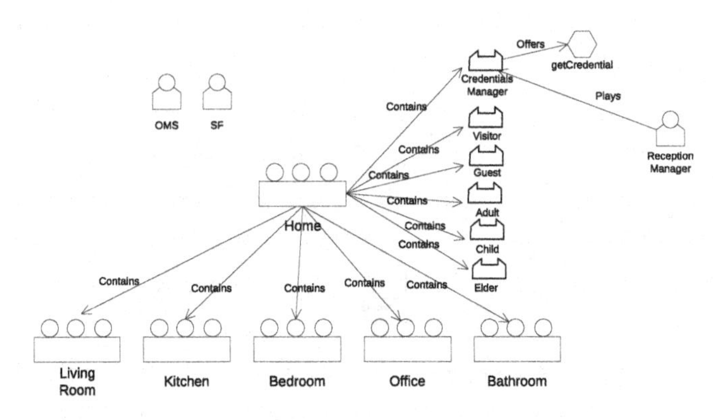

Fig. 2 Organizational view of a smart home

4 Applying Magentix2 to Smart-Homes

In order to illustrate the use of Magentix2 in a smart environment, we have considered smart-home an scenario where the smart-home is modeled as a service-oriented MAS based on virtual organizations. In the following sections, we describe the organizational view, the normative context, and the system dynamics of our proposal.

4.1 Organizational View

The organizational view describes the components of the system and their relationships. In this view, we define the agents, the organizational units, and the roles defined inside these organizational units. The organizational view of the MAS for smart-homes consists of the following entities (see Fig. 2):

- Agents: there is an agent that plays the organizational manager role *OMS* and an agent that plays the service facilitator role *SF*.
- Organizational Units (OU): we consider an organization unit called *Home*. Inside *Home OU*, we define a organizational unit for each of the rooms of the physical home.
- Roles: at the *Home OU* there is defined a set of organizational roles. These roles define the profiles that can be played by the agents inside the *Home OU*. We have considered a role for each type of inhabitant of the living space. The roles *Adult*, *Child*, and *Elder* represent the usual inhabitants of our smart home scenario. *Visitor* and *Guest* represent eventual inhabitants. Finally, the role *Credentials manager* represent the profile of an agent that manages who enters in the *Home OU* and which role must acquire an external agent to be a member of the organization.

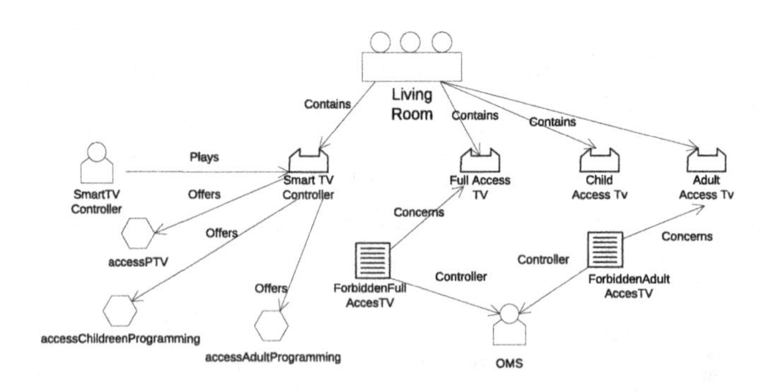

Fig. 3 Organizational view of the living room

Figure 3 describes with detail the *LivingRoom OU*. Inside the *LivingRoom OU* there is a set of roles defined. The *smartTVcontroller* role offers a set of services related to available TV programs in the TV. The roles *FullAccess* and *ChildAccessTv* are defined in order to control the access to the TV programs.

4.2 Normative Context

There is a set of norms that controls the acquirement of the required roles to access to the different services offered by the smart home. As an example, Fig. 4 shows the norm called *forbiddenAcquiereRole*. This norm is registered in the *Home OU* and prevents agents from acquiring roles directly. Thus, they only can acquire roles (credentials) through the *credendialsManager* agent. This norm is described using the Magentix2 normative language.

```
@forbiddenAquireRole[f, <agentName:_>, acquireRole(_,home,_),_,_]
```

Fig. 4 Forbidden norm registered in the *Home OU* that prevents agents from acquiring roles directly

In a similar way, Fig. 3 shows a set of norms that controls which agents can play the roles defined in the *Living Room OU*. Specifically, the norm *forbiddenFullAccessTV* (Fig. 5) avoids that agents which play *child* and *visitor* roles in *Home OU* buy films, because they cannot obtain the role required (Fig. 6).

```
forbiddenFullAccessTV[f,<agentName:Agent>,
acquireRole(fullAccessTV,livingRoom,Agent),isAgent(Agent)&
not playsRole(adult,home,Agent),_]
```

Fig. 5 Norms of the Living Room

4.3 System Dynamics

In order to illustrate the MAS functionality and how agents interact in the system, we present different scenarios that may occur in our smart-home context.

(a) Access to the Smart Home. An external agent a_i wants to get into the smart home system. The entry point is the *Home OU*. To be a member of the organization, the agent a_i must play at least one of the roles defined inside of *Home OU*. In order

to acquire one of these roles, the external agent contacts with a *ReceptionManager* agent that is inside the organization and plays the role *CredentialsManager*. The *ReceptionManager* agent offers a service (i.e., *getCredentials*) that allocates roles to newcomers based on the information contained in their profiles. The *ReceptionManager* interacts with the OMS agent to register members in the organization. After that, the newcomer agent is part of the *Home OU*.

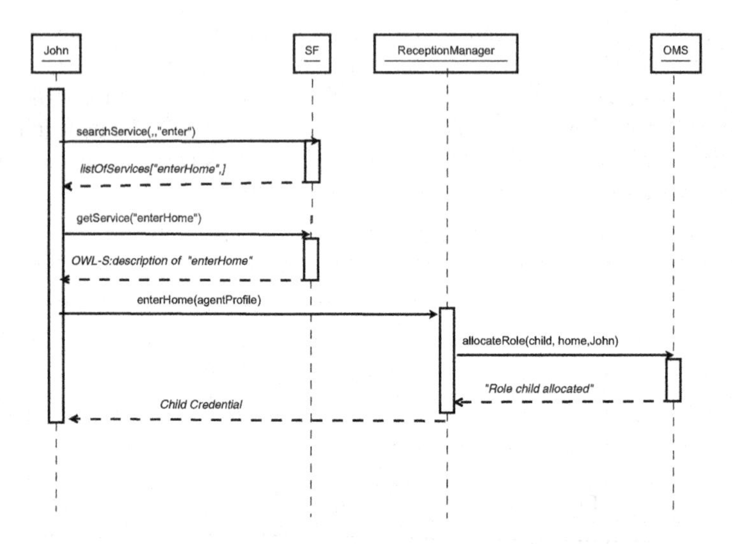

Fig. 6 Enter Home Scenario: a child named John gets his corresponding credential

(b) Asking for a service that you are not allowed to ask. An agent a_i that plays the role *Child* in *Home OU* enters in the organization *LivingRoom OU*. The agent wants to request a service in order to buy a film using the smart TV. The agent a_i asks the SF agent for a service of films and gets the information about services that offer pay-per-view movies and which are the provider agents. The agent a_i selects the Smart TV provider agent and asks it for the service. In order to provide the service to the agent a_i, the smart TV agent checks if a_i plays the required role. This verification is done by asking the OMS about the role that plays the requester. Based on this information the smart TV agent offers or not the service (Fig. 7).

(c) Asking for a service that you are allowed to ask. An agent a_i that plays the role *Adult* in *Home OU* enters in the *LivingRoom OU*. The agent a_i wants to request a service in order to buy a film. Therefore, it asks the SF agent for a service of films and then it gets the information about the available services and the agents that provide them. The agent a_i analyzes the information about the service profile of the provider agents and notices that to ask for the service, it is necessary to play the role *FullAccessTV*. Therefore, the agent a_i interacts with the OMS agent in order to acquire this role. Once the agent a_i plays this role, it asks the Smart TV agent for the service.

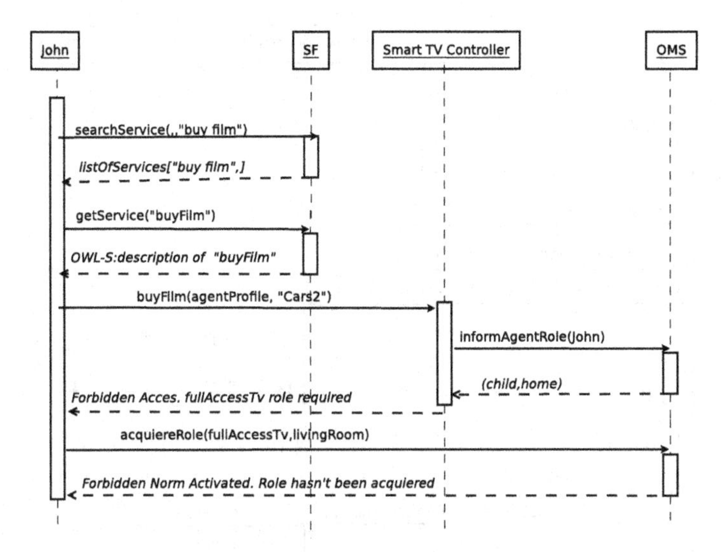

Fig. 7 Not Allowed Service Scenario: a child named John tries to access to the buyfilm service

(d) Access of a visitant (medical assistant). An external agent a_i, that represents a medical assistant, wants to get into the smart home system in order to control the health of one of the inhabitants. The entry point is the *Home OU*. To be a member of the organization, a_i must play at least one of the roles defined inside the *Home OU*. In order to acquire one of these roles, the external agent a_i contacts with the *ReceptionManager* agent that is inside the *Home OU*. As the external agent a_i has in its profile its occupation, the controller agent allocates the role *Visitor* and the role *HealthStaff*. The *ReceptionManager* agent interacts with the OMS agent for the registration process of the new member of the organization. Once the agent a_i is part of the organization and plays the role *HealthStaff* and *Visitor*, it asks the SF about services related to medical control (blood pressure, diabetic information, insurance information,...). With this information, the agent a_i can interact with the agents that provide the services in order to create an inform about the current situation of a home inhabitant (Fig. 8).

(e) Trace of events in the Smart Home An agent would be interested in tracing some activities performed by other inhabitants. For instance, the agent a_i that plays the role *Adult* can ask for tracing the services associated to the activity of other agent that plays the role *Elder* inside the *Home OU*. With this trace, the agent a_i controls if the medical assistant arrived at home. The trace can be also useful for controlling the activity of other inhabitants. For instance, if an agent that plays the role *Child* has tried to ask for a service that it is not allowed for agents that play the *Child* role.

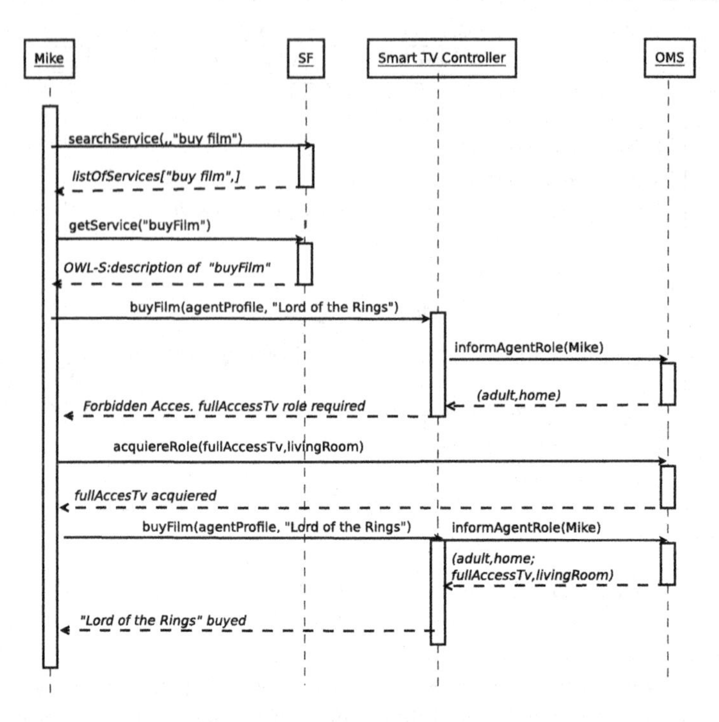

Fig. 8 Allowed Service Scenario: an adult named Mike asks for the buyfilm service

5 Conclusions

In this paper, we have presented a MAS for smart-home environments based on the Magentix2 platform. Our approach facilitates the management of multiple inhabitants in a living space. Through the use of organizational concepts, it is possible to control the activities of internal and external agents in an efficient way. We illustrate the applicability of our proposal through a set of scenarios in the context of smart-homes. As future work, we will extend Magentix2 to provide security services and consider norms in the Service Facilitator services. Moreover, we will compare empirical results of our proposal with other agent platforms that allow the definition and execution of norms.

Acknowledgments This work is supported by the Spanish government grants CONSOLIDER INGENIO 2010 CSD2007-00022, MINECO/FEDER TIN2012-36586-C03-01, TIN2011-27652-C03-01, and SP2014800.

References

1. Bajo J, Fraile JA, Pérez-Lancho B, Corchado JM (2010) The thomas architecture in home care scenarios: a case study. Expert Syst Appl 37(5):3986–3999
2. Cetina C, Giner P, Fons J, Pelechano V (2009) Autonomic computing through reuse of variability models at runtime: The case of smart homes. Computer 42(10):37–43
3. Cook DJ (2009) Multi-agent smart environments. J Ambient Intell Smart Environ 1(1):51–55
4. Crandall AS, Cook DJ (2010) Using a hidden markov model for resident identification. In: 6th international conference on intelligent environments, pp 74–79. IEEE
5. Criado N, Argente E, Botti V (2013) THOMAS: an agent platform for supporting normative multi-agent systems. J Logic Comput 23(2):309–333
6. Davidoff S, Lee MK, Zimmerman J, Dey A (2006) Socially-aware requirements for a smart home. In: Proceedings of the international symposium on intelligent, environments, pp 41–44
7. Grupo de Tecnología Informática e Inteligencia Artificial (GTI-IA) (2015). http://www.gti-ia.upv.es/sma/tools/magentix2/archivos/Magentix2UserManualv2.1.0.pdf. Magentix2 User's Manual v2.0
8. Loseto G, Scioscia F, Ruta M, di Sciascio E (2012) Semantic-based smart homes: a multi-agent approach. In: 13th Workshop on objects and agents (WOA 2012), vol 892, pp 49–55
9. Rodriguez S, Julián V, Bajo J, Carrascosa C, Botti V, Corchado JM (2011) Agent-based virtual organization architecture. Eng Appl Artif Intell 24(5):895–910
10. Rodríguez S, Paz JFD, Villarrubia G, Zato C, Bajo J, Corchado JM (2015) Multi-agent information fusion system to manage data from a WSN in a residential home. Inf Fusion 23:43–57
11. Such JM, Garca-Fornes A, Espinosa A, Bellver J (2012) Magentix2: a Privacy-enhancing Agent Platform. Eng Appl Artif Intell
12. Sun Q, Yu W, Kochurov N, Hao Q, Hu F (2013) A multi-agent-based intelligent sensor and actuator network design for smart house and home automation. J Sens Actuator Netw 2(3):557–588
13. Val E, Criado N, Rebollo M, Argente E, Julian V (2009) Service-oriented framework for virtual organizations. 1:108–114
14. Wu C-L, Liao C-F, Fu L-C (2007) Service-oriented smart-home architecture based on osgi and mobile-agent technology. IEEE Trans Syst Man Cybern Part C Appl Rev 37(2):193–205
15. Yin J, Yang Q, Shen D, Li Z-N (2008) Activity recognition via user-trace segmentation. ACM Trans Sens Netw (TOSN) 4(4):19

Analyzing Accelerometer Data for Epilepsy Episode Recognition

José R. Villar, Manuel Menéndez, Javier Sedano, Enrique de la Cal and Víctor M. González

Abstract Epilepsy is one of the main neurological disorders with high impact in the patient's everyday life. An incorrect treatment or a lack in monitoring might produce cognitive damage and depression. Therefore, developing a wearable device for epilepsy monitoring would eventually complete the anamnesis, enhancing the medical staff diagnosing and treatment setting. This study shows the preliminary results in epilepsy onset recognition based on wearable tri-axial accelerometers and simple fuzzy set learnt using genetic algorithms. A complete experimentation for learning the fuzzy set is detailed. According to the obtained results, some generalized feasible solutions are discussed. Results show a very interesting researching area that might be easily transferred to embedded devices and online health care systems.

1 Introduction

Epilepsy is one of the most common neurological disorders in the society with a high impact in the life quality, professional career and social behaviour. One of the consequences of suffering epilepsy is a lack in motorization, which is direct linked

J.R. Villar (✉) · E. de la Cal
Computer Science Department, University of Oviedo, Oviedo, Spain
e-mail: villarjose@uniovi.es

E. de la Cal
e-mail: delacal@uniovi.es

M. Menéndez
Morphology and Cellular Biology Department, Oviedo, Spain
e-mail: menendezgmanuel@uniovi.es

J. Sedano
Instituto Tec. de Castilla Y León, Burgos, Spain
e-mail: javier.sedano@itcl.es

V.M. González
Electrical Engineering Department, University of Oviedo, Oviedo, Spain
e-mail: vmsuarez@uniovi.es

Á. Herrero et al. (eds.), *10th International Conference on Soft Computing Models in Industrial and Environmental Applications*, Advances in Intelligent Systems and Computing 368, DOI 10.1007/978-3-319-19719-7_4

with some comorbidities, like cognitive damage and depression [1]. Epilepsy has a dramatic impact in the health care system's annual budgets [2]. Besides, an epilepsy crisis is a clinical manifestation that has its origin in an abnormal activity -either excessive or hypersyncronic- from a variable size group of brain neurons. This abnormal activity -or epileptic discharge- uses to occur suddenly, during a short and transitory period of time, that is, paroxysmal. Therefore, the symptoms and signs that characterize an epilepsy crisis also typically show paroxysmally.

The term epileptic syndromes must be used instead of epilepsy, thus it includes a wide range of epileptic symptoms, conditions, etiology, manifestations, treatments and trends. The percentage of wrong epilepsy diagnosis might be up to 20 % of the patients before attending an epilepsy unit for evaluation; consequently, the delay in the diagnose might be as long as 10 years [3]. The epilepsy diagnose is basically done using a clinical procedure: the anamnesis is the starting point for the initial epilepsy diagnose, although there are several tests to support the diagnose.

Basically, there are two main epileptic types of crisis: the generalized and the focal crisis [4]. In both of them there are subtypes with or without motor activity: those with motor activity are the most common cases. The motor activity can vary from the generalized tonic-clonic crisis -losing the consciousness, a short tonic stage followed by a prolonged generalized and repeated clonic movements of the whole body- to the focal mio-clonic crisis -repeated bursting movements of one limb, the upper and lower limbs of one body side or a combination of movements of the limbs and face. After the diagnose, the patient should keep a log of the suffered seizures, so the medical staff can evaluate the evolution of the patient and the efficiency of his/her treatment.

This research aims to design a model together with a wearable device for supporting patients in maintaining the log of seizures. This preliminary study analyzes the motor activity registered in the data gathered from tri-axial accelerometer (3DACC) placed on the dominant wrist. The idea is to determine if it is possible to identify an epilepsy seizure from realistic simulated data. This realistic data is based on a protocol defined by the medical staff, describing how a patient behaves during a certain type of epilepsy crisis. To identify the seizure, a simple fuzzy partition is learnt for the subject from the data using a genetic algorithm. In this stage, no generalized method is searched, but only the possibility of using Genetic Fuzzy Systems (GFS) to deal with this time series classification task. The aim of this study is to evaluate if GFS can cope with this problem or not, in order to design a feasible solution before dealing with data from real patients. The organization of this study is as follows. First of all, a revision of the literature is presented in the next section, while in Sect. 3 the preliminary approach is detailed. The experiments are described in Sect. 4. This study ends with the discussion on the results and the conclusion remarks.

2 Epilepsy Episode Recognition

The current trend in bio-engineering is introducing wearable sensors as the mechanism of gathering data from real or even controlled scenarios. Due to the fact that the majority of epilepsy episodes have a clear impact in the motor system, the main part of the literature focuses on reported studies that make use of 3DACC for their solutions.

As stated in [5], it is possible to observe epilepsy episodes using 3DACC when experts visualize the data gathered from experiments. However, the author have asserted that the problem is by any means solved [6]: there is not any feasible solution able to recognize epilepsy episodes out of controlled environments. As stated in [7], the characteristic movements vary according to the type of epilepsy crisis.

In [8], 3DACC bracelets were placed on the wrists of patients suffering from epilepsy. The wearable device were programmed for detecting the rhythmical movements in topic-clonic episodes: 7 out of 8 of the considered types were detected, but up to 204 false alarms were also arose. Although the approach is very interesting, its current status is not valid enough for deployment.

Not only the acceleration values have been considered, but also several different transformations. For instance, the amount of movement was used for generating alarms for epileptic episodes occurring during the sleeping periods of the patients has also been publish [9]. In this study, thresholds were defined as the limits of the amount of movement an individual can perform when sleeping: values higher than these limits were considered alarms of epilepsy crisis. Due to the use of thresholds, the study was limited to reduced movements activities.

The use of several sensors -3DACC among others- for epileptic episodes discovering is proposed in [10]. As in previous studies, the focus of the study is kept on discovering epileptic episodes while the subject is sleeping. Actually, very impressive results were reported as up to 95 % of the cases were discovered. Nevertheless, the cost of this approach as well as the restrictions in the type of activities considered represent the most relevant drawback of this approach. Interestingly, Schulc and colleagues studied different transformations for the data gathered from the sensors in order to recognize the epileptic episodes [11]. These transformations represent a valid starting point for discovering patterns on the data.

3 A Study for Epilepsy Identification

From now on, the terms raw acceleration (AC), body acceleration (BA) and gravity (G) refer to the acceleration data gathered from the 3DACC, the acceleration of the part of the body where the sensor is placed and the gravity acceleration, correspondingly. When a single subindex is included, it refers to the axis -e.g., AC_x or G_z-. A second subindex might be included for referencing the time stamp -e.g., $AC_{x,i}$-. To extract the BA and G from the ACC, a high pass filter or a low pass filter must be used [12].

One of the first steps when focusing the identification of a pattern is the visualization of the time series. Figure 1 shows the evolution of the acceleration components during a simulated tonic-clonic epilepsy episode. The duration of the episode is about 30 s long; then the subject remains still in the post-episode recovery. It was simulated by the medical staff following the behavior that is well documented in the literature. All the scales are the same for the sake of readability. The simulation started from a seated position, the data goes from a very steady signal to a high deviation signal in all the cases.

From the figure it might be concluded that using statistical data could be enough for identifying an epilepsy episode. However, if we consider the different activities a subject may carry during his/her everyday life, the problem becomes more complex. For instance, walking may have a similar variability of the signals but with smaller the mean values, while running have higher mean values. Furthermore, a feasible solution must consider the variability among the episodes and the differences performed by the different subjects.

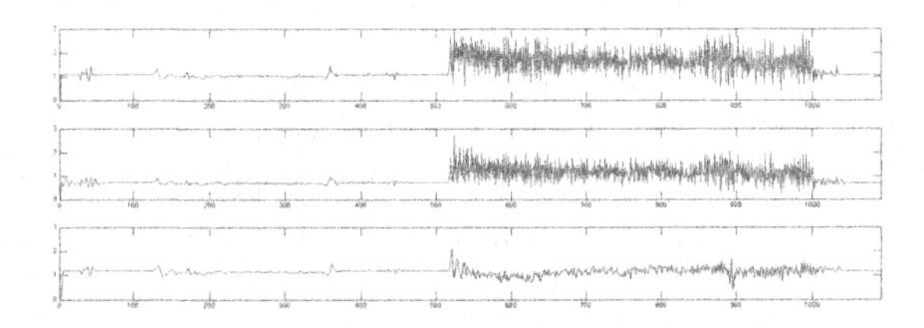

Fig. 1 Acceleration signals from a 2G 3DACC sampled at 16 Hz placed on the dominant wrist. Upper, central and lower parts depict the AC, BA and G, respectively

Different alternatives were applied to the accelerations but no relevant patterns were extracted from the data: using the acceleration data was not enough. In a previous study concerned with human activity recognition [12], several transformations of the different accelerations were analyzed. Additionally, some of these features were analyzed regarding epilepsy identification [11]. After a study of the most promising features, the *Signal Magnitude Area* -computed as $\mathbf{SMA} = \frac{1}{w}\sum_{i=1}^{w}(|b_{x,i}| + |b_{y,i}| + |b_{z,i}|)$-, was found the most representative. Still, the problem of obtaining a generalized solution can not be considered solved as different activities might have similar SMA values and because of the differences between subjects and episode intensities and behavior.

In order to make the problem easier to tackle, only one of the possible epilepsy episodes is considered in this study: the myoclonic type. This epilepsy episode is characterized by repeated contracting-elongating movements of the forearm, with a angle variation about 30°, with a frequency about 1 swing every 3 s [7]. An episode uses to last about 30 s with a pos-critic period of 30 s, in which the movements of the subject are somewhat erratic, without a clear pattern nor intention. The subject neither falls nor loses consciousness.

3.1 The Proposal for Identifying Tonic-Clonic Epilepsy Episodes

In order to tackle the above mentioned problem, a fuzzy set based solution has been developed. The idea underneath is to take advantage of the generalization capabilities that fuzzy sets provide in order to determine if a sequence of SMA values belong or not to the Epilepsy class. A single fuzzy set with trapezoidal membership function is proposed for determining the membership degree of current window's SMA value to the analyzed epilepsy episode. However, the fuzzy generalized solution can not be obtained from scratch but from a learning stage: a genetic algorithm (GA) is proposed to evolve the fuzzy membership function parameters.

The GA will evolve the parameters of the trapezoidal fuzzy membership function. The individual representation of the fuzzy function parameters is a vector of 4 double values $\hat{x} = [x(1), x(2), x(3), x(4)]$ with the corresponding restrictions. Furthermore, the following default options hold: random individual generation provided they accomplish with the restrictions, an elite subpopulation, rank fitness scaling, stochastic uniform selection, a fraction of the new population is generated by means of crossover, adaptive feasible mutation operator -generating random directions that are adapted to the last successful or unsuccessful generation according to the bounds and restrictions- and the scattered crossover operator. The stopping criteria includes the generation limit of 50 generations or whenever 20 generations are performed without fitness improvement is smaller than $1 \cdot 10^{-6}$.

To evaluate each individual a set of segmented files are given, reducing the problem to a two-class classification. The classification error, calculated as the number of misclassified examples, is used as the fitness function. A penalty factor is included for favoring those individuals with smaller width. This measurement has been chosen to avoid as much as possible the imbalanced problems that use to happen with time series classification. A sample is considered of the class if its membership value is better than 0.5.

In this study, a cross validation scheme based on 5×2 is introduced. Firstly, each data set file is tagged as {TRAIN, TEST or TRAIN-AND-TEST}. A file marked with TRAIN will be used for training exclusively; similarly, a file marked with TEST will be used in the validation of the approach only. A file marked with TRAIN-AND-TEST will be included in both the training and the validation stages. Using this 5×2 cross validation scheme has two main purposes. Firstly, it might keep the proportion the samples belonging or not to the analyzed class. Secondly, the generalization capabilities are strongly tested as the number of unseen files is increased. The mean value of the error measurement among the different files used for training or validating is used for aggregating the results for each fold.

4 Experiments and Results

4.1 Data Set Generation

The medical staff designed a protocol for the simulation of the tonic-clonic trials of myoclonic epilepsy episodes. These episodes are as briefly outlined in Sect. 3.

In this way, several simulations were carried out for a subject and, afterwards, realistic data from 10 believable fiction subjects were generated. Special care was taken to mimic the time series in the onset of the epilepsy. Half of the files generated for normal activities were considered to be tagged as TRAIN with probability 0.8; the remaining were tagged as TEST. Additionally, half of the epilepsy files were tagged as TRAIN, while the remaining were tagged as TRAIN-AND-TEST.

4.2 Tuning the GA Parameters

Basically, the number of generations and the stop conditions are kept constant, while the the performance of the rest of parameter sets are studied. During the experimentation, once a parameter set is found better it is kept for the remaining experimentation. More specifically, the following comparisons have been evaluated: (i) the use of subpopulations versus a single big population, (ii) the migration factor if subpopulation outperforms the single population results, (iii) the elite population size, (iv) the selection operator, (v) the fraction of individuals created using the crossover genetic operator and (vi) the crossover operator. Besides, the mutation operator does not change for the sake of the defined bounds and constraints.

For the sake of space, only a remark of the experimentation findings. Single population definitely performs better, with enhanced diversity. There were no variations in the results with the elite size. The uniform selection performs with a slightly better diversity and the default crossover fraction –0.75- seemed to be the best choice. In the remaining text the experimentation with the crossover operator is shown.

Four different crossover operators were compared: the scattered operator -randomly picking a gene from each parent-, the classic two points crossover, the heuristic crossover -generating an offspring from a linear combination of the parents- and the arithmetic crossover -a weighted combination of the parents is used-. Not only the fitness evaluation of the final population will be used for comparing the methods, but also the diversity at this stage. The results are depicted in Figs. 2 and 3, and in Table 1 as well. Small variations again are found in the phenotype, the main advantage is found in the diversity. According to this value, the most interesting operator is the heuristic operator.

The results shown in Fig. 3 are really interesting as that is what might be happening when deploying this model. The On/Off output depicted the figure is calculated as 1 when the value of membership of belonging to the class EPILEPSY is higher than 0.5. Clearly, the output of the model is not what is desired, as plenty of the

Table 1 Results for the crossover operators comparison. scat, heu, arth and 2pnt stand for the scattered, the heuristic, the arithmetic and the two points crossover operators, respectively

Op.	Dataset	Mean	Median	Std	Op.	Dataset	Mean	Median	Std
scat	Trn	98.5570	97.4742	4.0126	heu	Trn	96.8538	96.8036	1.8787
	Tst Mi	125.8407	122.4060	12.9749		Tst Mi	125.8407	122.4060	12.9749
	Tst Mn	173.4868	171.2261	16.9209		Tst Mn	173.4868	171.2261	16.9209
	Diversity	1.2314	1.1836	0.6073		Diversity	3.9885	4.1978	1.7753
arth	Trn	103.1752	99.5921	12.8982	2pnt	Trn	99.1747	98.4409	4.7719
	Tst Mi	125.8407	122.4060	12.9749		Tst Mi	125.8407	122.4060	12.9749
	Tst Mn	173.4868	171.2261	16.9209		Tst Mn	173.4868	171.2261	16.9209
	Diversity	3.1480	3.2235	2.1385		Diversity	1.5561	1.4894	0.6469

EPILEPSY samples were wrongly classified. Therefore, a more complex system is needed as simple linguistic variables or even thresholds are not valid.

The question is what kind of models can be valid for solving this type of problems. This question is partially solved by that Fig. 3 : learning rules from the data will lead to obtaining rules that might have an output like the On/Off depicted. It is possible to learn several rules and use ensembles in order to obtain a more continuous output.

Nevertheless, this behavior suggests that it may be interesting to learn models that keep track of the current state, and the rules acting as transitions between states. In other words, this may lead to Finite State Machines. There are several approaches that can be suitable, as the Genetic Fuzzy State Machines [13] or the Hidden Markov Models [14]. The idea, then, can include developing a suitable model for each type of epilepsy episode and make use of an ensemble scheme for the fusion of the different outputs. These techniques represent the most promising research lines in order to learn classifiers for the different types of epilepsy episodes.

In apart, the evaluation of the models needs further study. How to evaluate the fitness of a model according to the goodness of its output is the main concern. Typically, classification error based measurements have been used in the literature. After the experimentation stage, it seems that using distance measurements between the class and the output of the model instead of the classification error measurements can enhance the learning process. All of these issues need further study.

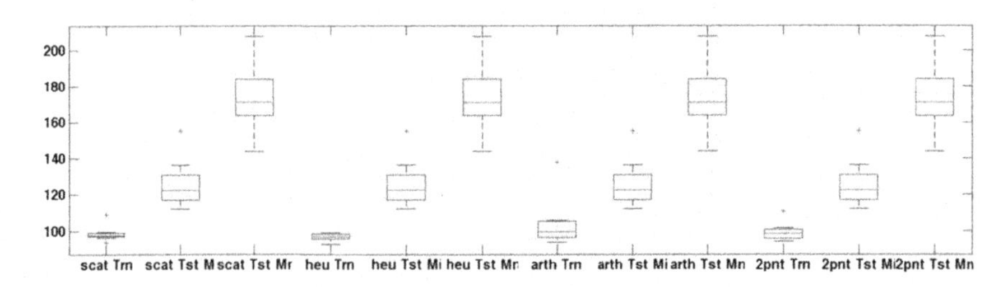

Fig. 2 The boxplot of the fitness results: each crossover operators has 3 box plots -best individual in training, best individual in test and mean individual in test-. From left to right: the scattered, the heuristic, the arithmetic and the two points crossover operators, respectively

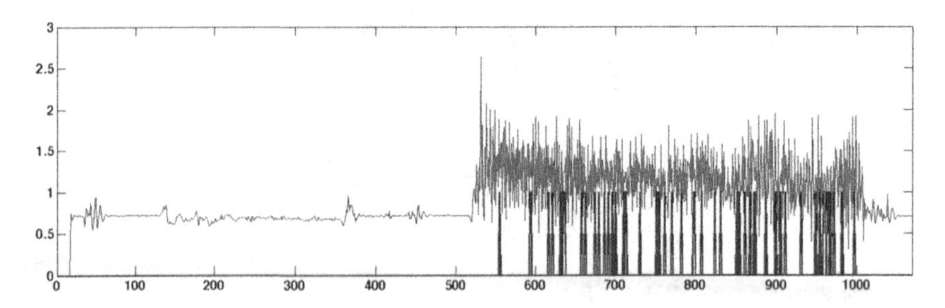

Fig. 3 Evaluation of the Fuzzy set with a simulation of an epilepsy episode the lower On/Off signal the discretized output when the membership function for the value of SMA is higher or equal to 0.5

5 Discussion and Conclusions

After all this experimentation, the best set of genetic parameters subset for the genetic algorithm evolving the fuzzy membership function characteristic points has been found. With this simple tool we have been able to correctly classify the time series with a test classification error that has been kept lower than 200 wrong classified samples when using the realistic data. However, this approach did not seems to be a generalized solution as far as the performance of the fuzzy set degrades when using the data from one subject to other. It was found that such a simple approach can not deal with the problem of identifying an epilepsy episode, as seen in Fig. 2. The analysis of the results suggests that considering the current state of the subject and learning Finite State Machines can lead to valid solutions. Nonetheless, generalized solutions should be considered because it would be practically impossible -and undesirable- to gather data from epilepsy episodes for a subject in order to learnt the solution model for him/her. These models should consider the variability of the behavior and the uncertainty in the data as well. Thus, it may be interesting to include Fuzzy Finite State Machines. Additionally, the use of Markov Hidden models and Fuzzy Finite State Machines represent an alternative to the previous mentioned approaches. Instead of transforming the time series into a, let's say, static data sets, these approaches aim to learn from the dynamic and to learn the causes of change of state.

Furthermore, it is expected that using some transformations apart from the SMA -like the amount of movement and the intensity of the movement- would eventually allow us to cluster the time series from an epilepsy crisis. The feature selection needs further study. Besides, a very related to the approach to that of the clusters is the learning of Fuzzy Association Rules. The data that can be gathered from patients will include quite a few time series; the higher the number of time series and the larger they are, the worse the clustering might perform. However, the learning of Fuzzy Association Rules have been successfully tested in large data sets and big data problems; the challenge will be to choose a suitable representation of the time series.

Acknowledgments This research has been funded by the Spanish Ministry of Science and Innovation, under projects TIN2011-24302 and TIN2014-56967-R, Fundación Universidad de Oviedo project FUO-EM-340-13, Junta de Castilla y León projects BIO/BU09/14 and SACYL 2013 GRS/822/A/13.

References

1. Marinas A, Elices E, Gil-Nagel A, Salas-Puig J, Sánchez JC, Carreno M, Villanueva V, Rosendo J, Porcel J, Serratosa J (2011) Socio-occupational and employment profile of patients with epilepsy. Epilepsy Behav 21(3):223–227
2. Villanueva V, Girón J, Martín J, Hernández-Pastor L, Lahuerta J, Doz M, Lévy-Bachelot MCL (2013) Quality of life and economic impact of refractory epilepsy in spain: the espera study. Neurologia 28(4):195–204

3. Smith D, Defalla B, Chadwick D (1999) The misdiagnosis of epilepsy and the management of refractory epilepsy in a specialist clinic. QJM 92(1):15–23
4. Engel J Jr (2001) International league against epilepsy (ilae). A proposed diagnostic scheme for people with epileptic seizures and with epilepsy: report of the ilae task force on classification and terminology. Epilepsia 42(6):796–803
5. Nijsen T, Arends J, Griep P, Cluitmans P (2005) The potential value of three-dimensional accelerometry for detection of motor seizures in severe epilepsy. Epilepsy and Behavior 7(1):74–84
6. Nijsen T, Cluitmans P, Arends J, Griep P (2007) Detection of subtle nocturnal motor activity from 3-d accelerometry recordings in epilepsy patients. IEEE Trans Biomed Eng 54(11):2073–2081
7. Silva CJP, Rémi J, Vollmar C, Fernandes J, Gonzalez-Victores J, Noachtar S (2013) Upper limb automatisms differ quantitatively in temporal and frontal lobe epilepsies. Epilepsy Behav 27(2):404–408
8. Lockman J, Fisher R, Olson D (2011) Detection of seizure-like movements using a wrist accelerometer. Epilepsy Behav 20(4):638–641
9. Bonnet S, Jallon P, Bourgerette A, Antonakios M, Guillemaud R, Caritu Y, Becq G, Kahane P, Chapat P, Thomas-Vialettes F, Gerbi D, Ejnes D (2011) An ethernet motion-sensor based alarm system for epilepsy monitoring. IRBM 32(2):155–157
10. de Vel AV, Cuppens K, Bonroy B, Milosevic M, Huffel SV, Vanrumste B, Lagae L, Ceulemans B (2013) Long-term home monitoring of hypermotor seizures by patient-worn accelerometers. Epilepsy and Behav 26(1):118–125
11. Schulc E, Unterberger I, Saboor S, Hilbe J, Ertl M, Ammenwerth E, Trinka E, Them C (2011) Measurement and quantification of generalized tonic-clonic seizures in epilepsy patients by means of accelerometry- an explorative study. Epilepsy Res 95(1–2):920–1211
12. Villar JR, González S, Sedano J, Chira C, Trejo JM (2013) Human activity recognition and feature selection for stroke early diagnosis. In: Pan JS, Polycarpou MM, Wozniak M, de Carvalho A, Quintián H, Corchado E (eds) Hybrid Artificial Intelligent Systems, vol 8073, Lecture Notes in Computer Science. Springer, Berlin Heidelberg
13. Alvarez-Alvarez A, Triviño G, Cordón O (2012) Human gait modeling using a genetic fuzzy finite state machine. IEEE Transactions on Fuzzy Systems 20(2):205–223
14. Cilla, R., Patricio, M.A., García, J., Berlanga, A., Molina, J.M.: Recognizing human activities from sensors using hidden markov models constructed by feature selection techniques. Algorithms **2**, 282–300

A Preliminary Cooperative Genetic Fuzzy Proposal for Epilepsy Identification Using Wearable Devices

E.A. de la Cal, J.R. Villar, P.M. Vergara, J. Sedano and A. Herrero

Abstract The epilepsy is one of the neurological disorders that affects people of all socioeconomic groups and ages. An incorrect treatment or a lack in monitoring might produce cognitive damage and depression. In previous work we presented a preliminary method for learning a generalized model to identify epilepsy episodes using 3DACC wearable devices placed on the dominant wrist of the subject. The model was based on a Fuzzy Finite State Machines to detect the epilepsy episodes in 3DACC time series. The learning model applied was a classical Genetic Fuzzy Finite State Machine. The goal of the present work is to adapt the previous learning scheme to a Cooperative Coevolutionary Genetic Fuzzy Finite State Machine to improve the classification results. The obtained results show that a Cooperative proposal outperform moderately the results of the original proposal.

Keywords Cooperative coevolutionary genetic fuzzy finite state machine · Time series classification · Human activity recognition · Epilepsy identification

1 Introduction

The epilepsy is one of the neurological disorders that affects people of all socioeconomic groups and ages [1]. Basically, there are two main epileptic crisis types: the generalized and the focal crisis. In both of them there are subtypes with or without motor activity: those with motor activity are the most common cases. The motor

E.A. de la Cal (✉) · J.R. Villar · P.M. Vergara
Computer Science Department, University of Oviedo, Oviedo, Spain
e-mail: delacal@uniovi.es

J. Sedano
Instituto Tecnológico de Castilla y León, Burgos, Spain
e-mail: javier.sedano@itcl.es

A. Herrero
University of Burgos, Burgos, Spain
e-mail: ahcosio@ubu.es

Á. Herrero et al. (eds.), *10th International Conference on Soft Computing Models in Industrial and Environmental Applications*, Advances in Intelligent Systems and Computing 368, DOI 10.1007/978-3-319-19719-7_5

activity can vary from the generalized tonic-clonic crisis to the focal mio-clonic crisis. Where the tonic-clonic episodes includes the following symptoms: losing the consciousness, a short tonic stage followed by a prolonged generalized and repeated clonic movements of the whole body, on the other hand the focal mio-clonic epilepsy typology involves repeated bursting movements of one limb, the upper and lower limbs of one body side or a combination of movements of the limbs and face [2].

Classical methods to treat epilepsy include medication, brain stimulation, surgery, dietary therapy or various combinations of the above, directed toward the primary goal of eliminating or suppressing seizures [3]. But roughly the 30 % of patients suffer from medically refractory epilepsy. These cases have lead researchers to investigate the mechanisms of seizures in refractory epilepsy using techniques from many scientific disciplines, like molecular biology, genetics, neurophysiology, neuroanatomy, brain imaging and computer modeling [4]. But computer modeling has been used as a general method to tackle not only refractory epilepsy cases. Traditionally in medicine fields Computer Modeling stands for computer aided mathematical techniques used to obtain neurological models based on scalp and intracranial ElectroEncephaloGram (EEG) signals [4, 5]. But there are other computer models based on other kind of measures like electrocardiography, accelerometry and motion sensors, electrodermal activity, and audio/video captures [6, 7]. Considering the current state of the wearable technology, it can be stated that the use of personal devices with tri-axial accelerometer (3DACC) is a cheap and promising medium to monitor some kind of epilepsy [7–10].

In our previous work [11], a method for recognizing focal mio-clonic epilepsy episodes based on analyzing the motor activity by means of a 3DACC placed on the dominant wrist was presented. The main hypothesis in the previous work was that it is possible to learn general models that may allow us to classify the current activity as an epilepsy onset. It worth to remark that when deploying Activity Recognition (HAR) solutions, it is always feasible to gather data for the specific user so the model can be tuned to fit the current user: it is possible to define a training routine that the user has to accomplish with at the very beginning of using the solution. However, this tuning stage is not longer feasible when the problem is focused on the recognition of an illness episode or onset. Therefore, obtaining generalized models is a main requisite that any proposal that aims to recognize an illness episode or onset should perform.

In the previous work a method based on Genetic Fuzzy Finite State Machines (hereafter referred as to as the original proposal) was proposed to the capture of the generic model. But two main problems of the original proposal are the poor classification results and the so heavy computation cost because of the large search space. Thus in this work it's proposed a Cooperative Coevolutionary (hereafter referred as to as the coevolutionary proposal) scheme [12–14] to outperform the results of the original proposal.

The work is structured as follows: in Sect. 2 a revision of the literature is presented, while in Sect. 3 the Cooperative Coevolutionary proposal is explained. Finally, numerical results and the conclusions and future work are presented.

2 State of Art

2.1 Detect Epilepsy Using 3DACC Devices

Epilepsy detection algorithms involve two main phases: first, apposite quantitative values or features, such as EEG signals, movements, or other biomarkers, must be obtained from the data, second a threshold or model-based criteria must be applied to the features to determine the presence or absence of a seizure [7].

Considering Epilepsy has an important impact in the motor system, it can be stated that a big part of the works in the literature use a 3DACC [7, 15, 16]. Different proposals on this topic have been present during the last decade: in [17] a review on the extraction of characteristic epileptic patterns from data is presented, and they conclude that typical movements differ on the kind of epilepsy crisis. An study about 8 types of epilepsy crisis using a 3DACC brazelets to detect rhythmical movements in topic-clonic episodes was presented in [9] but the results are not mature to be deployed.

Likewise, in the work [18], the transformation "amount of movement" is used for calculating alarms in sleeping epileptic episodes in a controlled environment. The heuristic is very simple, when the amount of movement surpass a threshold then an epileptic episode is identified. Due to the use of thresholds, the study was limited to reduced movements activities. Also, in [19] an study on detecting epilepsy episodes in sleeping phase is tackled.

An interesting work is [20] where they use a combination of sympathetically mediated electrodermal activity (EDA) and accelerometry measured using a novel wrist-worn biosensor. In this work data are gathered in half-controlled environment, where the patients can do wide range of activities of daily living such as emotionally and physically activating games.

A good reference work on transformations for the data gathered from the sensors to recognize the epileptic episodes can be found in [10].

2.2 Automatic Techniques to Detect Epilepsy Using 3DACC

The second phase to detected epilepsy, called classification, might be as simple as thresholding a value or might require models derived from modern machine learning algorithms.

Several proposal are based on time domain and frequency domain-based algorithms, [21–23] trying to capture the pattern of frequency during crisis episodes. Other works use simple analytic techniques such as: [24] based on linear threshold function to determine nocturnal seizures, in [25] they use standard deviations of moving epochs and uses moving average filter to detect nocturnal seizures. Also [9, 19] try to obtain motor patters using different mathematical techniques. And finally in [26], an algorithm based on a time series comparison using Dynamic time warping (DTW) is presented.

All these works have a common restriction, they output specific models for each patient. Besides it can be stated the lack of application of intelligent analysis of data in above problem. Thus, in our previous work [11] we presented a general model to detect epilepsy based on Genetic Fuzzy Systems, but the results can be improved using an new model based on coevolutionary techniques.

2.3 Coevolutionary GFFS to Detect Epilepsy

Our preliminary proposal was based on a Fuzzy Rule Base System (FRBS) using a Pittsburg approximation [27, 28], so the search space in this kind of solution is huge. Thus, as Casillas et al. recommend in [12] a coevolutionary scheme for this kind of problem is suitable. Let's review how a Cooperative Coevolutionary FRBS (Fuzzy Rule Base System) works [12]. First it's important to define what a FRBS is [29]: basically is a set of IF-THEN fuzzy rules, whose antecedents and consequents are composed of fuzzy statements, related with the dual concepts of fuzzy implication and compositional rule of inference. Here the aim is to derive an appropriate knowledge base (KB) about the problem being solved. The KB stores the available knowledge in the form of fuzzy IF-THEN rules. It consists of the rule base (RB), comprised of the collection of rules in their symbolic forms, and the data base (DB), which contains the linguistic term sets and the membership functions defining their meanings. One of the most used automatic methods to perform the derivate task is the Genetic Algorithms (GA) [13]. When only the derivation of the RB is tackled, methods operate in only one phase. And the DB is commonly obtained from the expert information. However, methods that design both RB and DB are preferable since the automation is higher. These methods presents two different approaches: simultaneous derivation (RB and DB are evolved at the same time) and sequential derivation (the learning process is divided in several stages). In most cases, a sequential process by firstly learning the RB and then tuning the DB is considered

When the RB and the DB are simultaneously derived, the high dependency of both components make derivation process considerably more complex because the search space grows and the selection of a suitable technique is key. The direct decomposition of the KB derivation process (thus obtaining two interdependent components, learning of the RB and DB) makes coevolutionary algorithms with a cooperative approach [30] very useful for this purpose.

3 A Cooperative Coevolutionary Genetic Fuzzy Finite Machine to Detect Epilepsy in Time Series

In our previous work [11], it was presented a preliminary method for learning a generalized model to identify epilepsy episodes using 3DACC wearable devices placed on the dominant wrist of the subject. The model used in this proposal was based on Fuzzy Finite State Machines using one-population Genetic Algorithm [31]. The goal of the present work is to adapt the previous learning scheme to a Collaborative Coevolutionary Genetic Fuzzy System [12, 32] to improve the former results.

3.1 The Original Work

First of all, it's important to review the parts of the previous methodology proposed. Our methodology follows the typical stages in the methods to detect epilepsy [7]:

- Feature computation
    - Preprocessing or filtering
    - Feature computation
    - Feature reduction or extraction
- Model learning

Feature computation The data gathered from a 3DACC sensor are the three acceleration components (known as raw data (RD)), and includes both the gravity acceleration (G) and the body acceleration (BA) components. The BA depicts the real acceleration that the part of the body where the sensor is located is being affected. The relevance of each of these components or even of the RD varies according to the current problem and whether the sensor is placed. When the sensor is placed on the limbs, the variable that gets the focus is the BA as long as it resembles the current body movements.

It is worth mentioning that these features have been mainly documented for HAR using 3DACC, with different sensor number and body's placement. The transformations include, among others, **(i)** the first to fourth moment statistics -mean, standard deviation, skewness and kurtosis-, **(ii)** the *Mean Absolute Deviation* ($MAD_t(s)$) computed for the RD, **(iii)** the *Root Mean Square* ($RMS_t(s)$) for the RD, **(iv)** the *sum of the absolute values* of the BA, **(v)** the *vibration of the sensor* ($VoS_t(s)$), **(vi)** the *tilt of the body* for the RD, **(vii)** the *Signal Magnitude Area* ($SMA_t(s)$) for the BA, **(viii)** the *Amount of Movement* ($AoM_it(s)$) for the BA, **(ix)** the *Delta coefficients* for estimating the first order time derivate of each of the G signal components, **(x)** the *Shifted Delta Coefficients* for estimating the first order time derivate of each of the BA signal components, **(xi)** the *Average Energy* using the FFT, **(xii)** the *correlation between axes*, **(xiii)** the *Intensity of the movement* ($IoM_t(s)$) as the mean first derivative of the raw acceleration data, **(xiv)** the *Time Between Peaks* measured for any possible variable as the number of peaks that occurs within the current window, and **(xv)** the *histogram* of the values of the considered acceleration for the current window. An interested readers can find studies of these transformations in [10, 33].

A problem that arises with all these transformations is that, for each concrete problem -where a number of sensors and their locations are specifically chosen-, only the most suitable transformations should be selected for the modeling stage. Thus, the feature domain reduction step is the very first stage that should be analyzed in Time Series (TS) classification.

Furthermore, we are interested on the movements of the limb, which is measured using the BA so, we only consider the modulus of BA and its related transformations.

The tilt of the body has no point as long as it only measures the relevance of two components, thus this feature is more interesting when the sensor is placed centered with the body. Similarly, the correlation between axes was disregarded as well. The VoS only considers the raw data acceleration and the G, both of them have been neglected for this problem. Furthermore, the sum of absolute values and the SMA represent the same transformation with different scaling factor.

After applying different selection criteria like disregarding by computational restrictions, histogram comparison and visual analysis, following three transformations based on BA were selected:

- Signal Magnitude Area (SMA).
- Amount of Movement (AoM).
- Time between peaks (TbP).

The model The problem to detect the state that better resembles the current sliding windows of an acceleration samples Time Series, can be tackled with a Fuzzy Finite State Machine (FFSM). Following the reason proposed in [31], it can be assumed the problem of identifying the class can be represented by means of a $FFSM = \{Q, X, f, Y, g\}$, where $Q = \{q_1, \cdots, q_p\}$ is the set of p fuzzy states, X is the set of input variables of the system, $f : Q \times X \rightarrow Q$ is the set of transition mapping functions, Y is the set of outputs, and $g : Q \times X \rightarrow Y$ is the output function. Let's define the elements of our FFSM:

- The fuzzy states to represent Q are {EPILEPSY, NO_EPILEPSY} and are defined as $S(t)$, where $S[t] = (s_1[t], \cdots, s_p[t]) = (s_1, \cdots, s_p)$ is the membership value of current state at time t to each of the available states q_i, with $s_k \in [0.0, 1.0]$ and $\sum_{k=1}^{p} s_k[t] = 1, \forall t$.
- X is the set of features that best describes the evolution of the class for the different time series. In our problem X is {SMA, AoM and TbP}.
- The transitions functions $f_r, \forall r \in \{1, \cdots, R\}$ are represented by means of Fuzzy rules (FR) with the form depicted in Eq. 1, where $\otimes$ is the t-norm for aggregation of the antecedents of the rule, $\oplus$ is the t-conorm for aggregation of the different fuzzy sets involved in the rule for an input variable.

$$\textbf{IF } (S[t] \text{ is } S_r^a) \otimes C_r^1 \otimes \cdots \otimes C_r^N \textbf{ THEN } S_r \tag{1}$$

In this Eq. 1, S_r^a and S_r are the state from which the rule r fires and the destination state, respectively, where C_r^n is the combination of fuzzy sets for variable x_n that are involved in rule r as an antecedent. We have decided to use 3 fuzzy trapezoidal partitions for each fuzzy set.
Here we follow the original proposal by Alvarez et al. [31] using the *min* function as t-norm, and the Lukasiewicz t-conorm ($\vee_{Luk}(a, b) = min(a + b, 1)$). Besides, we adopt the inference mechanism, the authors proposed computing the new state with their proposal of formula to calculate the strength of one rule.
- As for the original study, [31] $Y[t] = g(S[t], X[t]) = S[t]$

When a problem is focused using this FFSM representation several design decisions must be taken, like the specific input feature domain, the partitioning of the variables, the parameterization of the FR, etc. Once all these issues have been defined, it is possible to propose a method for learning either the partitioning, or the FR Base or both simultaneously. According to [31], the GFFSM has been found valid for learning TS classification models in the context of HAR. However, the approach should be adapted to the specific problem as explained in [33]. In our previous work a classical GFFSM scheme was used.

3.2 The New Proposal

In this new proposal, a Cooperative Coevolutionary Genetic Fuzzy Finite State Machine is applied following the proposal of Casillas et al. [12] but adapted to a FFSM. Let's define F_{ij} as the FRBS composed by the individuals i and j of the species 1 (RBs) and 2 (DBs) respectively. The fitness function is the Mean Absolute Error (MAE), which is calculated using Eq. 2. In this equation, N is the number of datasets in the corresponding fold, T is the number of examples in each data set, and si[t] and si*[t] are the degree of activation obtained from a designed FRBS F_{ij} and the expected degree of activation, respectively, for state qi at time step t = e. Let's $s_d[t] = F(t)$

$$MAE(F) = \frac{1}{N}\frac{1}{T}\sum_{d=1}^{N}\sum_{e=0}^{T} |s_d[t] - s_d^*[t]| \tag{2}$$

Each individual of species RB y DB are evaluated with corresponding fitness functions f1 and f2 (see Eq. 3) :

$$\begin{aligned} f1(i) &= min_{j \in C2} MAE(F_{ij}) \\ f2(j) &= min_{i \in C1} MAE(F_{ij}) \end{aligned} \tag{3}$$

where i and j are the individuals of species RB and DB respectively. And C_1 and C_2 are sets individuals selected at random from the previous population (generation = $t - 1$). The summary of fitnesses f1 and f2 evaluation is showed in Fig. 1.

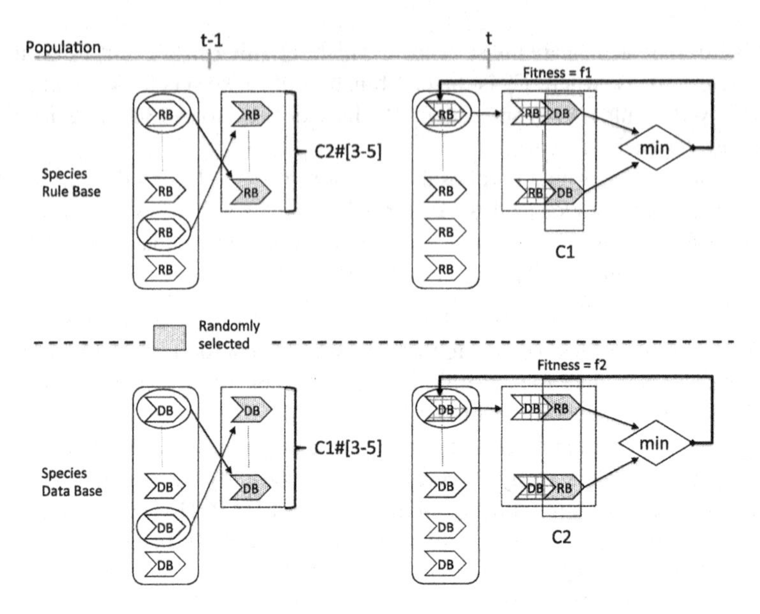

Fig. 1 Cooperative coevolutionary Fitness calculation scheme

The genotype of both species As three input variables are used, the SMA, the AoM and TbP, everyone considered linguistic variables with 3 labels each. Ruspini trapezoid membership functions are used, so we need to learn 4 parameters for each input variable. A GA evolves the partitions and the rules in a Pittsburg style: up to 36 binary genes coding the rules (the RB species) and 12 real-coded genes for the membership function parameters (the DB species).

Genetic operators: A generational scheme is followed in both species. Tournament selection procedure of 4 individuals together with an elitist mechanism (that ensures to select the best individual of the previous generation) are used. A two-point crossover for the RB species and BLX-α (α = 0.5) crossover for the DB species applied twice for obtaining two new pairs of chromosomes and a classical bitwise mutation for the rule base (RB species) and uniform mutation for the real-coded part (DB species). It has used a probability of crossover and mutation of 1.0 and 0.1 respectively.

4 Experimentations and Results

4.1 Experimentation

A set of experiments have been run in order to obtain the data for training and testing. To do so, a bracelet including a 3DACC has been delivered to every subject being involved in these experiments; each bracelet was placed on the subject's dominant wrist. The bracelets have wireless capabilities, transferring the data to a computer at a sample rate of 16 Hz, the range of the sensors fixed to $2 \times g$.

Nevertheless in current work a simple protocol with two similar activities (EPILEPSY and SAWING) simulated by a group of subjects has been defined in order to gather realistic data.

Up to 6 individuals with ages between 22 and 47 yeas, have successfully completed this experimentation, all of them healthy subjects. Every subject has carried out at least 10 runs per kind of activity.

To run the training and test computation process a 5×2 Cross Validation procedure has been performed with all datasets taken from all the subjects.

The main goal in this paper is to compare our Coevolutionary proposal against the original proposal from the previous work. Thus, the parameters population size (ps) and number of generations (g) in both learning schemes have been adjusted in order to perform the same number of fitness evaluations in the whole training process. So, let's define the number of fitness evaluations (NFE) for the whole generations in both learning schemes (see Eq. 4).

$$\begin{aligned} g_{\text{original}} &= \text{Number of generations (50)} \\ ps_{\text{original}} &= \text{Population Size (50)} \\ np_{\text{original}} &= \text{Number of populations(1)} \\ NFE_{\text{original}} &= g_{\text{original}} * ps_{\text{original}} * np_{\text{original}} \\ NFE_{\text{original}} &= 2500 \end{aligned} \tag{4}$$

Considering the $\text{NFE}_{\text{coevol}}$ must be equal to $\text{NFE}_{\text{original}}$ (2500), the g and the ps have to be adjusted. If ps is fixed to 20 then we have to calculate g_{coeval} (see Eq. 5).

$$\begin{aligned} g_{\text{coevol}} &= \text{Number of generations} \\ ps_{\text{coevol}} &= \text{Population Size(20)} \\ np_{\text{coevol}} &= \text{Number of populations(2)} \\ cn &= \text{Number of Cooperators(3,5)} \\ NFE_{\text{coevol}} &= g_{\text{coevol}} * ps_{\text{coevol}} * np_{\text{coevol}} * nc \\ NFE_{\text{coevol}} &= g_{\text{coevol}} * 20 * 2 * [3-5] \\ g_{\text{coevol}} * 20 * 2 * [3,5] &= 2500 \\ g_{\text{coevol}} &= [2500/(40*5), 2500/(40*3)] \\ g_{\text{coevol}} &= [12.5, 20.83] \end{aligned} \tag{5}$$

Consequently, we have three types of experiments attending to the number of generations: the original proposal with 50 generations and the coevolutionary proposal with 3 cooperators (20 generations) and 5 cooperators (12 generations).

On the other hand, since the search space in this problem is huge it has be decided to inject in the initial population an expert individual in order to guide the evolution of the genetic algorithm. This expert individual is a manually selected good solution but not the best. Besides, two percentage of elite individuals, 10 % and 50 %, have used in order to promote the expert individuals. The original proposal was run only with 50 % of elite individuals, while the new proposal was run with 10 % and 50 %.

Thus, we will compare different combination of genetic parameters for the coevolutionary proposal against the original proposal using a fixed set of parameters (see Table 1).

4.2 Numerical Results

Considering the 5 sets of schemes/parameters in Table 1, a 5×2 fold cross validation procedure has been run for each one using the 60 % of the available datasets. The remaining datasets have been kept to reduce the computation time.

Figure 2 depicts box plots for the error (MAE) of individuals with the best fitness in the testing partitions. The best fitness individual for a fold is the combination of the best sub-individual in each species (the Rule Base population and the Data Base population) in this fold.

Table 1 Experiments and configuration summary, where CV is Cross Validation, PS is Populations Size, CN is Cooperators Number, Gen is the number of generations, Cross./Mut P. are the crossover/mutation probability and Elite P. is the elite probability

Scheme	PS	CN	Gen.	Cross./Mut. P.	Elite P.
Original	50	-	50	1.0/0.1	50 %
Coevolutionary	20	3	20	1.0/0.1	10 %
Coevolutionary	20	3	20	1.0/0.1	50 %
Coevolutionary	20	5	12	1.0/0.1	10 %
Coevolutionary	20	5	12	1.0/0.1	50 %

We think that box plots don't allow to conclude statistically that coevolutionary proposals (Coevol. Cooper. box plots) are much better that the original proposal (Evolutionary labeled boxplot). May be the configuration Coevol. Cooper with 5 cooperators and 50 % of elite individuals shows better behavior having the best median of the whole configurations. Anyway the statistical results are not determining. it seems that a bigger number of cooperators improves lightly the results. Also we think that a new coevolutionary fitness function considering a mixture of random and elite selected cooperators could improve the results. In current coevolutionary proposal all the cooperators are selected at random.

Considering the best coevolutionary proposal is the one with 5 cooperators and 10 % of elite probability, let's compare the evolution of this proposal against the original proposal in training partitions. Figure 3 shows the evolution of the fitness for best individual in the original proposal (square marker) combined with the evolution of the fitness for two best sub-individuals in coevolutionary species (circle and triangle

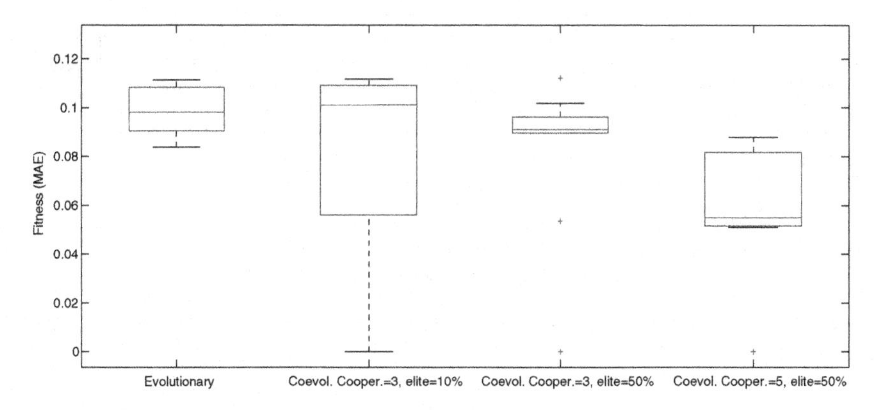

Fig. 2 Testing boxplots for 5 × 2 CV for original proposal (Evolutionary) vs coevolutionary proposal (Coevol. Cooper.)

markers). It can be stated that coevolutionary best solutions in both species outperform evolutionary solution during the 12 generations. The evaluation of fitness for the coevolutionary sub-individuals was performed taken the best combination with the cooperators in each generation (see Fig. 1).

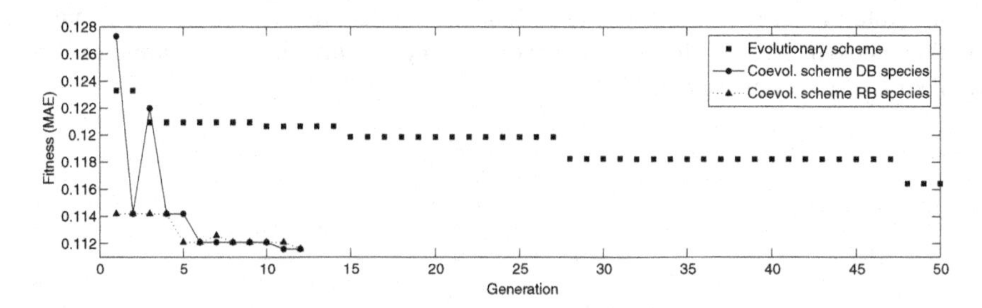

Fig. 3 Training best fitness evolution, original proposal (Evolutionary scheme, 50 generations) vs Coevol. Cooper. proposal (Coevolutionary scheme, 12 generations)

Finally, let's analyze the evolution of the coevolutionary proposal being evaluated with testing datasets. Figure 4 shows the fitness values for the coevolutionary proposal with the four sets of parameters. The fitness function has been calculated for the combination of the best sub-individuals of the both species in each generation using the testing datasets. It's not clear which configuration is better, may be the ones with 5 cooperators (solid line curves) with present a more regular behavior. Also, it can be observed that testing results are better than training results, and this is due to the distribution of EPILEPSY datasets. The datasets are not balanced.

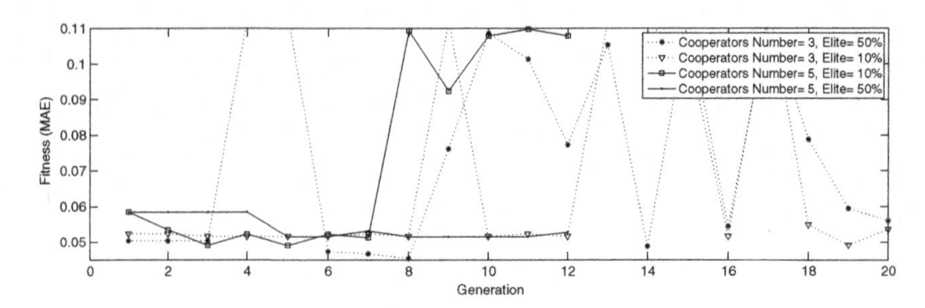

Fig. 4 Testing best fitness evolution, Coevolutionary proposal with 3 (20 generations) and 5(12 generations) cooperators

4.3 Time Performance

As it's said in previous Sect. 2.3, coevolutionary schemes are suitable for problems with a a huge space of search. So, if we take a look to Table 2 we can conclude that the extra time (see formula 6) spent by coevolutionary proposals (Coevol. Cooperators/Elite) respect to the original proposal (with speedups between 0.90 % and 0.93 %) is affordable. Also it can be observed that the reduction of generations allow to each coevolutionary scheme to invest more time in the fitness evaluation (Fit. Eval). Note that coevolutionary proposal has to evaluate the fitness for two populations considering that the fitness for coevolutionary individuals is more complex due to the use of the cooperators.

$$SpeedUp = \frac{T_{old}}{T_{new}} \tag{6}$$

Table 2 Time data obtained with the execution of the different proposals for 1 fold (Time measured in seconds). Fit. Eval. stands for the time invested in fitness evaluation in one generation, Total Fit. Eval = Generations*Fit. Eval, Remaining Ops. is the time invested in the genetic operations (initialize the population, crossover operations, etc.), but the Fitness function

Scheme	Generations	Fit.Eval	Total Fit.Eval.	Remaining Ops.	Total Time	SpeedUp
Evolutionary	50	285,00	14249,81	332,41	14582,22	-
Coevol 3/10	20	766,24	15324,79	753,30	16078,09	0,91
Coevol 3/50	20	748,31	14966,13	749,32	15715,45	0,93
Coevol 5/10	12	1228,66	14743,94	1217,54	15961,48	0,91
Coevol 5/50	12	1249,46	14993,57	1248,23	16241,79	0,90

5 Conclusions and Future Work

In a previous work we got a generalized model based on a GFFSM to identify epilepsy episodes using 3DACC wearable but the results could be improved. Thus, in current paper a Cooperative Coevolutionary proposal based on Genetic Fuzzy Systems is presented. It can be stated that the statistical results obtained for the whole configurations of new proposal don't outperform clearly the original proposal ones, since all the bloxplots are not determining (see Fig. 2). Only the best median of MAE error from the different coevolutionary configurations (5 cooperators and 50 % elite got a median of 5.8 %) outperforms the original proposal MAE of 9.8 %. The remaining coevolutionary boxsplots overlap the original proposal boxplot.

Besides, as we can see in Sect. 4.3, the time spent by the coevolutionary proposals don't surpass the original proposal in more than a 10 %, what is a modest overtime that can be assumed.

As current paper only analyze datasets from only two typologies of activities EPILEPSY and SAWING, more activities like WALKING and RUNNING must be included in future works in order to obtain a more general model. Also we think that a new coevolutionary fitness function considering a mixture of random and elite selected cooperators could improve the results; current coevolutionary proposal all the cooperators are selected at random. Besides a more extended study of the genetic parameters is needed.

Acknowledgments This research has been funded by the Spanish Ministry of Science and Innovation, under projects TIN2011-24302 and TIN2014-56967-R, Fundación Universidad de Oviedo project FUO-EM-340-13, Junta de Castilla y León projects BIO/BU09/14 and SACYL 2013 GRS/822/A/13.

References

1. Villanueva V, Girón J, Martín J, Hernández-Pastor L, Lahuerta J, Doz M, Lévy-Bachelot MCL (2013) Quality of life and economic impact of refractory epilepsy in spain: the espera study. Neurologia 28(4):195–204
2. Engel JJ (2001) International league against epilepsy (ilae). a proposed diagnostic scheme for people with epileptic seizures and with epilepsy: report of the ilae task force on classification and terminology. Epilepsia 42(6):796–803
3. Shorvon S (2010) Handbook of epilepsy treatment. Wiley-Blackwell
4. Stefanescu RA, Shivakeshavan R, Talathi SS (2012) Computational models of epilepsy. Seizure 21(10):748–759
5. Holt AB, Netoff TI (2013) Computational modeling of epilepsy for an experimental neurologist. Exp Neurol 244(0):75–86 (Special Issue: Epilepsy)
6. Becq G, Bonnet S, Minotti L, Antonakios M, Guillemaud R, Kahane P (2011) Classification of epileptic motor manifestations using inertial and magnetic sensors. Comput Biol Med 41(1):46–55
7. Ramgopal S, Thome-Souza S, Jackson M, Kadish NE, Fernndez IS, Klehm J, Bosl W, Reinsberger C, Schachter S, Loddenkemper T (2014) Seizure detection, seizure prediction, and closed-loop warning systems in epilepsy. Epilepsy Behav 37:291–307

8. Van de Vel A et al (2011) P26.3 accelerometers for detection of motor seizures during sleep in pediatric patients with epilepsy. Eur J Paediatr Neurol 15(Supplement 1(0)):S134
9. Lockman J, Fisher RS, Olson DM (2011) Detection of seizure-like movements using a wrist accelerometer. Epilepsy Behav 20(4):638–641
10. Schulc E, Unterberger I, Saboor S, Hilbe J, Ertl M, Ammenwerth E, Trinka E, Them C (2011) Measurement and quantification of generalized tonic-clonic seizures in epilepsy patients by means of accelerometry- an explorative study. Epilepsy Res 95(1–2):920–1211
11. Villar JR, Menéndez M, de la Cal E, González VM, Sedano J (2015) Obtaining general models for epilepsy episode recognitions. Inf Sci (2015 submitted)
12. Casillas Jorge, Cordón Óscar, Herrera Francisco, Merelo Juan Julián (2002) Cooperative Coevolution for Learning Fuzzy Rule-Based Systems. In: Collet Philippe, Fonlupt Cyril, Hao J-K, Lutton Evelyne, Schoenauer Marc (eds) EA 2001, vol 2310., LNCSSpringer, Heidelberg, pp 311–322
13. Herrera F (2008) Genetic fuzzy systems: taxonomy, current research trends and prospects. Evol Intell 1(1):27–46
14. Coello CAC, Lamont GB, Veldhuizen DAV (2007) Evolutionary algorithms for solving multi-objective problems (genetic and evolutionary computation). Springer
15. Patel S, Park H, Bonato P, Chan L, Rodgers M (2012) A review of wearable sensors and systems with application in rehabilitation. Journal of neuroengineering and rehabilitation 9 (April 2012) 21+
16. Cogan D, Pouyan M, Nourani M, Harvey J (2014) A wrist-worn biosensor system for assessment of neurological status. In: 36th annual international conference of the IEEE engineering in medicine and biology society 5748–5751
17. Silva CJP, Rémi J, Vollmar C, Fernandes J, Gonzalez-Victores J, Noachtar S (2013) Upper limb automatisms differ quantitatively in temporal and frontal lobe epilepsies. Epilepsy Behav 27(2):404–408
18. Bonnet S, Jallon P, Bourgerette A, Antonakios M, Guillemaud R, Caritu Y, Becq G, Kahane P, Chapat P, Thomas-Vialettes B, Thomas-Vialettes F, Gerbi D, Ejnes D (2011) An ethernet motion-sensor based alarm system for epilepsy monitoring. IRBM 32(2):155–157
19. Van de Vel A et al (2013) Long-term home monitoring of hypermotor seizures by patient-worn accelerometers. Epilepsy Behav 26(1):118–125
20. Poh M, Loddenkemper T, Reinsberger C, Swenson N, Goyal S, Sabtala M, Madsen J, Picard R (2012) Convulsive seizure detection using a wrist-worn electrodermal activity and accelerometry biosensor. Epilepsia 5(53):93–97
21. Nijsen T, Aarts R, Cluitmans P, Griep P (2010) Time-frequency analysis of accelerometry data for detection of myoclonic seizures. IEEE Trans Inf Technol Biomed 14:1197–1203
22. Beniczky S, Polster T, Kjaer T, Hjalgrim H (2013) Detection of generalized tonicclonic seizures by a wireless wrist accelerometer: a prospective, multicenter study. Epilepsia 4(54):e58–61
23. Kramer U, Kipervasser S, Shlitner A, Kuzniecky R (2011) A novel portable seizure detection alarm system: preliminary results. J Clin Neurophysiol 4(28):36–8
24. Nijsen T, Cluitmans P, Arends J, Griep P (2007) Detection of subtle nocturnal motor activity from 3-d accelerometry recordings in epilepsy patients. IEEE Trans Biomed Eng 54(11):2073–2081
25. Cuppens K, Lagae L, Ceulemans B, Van Huffel S, Vanrumste B (2009) Detection of nocturnal frontal lobe seizures in pediatric patients by means of accelerometers: a first study. In: Conference Proceedings IEEE Engineering Medicine and Biology Society 6608–11
26. Dalton A, Patel S, Chowdhury A, Welsh M, Pang T, Schachter S et al (2012) Development of a body sensor network to detect motor patterns of epileptic seizures. IEEE Trans Biomed Eng 59:3204–11
27. Tan CH, Yap KS, Yap HJ (2012) Application of genetic algorithm for fuzzy rules optimization on semi expert judgment automation using pittsburg approach. Appl Soft Comput 12(8):2168–2177

28. Tian J, Li M, Chen F (2010) Dual-population based coevolutionary algorithm for designing rbfnn with feature selection. Expert Syst Appl 37(10):6904–6918
29. Fernández A, López V, del Jesus M, Herrera F (2015) Revisiting evolutionary fuzzy systems: Taxonomy, applications, new trends and challenges. Knowl-Based Syst (In Press, Accepted Manuscript, February 2015)
30. Potter MA, Jong KAD (2000) Cooperative coevolution: an architecture for evolving coadapted subcomponents. Evol Comput 8:1–29
31. Alvarez-Alvarez A, Triviño G, Cordón O (2012) Human gait modeling using a genetic fuzzy finite state machine. IEEE Trans Fuzzy Syst 20(2):205–223
32. Peña Reyes CA, Sipper M (2001) Fuzzy coco: a cooperative-coevolutionary approach to fuzzy modeling. IEEE Trans Fuzzy Syst 9(5):727–737
33. Villar J, Gonzlez S, Sedano J, Chira C, Trejo-Gabriel-Galan J (2014) Improving human activity recognition and its application in early stroke diagnosis. Int J Neural Syst 10:1–20

Analysis of Knowledge Management in Industrial Sectors by Means of Neural Models

Álvaro Herrero, Emilio Corchado, Lourdes Sáiz-Bárcena and Miguel A. Manzanedo

Abstract It is required for an organization, before successfully applying a Knowledge Management (KM) methodology, to develop and implement a knowledge infrastructure, consisting of people, organizational and technological systems. Up to now, few approaches have been proposed for such technological systems supporting KM in organizations. Present paper advances previous work by proposing neural projection models for the analysis of the KM status of companies from two different industrial sectors. Exploratory methods are applied to real-life case studies to know and understand the structure of KM data. Subsequently, the application of such models generates meaningful conclusions that allow experts to diagnose KM from two different points of view: companies on the one hand and industrial sectors on the other hand.

Keywords Knowledge management · Unsupervised neural networks · Exploratory projection pursuit

Á. Herrero (✉) · L. Sáiz-Bárcena · M.A. Manzanedo
Department of Civil Engineering, University of Burgos, Avenida de Cantabria s/n, 09006 Burgos, Spain
e-mail: ahcosio@ubu.es

L. Sáiz-Bárcena
e-mail: lsaiz@ubu.es

M.A. Manzanedo
e-mail: mmanz@ubu.es

E. Corchado
Departamento de Informática y Automática, Universidad de Salamanca, Plaza de la Merced s/n, 37008 Salamanca, Spain
e-mail: escorchado@usal.es

Á. Herrero et al. (eds.), *10th International Conference on Soft Computing Models in Industrial and Environmental Applications*, Advances in Intelligent Systems and Computing 368, DOI 10.1007/978-3-319-19719-7_6

1 Introduction

Knowledge Management (KM) [1] means that organizations can capture and share the collective experience and the know-how (knowledge) of their employees and apply their knowledge in intelligent ways [2]. In an environment such as today's, where everything changes at great speed and almost nothing remains static, it could be said that knowledge emerges as the key factor in any economy [3]. A firm requires both "general" knowledge and "specific/singular" knowledge, which will permit the firm to pursue excellence alongside others. This class of first-level knowledge is held by a small number of people, without forgetting that more select and sophisticated knowledge is required, if possible, in pioneering firms, such as those analyzed in this study, in an area where the offer is, unfortunately, further and further away from satisfying demand.

Companies from different industrial sectors in Burgos (Spain) are analyzed in present study. These companies are in a dynamic environment characterized by: high levels of competitiveness; clients with increasing demands that know their needs and how to satisfy them; a need for personalized products and services; the existence of novel techniques that require professionals to have a knowledge and a good command; highly-qualified providers; disconcerting changes and new problems that must be addressed.

Under such circumstances, traditional sources for competitive advantages (such as physical, financial and technological assets, access to raw material or special markets, and list of clients) are not enough because these sources are available for the majority of companies, subject to the same conditions.

To effectively compete, companies must focus on those resources and capabilities that are truly valuable: difficult to get by other companies, with a positive effect on the business, being irreplaceable for the company, highly complementary to other resources/capabilities, low accessible to competitors. At the same time, it is crucial to promote and strengthen the development of the own knowledge, leading to a distinctive identity and personality in the processes and activities the companies carry out. KM studies these changes, and the forecasting of them, trying to respond to the above mentioned challenges by designing and developing concepts, tools and management models.

Present study aims at analyzing the effect (at company and industrial sector levels) of some KM practices on competitiveness, success and survival. The study at the industrial sector level is important because both opportunities and menaces are the same for the companies in the sector. Among the analyzed practices, it is worth mentioning the competitive advantages, critical capabilities and pieces of knowledge, used resources, formalize and document the available knowledge, needed ways of learning, which one of those are more profitable, how do they have to design efficient strategies for knowledge transfer, share, and capitalization. To sum up, the study intends to generate interesting guidelines about some questions that have not been addressed up to now in the KM field and lead company activity to take advantage of KM.

To do that, present research proposes the application of neural projection models [1, 4, 5] (described in Sect. 2) to generate intuitive visualizations of KM data by reducing their dimensionality. Thanks to such a visualization, the inner structure of the dataset is revealed. A comprehensive analysis of such structure by a KM expert, leads to interesting conclusions about the KM status of both companies and industrial sectors. To validate the proposed tool, as a technological support for KM, it has been validated against two real-life datasets, that are related with two different industrial sectors; on the one hand the Electrical and Telecommunications industries and, on the other hand, Timber industries.

The proposed neural approach is presented in Sect. 2, while details about the data and the experiments are provided in Sect. 3. Section 4 presents the derived conclusions and proposed future work.

2 Neural Visualization

Projection models [6] operate on the spatial coordinates of high-dimensional data, in order to project them onto lower dimensional spaces. The main goal is to identify the patterns that exist across dimensional boundaries by identifying "interesting" directions, in terms of any specific index or projection. Such indexes or projections are, for example, based on the identification of directions that account for the largest variance of a data set –i.e. Principal Component Analysis (PCA) [7, 8] - or the identification of higher-order statistics such as the skew or kurtosis index –i.e. Exploratory Projection Pursuit (EPP) [6]. Having identified the most interesting projections, the data is then projected onto a lower dimensional subspace, plotted onto two or three dimensions, which makes it possible to examine its structure with the naked eye.

The combination of projection techniques together with the use of scatter plot matrices is a very useful visualization tool to investigate the intrinsic structure of multidimensional data sets, allowing experts to study the relations between different components, factors or projections, depending on the technique that is applied.

The solution proposed in this research applies an unsupervised neural model called Cooperative Maximum Likelihood Hebbian Learning (CMLHL) [4]. It is based on Maximum Likelihood Hebbian Learning (MLHL) [4], and introduces the application of lateral connections [4] derived from the Rectified Gaussian Distribution [9]. This connectionist model has been chosen because it reduces the data dimensionality while preserving the topology in the original data set. Considering an N-dimensional input vector (x), and an M-dimensional output vector (y), with W_{ij} being the weight (linking input j to output i), then CMLHL can be expressed as:

1. Feed-forward step:

$$y_i = \sum_{j=1}^{N} W_{ij} x_j, \forall i. \tag{1}$$

2. Lateral activation passing:

$$y_i(t+1) = [y_i(t) + \tau(b - Ay)]^+. \quad (2)$$

3. Feedback step:

$$e_j = x_j - \sum_{i=1}^{M} W_{ij} y_i, \forall j. \quad (3)$$

4. Weight change:

$$\Delta W_{ij} = \eta . y_i . sign(e_j) |e_j|^{p-1}. \quad (4)$$

where: η is the learning rate, τ is the "strength" of the lateral connections, b the bias parameter, p a parameter related to the energy function and A a symmetric matrix used to modify the response to the data [4]. The effect of this matrix is based on the relation between the distances separating the output neurons. This neural projection model has been previously applied to the KM field [1].

3 Experiments & Results

As previously mentioned, neural visualization models have been applied to analyze data coming from companies in different industrial sectors. The analyzed data, as well as the obtained results are described in this section.

3.1 Dataset

For the proposed analysis, data coming from 44 companies in two different sectors (Electrical and Telecommunications industries, and Timber industries) have been collected. The headquarters of all the studied companies are located in Burgos (Spain). Many different features have been collected for each company, related to four main areas:

- Learning and Knowledge.
- KM Practices
- Contribution and Competitiveness
- Sector and company characteristics.

Data from those hour areas amount to 146 features. The complete analysis is too large for present paper so, only features related to the first area (learning and knowledge) are addressed. The 33 analyzed features are organized in some different subareas: competitive advantages, capabilities for competitive advantages, KM elements, ways of learning, knowledge documentation and given situations. Information about the features and their values is shown in Table 1.

Table 1 Data features for each one of the companies in the dataset

#	Feature name	Value range
Competitive advantages		
1	Product/service	0 (absent) and 1 (present)
2	Customer service	0 (absent) and 1 (present)
3	Innovation	0 (absent) and 1 (present)
4	Company's image/brand	0 (absent) and 1 (present)
5	Employees	0 (absent) and 1 (present)
6	Agility/adaptation	0 (absent) and 1 (present)
7	Technology	0 (absent) and 1 (present)
8	Company's management/organization	0 (absent) and 1 (present)
Capabilities for competitive advantages		
9	Human resources	0 (absent) and 1 (present)
10	Knowledge/experience/know-how	0 (absent) and 1 (present)
11	Innovation/Competitiveness/Design	0 (absent) and 1 (present)
12	Organization	0 (absent) and 1 (present)
13	Company's nature	0 (absent) and 1 (present)
KM elements		
14	Patents and brands	0 (absent) and 1 (present)
15	Processes	0 (absent) and 1 (present)
16	Experience	0 (absent) and 1 (present)
17	Technology	0 (absent) and 1 (present)
18	Innovation	0 (absent) and 1 (present)
Ways of learning		
19	Several employees working together	0 (absent) and 1 (present)
20	Experienced employees working with unexperienced ones	0 (absent) and 1 (present)
21	Some employees get deep knowledge in specialized areas	0 (absent) and 1 (present)
22	Internal and external training courses	0 (absent) and 1 (present)
23	New duties are assumed by employees on a regular basis	0 (absent) and 1 (present)
24	New employees, specialist in a certain issue, join the company	0 (absent) and 1 (present)
25	Collaboration with competitor companies	0 (absent) and 1 (present)
26	Opinion of clients and suppliers is taken into account	0 (absent) and 1 (present)
27	External staff is subcontracted	0 (absent) and 1 (present)
Knowledge documentation		
28	Level of knowledge documentation	1 (low) to 4 (high)
Given situations		
30	When a problem arises, the company knows who should be contacted	1 (no) to 4 (totally)
31	New ideas and proposals are welcomed	1 (no) to 4 (totally)
32	There are collaboration and knowledge exchange among employees, clients and suppliers	1 (no) to 4 (totally)
33	Experienced employees teach the other ones	1 (no) to 4 (totally)

3.2 Results

The obtained projections, together with the conclusions derived from them are described in this section. Each company in the dataset is depicted as a single point (clack square) in the figures below.

The PCA projection of the dataset is shown in Fig. 1. A clear structure of the dataset can not be identified from this projection as groups of data are not easily identified. However, the ordering of data in such a visualization has been analyzed by studying every single data. As a result, it can be said that companies in the right side of the projection are those with best positions regarding competitive advantages, capabilities, ways of learning and given situations. Complementarily, companies in the left side of the projection are those in the worst situations, getting worse and worse while moving down, to the bottom of the projection. These are companies with only one competitive advantage, only one capability, and the ways of learning and knowledge documentation are poor.

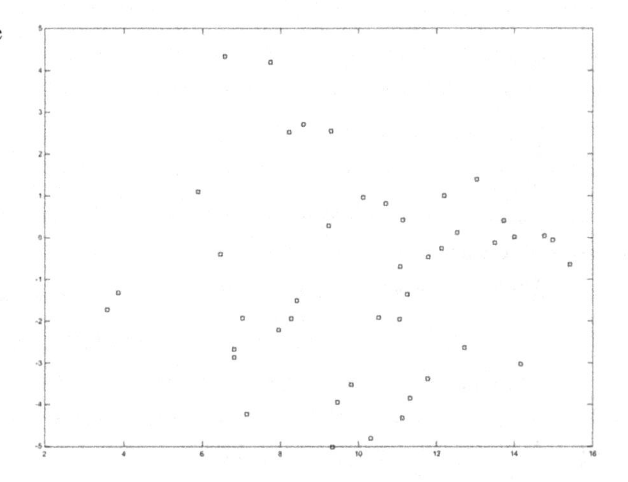

Fig. 1 PCA projection of the dataset

Figure 2 shows the projection of the dataset by CMLHL. According to the location of each data (company) in the projection, some conclusions arise, as described below.

From the perspective of the "competitive advantages" group of features, it has been checked that those data located in the first upper third of Fig. 2 represent companies with all the competitive advantages proposed in present study. In the middle of the projection, CMLHL places those companies with about 3 or 4 of the competitive advantages (product/service, customer service, employees, and company's management/organization). Those companies at the bottom only attain one competitive advantage: agility/adaptation (left side) or Innovation (right side). The meaning of this ordering is that the companies in the upper right third of Fig. 2 (groups 2.1 and 2.2) are those that have been able to develop and apply knowledge for the creation of an important amount of competitive advantages. From top to bottom, the companies are ordered according to this amount of competitive advantages. By applying CMLHL, it can be easily identified those companies that excel in applying knowledge to get competitive advantages. Additionally, there are some other companies in the way to get all the advantages or some other elements to improve their competitiveness.

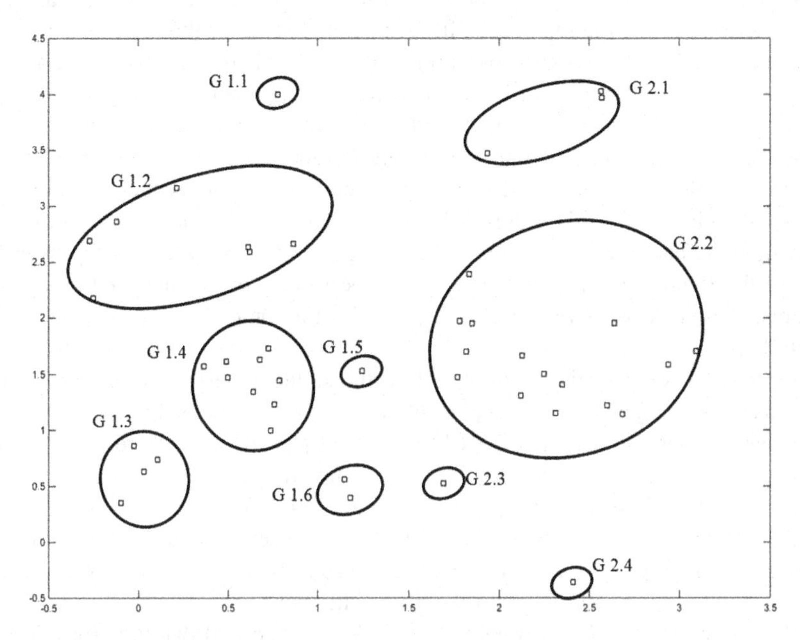

Fig. 2 CMLHL projection of the dataset

With regard to the "capabilities for competitive advantages" group of features, the ordering is similar to the one for "competitive advantages" but, in this case, it is more clearly defined. All the companies in the upper right third of Fig. 2 are those with the following capabilities: "Human resources", "Knowledge/experience/know-how" and "Organization". From the top to the bottom of the figure, some of these capabilities are not present ("Knowledge/experience/know-how"). At the very bottom, only the "Organization" capability is present. There is a similar situation in the left side of the figure; companies in the top left side are those with the "Knowledge/experience/know-how" and "Innovation/Competitiveness/Design" features. From the top to the bottom of the figure, some of these capabilities are not present. Most companies are in intermediate positions. However, 16 of them are different from the other ones because at least three capabilities are present. From a combined analysis of the "competitive advantages" and "capabilities for competitive advantages" features, it can be said that the more competitive advantages, the greater capabilities for the company. The preferably capabilities are "Human resources", "Knowledge/experience/know-how", and "Organization".

The third analyzed group of features is "KM elements", that is related to the meaning of KM for the company. The ordering is similar to that of the two previous groups of features (top to bottom and right to left). Companies in the upper right third of Fig. 2 consider that KM consists of at least four elements ("Processes", "Experience", "Technology", and "Innovation"). It is then easy to identify that companies in the right side of Fig. 2 relate the KM to more elements than those in the left side.

After analyzing the "Ways of learning" group of features, it can be said that once again, there is a clear ordering of data. It is similar to that for the three previous groups of features; all the different ways of learning, except "Collaboration with competitor companies" and "External staff is subcontracted", are present for companies in the upper right third of Fig. 2. In the intermediate region of the figure, the feature "New employees, specialist in a certain issue, join the company" is not present. At the bottom, only the "Opinion of clients and suppliers is taken into account" is present.

As a conclusion for this group of features, it can be said that those companies applying more and new ways of learning are located in better positions (upper right third of the figure). Those companies are the best ones for the analyzed groups of features: "Competitive advantages", "Capabilities for competitive advantages", and "KM elements".

For the "Given situations" group of features, the ordering is similar to the previous ones; companies in the upper right third of Fig. 2 get the highest scores for these features. On the contrary, CMLHL does not provide a clear ordering for the feature "Knowledge documentation" as companies in the best situations for such feature are located in opposite groups (1.1 and 2.4). Group 2.4 (with only one company) can be considered as a special case because its situation for the competitive advantages and capabilities is good but there are some weaknesses regarding the ways of learning and the given situations.

To sum up the conclusions related to Fig. 2, it can be said that the projection is coherent with regard to most of the analyzed groups of features. The companies with many competitive advantages, do have capabilities that support the advantages, what leads to innovative ways of learning to generate new ideas and apply new knowledge. The main goal is to improve/development of products, services, functions, processes, etc. which state the competitiveness and the survival of the company. In a more precise way, it can be said that the companies in the best situation from the KM perspective are gathered in groups 2.1 and 2.2, while those companies in the worst position are placed in group 1.3. All the other ones are in intermediate positions.

After the previous analysis (at a company level), a more general analysis (at an industrial sector level) has been carried out. To the best of the authors knowledge, this is the first time that neural models are applied to the analysis of KM along different industrial sectors. The CMLHL projection of the data has been modified to distinguish between companies belonging to one sector or the other one. Thus, in Fig. 3, Electrical and Telecommunications industries are depicted as black stars, while Timber industries are depicted as red circles.

From Fig. 3, it can be said that those companies belonging to the Electrical and Telecommunications sector are mainly located in the right side of the projection.

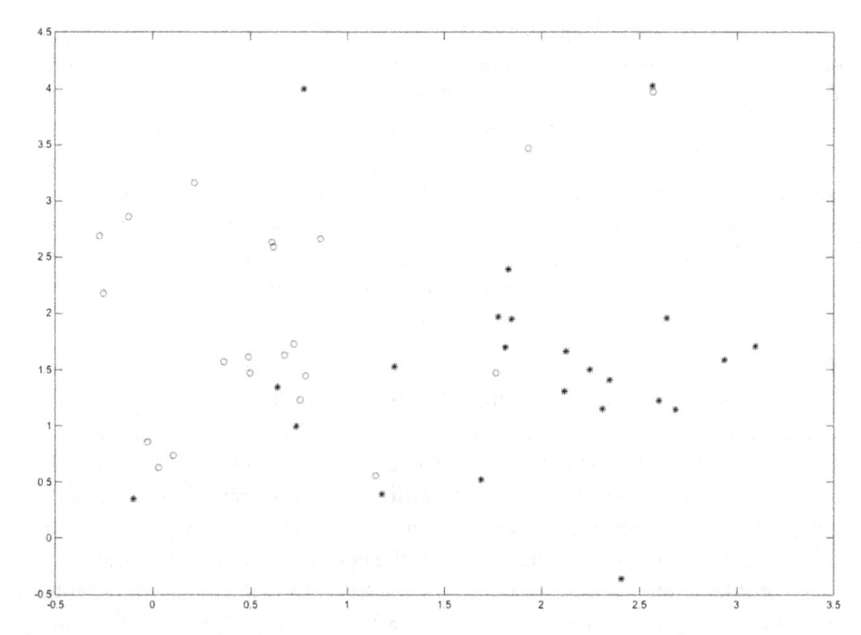

Fig. 3 CMLHL projection of the retention factor dataset

Those in best positions are located in group 2.2, standing out because of their competitive advantages, ways of learning and given situations. On the other hand, companies belonging to the Timber sector are mainly located in the left side of the projection. Group 1.2 contains companies that stand out because of the ways of learning and the behavior to employ and increase the knowledge. Similarly, group 1.4 contains those companies from the timber sector that excel in distinctive capabilities, KM elements, and the given situations.

Generally, it can be said that there are not strong differences between the two analysed industrial sectors. Taking into account all the companies belonging to each one of them, it can be said that their status is almost similar, although it could be defined as slightly better for the Electrical and Telecommunications sector. This can be easily seen in Group 2.2 that is one of those gathering companies in the best situations. All the companies in such groups, expect one, belong to the Electrical and Telecommunications sector.

4 Conclusions and Future Work

The main objective for present research was two-fold: on the one hand, to check neural projection models for the intuitive visualization of qualitative, abstract and disordered data associated to KM in different industrial sectors. More precisely, the analysis is focused on data related to Learning and Knowledge, for a significant real-life dataset in two industrial sectors: Electrical and Telecommunications industries and Timber industries. On the other hand, present research proposes technological support to bridge the gap between theoretical KM formulations to practical tools that support KM experts in diagnosis and decision taking.

This objective has been met as CMLHL has ordered and grouped the unknown dataset under analysis. After a thorough subsequent study, it has been proved that the obtained projection is coherent with the data and meaningful. Hence, it can support decision taking aimed at improving effectiveness of KM practices, ways of learning and behaviors to assure a proper KM in companies. Among the obtained results, the following ones are worth mentioning:

- The precise location of companies in CMLHL projection according to certain features determines the level of the companies according to their configuration of Learning and Knowledge, included in the KM. This is interesting to know the situation related to a certain feature for a company and take corresponding decisions. More precisely, obtained results let us know whether a certain company has competitive advantages, and which capabilities are supporting them. Similarly, the ways of learning that the company applies to guarantee and make profitable the knowledge it has or need to acquire, the behavior or attitude of employees in usual situations related to KM, such as problem solving, expert querying, encouragement of new ideas, or sharing, teaching and exchanging the existent knowledge.
- The convergence of some features, with certain values, in areas clearly identified, that represents industrial situations of a diverse nature, ranging from very good to extremely bad. These groups of data, located in a certain area of the projection, allow for not only informing about the situation of the company regarding Learning and Knowledge, but also, deploying a wide variety of actions for moving to the best locations and leaving those that are unrecommendable or inefficient.
- The support for cross-sector analysis of industrial companies in a way that general conclusions about a certain sector can be drawn from the projection of the companies belonging to it.

Consequently, the applied neural model supports the easy and meaningful analysis of the features associated to Learning and Knowledge. Furthermore, interesting conclusions can be obtained about how to evolve and progress towards the deployment and execution of efficient KM in industrial companies from the Electrical and Telecommunications and Timber sectors.

For future work, it is proposed to extend present study to cover some other features associated to KM, and some more sectors and companies. Additionally, some other neural projection models could be applied to obtain more interesting visualizations of data.

References

1. Herrero Á, Corchado E, Sáiz L, Abraham A (2010) DIPKIP: A connectionist knowledge management system to identify knowledge deficits in practical cases. Comput Intell 26:26–56
2. Durst S, Edvardsson IR (2012) Knowledge management in SMEs: a literature review. J Knowl Manag 16:879–903
3. Levy M (2011) Knowledge retention: minimizing organizational business loss. J Knowl Manag 15:582–600
4. Herrero Á, Corchado E, Jiménez A (2011) Unsupervised neural models for country and political risk analysis. Expert Syst Appl 38:13641–13661
5. Herrero Á, Corchado E, Gastaldo P, Zunino R (2009) Neural projection techniques for the visual inspection of network traffic. Neurocomputing 72:3649–3658
6. Friedman JH, Tukey JW (1974) A projection pursuit algorithm for exploratory data-analysis. IEEE Trans Comput 23:881–890
7. Hotelling H (1933) Analysis of a complex of statistical variables into principal components. J Educ Psychol 24:417–444
8. Pearson K (1901) On lines and planes of closest fit to systems of points in space. Phil Mag 2:559–572
9. Seung HS, Socci ND, Lee D (1998) The Rectified Gaussian Distribution. Adv Neural Inf Process Syst 10:350–356

Classification and Clustering Methods

Towards a Soft Evaluation and Refinement of Tagging in Digital Humanities

Gonzalo A. Aranda-Corral, Joaquín Borrego Díaz and Juan Galán Páez

Abstract In this paper we estimate the soundness of tagging in digital repositories within the field of Digital Humanities by studying the (semantic) conceptual structure behind the folksnonomy. The use of association rules associated to this conceptual structure (Stem and Luxenburger basis) allows to faithfully (from a semantic point of view) complete the tagging (or suggest such a completion).

1 Introduction

According to Wikipedia, Digital Humanities (DH) is an area of research and teaching at the intersection between computer science and humanities. DH embraces a variety of topics, from on line collections curation to data mining on large cultural data sets, where researchers use tools from Computing as Knowledge Extraction (KE), Machine Learning, Agent-Based Modeling techniques, as well as solutions from the Social Web. In order to bridge the gap between Humanities and Computing methodologies for Knowledge Organization, it is usual to provide humanists with services to self-organize digital content. Often resources are indexed and classified by categories. Also it is interesting to create tagging services for the community of researchers in order to enrich the content and provide a better navigation.

The reason of the success of tagging in the social web is that it does not have any kind of limitation. Tagging has several use cases in the social web [14]: personal information management (navigate through our selected and tagged resources), digital objects tagging helps to share and spread them, or even to improve user experience

This work was partially supported by TIC-6064 Excellence project (*Junta de Andalucía*) and TIN2013-41086-P project (Spanish Ministry of Economy and Competitiveness), co-financed with FEDER funds.

G.A. Aranda-Corral
Departamento de Tecnologías de la Información, Universidad de Huelva,
Crta. Palos de La Frontera S/n, 21819 Palos de La Frontera, Spain

J.B. Díaz · J.G. Páez (✉)
Departamento de Ciencias de la Computación e Inteligencia Artificial,
Universidad de Sevilla, Avda. Reina Mercedes S/n, 41012 Sevilla, Spain

Á. Herrero et al. (eds.), *10th International Conference on Soft Computing Models in Industrial and Environmental Applications*, Advances in Intelligent Systems and Computing 368, DOI 10.1007/978-3-319-19719-7_7

in e-commerce platforms. This allows the user take advantage of its *personomy* as well as other users personomies: an user can access resources uploaded and/or tagged by other users [11]. In the case of collaborative tagging, the full set of resources and tags represents a folksonomy which weakly represents an implicit ontology on community's knowledge. Tagging is an activity which produces folksonomies (inducing consensual vocabularies for the community), that can be understood as a kind of *emergent ontology* which facilitates the organization and navigation.

In general terms, these ontologies suffer of a number of limitations and deficiencies. On the one hand, since the vocabulary has user-dependent intentionality, semantic heterogeneity occurs: some tags represent distinct features for distinct users. Semantic heterogeneity is an intrinsic problem of tagging which prevents the user from exploiting other user's tagging with reliability. Another major drawback of existing social tagging systems is that social tags are used as keywords in keyword-based search. They focus on keywords and their interpretation by humans rather than on computer interpretable semantic knowledge [10]. In the case of DH it is usual to take into account that users share tag interpretation (to a certain extent). Despite these limitations, shared tagging is a potential solution to provide the community with semantically organized knowledge. However, this knowledge is not machine processable and the semantic heterogeneity is a common problem in multi-topic tagging services.

In [8] tagging is described as a task providing resources with sense and aims to categorize resources producing emergent meaning [15]. As a consequence of this, an individual tagging will not be really useful as a public one. Objects tagging made by the community can show the same problem, although in a different scale. However, this problem can be solved by means of Collective Intelligence: when the community tags collectively, they tend to unify the use of tags. Thus the most common tags set associated to an object provides a collective description of a certain concept [6]. In fact, these collective tagging are useful to build recommender systems [5].

The aim of this paper is to propose an (soft) estimate of the soundness of existent tagged digital objects in repositories, as well as to propose a rule set for the automatic refinement of existent tagging (semantics-based). The idea consists in estimating the topological structure of a conceptual network extracted from the tagging system by using Formal Concept Analysis, a mathematical theory which also provides reasoning tools useful for this second goal. The proposed methodology is applied to two tagged repositories relevant in Digital Humanities.

2 Formal Concept Analysis for Tagging Services

Formal Concept Analysis (FCA) [7] provides powerful semantic tools for classification, data mining and KE and Discovery (KD). Among these tools particularly interesting are concepts extraction and organization, and implication basis. The last one, represents a sound approach to rule extraction for classification. This task is a significant issue in KD where FCA applications in the Soft Computing field have been implemented (see for example [16]).

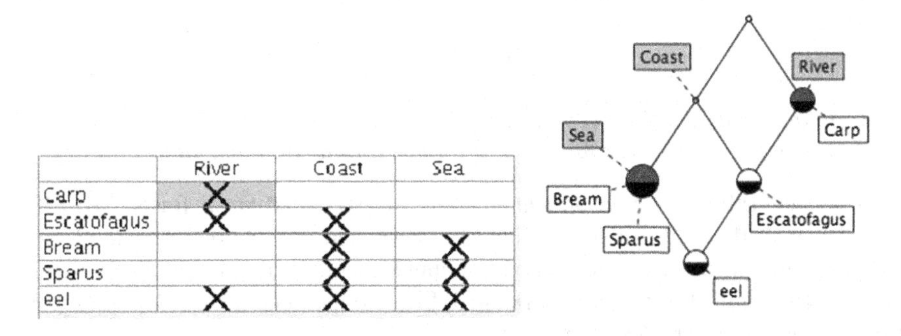

	River	Coast	Sea
Carp	X		
Escatofagus	X	X	
Bream		X	X
Sparus		X	X
eel	X	X	X

Fig. 1 Formal context on fish, and its associated concept lattice

A *formal context* $M = (O, A, I)$ consists of two sets, O (objects) and A (attributes), and a relation $I \subseteq O \times A$. Finite contexts can be represented by a 1-0-table (identifying I with a boolean function on $O \times A$). Given $X \subseteq O$ and $Y \subseteq A$, it defines $X' = \{a \in A \mid oIa \text{ for all } o \in X\}$ and $Y' = \{o \in O \mid oIa \text{ for all } a \in Y\}$

The classical method for defining a "concept" is actually twofold: The concept is defined *extensionally* by some set of objects that are instances of that concept. The concept is defined *intensionally* by a property that all the instances have in common but that is not possessed by any of the remaining objects. FCA mathematizes this philosophical understanding as a unit of thoughts composed of two parts: the extent and the intent. The extent covers all objects belonging to the concept, while the intent comprises all common attributes valid for all the objects under consideration [7]. The main goal of FCA is the computation of the set of concepts associated with the context.

A (formal) concept is a pair (X, Y) such that $X' = Y$ and $Y' = X$. The set of concepts of a context given M, $CL(M)$, can be endowed with the lattice structure by means of the "subconcept" relationship [7]. For example, the concept lattice from the formal context of fishes of Fig. 1, left (attributes are understood as "live in") is shown in Fig. 1, right. Each node is a concept, and its intension (or extension) is formed by the set of attributes (or objects) included along the path to the top (or bottom). For example, the bottom concept $(\{eel\}, \{Coast, Sea, River\})$ is the concept *euryhaline fish*.

Knowledge Bases (KB) in FCA are formed by *implications between attributes*. An implication is a pair of sets of attributes, written as $Y_1 \to Y_2$. We say that the implication is true with respect to $M = (O, A, I)$ according to the following definition: A subset $T \subseteq A$ *respects* $Y_1 \to Y_2$ if $Y_1 \not\subseteq T$ or $Y_2 \subseteq T$. $Y_1 \to Y_2$ is said to hold in M ($M \models Y_1 \to Y_2$ or $Y_1 \to Y_2$ is an implication of M) if for all $o \in O$, the set $\{o\}'$ respects $Y_1 \to Y_2$.

Definition 1 *Let $\mathcal{L}$ be a set of implications and L be an implication.*

1. *L follows from $\mathcal{L}$ ($\mathcal{L} \models L$) if each subset of A respecting $\mathcal{L}$ also respects L.*
2. *$\mathcal{L}$ is complete if every implication L*

$$M \models L \Rightarrow \mathcal{L} \models L$$

3. *$\mathcal{L}$ is non-redundant if for each $L \in \mathcal{L}$, $\mathcal{L} \setminus \{L\} \not\models L$.*
4. *$\mathcal{L}$ is a (implication) basis for M if $\mathcal{L}$ is complete and non-redundant.*

A particular basis is the *Duquenne-Guigues* or so called *Stem* Basis (SB) [9]. In order to work with formal contexts, stem basis and association rules, the Conexp[1]software has been selected. The reasoning system we use is a production system described in [3]. Initially it works with SB, and the entailment is based on the following result (see [3] for details):

Theorem 1 *Let S be a basis for M and $\{A_1, \ldots, A_n\} \cup Y \subseteq A$. The following statements are equivalent:*

1. *$S \cup \{A_1, \ldots A_n\} \vdash_p Y$ ($\vdash_p$ is the entailment by means of a production system).*
2. *$S \models \{A_1, \ldots A_n\} \to Y$*
3. *$M \models \{A_1, \ldots A_n\} \to Y$.*

In conditions of above definition, let define

$$S[\{A_1, \ldots, A_n\}] := \{a \in A \ : \ S \cup \{A_1, \ldots A_n\} \vdash_p a\}$$

In FCA, association rules are also implications between sets of attributes. Confidence and support are defined as usual in data mining. The analogous to Stem Basis for association rules is the Luxenburger basis [12]. The reasoning system for SB can be adapted for reasoning with Luxemburger basis [3]. Recall that Y is closed if $Y'' = Y$.

Definition 2 *Let be $M = (O, A, I)$ a formal context and $Y, Y_1, Y_2 \subset A$.*

- *Given Y_1, Y_2 closed, we denote $Y_1 \prec Y_2$ if there is not Y closed such that $Y_1 \subset Y \subset Y_2$.*
- *The support of an attribute set $Y \subseteq A$ is $supp(Y) = |Y'|$.*
- *The support of an implication $L = Y_1 \to Y_2$ is $supp(L) = |(Y_1 \cup Y_2)'|$*
- *The confidence of L is $conf(L) = \dfrac{supp(Y_1 \cup Y_2)}{supp(Y_1)}$*

Definition 3 *Given γ and δ, the Luxenburger basis of a context M with confidence γ and support δ, denoted by $\mathcal{L}(M, \gamma, \delta)$, is*

$$\mathcal{L}(M, \gamma, \delta) := \{L : Y_1 \to Y_2 \mid Y_1, Y_2 \text{ closed}, Y_1 \prec Y_2, conf(L) \geq \gamma, sup(L) \geq \delta\}$$

[1]http://sourceforge.net/projects/conexp/.

Implications from the Luxenburger basis can be interpreted as association rules from classic data mining, and therefore they allow reasoning under uncertainty. Conexp software provides association rules (and their confidence) associated to formal contexts. The subset of implications from the Luxenburger basis having confidence equal to one (those which are always true within the context) are the same than in the Stem Basis.

For the example from Fig. 1, the basis $\mathcal{L}(M, 0.8, 5)$ and $\mathcal{L}(M, 0.5, 2)$ are[2]

```
1 < 3 > Sea =[100%]=> < 3 > Coast;
2 < 5 > { } =[80%]=> < 4 > Coast;
```

```
1 < 3 > Sea =[100%]=> < 3 > Coast;
2 < 5 > { } =[80%]=> < 4 > Coast;
3 < 4 > Coast =[75%]=> < 3 > Sea;
4 < 3 > River =[67%]=> < 2 > Coast;
5 < 5 > { } =[60%]=> < 3 > River;
6 < 2 > River Coast =[50%]=> < 1 > Sea;
```

In general terms, when applying FCA to tagging systems, it is necessary to adapt its environment (formed by resources, tags, and users) to the format required by formal contexts. In this case our aim is to analyze the global structure of the whole tagged repository, thus it is not necessary to take into account which user tagged what resource. The general methodology to apply FCA on tagging is to consider tagged items as objects of the formal context and its tags as attributes on those objects. In this way, a formal context is associated to a folksonomy without taking users into account. Once the context is built, the associated concept lattice $CL(M_{\mathbb{F}})$ can be extracted. This concept lattice represents a concept hierarchy on the universe of the given folksonomy.

3 Meaning-Free Tagging Evaluation

In this section it is shown how to evaluate the semantic suitability of folksonomies. It should be noted that this analysis must be independent with respect to the topics of the repository and the field of study it belongs to. Therefore, any methodology used with this aim should take into account, only from a structural point of view, the structure of concepts $CL(M_{\mathbb{F}})$ obtained by means of the process formerly described.

An important feature in semantic networks is the degree distribution given by the connectivity of its nodes (concepts in this case, related by $\prec$), which have been deeply studied. It is expected that concept networks sharing a similar structure could share as well other properties, for instance, those related with its semantics and its suitability as knowledge representation in a certain domain. Therefore it is expected that the topological analysis of the $CL(M)$ shows a big picture of the semantics implicit in the folksonomy itself.

It should be noted that the $CL(M)$ is a complex network of semantic relationships that is not bounded by the self language, as in other semantic networks [13]. That is to say, there are concepts that are not represented by a single language term nor a

[2] the format of $L = Y_1 \to Y_2$ is $< \mathtt{supp}(Y_1) > Y_1 = \mathtt{conf(L)} => < \mathtt{supp}(Y_2) > Y_2$.

Table 1 Features of $CL(M_{\mathbb{F}})$ for case studies. *Density* is $|I|/|O \times A|$ and $< k >$ if the mean degree of the nodes (concepts) of $CL(M_{\mathbb{F}})$

	$\lvert\mathbb{O}\rvert$	$\lvert\mathbb{A}\rvert$	$\lvert\mathbb{I}\rvert$	Density	$\lvert CL\rvert$	$\langle k\rangle$
Baroque Art	11.062	221	74.993	3,067 %	17.817	7,949
Gothic Past	3.246	1.781	66.432	1,149 %	416.896	9.834

intelligible definition by the observer. Thus, is a task of the field specialist to interpret such concepts. This feature produces complex networks with extreme structural topology.

A scale-free network is one whose degree distribution follows a power law, at least asymptotically: the fraction $P(k)$ of nodes in the network having k connections to other nodes goes for large values of k as $P(k) \sim ck^{-\gamma}$ where c is a normalization constant and γ is a parameter whose value is typically in the range $2 < \gamma < 3$, although occasionally it may lie outside these bounds (as we will see below). It is more common for this behavior to appear from a certain threshold x_{min}. The *scale-free* residue of a *CL*(*M*) is the set of its nodes whose degree is greater than x_{min} (Table 1).

The analysis of the topology of Concept Lattices is a promising method for addressing the issue raised in the introduction, namely, whether sound qualitative modelizations (in our case, the Concept Lattices) share a similar structure. In [2] the following working hypothesis, called *Scale-Free Conceptualization Hypothesis* (SFCH) is stated, analyzed and experimentally validated:

Only if the attribute set selected to observe the System is computable, objective, and induces a Concept Lattice that provides a sound analysis of the CS, then its degree-distribution is scale-free.

This hypothesis (SFCH) has been tested in different experiments, In one of these experiments it was shown that random formal contexts do not respect the SFCH [1, 2]. In the case of the present work, regarding the analysis of folksonomies representing cultural complexity, the statement of the SFCH would be as follows: *A tags set is a suitable knowledge representation for a repository if the conceptual structure that it induces is an scale free network.*

4 Analyzing Tagging in DH Repositories

Two DH digital repositories have been chosen as example case study for the proposed methodology: *Baroque Art* from CulturePlex lab[3] and *Gothic Past.*[4]

The Hispanic Baroque: Complexity in the first Atlantic culture[5] is a multidisciplinary project carried out by a group of researchers from different universities and financed by the *Social Sciences and Humanities Research Council of Canada*. The

[3]http://baroqueart.cultureplex.ca/ in http://www.cultureplex.ca/ .

[4]http://www.gothicpast.com/.

[5]http://www.hispanicbaroque.ca/.

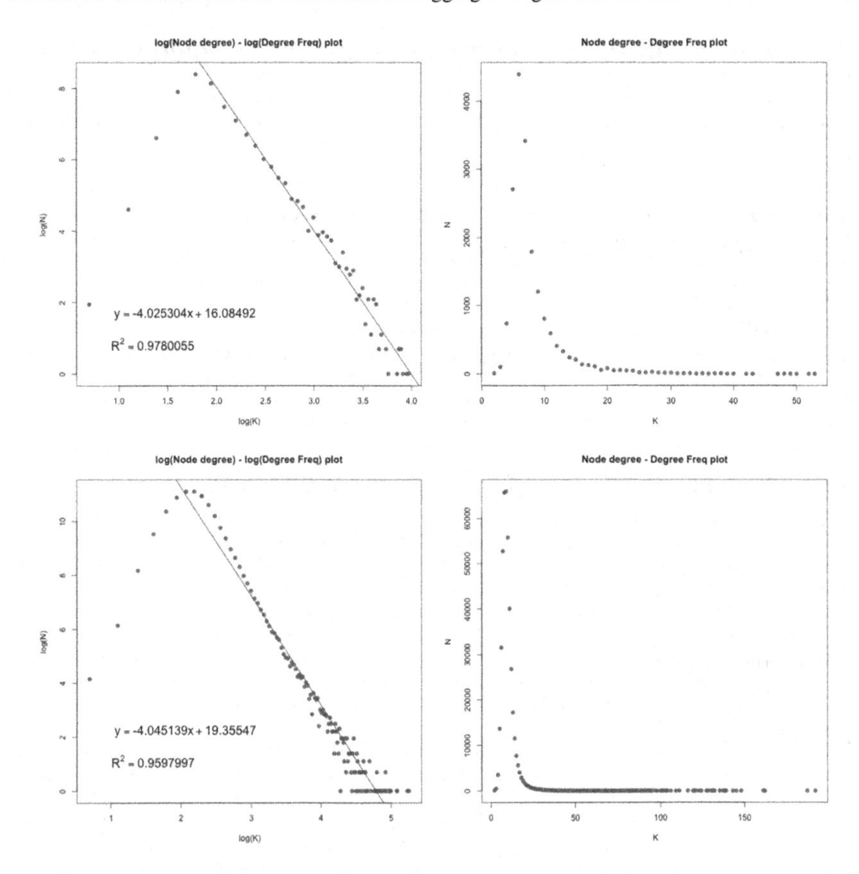

Fig. 2 Degree distribution of Concept Lattices associated to *Baroque Art* (up) and *Gothic Past* (down)

artwork repository selected, *Baroque Art*, belongs to this project. *Baroque Art* is a repository of artworks tagged by small and closed group of people and following common tagging rules and using as tags a preset common vocabulary (an ontology). The dataset consists of 11.000 artworks and 200 tags approximately.

Figure 2 (up) shows the degree distribution of the conceptual structure extracted from the tagged artworks, which presents an scale free distribution. According to the SFCH that means that the tags set used provides a sound and consistent knowledge representation for the artworks.

Gothic Past is a public on line repository for the study of the medieval architecture in Ireland. The repository provides for each element, different information items as pictures, tags, detailed descriptions, etc. In this case, the system allows other users to add new elements to the repository or to modify existing ones. Therefore the number of people involved in this tagging process is higher than in the former repository, possibly leading to more heterogeneous tagging criteria. Figure 2 (down) shows the degree distribution of the conceptual structure associated to the tagged repository on Irish Gothic monuments, which also presents a scale free distribution.

5 Luxenburger Basis for Automated Tagging Completion

By considering folksnomies as formal contexts it is possible to use (Luxenburger) Implication basis for suggesting new tags:

Definition 4 *Let $M_{\mathbb{F}}$ be the context associated to a folksonomy $\mathbb{F}$, $\mathcal{S}$ be a basis and r be a resource of the folksonomy (that is, an object of $M_{\mathbb{F}}$)*

- *The completion tagging of r, is $c(r, \mathbb{F}) := \mathcal{S}[\{r\}']$*
- *The suggested tagging for r is $s(r, \mathbb{F}) := c(r, \mathbb{F}) \setminus \{r\}'$*
- *The Luxenburger tagging with confidence γ and weight δ is $\mathcal{L}(M, \gamma, \delta)$ is*

$$c(r, \mathbb{F}, \gamma, \delta) := \mathcal{L}(M_{\mathbb{F}}, \gamma, \delta)[\{r\}']$$

- *The suggested Luxenburger tagging with respect to $\mathcal{L}(M, \gamma, \delta)$ is*

$$s(r, \mathbb{F}, \gamma, \delta) := s(r, \mathbb{F}, \gamma, \delta) \setminus \{r\}'$$

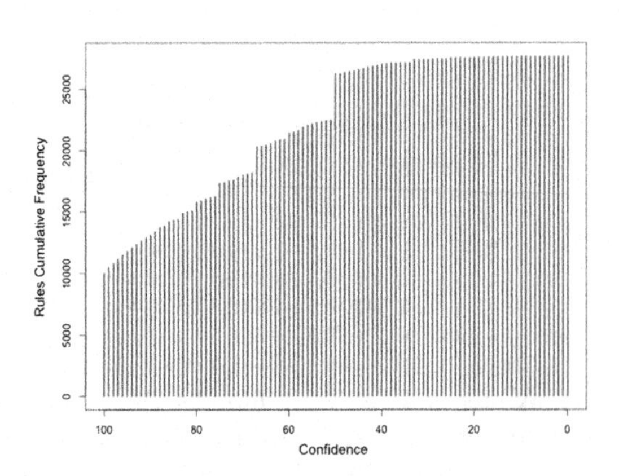

Fig. 3 Distribution of $L(M_{\mathbb{F}}, \gamma, \delta)$ according to rules' confidence, associated to *Baroque Art* repository

Proposition 1 *The completion tagging does not depend on the basis selected.*

Proof Let $r \in O$ and $\mathcal{S}_1, \mathcal{S}_2$ be two basis. If $a \in \mathcal{S}_1[\{r\}']$ then $\mathcal{S}_1 \cup \{r\}' \models a$ so $\mathcal{S}_1 \models \{r\}' \rightarrow \{a\}$ and thus $M \models \{r\}' \rightarrow \{a\}$. Therefore $\mathcal{S}_2 \models \{r\}' \rightarrow \{a\}$

Moreover, if the intent of all objects in the context is augmented by applying a Luxenburguer basis, then the implications of this turns to be true within the new context (thus they belong to a basis formed by implications with confidence 1 of the new context):

Proposition 2 *Let be* $M_{\mathbb{F}}^{\gamma,\delta} = (O, A, I^{\gamma,\delta})$ *where*

$$(o, a) \in I^{\gamma,\delta} \iff a \in \mathcal{L}(M_{\mathbb{F}}, \gamma, \delta)[\{o\}']$$

Then $\mathcal{L}(M_{\mathbb{F}}, \gamma, \delta) \subseteq \mathcal{L}(M_{\mathbb{F}}^{\gamma,\delta}, 1, 0)$

Fray Pedro Machado

Creator: Zurbarán, Francisco de
Dated In: 1630 - 1632
Original Location: Convento de la Merced Calzada
Current Location: Real Academia de Bellas Artes de San Fernando (Madrid, Spain)
It belongs to series: Retratos de personajes ilustres de la Orden Mercedaria-Zurbarán

General Description

[Objeto] → Clasificación → Género → Retrato
[Objeto] → Clasificación → Temática → Religioso
[Objeto] → Clasificación → Tipo → Pintura
[Objeto] → Propiedades Físicas → Color → Blanco
[Objeto] → Propiedades Físicas → Color → Negro
[Objeto] → Propiedades Físicas → Color → Rojo
[Objeto] → Propiedades Físicas → Material → Lienzo
[Objeto] → Propiedades Físicas → Tamaño: 193 x 122 cm
[Objeto] → Propiedades Físicas → Técnica → Óleo

Suggested tags

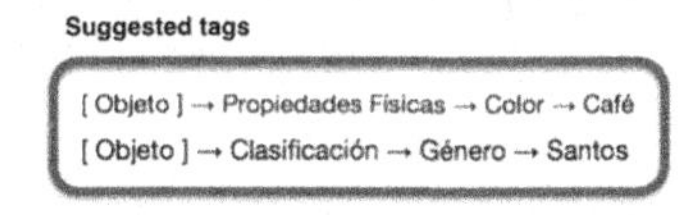

Fig. 4 Suggested tagging for the object $o353$, Zurbaran's artwork *Fray Pedro Machado*, by means of $\mathcal{L}(M_{\mathbb{F}}, 0.5, 30)[\{o353\}']$

Given a collaborative tagging service inducing a folksonomy $\mathbb{F}$, and a resource r, the tag set $c(r, \mathbb{F})$ extends the tagging $\{r\}'$ in order to allocate the object (the resource) in the most specific concept (as possible), according to its original tagging. However, $c(r, \mathbb{F}, \gamma, \delta)$ provides suggested tagging with a certain confidence degree. Thus, the user acceptability (or community of users) is important. Figure 3 shows the distribution of $|\mathcal{L}(M_{\mathbb{F}}, \gamma, \delta)|$ for *Baroque Art* repository. It is worthy to note that the tags set for a resource is very small with respect to the set of all tags. Therefore the computing of $\vdash_p$ (with confidence propagation [3]) is very fast. Particularly in the case of *Baroque*, the ontology-assisted tagging makes the basis $\mathcal{L}(M_{\mathbb{F}}, 1, 0)$ to have a relevant size: $|\mathcal{L}(M_{\mathbb{F}}, 1, 0)| = 10.007$ whilst $|\mathcal{L}(M_{\mathbb{F}}, 0.5, 0)| = 22.457$. As example, Fig. 4 shows some suggested tags (in red) for an artwork $o353$ (http://baroqueart.cultureplex.ca/artworks/353/) from *Baroque Art*. The tags belong to $\mathcal{L}(M_{\mathbb{F}}, 0.5, 30)[\{o353\}']$.

6 Conclusions and Related Work

Two uses of FCA within DH projects are described: the evaluation of the soundness of Knowledge Organization in tagging services and reasoning with implication basis to augment its tagging. Future work is focused on the use of *attribute exploration* [7] as a web service for accepting new tags (offered as plug-in). This idea (suggested in [4]) could be useful in cases where the repository is complete enough to extract useful knowledge from it, in the form of expert system.

In [11] authors study folksonomies by means of using triadic concepts, by considering the user as responsible of the tag. In our case, tagging in collaborative platforms, those are anonymized. It is also possible to exploit domain ontologies for suggesting tags (see for example [10]). In the first of the presented case studies (Baroque), the main tag vocabulary is provided by an ontology, thus it would be possible to expand or refine suggested tagging. In the second one this is not possible because, to the best of our knowledge, there is not a similar ontology.

The consensus a community can reach on collaborative tagging on a specific topic is different from personal information organization systems as for example Delicious (http://delicious.com/) or Diigo (https://www.diigo.com/). In this case it is necessary to reconcile their knowledge with other users to leverage their information (as in [4] by using FCA also).

References

1. Aranda-Corral GA, Borrego-Díaz J, Galán-Páez J (2013) Complex concept lattices for simulating human prediction. Sport J Syst Sci Complex 26(1):117–136
2. Aranda-Corral GA, Borrego-Díaz J, Galán-Páez J (2013) On the phenomenological reconstruction of complex systems-the scale-free conceptualization hypothesis. Syst Res Behav Sci 30(6):716–734
3. Aranda-Corral GA, Borrego-Díaz J, Galán J (2011) Confidence-based reasoning with local temporal formal contexts. In Proceedings 11th international conference artificial neural networks conference on advances in computational intelligence - volume part II (IWANN'11). Lecture notes in computer science 6692, pp 461–468 Springer
4. Aranda-Corral GA, Borrego-Díaz J, Giráldez-Cru J (2012) Agent-mediated shared conceptualizations in tagging services. Multimedia Tools Appl 65(1):5–28
5. Carmel D, Roitman H, Yom-Tov E (2010) Social bookmark weighting for search and recommendation. VLDB J 19(6):761–775
6. Halpin H, Robu V, Shepherd H (2007) The complex dynamics of collaborative tagging. In: Proceedings 16th international Conference WWW '07, pp 211–220
7. Ganter B, Wille R (1999) Formal concept analysis - mathematical foundations. Springer, Berling
8. Golder S, Huberman BA (2006) The structure of collaborative tagging systems. J Inf Sci 32(2):98–208
9. Guigues J-L, Duquenne V (1986) Familles minimales d' implications informatives resultant d'un tableau de donnees binaires. Math Sci Humaines 95:5–18
10. Hsu I-C (2013) Integrating ontology technology with folksonomies for personalized social tag recommendation. Appl Soft Comput 13(8):3745–3750

11. Jäschke R, Hotho A, Schmitz C, Ganter B, Stumme G (2008) Discovering shared conceptualizations in folksonomies. J Web Semant 6(1):38–53
12. Luxenburger M (1991) Implications partielles dans un contexte. Math Inf Sci Hum 11:335–55
13. Motter AE, de Moura APS, Lai Y, Dasgupta P (2002) Topology of the conceptual network of language. Phys Rev E 65
14. Smith G (2007) Tagging: people-powered metadata for the social web. First New Riders Publishing, Berkeley
15. Weick K-E, Sutcliffe K-M Obstfeld D (2005) Organizing and the process of sensemaking. Organ Sci 16(4):409–421
16. Jianping Y, Chen L, Wenxue H, Shaoxiong L, Deming M (2015) A new approach of rules extraction for word sense disambiguation by features of attributes. Appl Soft Comput 27:411–419

Neural Networks and Linear Predictive Coding Coefficients Used for European Starling Detection in Vineyards

Petr Dolezel and Martin Mariska

Abstract The use of feedforward multilayer artificial neural network to detect European starling in vineyards is presented in this paper. In the first paragraphs, the idea of whole system is outlined. Then, the method of detection is described and demonstrated, the process of neural network design is illustrated and, in the end, the neural network is validated.

Keywords Pest birds · Neural networks · Signal processing

1 Introduction

Together with a beginning of viniculture, people had to protect vineyards against many kinds of pests. Pest birds are considered as a special and indispensable inclusion of vermin for their amount of caused damages as well as their legal status. Therefore, many technologies (including mist nets, propane cannons, sound alarms, falconry etc.) have been introduced in last decades and comprehensive summary can be found in [1]. However, the most effective methods are quite annoying for residents living near vineyards since significant noise pollution is produced. For example, propane cannons, which are commonly used for pest birds frightening, create noise at between 115 and 130 decibels and operate nonstop from the fecond half of August till the beginning of November.

Thus, authors of this paper in cooperation with several vineyards situated in South Moravia, Czech Republic, try to develop more sophisticated system for pest birds control. The whole system is designed as a set of blocks which allows us to adapt the concept for each customer individually. The main blocks are Central Control

P. Dolezel (✉) · M. Mariska
University of Pardubice, Pardubice, Czech Republic
e-mail: petr.dolezel@upce.cz

M. Mariska
e-mail: mariska.martin@gmail.com

Á. Herrero et al. (eds.), *10th International Conference on Soft Computing Models in Industrial and Environmental Applications*, Advances in Intelligent Systems and Computing 368, DOI 10.1007/978-3-319-19719-7_8

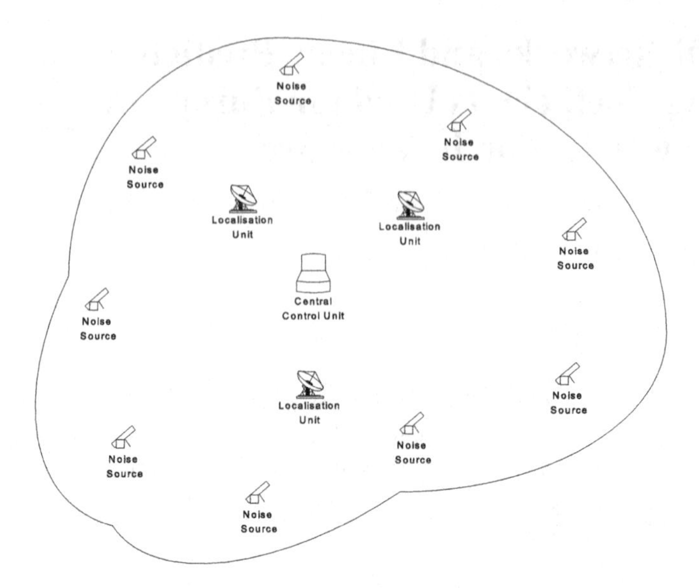

Fig. 1 Typical diagram of the system

Unit, Localization Unit and Noise Source. Typical diagram of the system applied in a vineyard can be found in Fig. 1 and data flow schedule is shown in Fig. 2.

This particular paper is focused on the concept of Localisation Unit, one its part, to be more specific. It is obvious that to work correctly, the Localisation Unit has to include strong pest birds detection device. So far, there have been published several possibilities of birds detection and recognition - some of them use sound samples [2], others radio detection [3] or image processing [4]. The aim of this paper is different. Since the majority of damages in Moravian vineyards is caused by European starling (Sturnus vulgaris) - see Fig. 3, the bird species to detect is known. However, it is necessary to determine whether the bird is present in vineyard or not and then to localize the position of the flock as exactly as possible.

2 Problem Formulation

As mentioned above, the problem especially consists in clear decision, whether the pest bird is present in vineyard. As a source of information, we selected sound samples - signal processing of bird sounds is less computationally demanding than image processing and does not require visual contact between emitter and receiver. However, birds produce a wide variety of sounds (songs, calls and mechanical sounds) and every type of sound should be applicable to pest bird detection. Thus, the detection unit is expected to work as shown in Fig. 4. The solution for each block of the referred flow chart is described in the following sections.

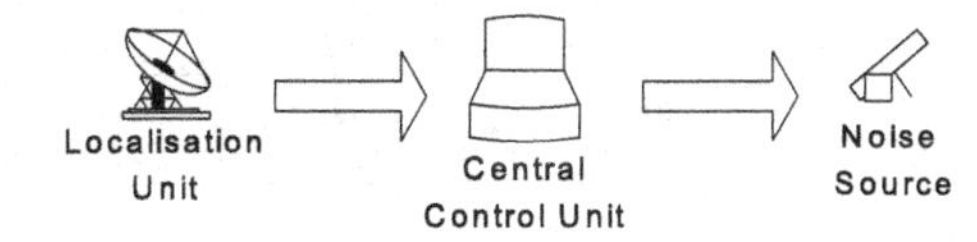

Fig. 2 Data flow between the blocks

Fig. 3 European starling

3 Data Preprocessing

The reason of data preprocessing is in the segmentation of the data into the frames of defined length and in data compression.

3.1 Sound Recording

The quality of recorded sound is determined especially by the field device used for the recording. For purposes of this paper, single-channel (mono) sounds with sample rate of 44100 Hz and double precision are used.

3.2 Sound Segmentation

Each recorded sound has to be divided into segments of constant length for further processing. It is a tricky procedure, since the length of the segment can significantly affect next steps. In various literatures [5–7], wide interval from 10 ms till 10 s is recommended and there is no approach to determine this value analytically. We tested several segment lengths and the results are posted in the next section of the paper.

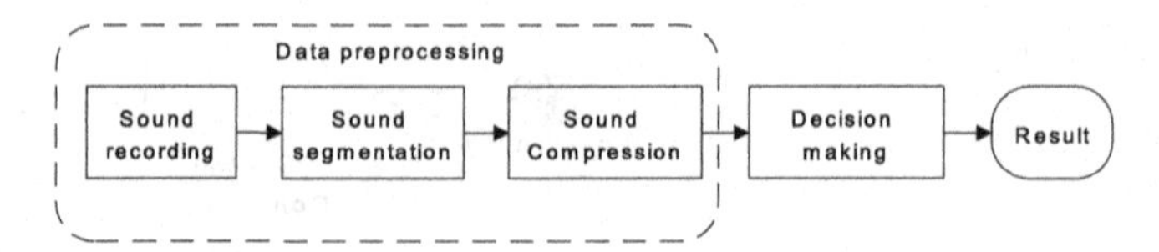

Fig. 4 Flow chart of the detection device

3.3 Sound Compression

As mentioned above, each second of the sound is represented by 44100 samples after data acquisition. This huge amount of values has to be reduced for further processing and many compression algorithms have been published so far. Let us mention at least Fast Fourier Transform [8], Linear Prediction Coding [9], Perceptual Linear Prediction [10] or Mel Frequency Cepstral Coefficients [11].

From options above, we chose Linear Prediction Coding (LPC). Firstly, it corresponds with the vocalizations of birds modeling by source-filter interaction [12]. Secondly, it has been many times successfully used in sound identification, recognition and compression. And thirdly, the usage of this method is not computationally demanding and can be performed by simple microprocessors in conformity with the general idea of the system.

The basic idea of LPC is that current sound sample $s(n)$ can be closely approximated as a linear combination of past samples, i.e.,

$$s(n) = \sum_{k=1}^{p} \alpha_k s(n-k) \tag{1}$$

where α is LPC coefficient and p is the number of LPC coefficients. Thus, limited number of LPC coefficients closely represents even thousands of samples. For our experiments, we decided to use 16 LPC coefficients for a 15th order LPC filter according to [12].

4 Decision Making

For the decision making, several approaches are suggested, most of them included in soft computing techniques (fuzzy theory, expert systems, neural networks, etc.). We decided to use neural networks, feedforward multilayer neural network, to be more specific. General diagram of used network is shown in Fig. 5 (numbers of inputs and neurons do not correspond with the actually used network).

The procedure of neural network design involves training and testing set acquisition, neural network training, pruning and validating. This sequence of processes is illustrated closely e.g. in [13, 14].

4.1 Data Acquisition

For training and testing set, songs, calls and other sounds produced by more than 30 birds were used (about 50 min of total time, where 6 min were produced by European starling). Furthermore, data were processed using some filtering and statistic tools to remove noise, silent segments and to balance the energies of all the recordings. Finally, the processed recordings were divided into samples of the defined length.

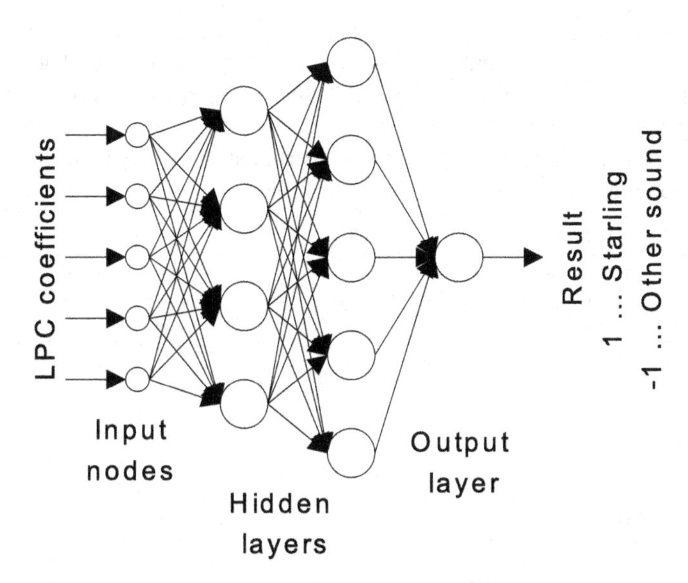

Fig. 5 General diagram of used neural network

4.2 Training Set

For training and testing set, each sample was compressed using LPC into the vector of 16 coefficients and these vectors were labeled either with -1 (non starling sounds) or with 1 (starling sounds). 70 % of the samples were placed into the training set, 15 % into the testing set and remaining 15 % were used for the validating.

4.3 Neural Network Training

Training means to find optimal weights and biases of the network. However, we used this procedure also to determine optimal length of sound segments derived in previous steps. To be specific, we trained the redundant feedforward multilayer neural network (hyperbolic tangent activation functions in all neurons) using Levenberg-Marquardt algorithm [15] with data gained from sounds divided into the segments of various lengths. For each length spread on the interval [10 ms; 2 s], 50 training procedures were performed and statistical data of the validation process were observed. Validation error was defined as follows:

$$E_{val} = \frac{1}{N}\sum_{i=1}^{N}[o(i) - y(i)]^2 \tag{2}$$

where $o(i)$ is the desired output, $y(i)$ is the actual output of the neural network and N is the amount of data. Data used for validation remained away from the training and testing set.

The statistics of the data gained from the experiments for some interesting lengths is concluded in the box graph in Fig. 6 and in Table 1.

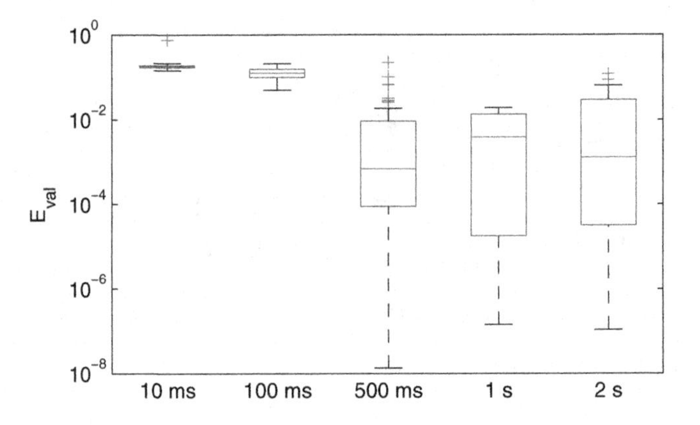

Fig. 6 The determination of the optimal sound length

Thus, we chose 500 ms as the length of the sound segments - we considered the median of E_{val} with respect to the minimal value of E_{val} as the main argument for this decision.

4.4 Neural Network Pruning

The aim of this step is to find optimal topology of the neural network. We simply trained a set of neural networks with different topologies (hyperbolic tangent activation functions in all neurons) using Levenberg-Marquardt algorithm fifty times and statistically evaluated the results. The resulting statistics is concluded in the box graph in Fig. 7 and in Table 2.

5 Results

According to the data in the previous section, we declared 5-1 topology (16 inputs, 5 neurons in one hidden layer and 1 output) as optimal - we considered the median of E_{val} with respect to the minimal value of E_{val} as the main argument for this decision. For the field use, the hyperbolic tangent activation function in output neuron was replaced with bipolar step function, so that the output is strictly bipolar (1 - starling is present, -1 - starling is not present). This adjusted network was validated once again with new data and results of that validation are summarized in Table 3 - the rates of correct answers seem to be very promising (more than 98 % of correct answers). In comparison to some works dealt with similar problematics, the rate is above average. In [16], the results are similar or worse according to the used technique, in [17], the experiments cannot be compared directly to our approach, however, in general, the rates published there are lower tha ours.

Table 1 The determination of the optimal sound length

Length	min(E_{val})	mean(E_{val})	std(E_{val})
10 ms	0.1451	0.1903	0.0837
100 ms	0.0501	0.1269	0.0380
500 ms	1.3270e-08	0.0156	0.0380
1 s	1.4035e-07	0.0062	0.0066
2 s	1.0700e-07	0.0175	0.0255

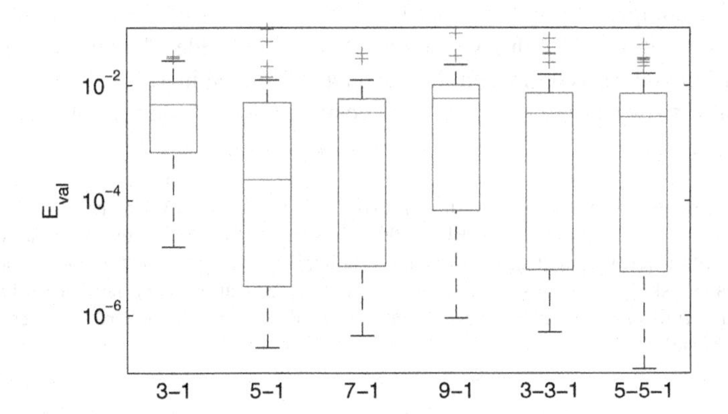

Fig. 7 Neural network pruning (X-axis represents the network topology to be examined)

Table 2 Neural network pruning

Topology	$\min(E_{val})$	$\mathrm{mean}(E_{val})$	$\mathrm{std}(E_{val})$
3-1	1.5478e-05	0.0078	0.0094
5-1	2.7444e-07	0.0057	0.0157
7-1	4.4267e-07	0.0045	0.0067
9-1	8.9090e-07	0.0077	0.0125
3-3-1	5.0886e-07	0.0076	0.0129
5-5-1	1.1902e-07	0.0069	0.0105

Table 3 Neural network validating

Number of sound segments	390
Rate of correct positive answers	18.5 %
Rate of correct negative answers	80.3 %
Total rate of correct answers	**98.8 %**
Rate of false positive answers	0.5 %
Rate of false negative answers	0.7 %
Total rate of false answers	**1.2 %**

6 Conclusion

The results discussed in this paper show that the developed neural network seems to be capable of detecting European starling. We introduced the approach which uses field sounds compressed by LPC based method. Compressed data are used then as the input to feedforward multilayer neural network to decide whether the pest birds are present in vineyards or not. Validation data mentioned in Table 3 show very low level of false answers - only 0.5 % of false positive answers and 0.7 % false negative answers.

Acknowledgments The work has been supported by the funds of the IGA, University of Pardubice, Czech Republic, project number SGSFEI2015006. This support is very gratefully acknowledged. In addition, this article was published within the sustainability of the project "Support of short term attachments and skilful activities for innovation of tertiary education at the Jan Perner Transport Faculty and Faculty of Electrical Engineering and Informatics, University of Pardubice, registration no. CZ.1.07/2.4.00/17.0107".

References

1. Bisho J, McKay H, Parrot D, Allan J (2003) Review of international research literature regarding the effectiveness of auditory bird scaring techniques and potential alternatives. Central Science Laboratories
2. Potamitis I, Ntalampiras S, Jahn O, Riede K (2014) Automatic bird sound detection in long real-field recordings: applications and tools. Appl Acoust 80:1–9

3. Klein L, Mino R, Hovan M, Antonik P, Genello G (2004) Mmw radar for dedicated bird detection at airports and airfields. In: Radar conference, 2004. EURAD. First European, pp 157–160
4. Pornpanomchai C, Homnan M, Pramuksan N, Rakyindee W (2011) Smart scarecrow. In: 2011 Third international conference on measuring technology and mechatronics automation (ICMTMA), vol 3, pp 294–297
5. Viswanathan V, Makhoul J, Schwartz RM, Huggins A (1982) Variable frame rate transmission: a review of methodology and application to narrow-band lpc speech coding. IEEE Trans Commun 30(4):674–686
6. Sun R, Marye Y, Zhao HA (2013) Fft based automatic species identification improvement with 4-layer neural network. In: 2013 13th International symposium on communications and information technologies (ISCIT), pp 513–516
7. Gardner WR, Rao B (1995) Theoretical analysis of the high-rate vector quantization of lpc parameters. IEEE Trans Speech Audio Process 3(5):367–381
8. Rader C, Brenner N (1976) A new principle for fast fourier transformation. IEEE Trans Acoust Speech Signal Process 24(3):264–266
9. Markel J, Gray A (1976) Linear prediction of speech. Springer, Berlin
10. Hermansky H (1990) Perceptual linear predictive (plp) analysis for speech recognition. J Acoust Soc Am
11. Davis S, Mermelstein P (1980) Comparison of parametric representations for monosyllabic word recognition in continuously spoken sentences. IEEE Trans Acoust Speech Signal Process 28(4):357–366
12. McIlraith A, Card H (1995) Birdsong recognition with dsp and neural networks. In: WESCANEX 95: communications, power, and computing. Conference proceedings., IEEE. vol 2, pp 409–414 vol 2
13. Haykin S (1999) Neural networks: a comprehensive foundation. Prentice Hall (1999) ISBN: 0132733501
14. Nguyen H, Prasad N, Walker C (2003) A first course in fuzzy and neural control. Chapman and Hall/CRC. ISBN: 1584882441
15. Hagan M, Menhaj M (1994) Training feedforward networks with the marquardt algorithm. IEEE Trans Neural Networks 5(6):989–993
16. Kiktova E, Lojka M, Pleva M, Juhar J, Cizmar A (2013) Comparison of different feature types for acoustic event detection system. In: Dziech A, Czyżewski A (eds) MCSS 2013, vol 368., CCISSpringer, Heidelberg, pp 288–297
17. Kiktova-Vozarikova E, Juhar J, Cizmar A (2003) Feature selection for acoustic events detection. Multimedia Tools Appl 1–21

Stroke-Based Intelligent Word Recognition Using a Formal Language

David Álvarez, Ramón Fernández and Lidia Sánchez

Abstract In this paper, we propose a new approach for off-line intelligent word recognition. Firstly, we segment a word into its single characters. Then, we tag every pixel of a character as vertical or horizontal and group them into vertical and horizontal strokes. We propose as main features the locations of the joints between a vertical stroke and each of their adjacent horizontal ones. These features can easily be obtained by dynamic zoning. After that, a Deterministic Finite Automaton (DFA) along with a regular grammar, let us generate a representative string of the above mentioned features. For the experiments we constructed sets of synthetic words from real characters written by two different authors to generate a knowledge base for each of the writers with the strings provided by the DFA. To achieve the recognition, we use an Inference Engine that matches representative strings of characters from unknown test words with those characters stored in the Knowledge Base. The experiments provide promising results.

Keywords Handwritten recognition · Word recognition · Stroke character representation · Inference engine · Knowledge base · Deterministic finite automaton

1 Introduction

One of the goals of document image processing and text recognition is automating the process of, both, reading previously scanned documents (off-line recognition) and identifying texts at the very time they are being written (on-line recognition)

D. Álvarez (✉) · R. Fernández · L. Sánchez
Department of Mechanical, Computer and Aerospace Engineerings, University of León, León, Spain
e-mail: etcdal00@estudiantes.unileon.es

R. Fernández
e-mail: ramon.fernandez@unileon.es

L. Sánchez
e-mail: lidia.sanchez@unileon.es

Á. Herrero et al. (eds.), *10th International Conference on Soft Computing Models in Industrial and Environmental Applications*, Advances in Intelligent Systems and Computing 368, DOI 10.1007/978-3-319-19719-7_9

[24]. Although there are solutions that work fairly well with printed documents [1, 12], handwritten text recognition is still a challenge [9].

On-line recognition generally means sampling the values of the pen position, velocity or acceleration as it is writing a character. Hence, there are on-line text recognizers that identify mathematical symbols [6], that verify signatures [15] or distinguish among different writers [26]. But perhaps, the main current application of on-line recognition involves text acquisition using smartphones and tablets as writing devices [3].

On the other hand, Off-line recognition starts with the acquisition of document images and applying some pre-processing techniques [8] such as thresholding [30], size normalization, rotation, slant correction [7] or noise reduction [13]. After that, all the parts of a document have to be identified (headings, paragraphs, lines), and that includes finding words in a line – word segmentation – and separating the single characters of each word – character segmentation [19, 29]. After that, the identification stage begins by generating a descriptor of each character from a set of its features. These features can be geometrical, based on projections, on histograms, based on the use of graphs or moments, or on the use of zoning [10, 18, 23, 31]. The descriptors may consist on simple vectors of features, but they can also be strings generated with the help of deterministic finite automata (DFA) and some formal grammar [28]. The last step uses some classifier (Hidden Markov Models [4], Neural Networks [21] or Fuzzy Logic [14]) to achieve the recognition. Finally, some dictionary or spell checker may be used to improve the accuracy of a recognizer [27]. The recognition rates are acceptable for printed documents [25], car plates [16] and even for figures on bank checks [20, 22] or ZIP codes [11]. But the recognition of handwritten characters is still considered an open problem [2, 17].

This paper is organized as follows: Sect. 2 details how we extract stroke based features with the use of dynamic zoning and how we use those features to construct a knowledge base. Section 3 describes the experiments and results obtained with a synthetic word repository made from two sets of single characters written by two different people. Finally, our conclusions and the future work are discussed in Sect. 4.

2 Methodology

This section describes our proposal for off-line intelligent word recognition: image preprocessing, character segmentation, feature extraction based on detecting vertical strokes and their joints with horizontal ones, and generation of character descriptors with the help of a deterministic finite automaton and a regular grammar.

After the image preprocessing and character segmentation steps, we propose the use of descriptors that codify the number of vertical strokes in a character as well as the relative positions of their connecting points with horizontal strokes. The descriptors of characters from a training set – i.e. a writer – help us make a knowledge base of that set. Then, an inference engine would be able to use that knowledge base in order to identify single characters from an unknown word, and a spell checker would improve the accuracy of the recognition task.

2.1 Pre-processing and Character Segmentation

First, a thresholding algorithm such as Otsu's [21] converts the image to binary format. If we consider that words are written with dark ink on clear paper, then Eq. 1 shows that foreground pixels become 0*s* in the binarized image, *Im*, while the background turns out into 1*s*.

$$Im(x, y) = \begin{cases} 1 & (x, y) \geq threshold \\ 0 & \text{otherwise} \end{cases} \tag{1}$$

After the thresholding, we have used Salvi et al. [26] algorithm to split a word image into its single characters (Fig. 1). Their proposal for character segmentation consists of the use of a graph model to, first, describe the possible locations for segmenting neighboring characters and, then, the application of an average longest path algorithm to find the globally optimal segmentation.

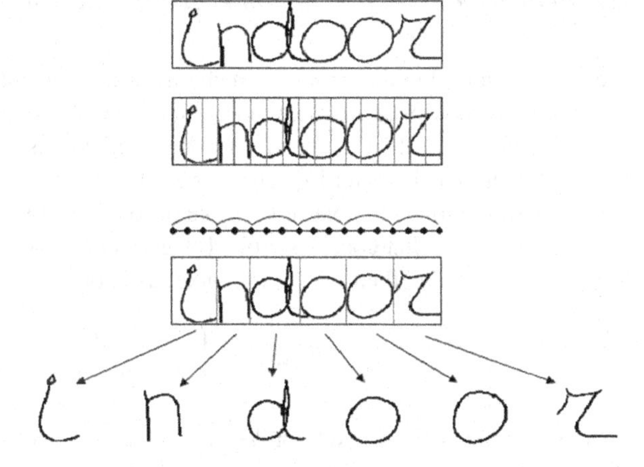

Fig. 1 Characters obtained after applying method in "indoor" word.

Binarized single character images, $M(x, y)$, may still need slant correction [28], that is, we have to find the angle, α, that would better transform a slanted character into a vertical one. First we choose a set of values $\alpha \in [-\frac{\pi}{4}, \frac{\pi}{4}]$ so that the skew transform provides several matrices $M_\alpha(x_\alpha, y_\alpha)$. Equation 2 transforms the original coordinates of pixel (x_s, y_s) into the new ones (x_α, y_α). The optimal α value would correspond to the matrix M_α whose vertical projection shows the highest peak [31], and that matrix M_α would be the slant corrected character (Fig. 2). However, this slant correction may create tiny holes inside the character strokes. Hence, the last pre-processing stage is a closing operation that removes those holes.

$$\begin{bmatrix} x_\alpha \\ y_\alpha \end{bmatrix} = \begin{bmatrix} 1 & -tg(\alpha) \\ 0 & 1 \end{bmatrix} \times \begin{bmatrix} x_s \\ y_s \end{bmatrix} \tag{2}$$

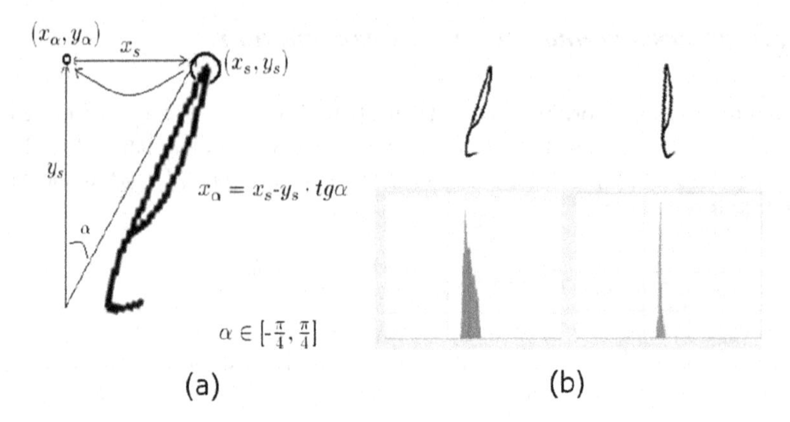

Fig. 2 **(a):** Horizontal traslation of pixels. **(b):** Character 'l' and its vertical projection before (left) and after (right) the slant correction.

2.2 Feature Extraction: Stroke Segmentation

We can consider that the strokes of a character may be either vertical or horizontal, and that vertical strokes are the main ones. Hence, the vertical strokes of a character and the locations of their connections to horizontal ones may be used as the features that represent that character [5]. Therefore, we have to tag every foreground pixel as v or h – i.e. it belongs to a vertical stroke or to a horizontal one. If the length of the vertical chord, vc, that contains a pixel is greater than or equal to the length of the horizontal chord, hc, then we tag that pixel as v; otherwise it would be h (Eq. 3).

$$P(x,y) = \begin{cases} v & vc \geq hc \\ h & \text{otherwise} \end{cases} \tag{3}$$

Sets of connected v-pixels or h-pixels constitute vertical strokes or horizontal strokes, respectively. However, as vertical strokes are the main ones, sets of h-pixels that overlap vertical strokes are re-labeled as v-pixels so that no horizontal stroke intersects a vertical one (Fig. 3).

To accomplish the feature extraction, we consider eight adjacency regions at both sides of a vertical stroke – regions 1 to 4 at the left side, and regions 5 to 8 at the right (Fig. 4). The regions where horizontal strokes touch a vertical one are the attributes of that vertical stroke.

2.3 Character Descriptors

The representation of a vertical stroke consists of a string that starts with the character V and is followed by the numbers that identify the adjacency regions that constitute its attributes. The descriptor of a character with two or more vertical strokes would be, therefore, the result of concatenating the strings that describe each of its vertical strokes. As vertical strokes are analyzed independently, ambiguities may appear due to horizontal strokes that connect several vertical strokes. That would result in equal representations for different characters as shown in Fig. 5. The disambiguation is achieved by duplicating the number that identify the adjacency region of a stroke that is connected to another vertical stroke. In this way, the horizontal stroke that connect two vertical strokes is differentiated from other horizontal strokes with a simple connection.

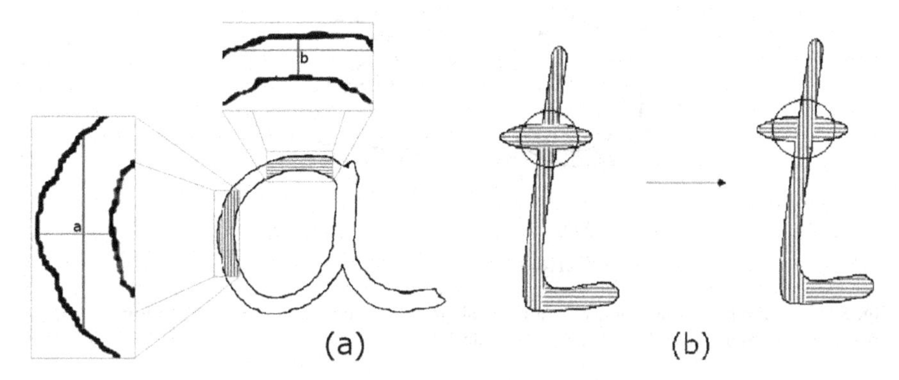

Fig. 3 **(a):** Comparison of chords lengths: a is tagged as v while b is labelled as h. **(b):** Horizontal strokes cannot overlap vertical ones.

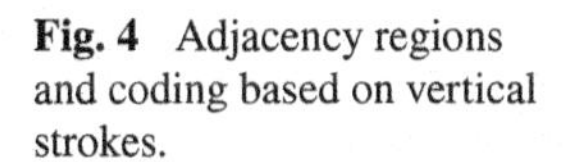

Fig. 4 Adjacency regions and coding based on vertical strokes.

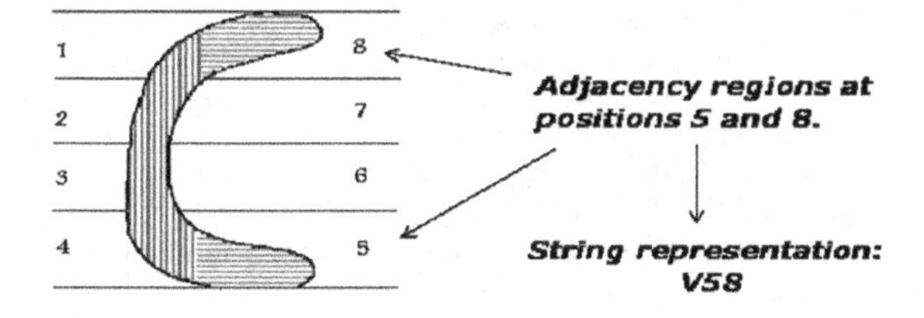

A deterministic finite automaton (DFA) is able to generate such character descriptors with the help of a regular grammar $G_{cr} = (T, N, S, P)$ that states the rules to generate valid strings (Fig. 6 shows the grammar and the DFA we have used).

2.4 Knowledge Based System

A Knowledge Base will store the strings that represent characters using XML, assuring the correctness of its structure with the help of a XML-SCHEMA.

The knowledge base consists of a tree whose root node is labeled as V and whose leaves store information of the characters in a training set. Each element of the string descriptor of a known character, c, is a inner node of a path that starts from the root and ends with a leaf that stores the value of c and also the number,n_c, of images of character c in the training set that resulted in the same path – i.e. that have the same descriptor. As different characters may result in the same descriptor, there may be several leaves at the end of a path. The value n_c let us compute the weight of the edge that reaches a leaf, that is the same as computing the probability of a path to correspond to character c (Fig. 7).

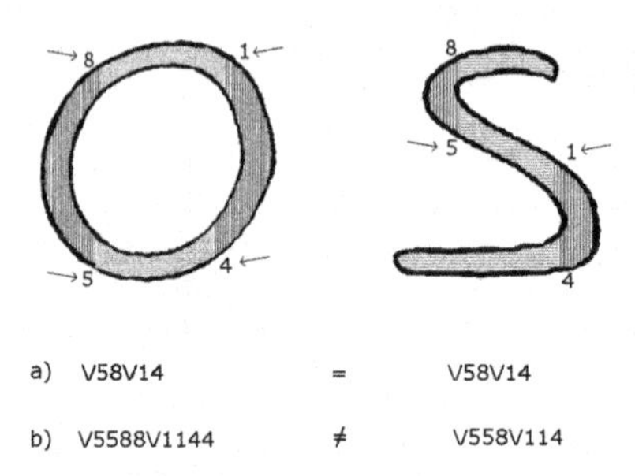

Fig. 5 (a) Ambiguities appear when horizontal strokes are connected to more than one vertical stroke. (b) Disambiguation by duplicating the labels.

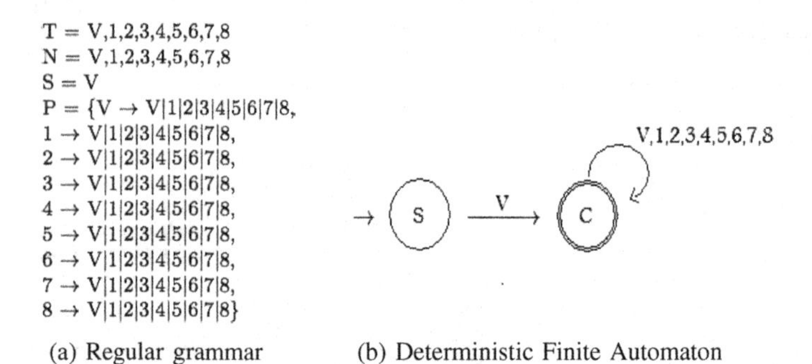

Fig. 6 The automaton uses the regular grammar that defines the rules for constructing the character representative string.

The recognition is achieved by means of an Inference Engine that matches the descriptor of one unknown character with one path in the tree that constitutes the knowledge base. If there is a matching path, their leaves provide the potential solutions, and the weights of the edges that reach them indicate the probabilities of the solutions to be correct. In case there is not any matching path,then the knowledge base would provide no solution –i.e. a rejection.

Among all the potential solutions for the characters of a word, the best one should correspond to a word that exists and whose characters have the highest probability of being correct. Hence, the recognizer generates a sorted list of all those words in descending order of probability. Then, a spell checker analyzes that list and chooses as solution the word with higher value of probability whose spelling is correct. If all the words in the list are incorrect, then we can consider that the word is not in the dictionary and the solution would be the word with the highest value of probability.

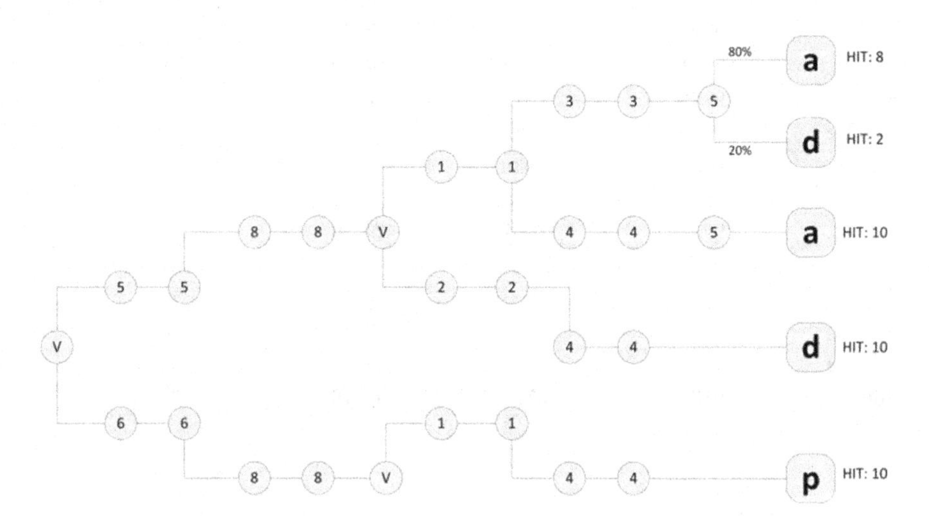

Fig. 7 Tree data structure with representation of some characters.

3 Experiments and Results

To accomplish the experiments we collected samples from two writers, each of one provided 20 images of every lower-case character ('a' to 'z'). We used 80 % of the samples – 16 samples of each character – to construct a knowledge base for each of the writers, and the remaining 20 % to generate the character test sets required to construct a synthetic word test set. We can generate a synthetic word image by picking a word randomly from a dictionary, choosing images of its characters randomly from the character test set, and changing some feature of each character image (we modified either the scale, the slant or the rotation).

We carried out 5 different tests with 5 sets of 10,000 synthetic words for every writer. Table 1 shows that a spell checker clearly improves the performance of the word recognizer. Note that without any spell checker, a mistake in just one character makes that the whole word recognition becomes a miss. But if the spell checker can correct the initial solution, it would become a hit.

Our results are comparable to others, as Kessentini et al. [17] which use the Hidden Markov Model paradigm with a multi-stream approach. For their experiments they use the IRONOFF database, which has a training set of 20,898 words and a

Table 1 Hit, Miss, reject and average rates in word recognition

Recognition Rates	with Spell Checker		without Spell Checker	
Writer 1	Hit (%)	Miss (%)	Hit (%)	Miss (%)
1	93.24	06.76	70.23	29.77
2	93.41	06.59	70.32	29.68
3	93.59	06.41	70.54	29.46
4	93.20	06.80	70,29	29.71
5	93.09	06.91	69.64	30.36
Average	93,31	06,69	70,20	29,80
Writer 2	Hit (%)	Miss (%)	Hit (%)	Miss (%)
1	90.12	09.88	58.86	41.14
2	90.16	09.84	58.16	41.84
3	90.02	09.98	57.76	42.24
4	90.00	10.00	58.21	41.79
5	90.43	09.57	57.76	42.24
Average	90.15	9.85	58.15	41.85
Total average	**91.73**	**8.27**	**64.18**	**35.83**

test set of 10,448 words. However, our experiments required only 1040 characters written by two different people, divided into 832 as training test for creating the custom knowledge base systems and 208 remaining characters for generating a test set of 100.000 synthetic words. The recognition rates are quite close: while we got 91,73 % hits, Kessentini et al. reached 89,8 %. Nevertheless, they used real words whereas that we used synthetic words.

4 Conclusions and Further Work

In this paper we propose a new methodology for word recognition. After the preprocessing and character segmentation steps, we locate the vertical and horizontal strokes of a character. With the vertical strokes and the locations of their connections with horizontal strokes we use a DFA and a grammar to generate a string that describes the character, and store such strings in a well-formatted XML knowledge base. Then, an Inference engine seeks new unknown characters in the knowledge base in order to achieve the recognition. XML makes the recognition faster, although our future work is focused on analyzing and reducing computation times.

Our results show that although the recognition rates are acceptable, especially when using a spell checker, they are very sensitive to the accuracy of the algorithm that decomposes characters into strokes. Hence, the next steps will include efforts to improve that algorithm. We also plan trying our method with other well known databases such as IRONOFF, IAM or CEDAR.

References

1. Rahman AFR, Fairhurst MC (1998) Machine-printed character recognition revisited: re-application of recent advances in handwritten character recognition research. Image Vis Comput 16(12–13):819–842
2. Alabau V, Martínez-Hinarejos C-D, Romero V, Lagarda A-L (2014) An iterative multimodal framework for the transcription of handwritten historical documents. Pattern Recogn Lett 35:195–203
3. Vicent A, Alberto S, Francisco C (2014) Improving on-line handwritten recognition in interactive machine translation. Pattern Recogn 47(3):1217–1228
4. AlKhateeb Jawad H, Olivier P, Jinchang R, Jianmin J (2011) Performance of hidden Markov model and dynamic Bayesian network classifiers on handwritten Arabic word recognition. Knowl Based Syst 24(5):680–688
5. Alonso A, Fernández RA, García I (2005) Recognition of merged characters based on vertical strokes and adjacency regions. In: V Congress of Hispalinux, pp 1–23
6. Álvaro F, Sánchez J-A, Benedí J-M (2014) Recognition of on-line handwritten mathematical expressions using 2D stochastic context-free grammars and hidden Markov models. Pattern Recogn Lett 35:58–67
7. Bertolami R, Bertolami R, Uchida S, Uchida S, Zimmermann M, Zimmermann M, Bunke H, Bunke H (2007) Non-uniform slant correction for handwritten text line recognition. In: Ninth International Conference on Document Analysis and Recognition (ICDAR 2007), pp 18–22
8. Caesar T, Gloger JM, Mandler E (1993) Preprocessing and feature extraction for a handwriting recognition system. In Proceedings of the Second International Conference on Document Analysis and Recognition, pp 408–411
9. Djeddi C, Siddiqi I, Souici-Meslati L, Ennaji A (2013) Text-independent writer recognition using multi-script handwritten texts. Pattern Recogn Lett 34(10):1196–1202
10. Dunn CE, Wang PSP (1992) Character segmentation techniques for handwritten text-a survey. In Pattern Recognition, 1992. Vol. II. Conference B: Pattern Recognition Methodology and Systems, Proceedings., 11th IAPR International Conference on, pp 577–580
11. Dzuba G, Filatov A, Volgunin A (1997) Handwritten zip code recognition. Analysis and Recognition, vol 1, pp 766–770
12. Adrià G, Ihab K, Jesús A-F, Alfons J (2014) Handwriting word recognition using windowed Bernoulli HMMs. Pattern Recogn Lett 35:149–156
13. Haji M, Bui TD, Suen CY (2012) Removal of noise patterns in handwritten images using expectation maximization and fuzzy inference systems. Pattern Recogn 45(12):4237–4249
14. Hanmandlu M, Murali Mohan KR, Chakraborty S, Goyal S, Roy Choudhury D (2003) Unconstrained handwritten character recognition based on fuzzy logic. Pattern Recogn 36(3):603–623
15. Jain AK, Griess FD, Connell SD (2002) On-line signature veriÿcation. 35:2963–2972
16. Jiang D, Mekonnen TM, Merkebu TE, Gebrehiwot A (2012) Car Plate Recognition System. In: 2012 Fifth International Conference on Intelligent Networks and Intelligent Systems, pp 9–12
17. Kessentini Y, Paquet T, Hamadou AMB (2010) Off-line handwritten word recognition using multi-stream hidden markov models. Pattern Recogn Lett 31(1):60–70
18. Liu C-L, Nakashima K, Sako H, Fujisawa H (2004) Handwritten digit recognition: investigation of normalization and feature extraction techniques. Pattern Recogn 37(2):265–279
19. Louloudis G, Gatos B, Pratikakis I, Halatsis C (2009) Text line and word segmentation of handwritten documents. Pattern Recogn 42(12):3169–3183
20. Morita ME, Letelier E (2000) Recognition of handwritten dates on bank checks using an HMM approach. Graphics and Image, pp 113–120
21. Il-Seok Oh, Suen CY (2002) A class-modular feedforward neural network for handwriting recognition. Pattern Recogn 35(1):229–244

22. Palacios R, Gupta A (2003) Training neural networks for reading handwritten amounts on checks. In: 2003 IEEE 13th Workshop on Neural Networks for Signal Processing, NNSP'03, pp 607–616
23. Zhang P, Chen L (2002) A novel feature extraction method and hybrid tree classification for handwritten numeral recognition. 23:45–56
24. Plamondon R, Srihari SN (2000) Online and off-line handwriting recognition: a comprehensive survey. IEEE Trans Pattern Anal Mach Intell 22(1):63–84
25. Namane RA, Guessoum A, Soubari EH, Meyrueis P (2014) CSM neural network for degraded printed character optical recognition. J Vis Commun Image Represent 25(5):1171–1186
26. Schlapbach A, Liwicki M, Bunke H (2008) A writer identification system for on-line whiteboard data. Pattern Recogn 41(7):2381–2397 July
27. Takahashi H, Itoh N, Amano T, Yamashita A (1990) A spelling correction method and its application to an OCR system. Pattern Recogn 23(3):363–377
28. Toselli AH, Juan A, Vidal E (2004) Spontaneous handwriting recognition and classification. ICPR 1:433–436
29. Zimmermann M, Bunke H (2002) Automatic segmentation of the IAM off-line database for handwritten English text. Object recognition supported by user interaction for service robots, 4(c)
30. Zou Y, Dong F, Lei B, Sun S, Jiang T, Chen P (2014) Maximum similarity thresholding. Digit Signal Proc 1:1–16
31. Trier ID, Jain AK, Taxt T (1996) Feature extraction methods for character recognition-a survey. Pattern Recogn 29(4):641–662

Link Similarity Reveals Multiscale Link Communities

Guishen Wang and Lan Huang

Abstract Overlapping community detection has been a hot topic in the research of complex network. In this paper, we proposed a novel link clustering method (NLC) for overlapping community detection. The method is consisted of two main steps. First step is a link similarity. The link similarity is to use a link similarity with a property of convergence to consider relationship of undirected links. The second step combines Markov Clustering Method with link similarity matrix got by first step with an extended measure of quality of modularity to determine the best partition of link communities. Extensive experiments on real world networks show our method is more reliable and reasonable than the other compared algorithms. Through varying parameters of our link similarity, our NLC method reveals multiscale link communities.

Keywords Multiscale link communities · Overlapping community detection · Link similarity

1 Introduction

As an important structure of complex network, overlapping community is important for understanding of complex network. Traditional overlapping community algorithms are mostly based on node. Early started in 2005, Palla uncovered the existence of overlapping community structure and put forward Clique Percolation Method (CPM) [1]. And then, with the rapid development in overlapping community, various algorithms came out continually [2–4].

For link communities has its own superiority, it considers the hierarchical and overlapping relationship of nodes simultaneously. Hence, lots of algorithms based on links are proposed in recent years. In 2010, Ahn et al. put forward Link

G. Wang · L. Huang (✉)
College of Computer Science and Technology, Jilin University, 130012 Changchun, China
e-mail: huanglan@jlu.edu.cn

Á. Herrero et al. (eds.), *10th International Conference on Soft Computing Models in Industrial and Environmental Applications*, Advances in Intelligent Systems and Computing 368, DOI 10.1007/978-3-319-19719-7_10

Clustering (LC) algorithm for link communities [5]. Later, Alex etc. realized the LC method and published R language pacakage "linkcomm" of LC [6]. In 2013, Huang et al. put forward a structural extended link similarity (ELS) and proposed Extended Link Clustering (ELC) algorithm for overlapping community detection [7]. In 2014, Lim etc. introduced DBScan clustering algorithm into link community detection and proposed Link Scan algorithm [8].

As representative methods for link community detection, LC algorithm uses a classical Jaccard similarity and ELC algorithm uses an extended link similarity (ELS) for evaluating relationships between links. But the Jaccard link similarity and ELS both have a main shortcoming that does not consider links with uncommon neighbors. Hence, from this point, we introduce the AH-Newman similarity for making up the shortcoming.

In rest parts of this paper, we will introduce our novel link clustering method in Sect. 2. Then experiments for testing our method's performance with other algorithm is discussed in Sect. 3. Finally, conclusion of this paper is in Sect. 4.

2 A Novel Link Clustering Method

In this section, we will introduce our method and steps in detail. Before discussing the steps, some basic definitions will be given first.

Definition 1: Let graph $G<V,E>$ be a network, where the nodes set V contains n nodes and the links set E contains m links. For any node i in V, its degree is $k(i)$. The adjacent matrix is A.

Definition 2 : Let graph $G_{link}<V_{link}, E_{link}>$ be a line graph that is a transformation of the original graph G. The transformation converts to the links set E into nodes set V_{link} and also transfer the nodes set V into the links set E_{link}. Hence, the elements in nodes set V_{link} are same to those in the links set E, namely $V_{link} = E$. However, the links set E_{link} is composed of the elements in V and $V \neq E_{link}$. Last, the adjacent matrix of G_{link} is denoted as A_{link}.

2.1 A Novel Link Clustering Method

Our Novel Link Clustering (NLC) method consists of four main steps as follows.

Iuput: link set E

Output: node communities of graph G

Steps:

1. Construct adjacent matrix A_{link}.
2. Use the improved link similarity to calculate the similarity between nodes in the line graph G_{link}, according to the adjacent matrix A_{link}. And get the similarity matrix S_{link}.

3. Use the MCL method on similarity matrix S_{link}. Through adjusting the parameters of MCL method, link communities of G with the best EQ value are identified.
4. Transform link communities back into node communities of G.

2.2 Improved Link Similarity

In 2006, Leicht et al. presented a similarity measure called AH-Newman for evaluating relationships between nodes [9]. The similarity measure is able to evaluating relationships between the nodes do not have common neighbor. Hence, in our paper, we introduce the AH-Newman similarity measure into evaluating relationships between links. The maximum value of AH-Newman similarity measure may exceed one. However, Leicht et al. proved the similarity converged. It makes the similarity effectively.

$$S_{link}(i,j) = \begin{cases} S_{unnormal-link}(i,j)/\max(S_{unnormal-link}), & i \neq j \\ 1, & i = j \end{cases} \tag{1}$$

With the α values smaller, similarity matrix becomes sparser and blocks are less obvious. Because with the increasing α value leads to the weaker relationships between links which have no common neighbors. Through varying with different values of α, the improved AH-Newman similarity measure fits more types of networks and multiscale link communities are revealed.

2.3 Markov Cluster Method

In this part, markov clustering method (MCL) is discussed in detail. As proposed by Stijin van Dongen in 2000, MCL method is a rapid extensible non-supervision graph clustering algorithm [10]. The method clusters based on the model of Markov chain. The method mainly contains expansion step and inflation step.

Input: link similarity matrix S_{link}
Output: link communities of G
Steps:

1. First normalized the similarity matrix S and then it will be as the start property matrix: $PS_{ij} = S_{ij}/\sum_{k=1}^{n} S_{ik}$.
2. Expansion Step: the step simulates the transfer matrix that randomly walk with e steps, namely $PS = PS^e$.

3. Inflation Step: After calculating the rth power, renormalize the PS matrix as follows: $PS_{ij} = PS_{ij}^{r} / \sum_{k=1}^{n} PS_{ik}^{r}$.
4. Convergence Step: While the iteration steps exceed the given steps or the energy function exceeds the threshold, the iteration will stop. The energy function is defined as follows:

$$Energy(m, s) = \max(m - s)$$

Among it, m is set to the maximum row element of PS matrix and s is set to the PS multiple itself.

2.4 EQ Evaluation Metrics

As an extension of modularity [11] for overlapping community detection, Extended Modularity (EQ) evaluation metrics was proposed by Shen et al. in 2009 [12] and has been a widely used evaluation metrics in the research of this area.

$$EQ = \frac{1}{2m} \sum_{l=1}^{c} \sum_{i \in C_l, j \in C_l} \frac{1}{O_i O_j} (A_{ij} - \frac{k(i) \cdot k(j)}{2m}) \tag{2}$$

3 Experiment

In this section, extensive experiments test on three real-world networks including NLC method, CPM method and LC method with three evaluation metrics.

3.1 Data Source and Evaluation Metrics

The details of three real-world networks are shown in the following:

(1) Zachary Karate Network (Karate) [13] is a social network that has two reference classes with 34 nodes and 78 links. The network represents the social relationship in a karate club at a US university and proposed by the sociologist Zachary in the 1970s.

(2) Dolphins Social Network (Dolphin) [14] is a social network that has two reference classes with 62nodes and 159 links. The network represents the

relationship between bottlenose dolphins in New Zealand and proposed by the biologist Lusseau etc. after their seven years researching.

(3) Books about US politics network (Politics) [15] consists of political books sold on the Amazon during 2004 and proposed by Krebs. The network has three reference classes with 105 nodes and 441 links.

Evaluation metrics we adopted are EQ, CN and NMI. EQ [12] and the number of communities (CN) are two evaluation metrics measures used for testing the performance of methods.

An extension of Normalized Mutual Information (NMI) is proposed by Lancichinetti et al. [4] and has become a most popular evaluation metrics used for the measuring quality of overlapping communities. NMI needs to know the ground-truth and ranges from 0 to 1. The higher value of NMI is, the closer the result is to the ground-truth.

3.2 Results

Results on three real world networks are presented in Table 1.

Results of three methods on three real-world networks with three different evaluation metrics are shown in Table 1. In the three networks, the values of α with the best division are 0.86, 0.58 and 0.95 respectively.

In the three real-world networks, communities identified by NLC method have the best EQ and NMI value than the other two methods. In Karate Network, communities found by LC method have higher EQ and NMI value than CPM method. However, in Dolphin Network and Political Network, the EQ values and NMI values of communities detected by CPM method are higher than LC method.

Table 1 Comparison with three methods on three real-world networks by three different evaluations[1]

Dataset	Karate			Dolphin			Political Books		
	CN	EQ	NMI	CN	EQ	NMI	CN	EQ	NMI
LC	8	0.2601	0.3089	13	0.0606	0.1455	32	0.1760	0.1022
CPM	3	0.1858	0.1745	4	0.3614	0.3810	4	0.4367	0.2470
NLC	3	**0.3170**	**0.4783**	3	**0.4015**	**0.5807**	3	**0.4496**	**0.2705**

[1]the bold data marked is the best value of each evaluation on the network

4 Conclusions

In this paper, we presented a novel link clustering method. The core steps of NLC method are link similarity step and MCL method step. The improved link similarity reveals multiscale link communities through varying parameter of it. After getting the similarity matrix, MCL method is used for identifying link communities with the best EQ value. In the tests of three real-world networks, the communities identified by NLC method have better modularity and more real meanings than CPM and LC method.

Acknowledgement This work is supported by the National Natural Science Fund Project of China (61472159) and the Science & Technology Development Projects of Jilin Province (20121805, 20140101180JC) and Graduate Innovation Fund of Jilin University (2014092).

References

1. Palla G, Derényi I, Farkas I, Vicsek T (2005) Uncovering the overlapping community structure of complex networks in nature and society. Nature 435(7043):814–818
2. Xie JR, Kelley S, Szymanski BK (2013) Overlapping community detection in networks: the state-of-the-art and comparative study. ACM Comput Surv 45(4):43
3. Lancichinetti A, Fortunato S, Kertesz J (2009) Detecting the overlapping and hierarchical community structure in complex networks. New J Phys 11(3):033015
4. Lancichinetti A, Radicchi F, Ramasco JJ, Fortunato S (2011) Finding statistically significant communities in networks. PLoS ONE 6(4):e18961
5. Ahn YY, Bagrow JP, Lehmann S (2010) Link communities reveal multi-scale complexity in networks. Nature 466(7307):761–764
6. Kalinka AT, Tomancak P (2011) linkcomm: an R package for the generation, visualization, and analysis of link communities in networks of arbitrary size and type. Bioinformatics 27 (14):2011–2012
7. Lan H, Guishen W et al (2013) Link clustering with extended link similarity and EQ evaluation division. PLoS ONE 8(6):e66005
8. Lim S, Ryu S, Kwon S, et al (2014) LinkSCAN*: overlapping community detection using the link-space transformation. In: 2014 IEEE 30th International Conference on Data Engineering (ICDE), pp 292–303
9. Leicht EA, Holme P, Newman MEJ (2006) Vertex similarity in networks. Phys Rev E 73 (2):026120
10. van Dongen SM (2000) Graph clustering by flow simulation
11. Newman MEJ, Girvan M (2004) Finding and evaluating community structure in networks. Phys Rev E 69(2):026113
12. Huawei S, Xueqi C, Kai C, Mao-Bin H (2009) Detect overlapping and hierarchical community structure in networks. Phys A: Stat Mech Appl 388(8):1706–1712
13. Zachary WW (1977) An information flow model for conflict and fission in small groups. J Anthropol Res 33:452–473
14. Lusseau D, Schneider K, Boisseau OJ, Haase P, Slooten E, Dawson SM (2003) The bottlenose dolphin community of Doubtful Sound features a large proportion of long-lasting associations. Behav Ecol Sociobiol 54(4):396–405
15. Newman MEJ (2006) Modularity and community structure in networks. Proc Nat Acad Sci 103(23):8577–8582

A Comparison of Clustering Techniques for Meteorological Analysis

Ángel Arroyo, Verónica Tricio, Emilio Corchado and Álvaro Herrero

Abstract Present work proposes the application of several clustering techniques (k-means, SOM k-means, k-medoids, and agglomerative hierarchical) to analyze the climatological conditions in different places. To do so, real-life data from data acquisition stations in Spain are analyzed, provided by AEMET (Spanish Meteorological Agency). Some of the main meteorological variables daily acquired by these stations are studied in order to analyse the variability of the environmental conditions in the selected places. Additionally, it is intended to characterize the stations according to their location, which could be applied for any other station. A comprehensive analysis of four different clustering techniques is performed, giving interesting results for a meteorological analysis.

Keywords Clustering techniques · K-means · SOM k-means · K-medoids · Agglomerative hierarchical clustering · Meteorology

Á. Arroyo (✉) · Á. Herrero
Department of Civil Engineering, University of Burgos, Burgos, Spain
e-mail: aarroyop@ubu.es

Á. Herrero
e-mail: ahcosio@ubu.es

V. Tricio
Department of Physics, University of Burgos, Burgos, Spain
e-mail: vtricio@ubu.es

E. Corchado
Departamento de Informática y Automática, University of Salamanca, Salamanca, Spain
e-mail: escorchado@usal.es

Á. Herrero et al. (eds.), *10th International Conference on Soft Computing Models in Industrial and Environmental Applications*, Advances in Intelligent Systems and Computing 368, DOI 10.1007/978-3-319-19719-7_11

1 Introduction

As they are usually perceived as similar issues, it is necessary to distinguish between meteorology and climatology. On the one hand, meteorology consists in the study of the atmosphere, the scientific study of phenomena and physical processes occurring in the atmosphere, and atmospheric effects on the weather. Meteorologists then produce forecasts that are intended to predict weather conditions over the short term. On the other hand, climatology is the study of atmospheric changes, that define average climates and their change over time, due to both natural and anthropogenic climate variability. Climatology studies the same parameters as the meteorology, but its purpose is different; not because it seeks to make immediate forecast, but to study the long-term climatic characteristics. Climatology employs a long-term perspective, analyzing models that are designed to predict changes in weather patterns in the years to come. Present study focuses on the study of the meteorology in four places in Spain for a certain period of time.

In Spain, a network of stations for meteorological data acquisition can be found at [1]. These measurement stations acquire data continuously and these data are available for further study and analysis.

Clustering can be defined as the unsupervised classification of patterns into groups [2]. Hence, clustering (or grouping) techniques divide a given dataset into groups of similar objects, according to several different "similarity" measures. These sets of techniques have been previously applied to meteorological data [3, 4].

Differentiating from previous work, in present paper several clustering methods are applied to a detailed case study, where four places with different climates are selected with the more significant variables. Results are analyzed in two ways: the meteorology of the four places selected and the comparison of the clustering techniques to establish the strengths of each method. The main idea of present study is to analyse data describing meteorology from a case study associated to four places in Spain. Firstly, Principal Component Analysis [5] is applied to reduce the dimensionality of analyzed data and get an intuitive visualization of their internal structure. By doing so, we can determine an approximate number of clusters. In a second step, clustering techniques are applied to the original data set in order to find the best possible clustering of data. Four relevant hierarchical [6] and partitional [7] clustering techniques are applied, combined with the most widely-used distance measures. The number of clusters identified in the first step is applied in the second step as some of the techniques do need that figure.

The rest of this paper is organized as follows. Section 2 presents the techniques and methods that are applied. Section 3 details the real-life case study that is addressed in present work, while Sect. 4 describes the experiments and results. Finally, Sect. 5 sets out the main conclusions and future work.

2 Clustering Techniques and Methods

This study checks the performance of several clustering techniques when analyzing meteorological variables (described in Sect. 3), in order to study the behavior of the climatology in different locations.

In order to analyze data sets with meteorological information, several clustering methods [2, 8] have been applied. Clustering is one of the most important unsupervised learning problems [9]. It can be defined as the process of organizing objects into groups whose members are similar in some way. A cluster is a collection of objects which are similar to those in the cluster and are dissimilar to those belonging to other clusters. Clustering techniques can be divided, in general terms, into two categories: partitional and agglomerative. Partitional clustering algorithms divide the data set into a specified number of clusters trying to minimize certain criteria [10]. On the contrary, agglomerative clustering algorithms begin with each pattern in a distinct (singleton) cluster, and successively merges clusters together until a stopping criterion is satisfied [2].

Those methods and measure distances are described in this section.

2.1 Partitional Clustering

2.1.1 *k*-Means

The well-known *k*-means [11] is an algorithm for grouping data into a given number of clusters. Its application requires two input parameters: the number of clusters (k) and their initial centroids, which can be chosen by the user or obtained through some pre-processing. Each data element is assigned to the nearest group centroid, thereby obtaining the initial composition of the groups. Once these groups are obtained, the centroids are recalculated and a further reallocation is made. The process is repeated until the centroids do not change. Given the heavy reliance of this method on initial parameters, a good measure of the goodness of the grouping is simply the sum of the proximity Error Sums of Squares (SSE) that it attempts to minimize:

$$SSE = \sum_{j=1}^{k} \sum_{x \in G_j} \frac{p(x_i, c_j)}{n} \quad (1)$$

where p is the proximity function, k is the number of the groups, c_j *are* the centroids and n the number of rows. In the case of Euclidean distance, the expression is equivalent to the global mean square error.

2.1.2 SOM *k*-Means

Traditional Self Organizing Maps (SOM) [12] cannot provide with precise clustering results, while traditional *k*-means depends on the initial value and it is difficult to find the centroid of cluster [13].

To overcome the limitations of both methods, SOM *k*-means [12] is proposed. It combines SOM and *k*-means in the following way: when the SOM training finishes, the *k*-means algorithm is applied to refine the weights obtained by the SOM. When the SOM clustering finishes, *k*-means is also applied to refine the final result of clustering.

2.1.3 *k*-Medoids

The objective function of *k*-medoids (partitioning around medoids) algorithm is to partition a given dataset (X) into c clusters. The input and output arguments are the ones that *k*-means uses [11]. The main difference between the two methods consists in the way cluster centers are calculated; in *k*-medoids, the new cluster center is the nearest data point to the mean of the cluster points [14]. The algorithm generates random cluster centres, and not a partition matrix for initialization.

2.2 Agglomerative Hierarchical Clustering

Algorithms in this category generate a cluster tree also called dendrogram by using heuristic techniques. The most popular algorithms that use merging to generate the cluster tree are called agglomerative. There are many implementations of agglomerative hierarchical algorithms [15].

2.3 Measure Distances

The above mentioned clustering techniques, take into account distance in order to cluster the data. Different distance criteria are defined. The distance measures applied in present study are described in this subsection.

2.3.1 Euclidean Distance

This is the most common metric, where each centroid is the mean of the points in that cluster:

$$d(x-c)=(x-c)(x-c)' \tag{1}$$

where d is the distance from the point x to the centroid c.

2.3.2 Seuclidean Distance

In Standardized Euclidean metric (Seuclidean), each coordinate difference between rows in X is scaled by dividing it by the corresponding element of the standard deviation:

$$d(x-c)=(x-c)V^{-1}(x-c)' \tag{2}$$

where V is the n-by-n diagonal matrix.

2.3.3 Cityblock Distance

In this case, each centroid is the component-wise median of the points in that cluster.

$$d(x-c)=1-\sum_{j=1}^{p}\left|x_j-c_j\right|' \tag{3}$$

where the exponent P is a scalar positive value and j an observation in the vector X.

2.3.4 Cosine Distance

This is defined as one minus the cosine of the included angle between points (treated as vectors). Each centroid is the mean of the points in that cluster, after normalizing those points to unit Euclidean length:

$$d(x-c)=1-\frac{xc'}{\sqrt{(xx')(cc')}} \tag{4}$$

2.3.5 Correlation Distance

In this case, each centroid is the component-wise mean of the points in that cluster, after centering and normalizing those points to zero mean and unit standard deviation.

$$d(x-c)=1-\frac{(x-\bar{x})(c-\bar{c})}{\sqrt{(x-\bar{x})(c-\bar{c})}\sqrt{(c-\bar{c})(c-\bar{c})}} \tag{5}$$

2.3.6 Minkowski Metric

The Minkowski distance is a metric in a normed vector space which can be considered as a generalization of both the Euclidean distance and the Manhattan distance, as defined by:

$$d(x-c)=\sqrt[p]{\sum_{j=1}^{n}\left|x_{sj}-x_{tj}\right|^{p}} \tag{6}$$

where p is a scalar positive value of the exponent, s and t are the indexes of the rows of the vector x and j is index of the column of the same vector x.

3 Real-Life Case Study

This study is focused on the analysis of meteorological data recorded in four different places in Spain, which is a country with a noticeable climatic diversity. As it was described in Sect. 1, the data were provided by the Spanish Meteorological Agency (AEMET) [1, 16]. From the database of AEMET, the following four stations were selected for present analysis, based on their very different meteorology and sparse geographical location:

1. **Burgos Airport**. Geographical coordinates: 42°21’22”N; 03°37’17”W; 891 meters above sea level, moderate Continental climate. Labelled as **BU**.
2. **Santiago de Compostela Airport**. Geographical coordinates: 42°53’51”N; 08°24’38’W; 370 meters above sea level, Atlantic climate. Labelled as **SA**.
3. **Almería Airport**. Geographical coordinates: 36°50’47”N; 02°21’25”W; 21 meters above sea level, Mediterranean dry climate. Labelled as **AL**.
4. **Palma de Mallorca Port**. Geographical coordinates: 39°33’12”N; 02°37’31”E; 3 meters above sea level, typical Mediterranean climate. Labelled as **PM**.

From the timeline point of view, data are selected from years 2004 and 2005. Year 2003 was characterized by extreme values, particularly a heat wave during the month of August, in three of the analyzed locations. During years 2004 and 2005 normal values were again recorded for the analyzed variables. This fact, together with the intention of analyzing subsequent years on these studies, are the reasons for selecting years 2004 and 2005 in present study There are a total of 2,924 samples as data are collected on a daily basis (365 days for 2004 and 2005), that is

730 samples for each one of the 4 stations and one sample per day (daily average). The following parameters (six meteorological variables) are gathered:

1. Maximum absolute temperature: maximum temperature in the whole day (C°).
2. Minimum absolute temperature: minimum temperature in the whole day (C°).
3. Wind speed: air maximum gust recorded in the whole day (m/s).
4. Number of hours of sunshine in the day (hours).
5. Maximum absolute atmospheric pressure in tenths of hectopascal in the whole day (hPa).
6. Minimum absolute atmospheric pressure in tenths of hectopascal in the whole day (hPa).

4 Experiments and Results

As previously stated, Principal Component Analysis (PCA) [17] has been firstly applied to the dataset. In this step, the inner structure of the dataset is searched. In this case study, three main clusters of data are identified. This figure is used as an approximation to the number of clusters to be selected in the subsequent experiments. Once the initial approximate number of clusters is obtained, several clustering techniques are compared, namely: *k*-means, SOM *k*-means, *k*-medoids, and agglomerative hierarchical. For present study, the Matlab [18] implementations of such methods have been applied.

The results obtained by those techniques (after twenty valid runs for *k*-means, *k*-medoids and agglomerative hierarchical methods and ten valid runs for SOM *k*-means) are listed and described in this section: Tables 1, 2, 3 and 4 shows the parameter values of the applied technique and the allocation of data (according to the meteorological station they come from: BU, AL, SA, and PM) to the defined clusters (K). Additionally, execution time is also gathered to compare the different methods.

In Table 1, column k represents the number of clusters specified for the algorithm in advance, Distance is the measure distance applied (see Sect. 2), Time is the execution time (in seconds) and the Cluster Samples Allocation represents the percentage of samples from each one of the stations (BU, AL, SA and PA) that are allocated to each one the clusters; e.g. [100 0] represents 100 % of samples allocated to the first cluster and 0 % allocated to the second one.

From the data in Table 1, two main aspects can be highlighted. Firstly, the big difference between the meteorology of Burgos and the one of the other three locations, as well as the similar Mediterranean conditions in Almería and Palma de Mallorca. This can be seen in the following tendency; as the number of clusters is increased, the samples belonging to Burgos tend to remain together (in the same cluster), while the subdivision of samples in different clusters is more usual for the locations Almería and Palma de Mallorca. It is important to highlight that in all cases, samples from Mallorca and Almería are included in the same clusters. It is

Table 1 *K*-means clustering results

			Cluster samples allocation (%)			
k	Distance	Time (s)	BU	AL	SA	PM
2	Seuclidean	0.051601	[0 100]	[100 0]	[98 2]	[100 0]
2	Cityblock	0.045175	[100 0]	[0 100]	[9 91]	[0 100]
2	Cosine	0.045913	[63 37]	[40 60]	[67 33]	[44 56]
2	Correlation	0.631190	[29 71]	[69 31]	[41 59]	[75 25]
3	Seuclidean	0.039746	[100 0 0]	[0 100 0]	[0 0 100]	[0 100 0]
3	Cityblock	0.071559	[100 0 0]	[0 100 0]	[0 0 100]	[0 99 1]
3	Cosine	0.054590	[28 42 30]	[48 3 49]	[51 29 20]	[44 8 48]
3	Correlation	0.839933	[47 9 44]	[11 45 44]	[29 19 52]	[29 19 52]
4	Seuclidean	0.051418	[0 0 100 0]	[59 0 0 41]	[0 100 0 0]	[51 0 0 49]
4	Cityblock	0.080244	[52 0 0 48]	[0 100 0 0]	[0 0 100 0]	[0 100 0 0]
4	Cosine	0.058141	[18 28 39 15]	[34 27 2 38]	[45 27 21 7]	[38 20 5 37]
4	Correlation	0.084806	[39 5 40 15]	[24 44 7 26]	[31 15 18 36]	[21 50 6 23]
5	Seuclidean	0.060528	[0 49 0 51 0]	[0 0 59 0 41]	[100 0 0 0 0]	[0 0 51 0 49]
5	Cityblock	0.063433	[0 0 100 0 0]	[0 0 0 52 48]	[36 64 0 0 0]	[0 0 0 45 54]
5	Cosine	0.078143	[22 35 37 5 1]	[29 7 1 29 33]	[43 10 18 26 4]	[31 0 4 29 35]
5	Correlation	0.079106	[39 27 1 14 18]	[6 1 44 21 28]	[15 14 11 33 27]	[6 0 51 17 26]
6	Seuclidean	0.069633	[0 100 0 0 0 0]	[0 0 48 30 23 0]	[60 0 0 0 0 40]	[0 0 46 26 28 0]
6	Cityblock	0.081311	[31 0 6 19 27 17]	[2 34 25 12 1 27]	[10 5 21 24 14 27]	[2 44 18 10 0 25]
6	Cosine	0.104866	[34 34 17 10 3 1]	[5 1 15 20 26 33]	[11 14 22 28 22 3]	[0 3 9 26 29 33]
6	Correlation	0.082790	[16 3 23 1 31 26]	[18 33 9 35 3 1]	[29 19 18 8 11 14]	[12 35 6 43 4 0]

also worth mentioning the great influence of the measure distance applied. While 'cosine' and 'correlation' usually split samples from a location in different clusters, 'seuclidean' and 'cityblock' generally keep the samples from the same location in the same cluster. This is because 'cosine' and 'correlation' measures the difference in the angle between two vectors and not the difference in the magnitude between two vectors [10]. Finally, regarding the elapsed time in executing the *k*-means algorithms, it could be say that 'Correlation' provides the lowest response when k equals 1, 2, and 3. A similar response is obtained when applying the other measure distance when k equals 4, 5 and 6.

In Table 2, the results obtained by SOM *k*-means are shown. In this table, Type is the type of applied algorithm in the neurons initialization process (it can be sequential or batch). Additionally Err shows the total quantization error for the data set, according to the distance from any given data point to a cluster center weighted by that data point's membership grade.

One of the first conclusions that can be drawn from Table 2 is that, as expected, the 'seq' algorithm is slower than the 'batch' one. Both are iterative algorithms, but the batch version is much faster in Matlab since matrix operations can be utilized efficiently [19]. In relation to the cluster sample allocation process, SOM *k*-means offer similar results to *k*-means (Table 1) when applying 'seuclidean' distance; this is because SOM k-means uses also a simple distance measure. The same patterns in the process of cluster samples allocation are repeated respect to *k*-means (Table 1).

In Table 3 the results of applying *k*-medoids to the original data set are shown.

By applying *k*-medoids (Table 3), the cluster sample allocation is similar to that obtained by *k*-means (Table 1) when the measure distance is 'seuclidean'. However, it is quite different when 'cosine' and 'correlation' distances are applied. In these cases *k*-means makes a wider division of the samples into clusters. The main drawback of *k*-medoids respect to *k*-means (Table 1), is the computational cost; as can be seen, when the number of clusters is increased, the execution time is much bigger in the case of *k*-medoids.

Table 4 shows the very different results of applying agglomerative hierarchical clustering technique to the original dataset from those obtained in partitional methods, (Tables 1, 2 and 3).

The main difference between agglomerative hierarchical clustering and the other three methods is that the former allocates, with almost 100 % of accuracy, the samples to clusters according to the location of the stations. Increasing the number of cluster is not then useful in the case of agglomerative hierarchical clustering because most of the additional clusters are empty (no samples are allocated to them). Not a very reliable response is detected in the samples cluster allocation process when k equals 3, 4, 5, and 6 and when the measure distance selected is 'seuclidean', the total of the samples are allocated in the same cluster. Additionally, when k equals 3 and 6 and selected distance is 'cityblock', the samples of Santiago de Compostela, Almería and Palma de Mallorca are allocated in the same cluster. Agglomerative hierarchical clustering is able to distinguish with almost 100 % accuracy three groups of data: Burgos, Santiago de Compostela and Almería together with Palma de Mallorca, when (k equals 5 and 6) and for 'euclidean' or

Table 2 SOM *k*-means clustering results

				Cluster samples allocation (%)			
K	Type	Err	Time (s)	BU	AL	SA	PM
2	Seq	3.74	1.934958	[0 100]	[100 0]	[98 2]	[100 0]
2	Batch	3.73	0.025540	[0 100]	[100 0]	[98 2]	[100 0]
3	Seq	0.99	1.923084	[0 0 100]	[0 100 0]	[100 0 0]	[0 100 0]
3	Batch	0.99	0.028748	[0 0 100]	[0 100 0]	[100 0 0]	[0 100 0]
4	Seq	0.83	1.928557	[0 0 0 100]	[59 0 40 0]	[0 100 0 0]	[51 0 49 0]
4	Batch	0.87	0.032140	[0 0 51 49]	[0 100 0 0]	[100 0 0 0]	[0 100 0 0]
5	Seq	0.74	1.917720	[0 100 0 0 0 0]	[0 0 28 24 48 0]	[99 0 1 0 0 0]	[0 0 25 29 46 0]
5	Batch	0.77	0.034465	[0 49 51 0 0 0]	[57 0 0 0 43 0]	[0 0 0 99 0 0]	[69 0 0 0 31 0]
6	Seq	0.66	1.943915	[0 100 0 0 0 0]	[0 0 48 29 0 23]	[32 0 0 0 68 0]	[0 0 46 26 0 29]
6	Batch	0.66	0.037405	[0 0 0 100 0 0]	[48 0 30 0 23 0]	[0 40 0 0 0 60]	[46 0 26 0 28 0]

Table 3 *K*-medoids clustering results

			Cluster samples allocation (%)			
K	Distance	Time (s)	BU	AL	SA	PM
2	Euclidean	0.477961	[0 100]	[100 0]	[93 7]	[100 0]
2	Seuclidean	0.624595	[100 0]	[0 100]	[50 50]	[1 99]
2	Cosine	0.590097	[0 100]	[100 0]	[92 8]	[100 0]
2	Correlation	0.461034	[0 100]	[100 0]	[92 8]	[100 0]
3	Euclidean	0.835087	[0 100 0]	[100 0 0]	[0 0 100]	[100 0 0]
3	Seuclidean	1.092403	[0 0 100]	[50 50 0]	[15 70 15]	[50 50 0]
3	Cosine	0.623375	[0 100 0]	[0 0 100]	[100 0 0]	[0 0 100]
3	Correlation	0.633741	[0 0 100]	[100 0 0]	[0 100 0]	[100 0 0]
4	Euclidean	1.352807	[0 0 100 0]	[0 50 0 50]	[100 0 0 0]	[0 47 0 53]
4	Seuclidean	1.420630	[0 45 0 55]	[50 0 50 0]	[76 1 15 9]	[50 0 50 0]
4	Cosine	1.063674	[100 0 0 0]	[0 51 0 49]	[0 0 100 0]	[0 46 0 54]
4	Correlation	1.134566	[0 100 0 0]	[0 0 50 50]	[100 0 0 0]	[0 0 46 54]
5	Euclidean	1.293159	[0 0 57 43 0]	[0 50 0 0 50]	[100 0 0 0 0]	[0 46 0 0 54]
5	Seuclidean	1.676479	[0 45 0 55 0]	[48 0 39 0 13]	[15 1 24 1 60]	[47 0 41 0 11]
5	Cosine	1.309666	[0 0 0 44 56]	[0 51 49 0 0]	[100 0 0 0 0]	[0 46 54 0 0]
5	Correlation	1.674484	[0 100 0 0 0]	[31 0 0 25 44]	[0 0 100 0 0]	[26 0 0 31 43]
6	Euclidean	1.946891	[0 57 0 43 0 0]	[0 0 25 0 44 31]	[100 0 0 0 0 0]	[0 0 31 0 43 26]
6	Seuclidean	1.544820	[0 55 0 0 0 45]	[30 0 29 4 38 0]	[15 2 11 58 13 0]	[37 0 10 10 42 0]
6	Cosine	1.947368	[0 100 0 0 0 0]	[46 0 0 27 0 28]	[0 0 66 0 34 0]	[41 0 0 23 0 35]
6	Correlation	1.974048	[0 57 0 43 0 0]	[50 0 50 0 0 0]	[0 0 0 0 39 61]	[47 0 53 0 0 0]

Table 4 Agglomerative hierarchical clustering results

			Cluster samples allocation (%)			
K	Distance	Time (s)	BU	AL	SA	PM
2	Euclidean	31.217975	[0 100]	[100 0]	[100 0]	[100 0]
2	Seuclidean	31.722818	[0 100]	[0 100]	[0 100]	[0 100]
2	Cityblock	30.559558	[100 0]	[0 100]	[0 100]	[0 100]
2	minkowsky	30.697608	[0 100]	[100 0]	[100 0]	[100 0]
3	Euclidean	31.083268	[0 100 0]	[0 0 100]	[0 0 100]	[0 0 100]
3	Seuclidean	31.626975	[0 100 0]	[0 100 0]	[0 100 0]	[0 100 0]
3	Cityblock	31.734806	[0 0 100]	[0 100 0]	[0 100 0]	[0 100 0]
3	minkowsky	31.577912	[0 100 0]	[0 0 100]	[0 0 100]	[0 0 100]
4	Euclidean	31.163991	[0 0 0 100]	[0 100 0 0]	[0 100 0 0]	[0 100 0 0]
4	Seuclidean	30.652015	[0 100 0 0]	[0 100 0 0]	[0 100 0 0]	[0 100 0 0]
4	Cityblock	31.023600	[0 100 0 0]	[0 0 0 100]	[0 0 0 100]	[0 0 0 100]
4	minkowsky	30.651821	[0 0 0 100]	[0 100 0 0]	[0 100 0 0]	[0 100 0 0]
5	Euclidean	30.652530	[0 0 0 0 100]	[0 100 0 0 0]	[100 0 0 0 0]	[0 100 0 0 0]
5	Seuclidean	30.602942	[0 100 0 0 0]	[0 100 0 0 0]	[0 100 0 0 0]	[0 100 0 0 0]
5	Cityblock	30.619702	[0 0 0 100 0]	[0 100 0 0 0]	[0 100 0 0 0]	[0 100 0 0 0]
5	minkowsky	30.623085	[0 0 0 0 100]	[0 100 0 0 0]	[100 0 0 0 0]	[0 100 0 0 0]
6	Euclidean	30.643442	[0 0 0 0 0 100]	[0 100 0 0 0 0]	[0 1 99 0 0 0]	[0 100 0 0 0 0]
6	Seuclidean	30.694775	[0 100 0 0 0 0]	[0 100 0 0 0 0]	[0 100 0 0 0 0]	[0 100 0 0 0 0]
6	Cityblock	30.762574	[0 100 0 0 0 0]	[0 0 0 100 0 0]	[0 0 0 100 0 0]	[0 0 0 100 0 0]
6	minkowsky	30.737657	[0 0 0 0 0 100]	[0 100 0 0 0 0]	[0 1 99 0 0 0]	[0 100 0 0 0 0]

'minkowsky' distance. This is because 'euclidean' distance is a particular case of 'mikowski' distance. Another drawback is that this technique is highly computationally demanding, regardless of the number of selected clusters or the distance metric applied.

5 Conclusions and Future Work

Main conclusions derived from previously explained results (see Sect. 4) can be divided into two groups, firstly, those regarding the analysis of meteorological conditions in the analyzed case study. Secondly, those related to the behaviour of the different clustering techniques applied in the case study.

Talking about the meteorological conditions in the four selected places, a clear conclusion is the big difference between the climatology in Burgos and that in the other three places. Also, in Santiago de Compostela it is appreciated a different climatology from the other three places, but not as pronounced as in the case of Burgos. However, the climate in Palma de Mallorca and Almería are very similar between them, as none of the applied methods has been able to split those samples in different clusters.

Regarding the applied clustering techniques, it should be emphasized the different results offered by the hierarchical agglomerative method compared with the partitional methods, and also the differences between the application of different measures of distance. In many cases, the agglomerative hierarchical clustering do not show a reliable response, not being able to allocate samples from different places in different clusters.. *K*-means, SOM *k*-means and *k*-medoids attain similar results, and the selected measure distance selected is a key factor. *K*-means is the best method in terms of computational load. None of the techniques has been able to separate the samples of the four locations in four different clusters, as it was initially detected through PCA.

Future work will consists on expanding the time window to analyze the temporal evolution of meteorological data. It will also include the application of probabilistic methods and different evaluating techniques, in order to extend the comparison.

References

1. National Network of meteorological stations—Spanish Agency of Meteorology. http://www.aemet.es/es/eltiempo/observacion/ultimosdatos
2. Jain AK, Murty MN, Flynn PJ (1999) Data clustering: a review. ACM Comput Surv (CSUR) 31(3):264–323
3. Lu Y, Ma T, Yin C, Xie X, Tian W, Zhong S (2015) Implementation of the fuzzy C-means clustering algorithm in meteorological data. Int J Database Theory Appl 6:1–18
4. Tian W, Zheng Y, Yang R, Ji S, Wang J (2015) A survey on clustering based meteorological data mining. Int J Grid Distributed Comput 7:229–240

5. Hotelling H (1933) Analysis of a complex of statistical variables into principal components. J Educ Psychol 24:417–444
6. Michie S, Richardson M, Johnston M, Abraham C, Francis J, Hardeman W, Eccles MP, Cane J, Wood CE (2013) The behavior change technique taxonomy (v1) of 93 Hierarchically clustered techniques: building an international consensus for the reporting of behavior change interventions. Ann Behav Med 46(1):81–95
7. Aparna K, Nair MK (2015) Comprehensive study and analysis of partitional data clustering techniques. Int J Bus Anal (IJBAN) 2:23–38
8. Anil K (2010) J.: Data clustering: 50 years beyond K-means. Pattern Recogn Lett 31:651–666
9. Barlow H (1989) Unsupervised learning. Neural Comput 1:295–311
10. Jain AK, Maheswari S (2013) Survey of recent clustering techniques in data mining. J Curr Comput Sci Technol 3
11. Ding C, He X (2004) K-means clustering via principal component analysis. In: Proceedings of the twenty-first international conference on Machine learning, vol 29 (2004)
12. Kohonen T (1990) The self-organizing map. Proc IEEE 78:1464–1480
13. Napoleon D, Pavalakodi S (2011) A New method for dimensionality reduction using K means clustering algorithm for high dimensional data set. Int J Comput Appl 13:41–46
14. Park HS, Jun CH (2009) A simple and fast algorithm for K-medoids clustering. Expert Syst Appl 36:3336–3341
15. Day WHE, Edelsbrunner H (1984) Efficient algorithms for agglomerative hierarchical clustering methods. J Classif 1:7–24
16. Hotelling H (1933) Analysis of a complex of statistical variables into principal components. J Educ Psychol 24:498–520
17. Mathworks. http://es.mathworks.com/products/matlab/?refresh=true (2015)
18. Vesanto J, Himberg J, Alhoniemi E, Parhankangas J (1999) Self-organizing map in Matlab: the SOM toolbox. In: Proceedings of the Matlab DSP Conference, vol 99, pp 16–27

Evolutionary Computation and Optimization

Evolutionary Battery Scheduling Optimization Under Variable Electricity Prices in Micro-Grids with Renewable Generation

R. Mallol-Poyato, S. Salcedo-Sanz, S. Jiménez-Fernández and P. Díaz-Villar

Abstract In this paper we propose an Evolutionary Algorithm (EA) to tackle the Battery Scheduling Optimization Problem (BSOP) in MicroGrids (MG). Specifically, we consider a MG that includes renewable generation, different load profiles, and is equipped with an energy storage device (battery) to address its scheduling (charge/discharge duration and occurrence) in a real scenario of variable electricity prices. We fully describe the proposed evolutionary algorithm, including its initialization and the different operators implemented to guide the search. Experiments in a MG with residential and industrial consumption profiles, photovoltaic and wind power generators' profiles, together with an energy storage device are carried out. To show the good battery scheduling performance of the proposed EA, we have compared the results with what we called a deterministic scenario. It is defined as a fixed way of using the energy storage device, that only depends on the pattern of load and generation profiles considered. Hourly values of both generation and consumption have been considered, and the good performance of the proposed EA is shown for four different weeks of the year (one per season), where the effect of the battery scheduling optimization obtains savings up 10 % of the total electricity cost in the MG, when compared with the deterministic approach.

1 Introduction

MicroGrids (MGs) are defined as the coordinated operation and control of distributed energy resources, involving different technologies, together with controllable or non-controllable loads and energy storage systems, operating connected to the utility grid and capable of islanding [1, 2]. MGs are often considered as the next evolution of the current electricity distribution systems, since they allow a high penetration of low emissions generation, a reduction of electricity transportation lines

R. Mallol-Poyato · S. Salcedo-Sanz (✉) · S. Jiménez-Fernández · P. Díaz-Villar
Department of Signal Processing and Communications,
Universidad de Alcalá, Alcalá de Henares, Madrid, Spain

Á. Herrero et al. (eds.), *10th International Conference on Soft Computing Models in Industrial and Environmental Applications*, Advances in Intelligent Systems and Computing 368, DOI 10.1007/978-3-319-19719-7_12

losses, the reuse of waste heat to service thermal loads, etc. MGs offer other important advantages such as a larger robustness against extreme weather events or attacks, an improved reactive power support and a save in deployment time of the systems [1].

In its basic configuration, a MG is a medium or low voltage network with distributed energy generators, multiple electrical loads and, optionally, energy storage devices. MGs are connected to the utility grid via the Point of Common Coupling, although their main feature is their capability to disconnect from the utility grid to operate in islanding mode in case of faults in the upstream network. Constraints about this islanded operation in regulations, on the basis of security and control of the grid, have limited the deployment of MGs so far. However, the development of fast and safe power electronic inverters and the consequent regulation revision may increase the number of MGs in operation. Like Super Grids at transmission level or Virtual Power Plants at software and communication one, MGs can be related with the global and opened concept of Smart Grids as one of the multiple and complementary options to develop the electricity network of the future [3]. As stated by the EU SmartGrid Platform, Smart Grid is an "electricity network that can intelligently integrate the actions of all users –generators, consumers and those that do both– in order to efficiently deliver sustainable, economic and secure electricity supply". Deployment of decentralized architectures, improve management and control techniques, efficiently integrate intermittent generation systems and enhance the role of the demand side are mentioned as key challenges for Smart Grids. MGs have to cope with these issues at the distribution level.

The research on different aspects of MGs, both theoretical (MG design and planning, control, renewable energies integration, etc.), and practical (rural electrification and stand alone systems, energy management, real application in smart cities) has been massive in the last few years [3]. Some of these works apply advanced computational methodologies, such as neural computation [4], evolutionary computation [5] or computational intelligence techniques in order to obtain good quality solutions when tackling difficult problems related to MG. For example, one of the key issues concerning MG technology deployment is related to the integration of distributed energy resources into distribution networks. In the last few years there have been many different works dealing with optimal location and sizing of distributed generation in MG [6–11]. Other aspects of MG design such as topology design [12], control [13, 14], load prediction [15] or back-up systems design [16, 17] have been recently studied.

This paper is focused on the management of an energy storage device in MGs within an environment of variable electricity prices, taking into account MG with renewable generation. Specifically, we consider the problem of battery scheduling in a MG for modifying the main grid consumption profile, in such a way that the electricity cost is reduced. We will refer to this problem as Battery Scheduling Optimization Problem (BSOP). We propose to apply an Evolutionary Algorithm (EA) to this task, and evaluate its performance taking into account different load and generation profiles through the year. An specific example based on the Spanish regulation for energy prices is considered and discussed in the experimental part of the paper.

The rest of this work is structured in the following way: next section details the problem definition considered in this paper. Section 3 presents the EA used to solve the problem. Section 4 presents the experimental part of the paper, where the performance of the proposed EA in a problem of battery scheduling under variable prices is evaluated.

2 Problem Definition

Let us consider, without lost of generality, a model of MG consisting of different loads, renewable energy generators and a battery (energy storage device) connected together to the main utility grid at the same point (see Fig. 1 as reference). We will focus on establishing the optimal battery scheduling for a given period (T hours). For that, we consider typical (or predicted) wind generation $\mathcal{W} = \{\mathcal{W}_1, \ldots, \mathcal{W}_T\}$ and photovoltaic generation $\mathcal{F} = \{\mathcal{F}_1, \ldots, \mathcal{F}_T\}$ annual profiles. We also consider two different load profiles $\mathcal{L}^1 = \{\mathcal{L}_1^1, \ldots, \mathcal{L}_T^1\}$ corresponding to a residential consumption, and $\mathcal{L}^2 = \{\mathcal{L}_1^2, \ldots, \mathcal{L}_T^2\}$, that corresponds to an industrial consumption. The battery scheduling is defined by a vector $\mathcal{B}$ that stands for the battery charging ($\mathcal{B} > 0$) or discharging ($\mathcal{B} < 0$) power. Note that if an annual scheduling is considered, then $\mathcal{B}$ is a 8760-length vector, whereas in a weekly profile the encoding of $\mathcal{B}$ would be a 168-length vector. A vector $\mathcal{P}$ is defined as the power exchanged between the MG and the main grid, considering the effect of all profiles described above acting together at the same node, as follows: $\mathcal{P} = \mathcal{L}^1 + \mathcal{L}^2 - \mathcal{F} - \mathcal{W} + \mathcal{B}$

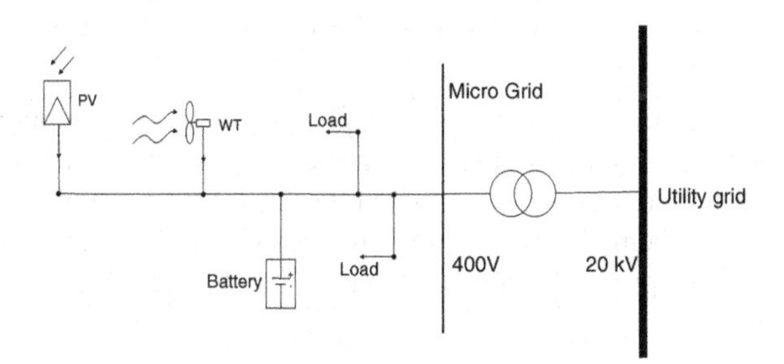

Fig. 1 MicroGrid structure used in this work

In this work, battery power values are limited, and minimum % SOC (State of Charge) value is set to 20 % to achieve longer battery life time (according to a simplified battery model). The simplest way for handling the battery is named here as *Deterministic* battery use. In that case, the battery is charged (with the maximum possible power) every period of time in which the generation is larger than the load's demand. If this power is within battery limits, there is no energy exported to the main grid. On the other hand, for the periods of time in which the load's demand is larger than the generation, the battery is discharged with the maximum possible power, always avoiding exportation to the grid.

Finally, an objective function $g(\mathcal{B}, \mathcal{W}, \mathcal{F}, \mathcal{L}^1, \mathcal{L}^2)$ is considered, where g stands for the total price paid for the electricity that the MG consumes from the main grid. Specifically, we have defined g as the electricity cost. This cost adds up two terms: an energy term (ET, the price for the energy consumption), and a power term (PT, the price for the availability of electric energy at our site), $g = ET + PT$.

In this work we have recreated the Spanish scenario for SMEs (Small and Medium-sized Enterprises) [18]. Precisely, we have implemented access tariff 3.1 for high voltage and power supplies up to 450kW. This scenario specifies three access tariff periods: P_1 (corresponding to high-priced hours, and with a duration of 4 hours), P_2 (mean-priced hours, and a duration of 12 hours), and P_3 (low-priced hours, and a duration of 8 hours). This means that the energy term is obtained as $ET = \sum_{j=1}^{3} \beta_j \cdot E_j$, where β_j is the access tariff price for the P_j period of time, and E_j is the energy consumption during the P_j period. Regarding the power term, and also for the Spanish scenario, an estimation of the maximum's power consumption for each of the three access tariff periods, HP_j (Hired Power in period P_j), has to be specified when signing the contract. Clients have to be meticulous when specifying and agreeing the HP_j, as the Electric Company will penalize them when their power consumption exceeds the agreed HP_j, and benefit them when their power consumption is limited to a certain percentage of that HP_j. PT is obtained as $PT = \sum_{j=1}^{3} \alpha_j \cdot IP_j$, where α_j is the power term price, and IP_j is the invoiced power during the P_j period. According to the above-mentioned benefit/penalty policy, IP_j may have three different values, depending on the maximum value of the power consumed in period j, $\mathcal{M}$:

$$IP_j = \begin{cases} HP_j & if \;\; 0.85 \cdot HP_j < \mathcal{M} < 1.05 \cdot HP_j \\ 0.85 \cdot HP_j & if \;\; \mathcal{M} < 0.85 \cdot HP_j \\ \mathcal{M} + 2 \cdot [\mathcal{M} - HP_j] & if \;\; \mathcal{M} > 1.05 \cdot HP_j \end{cases} \quad (1)$$

where $\mathcal{M} = \max(PC_j)$ and PC_j is the Power Consumption during the P_j period. In this scenario, it is obvious that the client needs to avoid the penalty region, and to restrain power consumption within the other two regions.

Mathematically the problem consists of obtaining the optimal battery scheduling $\mathcal{B}^*$ that produces the lowest value of $g(\mathcal{P})$ given $\mathcal{L}^1$, $\mathcal{L}^2$, $\mathcal{W}$ and $\mathcal{F}$.

3 Evolutionary Algorithm

Evolutionary algorithms (EAs) [19], are robust problems' solving techniques based on natural evolution processes. They are population-based techniques which codify a set of possible solutions to the problem, and evolve it through the application of the so called *evolutionary operators* [19]. In this section the main characteristics of the evolutionary algorithm proposed in this paper for the BSOP are described, including the algorithm's initialization selection, crossover and mutation operators proposed.

1. Let t be a counter for the number of generations, set it to $t = 1$. The first step of the algorithm consists then of generating an initial population of μ individuals (tentative solutions to the problem). As encoding, we consider the percentage of battery power charging or discharging respect to the battery maximums, i.e., a 168-length vector in the range $[-100, 100]$. Note that, using this encoding, the battery limits considered are needed for the calculation of $\mathcal{B}$. We initialize the population of the algorithm with five types of individuals: randomly generated solutions, solutions where the battery is not used ($\mathcal{B} = 0$), solutions obtained from slight random modifications of the $\mathcal{B} = 0$ case, solutions obtained from always charging and discharging the battery with the maximum available power (Deterministic way of using the battery) and finally solutions obtained from slight random modifications of the latter case.
2. Evaluate the fitness value for each individual of the population using the problem's objective function g (see Sect. 2).
3. Generate an offspring population, of length μ, by applying a multi-point crossover and mutation operators. The crossover operator applied in this work is carried out by randomly choosing two individuals, and selecting a fixed number of crossover points also at random (10 crossover points have been chosen in this case). A binary template is then randomly generated, and one offspring individual is obtained by assigning the 1 s to parts from one parent individual and the 0 s to parts from the other parent. Figure 2 shows an example of the crossover operator implemented. The mutation operator is applied with a small probability to each offspring individual (1 %). It consists of including random changes in the interval $[-100,100]$ that substitute previous values of the individual. The final fitness of the offspring population is obtained by means of function g.
4. Selection: we have chosen a ranking-based selection mechanism in this work. First, we get the union of parents and offspring individuals into just one population of length 2μ individuals. We sort the individuals in decreasing order of fitness value, and we discard the less fitted individuals, keeping the best μ ones. Note that this selection mechanism is elitist: the best solution found so far in the evolution is always kept in the population.
5. Stop if the stopping criterion is satisfied, and if not, set $t = t+1$ and go to Step 3. In this case, the stopping criterion established as a maximum number of generations (*max_ite*) has been reached.

4 Experiments and Results

In order to show the performance of the evolutionary algorithm proposed for the BSOP we have carried out a number of experiments in a MG equipped with micro-wind and micro-photovoltaic generation. We consider two different loads, one of them stands for a residential profile (standard profile published by the Spanish system operator, REE, [20]) characterized by an energy demand of 162500 kWh/year [21], that corresponds to 50 typical homes in Spain, with 3250 kWh/year each ($\mathcal{L}^1$).

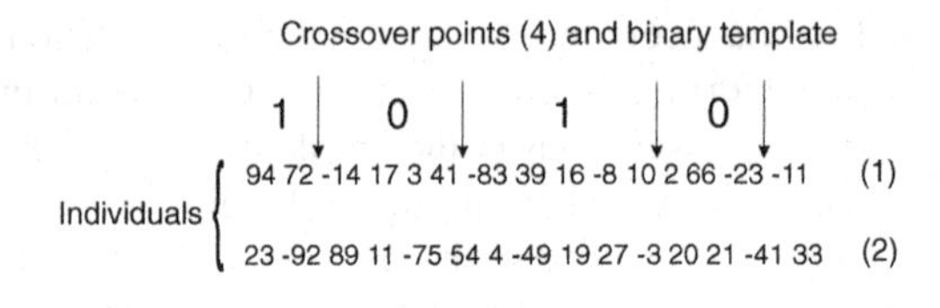

Fig. 2 Example of the crossover used in the EA: 15-length individuals, with 4 points crossover and binary matrix of length 4

Second, we consider an industrial consumption profile ($\mathcal{L}^2$), normalized in such a way that the annual energy consumption is 200000 kWh. Regarding the generation profiles, we have considered a 100 kW photovoltaic generator ($\mathcal{F}$) which provides 165000 kWh/year and a 100 kW wind power generator ($\mathcal{W}$) providing 140000 kWh/year. According to [22] we have considered a 300 kWh capacity battery that, in average, represents 1/3 of the daily energy demand. However, note that if we consider the energy provided by the generators, those 300 kWh would represent the energy consumed from the main grid during two days. Regarding α parameters of the objective function, we have considered $\alpha = [59.1735, 36.4907, 8.3677]$ €/kW, values periodically published by the Spanish Ministry of Industry, Energy and Tourism ([23]). The values of the β parameters contemplated are $\beta = [0.1044496, 0.089868, 0.065655]$ €/kW, published on a yearly basis by the electric companies (e.g. values valid for Endesa Energía's clients can be found at [24]). Finally, the values considered for HP_j are $HP_j = [72, 66, 58]$ kW, corresponding to the maximum power demanded in each period PC_j obtained with the deterministic solution.

In this work we have considered hourly-defined generation and consumption profiles for the whole year (52 weeks). These data have been applied to three different scenarios: (1) No battery use, (2) Deterministic battery use (as explained in Sect. 2), and (3) Battery scheduling optimization (proposed EA). The first two are baseline cases for comparison purposes, while the latter represents our proposed methodology. For clarification purposes, Table 1 presents the results for a randomly selected week out of every season (winter, spring, summer, and autumn) to depict different consumptions and generations. Finally, total values for the whole year are provided. In all cases, improvements obtained over case 1 and 2 are also displayed in the table.

To show some more specific details on the effect of evolutionary optimization when tackling battery scheduling in the BSOP, let us consider the case of one randomly selected week in the winter season. Figure 3 shows the generation profiles considered (wind and photovoltaic), as well as the load profiles (residential and industrial). Finally, the main grid consumption ($\mathcal{P}$) without battery, obtained by the addition of the abovementioned profiles, can be found. Figure 4 shows a comparison between the Deterministic approach and the proposed EA, in terms of consumption from the main grid, battery power scheduling and % SOC of the battery. First, note that an optimal power scheduling for an energy storage device on the MG provides a significant cost reduction over the Deterministic scheduling, providing a shorter

Table 1 BSOP results for four randomly selected weeks, each corresponding to one different season

	PT (€)	*ET* (€)	Cost/*fitness* (€)	Improvement over No battery	Improvement over Deterministic
Winter week					
No battery	137.20	366.88	504.09	0.00 %	–
Deterministic	137.20	290.48	427.68	15.16 %	0.00 %
Proposed EA	118.02	264.57	382.59	24.10 %	10.54 %
Spring week					
No battery	116.62	140.05	256.67	0.00 %	–
Deterministic	116.62	50.76	167.39	34.79 %	0.00 %
Proposed EA	116.62	48.67	165.29	35.60 %	1.25 %
Summer week					
No battery	124.95	313.52	438.46	0.00 %	–
Deterministic	124.95	234.38	359.33	18.05 %	0.00 %
Proposed EA	118.02	219.08	337.10	23.12 %	6.19 %
Autumn week					
No battery	128.88	316.98	445.86	0.00 %	–
Deterministic	116.62	211.44	328.06	26.42 %	0.00 %
Proposed EA	116.62	195.09	311.71	30.09 %	4.98 %
Total. Considering 52 weeks in a year					
No battery	6367.69	13498.37	19866.05	0.00 %	–
Deterministic	6281.03	8527.53	14808.55	25.46 %	0.00 %
Proposed EA	6078.37	7954.98	14033.35	29.36 %	5.23 %

recovery time of investment. The main goal of using an energy storage device is to avoid energy waste at the moments when the energy produced by the generators is larger than the energy demanded by the loads attached to the MG. However, we have shown that the use of the battery in a Deterministic way, taken only into account one single instant of load and generation, and ignoring the future instants predicted on its weekly profiles, is not an optimal procedure for performing the scheduling. Instead, the use of a meta-heuristic approach as the proposed EA is able to provide a solid battery scheduling, with a reduced consumption from the main grid.

An important characteristic of the scheduling obtained using the proposed EA, is that the periods of time in which the battery is charged may not be those in which an excess of generation occurs (see Fig. 4, main grid consumption). In this case, battery scheduling allows reducing the electricity consumption in the most expensive periods of time, shifting it towards the cheapest ones, which means a better battery use.

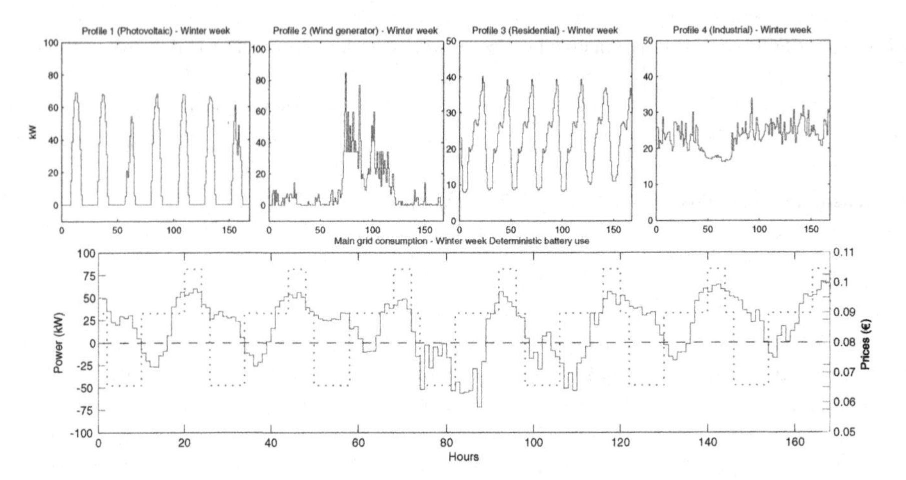

Fig. 3 For the winter week showed in Table 1: Generation and load profiles considered (*Above*). Consumption from the main grid ($\mathcal{P}$) occurred in the MG without the use of any battery in the system (baseline algorithm for comparison purposes, $\mathcal{B} = 0$). Dotted green line stands for the values of β in €/*kWh* (*Below*)

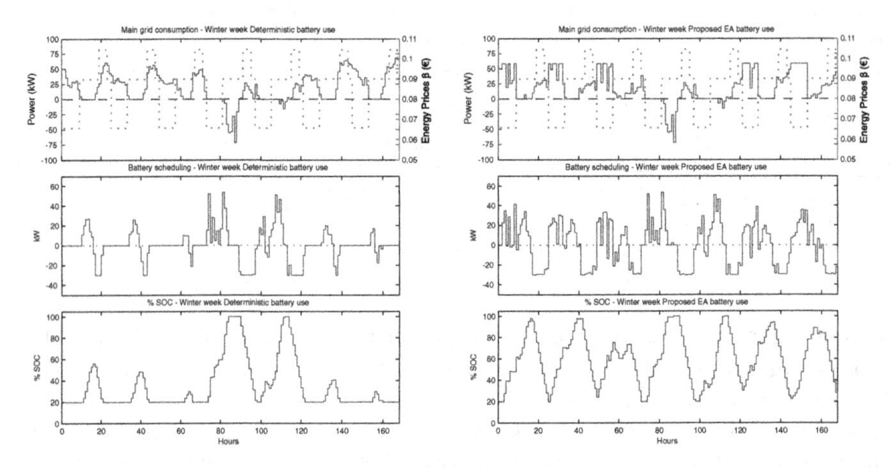

Fig. 4 Comparison between deterministic and proposed EA results: Consumption from the main grid; Battery power scheduling; and % SOC

5 Conclusions

This paper has presented the application of an evolutionary algorithm to a problem of battery scheduling optimization (BSOP) in MicroGrids (MG), in an environment of variable electricity prices. First, the problem of battery scheduling in MGs has been detailed, including a complete mathematical formulation of the problem. Then, the evolutionary algorithm proposed in this paper to solve the problem has been described, including the specific problem encoding, initialization and operators applied to evolve the population towards good quality solutions. In the experimental part of the work, we have shown the excellent performance of the proposed algorithm in a realistic BSOP, considering a MG with photovoltaic and wind generations and two loads type profiles (residential and industrial). Savings up to 35.6 % in the spring week (outperforming in 1.25 % the Deterministic use of the battery), around 24 % in winter (10.5 %), 23 % in summer (6 %) and 30 % in autumn week (5 %), are obtained.

Acknowledgments This work has been partially supported by Comunidad de Madrid, under project number S2013/ICE-2933, "PRICAM: Programa de redes eléctricas inteligentes en la Comunidad de Madrid."

References

1. Jiayi H, Chuanwen J, Rong X (2008) A review of distributed energy resources and MicroGrid. Ren Sust Energy Rev 12:2472–2483
2. Berry A, Platt G, Cornforth D (2010) Minigrids: analyzing the state-of-play. In: Proceedings of the IEEE International Power Electronics Conference, pp 710–716
3. Asmus P (2010) Microgrids, virtual power plants and our distributed energy future. Electr J 23(10):72–82
4. Xu FY, Zhou L, Lai LL (2010) Application of artificial neural network in electrical analysis of micro-grid load. In: Proceedings of the IEEE Power and Energy Society General Meeting, pp 1–5
5. Bajpai P, Dash V (2012) Hybrid renewable energy systems for power generation in stand-alone applications: a review. Renew Sustain Energy Rev 16:2926–2939
6. Moradi MH, Abedini M (2012) A combination of genetic algorithm and particle swarm optimization for optimal DG location and sizing in distribution systems. Int J Electr Power Energy Syst 34:66–74
7. Doagou-Mojarrad H, Gharehpetian GB, Rastegar H, Olamaei J (2013) Optimal placement and sizing of DG (distributed generation) units in distribution networks by novel hybrid evolutionary algorithm. Energy 54:129–138
8. Gözeland T, Hocaoglu MH (2009) An analytical method for the sizing and siting of distributed generators in radial systems. Electr Power Syst Res 79:912–918
9. Taher SA, Hasani M, Karimian A (2011) A novel method for optimal capacitor placement and sizing in distribution systems with nonlinear loads and DG using GA. Commun Nonlinear Sci Numer Simul 16:851–862
10. Devi S, Geethanjali M (2014) Application of modified bacterial foraging optimization algorithm for optimal placement and sizing of distributed generation. Expert Syst Appl 41:2772–2781
11. Ghosh S, Ghoshal SP, Ghosh S (2010) Optimal sizing and placement of distributed generation in a network system. Int J Electr Power Energy Syst 32:849–856
12. Zeng Z, Yang H, Zhao R, Cheng C (2013) Topologies and control strategies of multi-functional grid-connected inverters for power quality enhancement: A comprehensive review. Renew Sustain Energy Rev 24:223–270
13. Planas E, Gil de Muro A, Andreu J, Kortabarria I, Martínez de Alegría I (2013) General aspects, hierarchical controls and droop methods in microgrids: a review. Renew Sustain Energy Rev 17:147–159
14. Al-Saedi W, Lachowicz SW, Habibi D, Bass O (2013) Power flow control in grid-connected microgrid operation using Particle Swarm Optimization under variable load conditions. Int J Electr Power Energy Syst 49:76–85

15. Haesen E, Espinoza M, Pluymers B, Goethals I, Thong VV et al (2005) Optimal placement and sizing of distributed generator units using genetic optimization algorithms. Electr Power Qual Util J XI:97–104
16. Tan X, Li Q, Wang H (2013) Advances and trends of energy storage technology in Microgrid. Int J Electr Power Energy Syst 44(1):179–191
17. Moghaddam AA, Seifi A, Niknam T, Pahlavani MR (2011) Multi-objective operation management of a renewable MG (micro-grid) with back-up micro-turbine/fuel/battery hybrid power source. Energy 36:6490–6507
18. Real Decreto 1164/2001, October 26th, that establishes access tariffs for electric energy transport and distribution. https://www.boe.es/boe/dias/2001/11/08/pdfs/A40618-40629.pdf
19. Eiben AE, Smith JE (2003) Introduction to evolutionary computing. Springer, Berlin
20. Reference power demand and consumption profiles. Red Eléctrica de Espana, 2014. http://www.ree.es/es/actividades/operacion-del-sistema/medidas-electricas
21. Tariff 2.0 prices. BOE-A-2013-13803 (2013) Spanish Official Bulletin, no. 313, pp 106840–107066
22. Velik RM (2013) The influence of battery storage size on photovoltaics energy self-consumption for grid-connected residential buildings. Int J Adv Renew Energy Res 2(6):1–7
23. BOE-A-2014-1052 (2014) Spanish Official Bulletin, no. 28, pp 7147–697
24. Tariff 3.1. prices (2015) Endesa Energía, http://www.endesaonline.com/es/empresas/luz/tarifas_electricas_empresas_alta_tension/optima/precios/index.asp

Reliability-Based Design Optimization of Structures Combining Genetic Algorithms and Finite Element Reliability Analysis

Luis Celorrio

Abstract Structural optimization has undergone a substantial progress. Most of these research efforts deal with deterministic parameters. However, in a realistic structural design, it is necessary to consider inherent uncertainties in geometric variables and material properties to ensure safety and quality. Then, design constraints are formulated in terms such as probability of failure or reliability index. The process of design optimization enhanced by the addition of reliability constraints is referred as Reliability-Based Design Optimization (RBDO). Most of RBDO methods use classical mathematical optimization algorithms and require computing gradients of objective function and constraints. This task sometimes can be cumbersome and very hard because reliability constraints are implicit functions of design variables. However, the increased power of computers has made possible to apply heuristic methods, especially Genetic Algorithms in RBDO problems. In this paper Genetic Algorithms have been combined with Nonlinear Finite Element Reliability Analysis software, named OpenSees, to solve RBDO problems. A numerical example shows the performance of the implementation.

Keywords Reliability based design optimization · Nonlinear finite elements · Opensees · Genetic algorithms · Discrete design variables

1 Introduction

Structural optimization has undergone a substantial progress. Design and analysis software programs based in finite elements have recently added optimization and sensitivity assessment capabilities. Design methods for daily practice deal only with

L. Celorrio (✉)
Mechanical Engineering Department, Universidad de La Rioja, Logroño, La Rioja, Spain
e-mail: luis.celorrio@unirioja.es

Á. Herrero et al. (eds.), *10th International Conference on Soft Computing Models in Industrial and Environmental Applications*, Advances in Intelligent Systems and Computing 368, DOI 10.1007/978-3-319-19719-7_13

deterministic parameters. However, uncertainties are inherent in design variables and parameters such as material properties, loading and geometry parameters and, therefore, these uncertainties have to be considered in the design of any engineering system to ensure safety and quality. Traditionally, these uncertainties have been considered through partial safety factors in structural optimization methods. These partial safety factors are established in structural design codes for buildings. However, partial safety factors do not provide a quantitative measure of the safety margin in design and are not quantitatively linked to the influence of different design variables and their uncertainties on the overall system performance. Therefore, it is crucial to account for uncertainties explicitly in structural design.

Reliability Based Design Optimization (RBDO) is a set of methods to design optimal structures subject to reliability constraints. Each reliability constraint is written in probabilistic terms, such as, probabilities of failures or reliability indexes. These terms are computed using structural reliability methods.

In structural engineering, failures are formulated in mathematical form as limit state functions, which are expressed in terms of stresses and displacements. Violations of limit state usually occur when structures support extreme natural actions like large wind loads, earthquakes, wave loads, etc. In these cases linear structural analysis could be very inaccurate and material and geometrical nonlinearities must be considered. In this paper a power software program named OpenSees with excellent capabilities to nonlinear finite element analysis and structural reliability analysis has been used [1, 2].

Classical RBDO methods consist in a double loop procedure. At the outer loop, design optimization is carried out by mathematical methods and requires gradients of the objective functions and reliability constraints. Sequential Quadratic Programming (SQP), Sequential Linear Programming (SLP), and other nonlinear constrained optimization methods are often used. Reliability constraints and their gradients are evaluated at the inner loop.

Other types of mathematical methods have been developed to solve RBDO problems, like single loop methods and decoupled methods. Several reviews can be found in the literature about these methods [3, 4].

Recently the increased power of computers makes it possible to develop RBDO methods based in heuristics, like the Evolutionary Computation methods. In this paper Genetic Algorithms are proposed to solve RBDO problems. There is an extensive list of references about the application of GAs in deterministic structural optimization [5, 6]. However, relatively few efforts have been done to apply GA to RBDO problems involving frame and truss structures. Dimou and Koumousis [7] consider multiple-population GA in the RBDO of planar trusses. Multiple populations contend among themselves for computational resources and a meta-GA assigns these resources to the most-fit sub-populations. Shayanfar et al. [8] develop a GA-based RBDO method and apply it to analytical and structural problems. They only consider linear elastic structures. However nonlinear analysis has to be considered when extreme loads are applied and that is the case when reliability analysis is worthy. This paper proposes a GA-based RBDO method applied to nonlinear structures.

Some Multiobjective Genetic Algorithms (MOGA) have been used to solve multiobjective RBDO problems [9, 10]. In these problems, the weight of the structure and the reliability indexes of the probability constraints are defined as objective functions.

Next section explains the formulation of RBDO problems. Section 3 describes same features of GA based optimization. A communication transmission tower truss subjects to random loads is optimized applying the proposed method in Sect. 4. Finally, conclusions are outlined in Sect. 5.

2 Formulation of RBDO Problem

An RBDO problem can be formulated as minimization of an objective function subject to reliability and deterministic constraints. Also lateral constraints are considered for the design variables. The objective function is the cost of the structure and can include the cost of construction and costs along its life cycle, such as maintenance cost, reparation cost when failure occurs and demolition cost. Usually the probability of failure is very little and reparation cost can be negligible with respect to the construction cost. Constraints describe design conditions stated in codes of structural design and are written as limit state functions.

The most common mathematical formulation of a RBDO problem is:

$$\min_{\mathbf{d}, \boldsymbol{\mu}_{\mathbf{X}}} f(\mathbf{d}, \boldsymbol{\mu}_{\mathbf{X}}, \boldsymbol{\mu}_{\mathbf{P}})$$

$$s.t.\ P_{fi} = P[g_i(\mathbf{d}, \mathbf{X}, \mathbf{P}) \leq 0] \leq P_{fi}^t, i = 1, \ldots, n \tag{1}$$

$$\mathbf{d}^L \leq \mathbf{d} \leq \mathbf{d}^U, \boldsymbol{\mu}_{\mathbf{X}}^L \leq \boldsymbol{\mu}_{\mathbf{X}} \leq \boldsymbol{\mu}_{\mathbf{X}}^U$$

where $\mathbf{d} \in R^k$ is the vector of deterministic design variables. $\mathbf{d}^L$ and $\mathbf{d}^U$ are the upper and lower bounds of vector, respectively. $\mathbf{X} \in R^m$ is the vector of random design variables, that is, random variables whose mean values, $\boldsymbol{\mu}_{\mathbf{X}}$ are design variables. $\boldsymbol{\mu}_{\mathbf{X}}^L$ and $\boldsymbol{\mu}_{\mathbf{X}}^U$ are the upper and lower bounds of vector $\boldsymbol{\mu}_{\mathbf{X}}$. $\mathbf{P} \in R^q$ is the vector of random parameters. $\boldsymbol{\mu}_{\mathbf{P}}$ is the mean value of $\mathbf{P}$. $f(\cdot)$ is the objective function, n is the number of reliability constraints, k is the number of deterministic design variables, m is the number or random design variables and q is the number of random parameters. $g_i(\mathbf{d}, \mathbf{X}, \mathbf{P})$ is the i-th limit state function. P_{fi} is the probability of violating the i-th probabilistic constraint and P_{fi}^t is the target probability of failure for the i-th probabilistic constraint.

Reliability constraints are written in terms of nodal displacements and in terms of internal forces in any section of a member. If these displacements or internal forces efforts exceed a limit value a failure occurs. Limit state functions are defined in a way that $g_i(\cdot) \leq 0$ represents the failure domain. Then, the probability of failure for the i-th limit state function P_{fi} could be computed using the multivariate integral:

$$P_{fi} = \int_{g_i(\mathbf{d},\mathbf{X},\mathbf{P}) \leq 0} f_{\mathbf{X}}(\mathbf{x}) d\mathbf{x} \quad (2)$$

where $f_{\mathbf{X}}(\mathbf{x})$ is the joint probability density function of vector of random variables $\mathbf{X}$. The close-form solution of this integral is not usually available. Numerical solutions have been obtained until dimension 4 or 5. Because that, approximate reliability methods and simulation methods are available. Simulation methods as MonteCarlo Simulation (MCS) and Importance Sampling based MCS need a large computing time and generally are not used. The approximate First Order Reliability Methods (FORM) provides accurate reliability approximation. FORM is used in this research to compute P_{fi} of each reliability constraint. Alternatively the reliability of a limit state function can be written in terms of the reliability index. In FORM, these values are related by this equation:

$$P_{fi} \approx \Phi(-\beta_i) \quad (3)$$

where Φ is the normal standard CDF.
Now, if FORM is used to evaluate the reliability constraints the RBDO problem is expressed with the next equation:

$$\begin{gathered} \min_{\mathbf{d},\boldsymbol{\mu}_{\mathbf{X}}} f(\mathbf{d}, \boldsymbol{\mu}_{\mathbf{X}}, \boldsymbol{\mu}_{\mathbf{P}}) \\ s.t.\ \beta_i \geq \beta_i^t, i = 1, \ldots, n \\ \mathbf{d}^L \leq \mathbf{d} \leq \mathbf{d}^U, \boldsymbol{\mu}_{\mathbf{X}}^L \leq \boldsymbol{\mu}_{\mathbf{X}} \leq \boldsymbol{\mu}_{\mathbf{X}}^U \end{gathered} \quad (4)$$

where β_i^t is the target or admissible reliability index corresponding to the i-th reliability constraint.

The first step in FORM is to transform the limit state function from the space of original random variables to the space of uncorrelated standard normal random variables using an isoprobabilistic transformation. Then, each limit state functions is approximate by a first order function using the Taylor series around a determinate point in standard normal random space. FORM iteratively compute the reliability index moving from an initial point to a point named design point or Most Probable Point (MPP). The value of the limit state functions and its gradients with respect to the random variables are computed in each iteration. In structural design, the limit state functions are not explicit functions of random design variables, they rather than depend implicitly of design variables. An iteration in FORM implies a call to finite element software with capabilities to compute the gradients of the structural response with respect to random variables.

In this paper, the Finite Element software OpenSees is used to determine the response of the structural analysis. OpenSees is an open source code developed in University of California at Berkeley. OpenSees has capabilities of sensitivity and structural reliability analysis. Two methods are implemented to compute gradients: the Forward Finite Difference method (FFD) and the Direct Differentiation Method (DDM). As DDM is more efficient than FFD, it is used in this research.

3 GA Based Optimization

GAs are well known algorithms that have been applied extensively in optimization and classification problems. As other Evolutionary Computation procedures, they are inspired in the principles of natural selection and simulate the adaptive behavior of live beings. Basically, GAs start from an initial population of potential solutions that evolves by some genetic operators which act in a very similar manner than natural selection laws [11].

More important operators are:

Selection.

This operator is used to select individuals for reproduction in the current population. These individuals are named parents and form the mating-pool and generate children to form the next generation. There are several types of selection operators. The GA usually selects the individuals that have better fitness values. This value is the value of the objective function in unconstrained optimization problems or is a modified fitness that results adding to the objective function a penalty function in constrained optimization problems. Selection operators are stochastic, because are based in random numbers generators. This fact explains the capability of GAs to explore very complex search space of design variables. In the RBDO problems formulated here a tournament selection operator has been chosen.

Elitism.

Genetic Algorithms often maintain the best solution/s from a previous population through a mechanism termed elitism. In simplest terms, elitism is simply taking the best solution/s from the current population and simply carrying it/them over to the next generation.

Crossover.

This operator takes pairs of parents of the mating-pool and combines or interchanges the genes of these parents generating children. There are various types of crossover operators: mask, single point, two point, heuristic, etc.

Mutation.

This operator works after crossover and makes small random changes in single parents of the mating-pool. These changes area applied in stochastically selected genes. Mutation operator provides genetic diversity and enables the genetic algorithm to search a complex search space.

GAs present some advantages with respect to classical optimization methods. Unlike mathematical optimization methods which perform optimum search considering only a single solution at a time, GAs work with a population of individuals in each generation. Each individual is a codified version of the design variables of the structure. Objective and constraints functions are evaluated for all these individuals. GA is a free-gradient optimization method. This feature makes possible to consider objective and constraints functions with mathematical difficulties such as discontinuous or non-derivable functions [11].

Initial population and offspring in next generations are generated stochastically in GAs. This permits to maintain several search points and to explore complex search space. Therefore, the convergence of the optimization to local minima is prevented, although starting point has been poorly chosen. All these aspects result in more chances of finding the optimal solution than gradient based methods, even on problems having hard search spaces with multiple local minima.

The main drawback of these methods is the extreme computational effort required. Classical RBDO problems require a large quantity of objective and reliability constraints evaluations. In the case of structural applications, for each reliability constraint assessment some runs of structural analysis software as, for example, a finite element package, have to be carried out. The replacement of mathematical optimization methods by GAs increases even more the computational time because several design solutions are computed at the same time. However, parallel computing resources and High Performance Computing techniques allow dealing with complex GA-based RBDO problems.

4 Numerical Example: Space Truss

This example deals with the design of the space truss shown in Fig. 1. This structure has been frequently used as a transmission tower and has been proposed in the literature for evaluate optimization methods. This truss contains 25 bars and 10 nodes. It is subjected to the nodal loads represented also in Fig. 1. The relation stress – strain of the steel is represented with a bilinear model, represented in the Fig. 2. Elastic modulus, E, and yield stress, f_y, are considered random variables. The strain hardening ratio b is deterministic. Bar elements are modeled as nonlinear finite elements using the corotational transformation to consider geometric nonlinearity. The 25 bars are grouped in 8 groups. Bars in the same group have the same cross sectional area and these areas are the random design variables of the problem. Also, nodal loads are lognormal random variables. The random variables of the problem are registered in Table 1.

This, the RBDO problem is formulated as:

$$\min_{\mu_{A1}\cdots\mu_{A8}} V(\boldsymbol{\mu}_{\mathbf{A}}, \boldsymbol{\mu}_{\mathbf{P}}) = \sum_{i=1}^{8} \mu_{Ai} \sum_{j=1}^{nb_i} l_i$$

$$s.t. \quad \beta_{ij} \geq \beta_{ij}^{t} = 3.7 \; i = 1,2 \quad j = x, y \tag{5}$$

Table 1 Random variables in the tower truss example

Random variable	Description	Dist.	Mean	CoV
$A_1, \ldots, A_8$	Cross section area	N	Design Variable	0.1
E	Elastic modulus	LN	210E + 6 kPa	0.05
f_y	Yield stress	LN	250E + 3 kPa	0.05
b	Strain hardening ratio		Deterministic	0.00
P_1	Lateral load	LN	60 kN	0.1
P_2	Lateral load	LN	100 kN	0.1
P_3	Vertical load	LN	200 kN	0.1
P_4	Lateral load	LN	60 kN	0.1

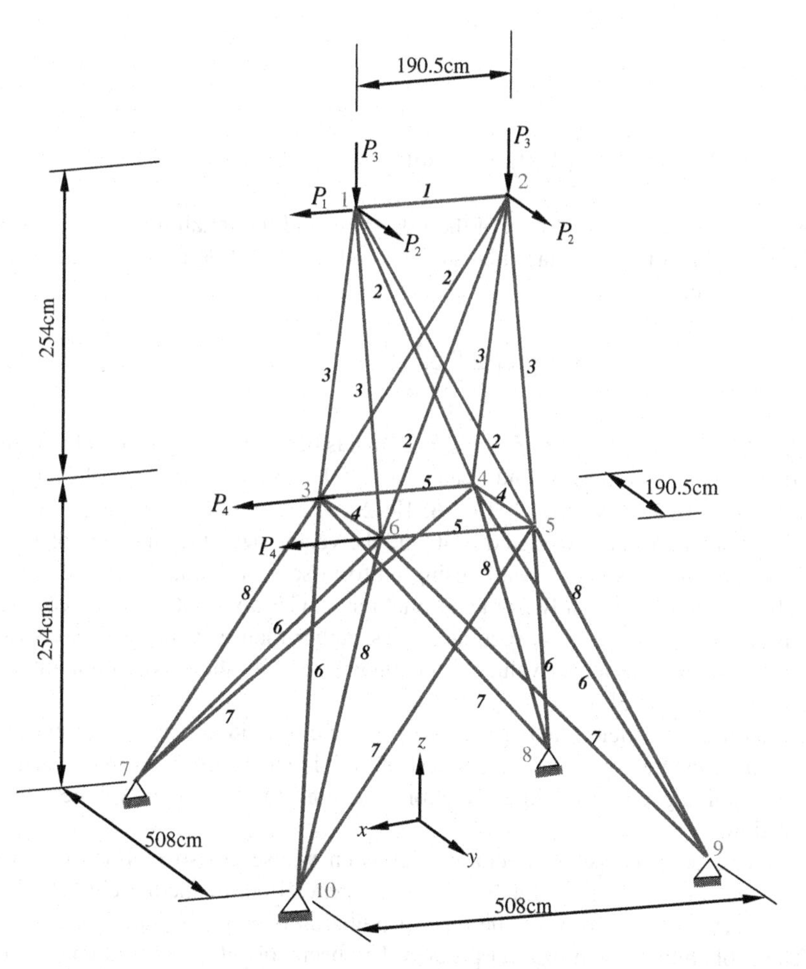

Fig. 1 Transmission Tower Truss

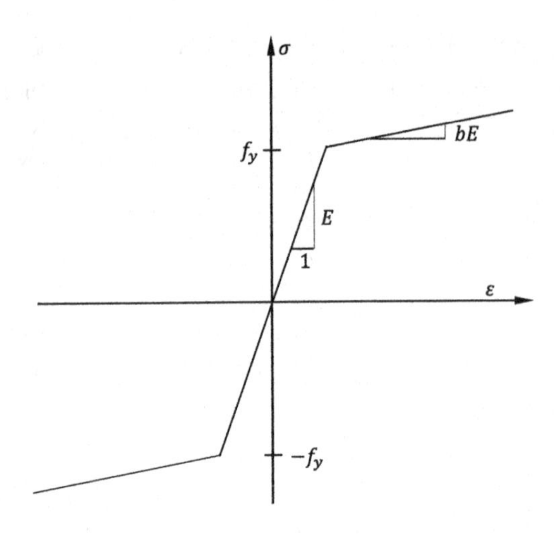

Fig. 2 Nonlinear behavior model for steel

$$P\left[g_{ij}(\mathbf{A},\mathbf{P})\leq 0\right]<\Phi\left(-\beta_i^t\right)i=1,2 \quad j=x,$$

where nb_i is the quantity of bars of the i-th group and the reliability constraints are written in terms of the displacements of the nodes 1 and 2 on coordinates x and y. These constraints are:

$$g_i(\mathbf{A},\mathbf{P})=1-\frac{u_{ij}(\mathbf{A},\mathbf{P})}{u_{max}} \; i=1,2 \quad j=x,y \tag{6}$$

Here the allowed displacement $u_{max}=2\,\mathrm{cm}$ has been set. Design variables μ_{A_i} are discrete design variables and can take 20 different values of the set [1, 2, 3, 4, 5, 6, 7, 8, 9, 10, 12, 14, 16, 18, 20, 22, 24, 26, 28, 30] cm^2.

The population size considered is 40 and the vector size of desing variables is 8. Initial population has been created using uniform selection, that is, all individuals have been selected randomly from an initial range. GA does not work directly with the mean values of the cross sectional areas, rather than it works with a codified version consisting in integer values in the range [1, .., 20]. The maximum number of generations used is 300.

Elistism is considered taking the two best fitness inviduals of the current population over the next generation. The rest of individuals for the next generation are formed applying crossover and mutation operators to the parents of the current population.

A single point crossover operator is chossen whose crossover fraction is 0.8. This is the fraction of individuals of the next generation, deducting elite children, that are created by crossover. The type of mutation operator is gaussian.

The probabilistic optimization problem has been solved four times considering different analysis models and various types on material behavior. This shows the flexibility of the GA based method.

In the first case, geometric and material nonlinearities are considered. The convergence is reached in 76 generations. The set of design variables with best fitness in the last generation is considered as the probabilistic optimum design. The optimum volume of the truss is 52901.8 cm^3, and the mean values of the cross sectional areas are: $\mu_{A1}, \ldots, \mu_{A8} = 1, 10, 7, 1, 3, 3, 2, 16\,\text{cm}^2$. In this probabilistic optimum the four reliability constraints are verified and the penalty function is equal to the fitness function. The values of the reliability indexes for the optimum design are: $\beta_{1x} = 6.060, \beta_{1y} = 3.781, \beta_{2x} = 6.064, \beta_{2y} = 4.895$. These results are summarized in Table 2.

Table 2 Results for the tower truss example

Case	$\mu_{A1} \ldots \mu_{A8}$ [cm^2]	Volume [cm^3]	$\beta_{1x}, \beta_{1y}, \beta_{2x}, \beta_{2y}$
1	1, 10, 7, 1, 3, 3, 2, 16	52901	6.060; 3.781; 6.064; 4.895
2	8, 14, 3, 2, 2, 2, 1, 14	49129	14.075; 3.812; 14.294; 4.984
3	1, 14, 3, 1, 2, 2, 2, 16	51967	5.246; 3.713; 5.156; 4.165
4	4, 14, 2, 2, 3, 4, 1, 16	54136	8.564; 3.752; 8.381; 4.570

It is worth to note that the second constraint that refers to the displacement of node 1 along y axle is more critical than the others constraints. Also, its value 3.781 is not exactly equal to the admissible value for the reliability index. The reason of this different is that optimization problem deal with discrete design variables.

Case 2 corresponds to consider geometric nonliearity and linear elastic material. Case 3 corresponds to consider geometric linearity and nonlinear material. And case 4 corresponds to consider geometric and material linearities. Resuts for these three cases are also registered in Table 2.

Although this numerical example involves a reduced quantity of random design variables, each probabilistic optimization took approximately an hour to provide the results. This period of time is about four times the computational time needed with gradient based RBDO methods.

MonteCarlo Simulation with Importance Sampling has been used to verify the results obtained by the GA-based RBDO method. Reliability indexes obtained with simulation were very close to reliability indexes obtained with the proposed method.

The proposed method has been applied to frames. However, due to space limitations a numerical example cannot be provided here. The interested reader is referred to coming works.

5 Conclusions

In this work a RBDO method combining GAs with Nonlinear Finite Element Reliability Analysis software OpenSees has been carried out. The proposed method has been applied to engineering structures subjected to extreme random loads caused by natural actions. In these cases large strains and displacements occur and geometric nonlinearity must be considered. The implemented method can be used to design structures when traditional methods prescribed in construction codes are not applicable or are imprecise.

In addition to the case with geometric and material nonlinearities, other three types of structural analysis have been implemented: geometric nonlinearity with linear elastic material, geometric linearity with nonlinear material behavior and both geometric and material linearity. The designer have to decide what is the adequate analysis and then, to select the values of the cross-sectional areas or the standardized steel profiles.

GA can deal with several types of design variables: continuous, integer or mixed. This allows the design of structures using standardized commercial steel sections. Also a near global optimum design can be reached.

The numerical example shows that precise results can be obtained in moderate period of time thanks to the increased power computing available even in home computers.

Advanced evolutionary algorithms such as distributed GA, MOGA, coevolutionary algorithm and others could be applied to solve complex RBDO problems and research efforts are focused on them.

References

1. OpenSees: open system for earthquake engineering simulation. http://opensees.berkeley.edu
2. Haukaas T, Scott MH (2006) Shape sensitivities in the reliability analysis of nonlinear frame structures. Comput Struct 84(15–16):964–977
3. Celorrio-Barragué L (2010) Metodología eficiente de optimización de diseño basada en fiabilidad aplicada a estructuras. Ph. D Thesis. Universidad de La Rioja, Logroño
4. Celorrio-Barragué L (2012) Development of a reliability-based design optimization toolbox for the FERUM software. In: 6th International Conference Scalable Uncertainty Management, 2012, LNAI, vol 7520, pp 273–286. Springer, Berlin
5. Pezeshk S, Camp CV (2002) State of the art on the use of genetic algorithms in design of steel structures. In: Burns S (ed) Recent Advances in Optimal Structural Design. American Society of Civil Engineers, Reston
6. Foley CM (2007) Structural optimization using evolutionary computation. Optimization of structural and mechanical systems. World Scientific, Singapore Chap. 3, pp 59–120
7. Dimou CK, Koumousis VK (2003) Competitive genetic algorithms with application to reliability optimal design. J Adv Eng Softw 34(11–12):773–785
8. Shayanfar M, Abbasnia R, Khodam A (2014) Development of a GA-based method for reliability-based optimization of structures with discrete and continuous design variables using OpenSees and Tcl. Finite Elements Analy Des 90:61–73
9. Mathakari S, Gardoni P, Agarwal P, Raich A, Haukaas T (2007) Reliability-based optimal design of electrical transmission towers using multi-objective genetic algorithm. Comput-Aided Civil Infrastruct Eng 22(4):282–292
10. Deb K, Gupta S, Daum D, Branke J, Mall AK, Padmanabhan D (2009) Reliability-based optimization using evolutionary algorithms. IEEE Trans Evol Comput 13(5):1054–1074
11. Golberg DE (1989) Genetic algorithm in search, optimization, and machine learning. Addison-Wesley Professional, Reading

Improved Energy-Aware Stochastic Scheduling Based on Evolutionary Algorithms via Copula-Based Modeling of Task Dependences

Zorana Banković and Pedro López-García

Abstract In this work we apply the copula theory for modeling task dependence in a stochastic scheduling algorithm. Our previous work, as well as the majority of the existing related works, assume independence between the tasks involved, but this is not very realistic in many cases. In this paper we prove that, when task dependence exists, better results can be obtained when it is modeled. Our results show that the performance of the stochastic scheduler is significantly improved if we assume a certain level of task dependence: on average 18 % of the energy consumption can be saved compared to the results of the deterministic scheduler, along with 81 % of improved test cases, versus 2.44 % average savings when task independence is assumed, along with 50 % of improved test cases.

1 Introduction

The most common approach when dealing with a system which depends on a group of random variables is to assume that the variables are independent, mainly because the mathematical apparatus becomes too complex, or simply because it is not possible to mathematically describe the underlying dependence. However, this simplification often results in assuming an initial scenario which is very different from reality, which limits the usefulness of the final result. An example of this approach is stochastic scheduling, where the relevant characteristics of the tasks are represented as random variables with the corresponding distribution function, and, as far as we know, in the majority of existing works, the variables describing different tasks are considered to be independent, or modelled with normal distributions [6], which is often not realistic.

Z. Banković (✉)
IMDEA Software Institute, Madrid, Spain
e-mail: zorana.bankovic@imdea.org

P. López-García
Spanish Council for Scientific Research (CSIC), Madrid, Spain
e-mail: pedro.lopez@imdea.org

Á. Herrero et al. (eds.), *10th International Conference on Soft Computing Models in Industrial and Environmental Applications*, Advances in Intelligent Systems and Computing 368, DOI 10.1007/978-3-319-19719-7_14

The main objective of our work is to minimize the energy consumption in multi-threaded multicore Dynamic Voltage and Frequency Scaling (DVFS) enabled platforms through optimal task scheduling. In general, the problem of scheduling and allocation is NP-hard. Thus, the problem has been addressed with different heuristic algorithms that are capable of obtaining sub-optimal solutions in real time due to their fast exploration of the search space. For this reason, our scheduler is based on an Evolutionary Algorithm (EA). Since DVFS reduces energy, but increases execution time, these two magnitudes are clearly in conflict. For this reason, we use a multi-objective optimisation approach in order to find a trade-off between energy consumption and execution time. We also provide an appropriate representation of solutions that captures two levels of parallelism, processor (core) and thread level parallelism, and at the same time performs allocation and scheduling and identifies appropriate voltage and frequency settings (i.e., (V,f) pairs), exploring in this way the entire search space.

This work is a part of a bigger tool for scheduling based on EAs [3], to which we want to add the possibility of modelling dependence between the tasks. For this reason, in this work we experiment with modeling dependence between the execution time and power of different tasks using copulas [12], in particular Archimedean copulas [11]. However, if we wanted a stand-alone implementation on copula-supported scheduling, a promising approach would be to use Estimation of Distribution Algorithms [8]. The main advantage of copulas when modeling dependence is the fact that they do not depend on the marginal distributions. In this work we have decided to study the applicability of Archimedean copulas for two important reasons: they have been extensively studied and the mathematical apparatus for their manipulation is quite mature, and they are known to properly model the existing tail dependency between two variables, i.e., the possibility of achieving extreme values at the same time, which can be expected in our case. In particular, in this work we experiment with Gumbel copulas [11] due to their proper modeling of positive right-tail dependence, e.g., if one task takes more time due to a prolonged memory access, it will lead to longer execution time of all the tasks that are related to it, as well as energy consumption, which is important when dealing with a time and/or energy budget. However, it does not model negative dependence, i.e., achieving small values at the same time, which is not important in our case since it does not affect the budgets mentioned above.

The rest of the paper is organised as follows. In Sect. 2 we give a short overview of the copula theory, necessary for understanding and reproducing the results of our work, while in Sect. 3 we detail our proposed approach. Section 4 presents and discusses the obtained results. In Sect. 5 we list the most important related work. Finally, some conclusions are drawn in Sect. 6.

2 Copulas for Modeling Dependency

In this section we give a short survey on copulas [12]. In essence, the copula theory gives us a mathematical framework for describing dependence between the variables irrespectively of their underlying distribution functions.

Sklar's Theorem (1959) Let H be a *continuous* two-dimensional distribution function with marginal distribution functions F and G. Then there exists a copula C such that

$$H(x,y) = C(F(x), G(y)) \implies C(x,y) = H(F^{-1}(x), G^{-1}(y)) \tag{1}$$

Thus, for any two distribution functions F and G and copula C, the function H is a two-dimensional distribution function with marginals F and G.

Archimedean Copulas. An important group of copulas are the Archimedean copulas, where the dependence level depends on one parameter. They are defined in the following way: let ϕ be a continuous strictly decreasing function from $\mathbf{I}$ to $[0, \infty]$ such that $\phi(1) = 0$, and let $\phi^{[-1]}$ denote the "pseudo-inverse" of ϕ:

$$\phi^{[-1]}(t) = \phi^{-1}(t) \text{ for } t \in [0, \phi(0)] \text{ and } \phi^{[-1]}(t) = 0 \text{ for } t \geq \phi(0) \tag{2}$$

Then, if ϕ is convex, the function

$$C(u,v) = \phi^{[-1]}(\phi(u) + \phi(v)) \tag{3}$$

is an *Archimedean* copula and ϕ is called its *generator function.* In the case of the Gumbel copula used in this work the generator function is the following:

$$\phi(t) = (-\ln t)^{\theta} \text{ where } \theta \in [1, \infty). \tag{4}$$

Monte Carlo for Copula-based Models. Since in our work we use Monte Carlo simulation to calculate the expected value of random variables, in the following we show how it is integrated in the copula model [10].

If we want to find the expected value of a function $g : \mathbb{R}^d \to \mathbb{R}$ applied to a random vector $(X_1, X_2, ..., X_d)$ whose cumulative distribution function (*cdf*) is H, the expected values we need are calculated in the following way:

$$\mathbb{E}[g(X_1, X_2, ..., X_d)] = \int_{\mathbb{R}^d} g(x_1, x_2, ..., x_d) \mathrm{d}H(x_1, x_2, ..., x_d) \tag{5}$$

If H is given by its copula model expressed with formula 1, formula 5 can be written in the following way:

$$\mathbb{E}[g(X_1, X_2, ..., X_d)] = \int_{[0,1]^d} g(F_1^{-1}(u_1), ..., F_d^{-1}(u_d)) \mathrm{d}C(u_1, ..., u_d) \tag{6}$$

If copula C and marginals $F_1, .., F_d$ are known or estimated, the expected value can be approximated using the following Monte Carlo algorithm:

1. Draw a sample $(U_1^k, U_2^k, ..., U_d^k) \sim C, k = (1, 2, .., n)$ of size n from copula C.
2. Calculate a sample of $(X_1, X_2, ..., X_d)$ by applying the inverse cdf of marginal functions:

$$(X_1^k, X_2^k, ..., X_d^k) = (F_1^{-1}(U_1^k), F_2^{-1}(U_2^k), ..., F_d^{-1}(U_d^k)) \sim H(k = 1, ..., n)$$

3. Approximate $\mathbb{E}[g(X_1, X_2, ..., X_d)]$ with its empirical value:

$$\mathbb{E}[g(X_1, X_2, ..., X_d)] \approx \frac{1}{n} \sum_{k=1}^{n} g(X_1^k, X_2^k, ..., X_d^k)$$

Archimedean Copula Simulation. In order to draw a sample from Archimedean copulas (step 1 of the previous Monte Carlo algorithm), we follow the algorithm for Laplace transform Archimedean copulas, which are all the copulas whose generator function ϕ is a Laplace transform of some function G. The algorithm has been proven correct in [10] and consists on the following steps:

1. Generate a pseudorandom variable V whose *cdf* is G.
   - For the Gumbel copula used in this work, V is a stable distribution (class of probability distributions allowing skewness and heavy tails), $St(1/\theta, 1, \gamma, 0)$, with $\gamma = (cos(\pi/2/\theta))^\theta$ and $\hat{G} = exp(-t^{1/\theta})$.
2. Generate independent and identically distributed random variables $(X_1, X_2, ..., X_d)$.
3. Return $U_i = \hat{G}(-\frac{\ln X_i}{V}), i = 1, ..., d$.

3 Our Proposed Approach Based on EAs

In the following we present the main aspects of our EA implementation. Our multiobjective EA is based on the NSGA-II implementation [5], since in our previous work [3] the best results were obtained when this technique was applied.

Individuals. An individual, as a representation of a solution to the problem, has to contain information about temporal and spatial allocation of each task. A solution to the scheduling, i.e., temporal aspect of the problem is a permutation of the task identifiers (IDs), where their order also stands for the order of their temporal execution, assuming that each task has a unique ID. In order to solve the allocation problem, i.e., on which thread (and core) each task is executed, we add delimiters to the permutation of the task IDs that define where the tasks are being executed, i.e., core, thread and (V,f) setting (the tasks between two delimiters are executed on the right-side one). In order to be able to distinguish the delimiters from the tasks, delimiters are coded as negative three-digit numbers, where the first digit stands for the core, the second one for the thread on that core, and the third one for the core (V,f) setting (assuming that there is a finite number of settings, which is realistic). As an example, a part of an individual is depicted in Fig. 1, where tasks with IDs 1, 2, 5 and 7 are executed in that order on thread 4 of core 2, with the (V,f) setting coded as 4. In the most general case, the order of delimiters is random. However, if two consecutive delimiters that belong to the same core have different (V,f) settings, this means that they are not being executed in parallel, since the voltage and/or frequency have to be changed. Representing individuals in the way described above has provided us with a relatively simple approach, which does not introduce great overhead when executing the EA.

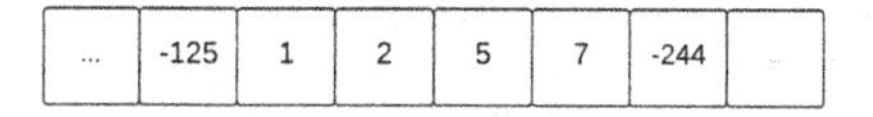

Fig. 1 An example of (part of) a solution (i.e., individual) representation

Population Initialisation. Since our problem setting in this work is simple (two cores with two thread each), individuals in the initial population are created by randomly assigning tasks to random threads in random (V,f) settings. However, in more complicated problem settings with more cores and threads per core, with task deadlines etc., we will have to consider adding a heuristic in order to provide a feasible solution in each run.

Solution Perturbations. Given that all the tasks and all the delimiters are different, different solutions are always permutations of the set of tasks and the set of delimiters. This gives us the opportunity to use some of the permutation-based crossover operators, and in this case we are using partial match crossover, since it performed better in terms of the objective function than cycle crossover, and slightly better than order crossover in terms of the objective function and the execution time. Since the order of delimiters is not important in the most general case, this operator provides at the same time variety in consecutive changes of (V,f) settings, and the capability of moving tasks from one thread to another. Regarding mutation, it is implemented in a way that two random threads exchange two random tasks with certain (low) probability.

Objective Functions. The objective of the scheduler is to minimize the total energy consumption, as well as the execution time. Since in this work we apply DVFS, which decreases energy, but increases time, these two values are clearly in conflict, and thus the use of multiobjective optimisation is justified. In general, these values are expressed with the following formulas for a given set of n heterogenous machines and a set of k tasks, for a particular machine-task assignment χ:

$$
\begin{aligned}
E(\chi) &= \sum_{1 \le i \le n} (P_{st,i} \cdot T(\chi) + \sum_{1 \le j \le k} x_{i,j} \cdot p_{i,j} \cdot \tau_{i,j}) \\
T(\chi) &= \max_{1 \le i \le n} \{ \sum_{1 \le j \le k} x_{i,j} \cdot \tau_{i,j} \}
\end{aligned} \tag{7}
$$

where $P_{st,i}$ is the static power of the machine i, $x_{i,j}$ is a binary value, $x_{i,j} \in \{0, 1\}$, that represents whether the task j is executed on the machine i ($x_{i,j} = 1$) or not ($x_{i,j} = 0$), $p_{i,j}$ is the (dynamic) power consumption of the task j on the machine i, and $\tau_{i,j}$ is the execution time of the task j on the machine i.

The objective of the stochastic scheduler is to minimize the expected value of these formulas:

$$
\begin{aligned}
&\min_{\chi \in \pi} \{\overline{E}(\chi)\} \\
&\min_{\chi \in \pi} \{\overline{T}(\chi)\}
\end{aligned} \tag{8}
$$

As in our previous work, these values are approximated using the Monte Carlo method, but here we have to introduce the necessary changes to account for the copula-based dependence:

- *Total Energy:* estimated in the Monte Carlo approximation presented in Sect. 2, taking $g(X_1, .., X_d) = \sum_{i=0}^{d} X_i$, where d is the total number of tasks and X_i are random variables representing the energy of each task.
- *Execution Time:* for each core approximated in the same way as the total energy, after that the maximum value between all the cores is taken.

In our implementation, we use the Mathematica system [1] to calculate the inverseCDF function (step 2 of the Monte Carlo method presented in Sect. 2) due to its capability to deal with all the probability functions used in this work. For this purpose, C++ executes Mathematica as an external program in MathLink mode [9].

4 Results and Discussion

4.1 Input Data

The input data to our scheduling algorithm consists of a set of tasks whose power consumption and execution time are given as random variables with a known probability density function. The following density functions are available at the moment: Uniform, Constant, Exponential, Normal Chi-squared, Gamma, Pareto, Poisson, Binomial, Negative Binomial, and any combination of the previous ones expressed as a sum of products. A sampling from all the above density functions can be obtained by using a package implemented by Robert Davies [4]. In different (V,f) settings, the power consumption is scaled with V^2 and f, and the corresponding execution time is scaled with f. Finally, the energy of a task is a random variable obtained as the product of their corresponding execution time and power random variables.

4.2 Obtained Results and Discussion

In this work we experiment with controlled dependency in synthetic data. Dependency is introduced in a way that some of the random variables which describe execution time and power of the tasks have the same fixed distribution. As we have previously mentioned, dependence in Archimedean copulas is controlled with the θ parameter, which in the case of the Gumbel copula belongs to the interval $[1, \infty)$, where $\theta = 1$ stands for independence, while $\theta \to \infty$ stands for comonotonicity, i.e., maximal positive dependence. However, $\theta \geq 10$ is already considered as a significant level of dependence.

In order to check the possibility of improving the results of the stochastic scheduling by introducing copulas for modeling dependence, we have created an experiment where we fix the testing scenario, start with independence assumption, i.e., $\theta = 1$, and then increase the θ parameter in order to increase the level of dependence. Since the main claim of our work is that the stochastic scheduling may improve its deterministic counterpart, we check how the results are being improved as the level of dependence increases. In particular, we repeat the simulation for three different levels of dependence: $\theta = 1$, which is equivalent to independence, $\theta = 5$ and $\theta = 10$, which assumes a high level of dependence.

Table 1 Summary of average improvements

Test case (θ)	Avg.% of Improved test cases	Avg. improv. (%)	Avg. min. Improv. (%)	Avg. max. Improv. (%)
1 *(1)*	50	2.46	−4.58	12.1
2 *(5)*	81.66	18	−3.51	31.96
3 *(10)*	68.75	8.42	−3.95	22.75

After performing the training and obtaining the Pareto front, in all the cases we took the solution with minimal energy consumption from all the solutions belonging to the Pareto front. The testing was performed on different sets of test cases, each having 10 test cases generated randomly. The results are summarised in Table 1, where we can observe (from left to right) average % of test cases where the stochastic scheduler improves the deterministic one, along with average improvement, and average minimal and maximal improvement in all test cases. Figure 2 illustrates the evolution of different test cases (1–6 are different test case sets, each having 10 different test

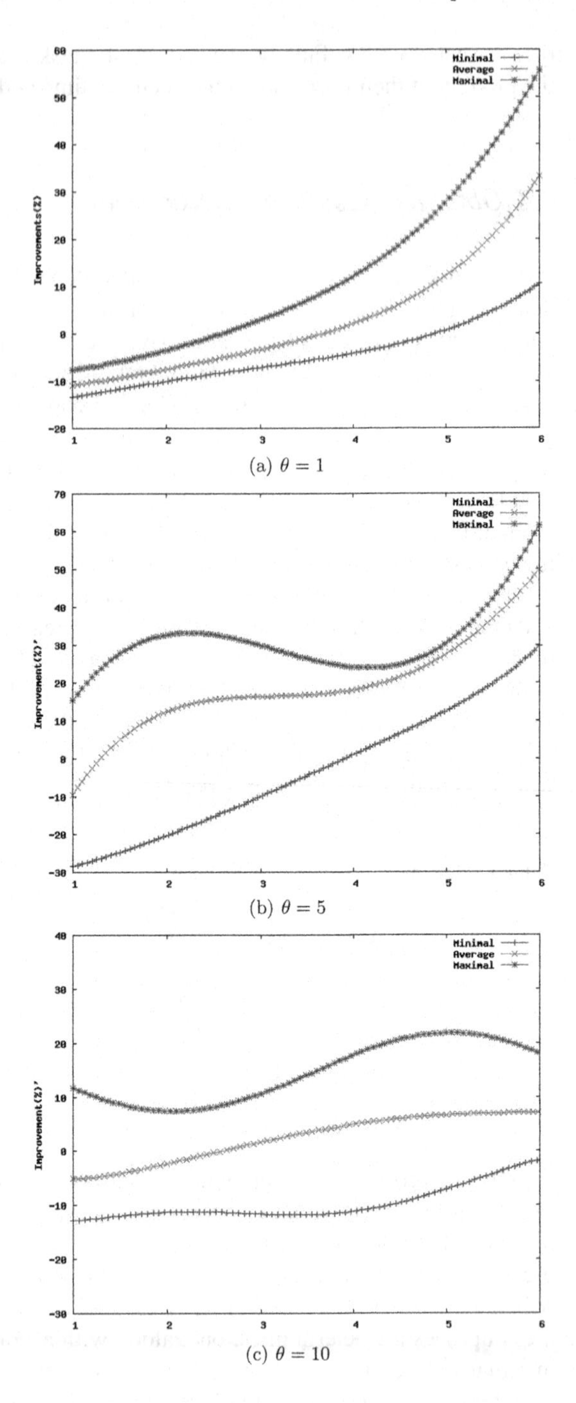

Fig. 2 Evolution of minimal, average and maximal improvements (%) for $\theta = 1, 5, 10$

cases), sorted by their improvement. Note that negative numbers actually mean that the stochastic scheduler worsened the performances.

In the current implementation we rely on Mathematica to calculate the inverse cdf, which significantly increases simulation time, since it takes 6–8 hours on a 2.5 GHz Intel Core i5 with 4 GB DDR3.

From the table and the figures we can draw a few interesting conclusions:

- The stochastic scheduler improves a significant number of test cases, even when assuming independence ($\theta = 1$).
- The best obtained results correspond to $\theta = 5$, which assumes certain level of dependence, although less than maximal positive dependence, which is close to $\theta = 10$. This confirm our expectation, since in our test cases some of the variables describing power and/or time are correlated, but not all of them.
- The maximal observed energy improvement achieved in a particular case is around 62 %, while the maximal observed performance (in terms of energy savings) decrease is 28 %, both belonging to the case $\theta = 5$ (Fig. 2).
- The test cases that obtained better performances with the deterministic scheduler, i.e., whose performances were decreased after applying the stochastic scheduler (and which can be observed as "negative" improvement in both Table 1 and Fig. 2), had values which were close to the average values of the corresponding distributions, which were used to design the deterministic scheduler. This behaviour was also expected, since the main idea of the stochastic scheduler is to improve the scheduling when the real data deviate significantly from the expected ones used to create the deterministic scheduler. However, from this testing scenario we were not able to properly decide the threshold level which would tell us when it is beneficial to start using the stochastic scheduler.

5 Related Work

Stochastic scheduling has gained lots of interest over the years, since many different cases include uncertainty. In general, approaches to optimisation under uncertainty include various modeling philosophies, the most important being the following ones:

- Expectation minimisation.
- Minimisation of deviations from goals.
- Minimisation of maximum costs.
- Optimisation over soft constraints.

Our approach clearly belongs to the first group. The solution presented in [7] is in the same group, however, it solves the stochastic scheduling problem by reducing it to the deterministic case. The benefit of this approach is the lower execution time, yet at the cost of decreased accuracy.

Copulas have been used in different versions of scheduling-related problems. For example, in [2] they are used in power supply system scheduling to model the presence of uncertain renewables, such as wind and solar energy. Another work given in [6] assesses the schedule reliability of airport schedules using copulas. One more example is given in [13], where the authors use copulas to assess the schedule risk of a software development project. However, all of them use Gaussian or Student-t copulas, which can be applied only if the marginal distributions are either normal or Student-t, while in our work there is no restriction on the marginal distributions. As far as we know, there were no attempts to use copulas in the way presented in this work.

6 Conclusions

In this work we have studied the possibility of introducing dependence of both power consumption and execution time of the tasks which run on the same platform in order to improve the results of optimal task scheduling. Power consumption and execution time of the tasks are represented as random variables with known distribution functions, while their dependence is modeled with Gumbel copulas, whose θ parameter is varied in order to simulate different levels of dependence. As far as we know, this is the first work that uses copula-based dependence in the context of stochastic scheduling where the important aspects of the tasks are modeled using random distribution functions. If there is dependence present in the underlying data, our results show that the performance of the stochastic scheduler is significantly improved after assuming certain level of dependence: on average 18 % of energy consumption can be saved compared to the results of the deterministic scheduler, along with 81 % of improved test cases, versus 2.44 % average savings when task independence is assumed, along with 50 % of improved test cases. The obtained rates demonstrate the potential of the approach in improving results of stochastic scheduling when dependence between the tasks is present. In the future we plan to devote additional effort to providing faster simulation time, which would enable a real world implementation of the approach. Furthermore, we will test the approach on real world data.

Acknowledgments The research leading to these results has received funding from the European Union 7th Framework Programme under grant agreement 318337, ENTRA - Whole-Systems Energy Transparency, Spanish MINECO TIN'12-39391 *StrongSoft* and TIN'08-05624 *DOVES* projects, and Madrid TIC-1465 *PROMETIDOS-CM* project.

References

1. Mathematica. http://www.wolfram.com/mathematica/
2. Abedi S, Riahy GH, Hosseinian SH, Farhadkhani M (2013) Improved stochastic modeling: an essential tool for power system scheduling in the presence of uncertain renewables
3. Banković Z, Lopez-Garcia P (2014) Stochastic vs. deterministic evolutionary algorithm-based allocation and scheduling for XMOS Chips. Neurocomputing 82–89
4. Davies R (2013) Random distributions. http://www.robertnz.net/

5. Deb K, Pratap A, Sameer A, Meyarivan T (2000) A fast elitist multi-objective genetic algorithm: Nsga-ii. IEEE Trans Evol Comput 6:182–197
6. Diana T (2011) Improving schedule reliability based on copulas: an application to five of the most congested US airports. J Air Transp Manag 17(5):284–287
7. Gong C, Wang X, Xu W, Tajer A (2013) Distributed real-time energy scheduling in smart grid: stochastic model and fast optimization. IEEE Trans Smart Grid 4(3):1476–1489
8. Gonzalez-Fernandez Y, Soto M (2014) copulaedas: an r package for estimation of distribution algorithms based on copulas. J Stat Softw 58(9):1–34
9. Mathematica. MathLink Reference Guide, 1993
10. McNeil AJ, Frey R, Embrechts P (2010) Quantitative risk management: concepts, techniques, and tools. Princeton Series in Finance. Princeton University Press, Princeton
11. McNeil AJ et al (2009) Multivariate archimedean copulas, d-monotone functions and l_1-norm symmetric distributions
12. Nelsen RB (2003) Properties and applications of copulas: a brief survey. In: First Brazilian conference on statistical modelling in insurance and finance, pp 10–28
13. Wu D, Song H, Li M, Cai C, Li J (2010) Modeling risk factors dependence using copula method for assessing software schedule risk. In: 2010 2nd International conference on software engineering and data mining (SEDM), pp 571–574

Relational Crossover in Evolutionary Ontologies

Oliviu Matei, Diana Contraş and Honoriu Vălean

Abstract In this paper we introduce a new genetic operator, namely relational crossover applied to evolutionary ontologies recently defined. Also we demonstrate that the relations preserve properties like reflexivity, irreflexivity, symmetry, antisymmetry, asymmetry, transitivity after applying relational crossover operator. Applying such an operator in the evolutionary process induces an important variation in the population, which is relevant for a better exploration of the ontological space.

1 Introduction

Introduced very recently by Matei et. al. [1], evolutionary ontologies (EO) are genetic algorithms in which the individuals are represented by the elements of an otology instead of numbers, binary strings or programs. EO play a major role in evolutionary computation because they fill the gap between subsymbolic and symbolic intelligence [2].

As well, the evolutionary process is mainly the same: the population undergoes certain genetic operators: selection, crossover, mutation generation after generation, until an ending condition is fulfilled. However, the operators have their own specific, adapted to the ontological character of the chromosomes.

O. Matei
Department of Electrical Engineering, Technical University of Cluj-Napoca,
North University Centre of Baia Mare, Baia Mare, Romania
e-mail: oliviu.matei@holisun.com

D. Contraş (✉) · H. Vălean
Department of Automation, Technical University of Cluj-Napoca,
Cluj-Napoca, Romania
e-mail: pdia17@yahoo.com

H. Vălean
e-mail: honoriu.valean@aut.utcluj.ro

Á. Herrero et al. (eds.), *10th International Conference on Soft Computing Models in Industrial and Environmental Applications*, Advances in Intelligent Systems and Computing 368, DOI 10.1007/978-3-319-19719-7_15

The general concepts of genetic operators in EO, including crossover, were presented in [1]. However, as the complexity of recombination applied to properties is very high, we address the subject in this article.

The rest of the paper is organized as follows: Sect. 2 contains references to the elements of an ontology and to genetic crossover operator, but we have to point out that the two concepts are first used together in this article. Section 3 introduces a new genetic operator, namely relational crossover, demonstrating under what circumstances it can be applied on an ontology's relations. In Sect. 4 we demonstrate that relations retain their properties like reflexivity, irreflexivity, symmetry, antisymmetry, asymmetry, transitivity after being involved in the process of recombination. The experimental results presented in Sect. 5 highlight the importance of relational crossover operator in EO. Finally, the conclusions are marked in Sect. 6.

2 Related Work

According to Lord [3], these components of an ontology can be divided usually in two parts: those describing the entities of the domain and those describing the ontology itself. The most common components are the classes, individuals and properties, either as relations or as attributes.

Recombination of the individuals by means of mating is a critical part of evolution, according to [4]. Mimicking the biological model, in case of crossover two individuals and a crossover point are chosen, after which the right portions of that point are swapped to create offspring [4]. Based on this concept, have been developed many types of crossover operators, according to data coding method and the requirements of the problem [5].

Thus, for binary encoding there are four appropriate types of crossover operator: single point, two point, uniformly and arithmetically crossover [5]. In [6–9] is mentioned the single point crossover. In [9, 10] it is specified that uniform crossover interchange information between parents.

Permutation encoding is usualy used in ordering problems, acording to [11, 12]. Standard crossover operators applied on permutation encoding individuals can generate invalid permutation chromosomes [13], therefore specific operators were developed. According to [14], the crossover operators for permutation encoding can be classified in three classes: a class that (a) preserves the relative order, (b) that respects the absolute position and (c) that tends to preserves the adjacency.

If binary encoding is difficult to use, then value encoding is required where every chromosome is a sequence of some values which can be real numbers, chars or objects, according to [5, 15, 16]. In case of value encoding can be used the same operators as in the case of binary technique [5].

Tree encoding is used in genetic programming for evolving programs or expressions [16]. For tree encoding single point crossover is required [5] and subtree or node will be exchanged to accomplish the crossover operator [17]. In [18] the crossover is considered as a special case of a general form of interpolation used to create designs.

Recently, Matei et al. introduced the concept of evolutionary ontologies in [1] and showed that it is a completely different field in evolutionary computation [19]. The complexity is given by the relationships between the entities of the ontology which undergo evolution. Therefore this paper is focused on the specificity of crossover with respect to these relationships.

3 Relational Crossover

For the sake of simplicity and understanding, we will commit ourselves to binary relations, in this article. According to Woronowicz et al. [20], a binary relation R is an ordered triple (X, Y, R) where X and Y are arbitrary sets (or classes), and R is a subset of the cartesian product $X \times Y$. The set X is called the domain, and Y is called the codomain. R is the relation (or the graph). The pair $(x, y) \in R$ can be interpreted as *x is R-related to y*, and is denoted by xRy or R(x, y).

In [21, 22] the intersection of two classes in an ontology is defined as a class that contains all common individuals of initial classes.

Are considered two randomly selected relations (X_1, Y_1, R_1) and (X_2, Y_2, R_2). Having two individuals $x_1R_1y_1$ and $x_2R_2y_2$ with $x_1 \in X_1, y_1 \in Y_1, x_2 \in X_2$ and $y_2 \in Y_2$ the crossover operator would generate two offspring, such that:

$$x_iR_iy_i. \tag{1}$$

$$x_kR_ky_k. \tag{2}$$

where $x_i, x_k \in \{x_1, x_2\}$; $R_i, R_k \in \{R_1, R_2\}$ and $y_i, y_k \in \{y_1, y_2\}$, depending on the choice of crossing point.

The necessary and sufficient condition for the crossover to occur is that the two codomains or the two domains must have common elements if the crossover point is chosen at the individual level or both domains and codomains must have non-empty intersection if the crossover point is chosen at the relation level.

Further on we demonstrate the previous statement.

If the crossover point is chosen at domain individual level are obtained two offspring $x_1R_1y_2$ and $x_2R_2y_1$ with $x_1 \in X_1, y_2 \in Y_1, x_2 \in X_2$ and $y_1 \in Y_2$. From $y_1 \in Y_1$ and $y_1 \in Y_2$ results $y_1 \in Y_1 \sqcap Y_2$. From $y_2 \in Y_2$ and $y_2 \in Y_1$ results $y_2 \in Y_1 \sqcap Y_2$. The results $y_1 \in Y_1 \sqcap Y_2$ and $y_2 \in Y_1 \sqcap Y_2$ lead to

$$Y_1 \sqcap Y_2 \neq \bot. \tag{3}$$

If the crossover point is chosen at codomain individual level are obtained two offspring $x_2R_1y_1$ and $x_1R_2y_2$ with $x_2 \in X_1, y_1 \in Y_1, x_1 \in X_2$ and $y_2 \in Y_2$. From

$x_1 \in X_1$ and $x_1 \in X_2$ results $x_1 \in X_1 \sqcap X_2$. From $x_2 \in X_2$ and $x_2 \in X_1$ results $x_2 \in X_1 \sqcap X_2$. The results $x_1 \in X_1 \sqcap X_2$ and $x_2 \in X_1 \sqcap X_2$ lead to

$$X_1 \sqcap X_2 \neq \bot. \tag{4}$$

If the crossover point is chosen at relation level are obtained two offspring $x_1R_2y_1$ and $x_2R_1y_2$ with $x_1 \in X_2, y_1 \in Y_2, x_2 \in X_1$ and $y_2 \in Y_1$. From the given individuals and the above relations result that (3) and (4) both have to occur when crossover point is chosen at relation level.

In particular, for a binary relation R, (X, Y, R), if the domain and the codomain coincide, i.e. X = Y, is stated that the binary relation is over X [23]. In these conditions, are considered two randomly selected binary relations (X_1, X_1, R_1) and (X_2, X_2, R_2). According to (3) and (4) the necessary and sufficient condition for applying the crossover is

$$X_1 \sqcap X_2 \neq \bot. \tag{5}$$

The question is whether $X_1 \sqcap X_2 = \bot$ the crossover operator can not be applied? According to [24] if R is a binary relation and p is a property that can be satisfied by adding pairs of elements to R, then the *p closure of R* is the smallest binary relation containing R that has property p. Based on the previous definition we introduce *the* X_2 *closure of* R_1, by adding to R_1 pairs of elements from X_2 and *the* X_1 *closure of* R_2, by adding to R_2 pairs of elements from X_1. We use the notations $C_{X_2}(R_1)$ for *the* X_2 *closure of* R_1, respectively $C_{X_1}(R_2)$ for *the* X_1 *closure of* R_2. Occurs $C_{X_2}(R_1) \sqcap C_{X_1}(R_2) \neq \bot$ and according to (5) the crossover can be applied.

4 Properties of Relations with Respect to Crossover

Be a binary relation R over X, i.e. (X, X, R). Woronowicz et al. [20] defined some important properties of the binary relation R over the set X as follow:

The binary relation R is **reflexive** if

$$\forall x \in X \implies xRx. \tag{6}$$

The binary relation R is **irreflexive** or **strict** if

$$\forall x \in X \implies \text{not } xRx. \tag{7}$$

The binary relation R is **symmetric** if

$$\forall x_1, x_2 \in X, x_1Rx_2 \implies x_2Rx_1. \tag{8}$$

The binary relation R is **antisymmetric** if

$$\forall x_1, x_2 \in X, x_1 R x_2 \text{ and } x_2 R x_1 \implies x_1 = x_2. \quad (9)$$

The binary relation R is **asymmetric** if

$$\forall x_1, x_2 \in X, x_1 R x_2 \implies \text{not } x_2 R x_1. \quad (10)$$

Moreover, the binary relation R is defined **transitive** as follow

$$\forall x_1, x_2, x_3 \in X, x_1 R x_2 \text{ and } x_2 R x_3 \implies x_1 R x_3. \quad (11)$$

In [25], a relation is said to be **coreflexive** if

$$\forall x_1, x_2 \in X, x_1 R x_2 \implies x_1 = x_2. \quad (12)$$

Are considered two binary relations (X_1, X_1, R_1) and (X_2, X_2, R_2) with $x_{11}R_1x_{12}$ and $x_{21}R_2x_{22}$. If crossover operator is applied on R_1 and R_2, according to (1) and (2), are obtained $x_{11}R_1x_{22}$ and $x_{21}R_2x_{12}$.

It is considered that the two relations, R_1 and R_2, are **reflexive**, therefore, according to (6), $x_{12}R_1x_{12}$ and $x_{22}R_2x_{22}$. If crossover operator is applied on $x_{11}R_1x_{22}$ and $x_{22}R_2x_{22}$ are obtained $x_{22}R_1x_{22}$ and $x_{11}R_2x_{22}$. Moreover, if crossover operator is applied on $x_{12}R_1x_{12}$ and $x_{21}R_2x_{12}$ are obtained $x_{21}R_1x_{12}$ and $x_{12}R_2x_{12}$. The results $x_{22}R_1x_{22}$ and $x_{12}R_2x_{12}$ demonstrate that for the new elements appeared after applying crossover operator, the relations R_1, R_2 keep (6), thus both relations preserve the property of reflexivity after crossover.

It is considered that the two relations, R_1 and R_2, are **irreflexive**, thus, according to (7), not $x_{12}R_1x_{12}$ and not $x_{22}R_2x_{22}$. In order to demonstrate that relations R_1, R_2 remain irreflexive after applying crossover operator must be demonstrated that occur not $x_{22}R_1x_{22}$ and not $x_{12}R_2x_{12}$. It is supposed that $x_{22}R_1x_{22}$, which, according to previous demonstration, does not take place unless the relation R_1 is reflexive, that contradicts the hypothesis statement not $x_{12}R_1x_{12}$. Thus it can be concluded that not $x_{12}R_1x_{12}$ is a valid result.

Likewise it is supposed that $x_{12}R_2x_{12}$, which, according to previous demonstration, does not take place unless the relation R_2 is reflexive, that contradicts the hypothesis statement not $x_{22}R_2x_{22}$. Therefore not $x_{12}R_2x_{12}$ takes place.

It is considered that the two relations, R_1 and R_2, are **symmetric**, hence, according to (8), $x_{12}R_1x_{11}$ and $x_{22}R_2x_{21}$. To demonstrate that relations remain symmetric after applying the crossover operator, must be proved that occur $x_{22}R_1x_{11}$ and $x_{12}R_2x_{21}$. Applying crossover on $x_{12}R_1x_{11}$ and $x_{22}R_2x_{21}$ are obtained $x_{22}R_1x_{11}$ and $x_{12}R_2x_{21}$, so the assertion that relations are symmetric after applying crossover operator is true.

It is considered that the two relations, R_1 and R_2, are **antisymmetric**, thus, according to (9)

$$x_{11}R_1x_{12} \text{ and } x_{12}R_1x_{11} \implies x_{11} = x_{12}. \quad (13)$$

$$x_{21}R_2x_{22} \text{ and } x_{22}R_2x_{21} \implies x_{21} = x_{22}. \quad (14)$$

To demonstrate that relations remain antisymmetric after applying the crossover operator must be shown that occur

$$x_{11}R_1x_{22} \text{ and } x_{22}R_1x_{11} \implies x_{11} = x_{22}. \quad (15)$$

$$x_{21}R_2x_{12} \text{ and } x_{12}R_2x_{21} \implies x_{21} = x_{12}. \quad (16)$$

Crossover operator is applied on $x_{11}R_1x_{22}$ and $x_{21}R_2x_{22}$ and are obtained $x_{21}R_1x_{22}$ and $x_{11}R_2x_{22}$. Crossover operator is applied on $x_{22}R_1x_{11}$ and $x_{22}R_2x_{21}$ and are obtained $x_{22}R_1x_{21}$ and $x_{22}R_2x_{11}$. It is supposed that relation R_1 is not anti-symmetric after applying the crossover operator, meaning that for a new element introduced in X_1 not exists an element in X_1 thereby the two elements are equal and R_1-related . Be x_{22} this new element. From this assumption, $x_{21}R_1x_{22}$ and $x_{22}R_1x_{21}$ result that $x_{21} \neq x_{22}$, which is in contradiction with (14). Thus the assumption is false.

Further, crossover operator is applied on $x_{11}R_1x_{12}$ and $x_{21}R_2x_{12}$ and are obtained $x_{21}R_1x_{12}$ and $x_{11}R_2x_{12}$. Crossover operator is applied on $x_{12}R_1x_{11}$ and $x_{12}R_2x_{21}$ and are obtained $x_{12}R_1x_{21}$ and $x_{12}R_2x_{11}$. It is supposed that relation R_2 is not anti-symmetric after applying the crossover operator, meaning that for a new element introduced in X_2 not exists an element in X_2 thereby the two elements are equal and R_2-related. Be x_{12} this new element. From this assumption, $x_{11}R_2x_{12}$ and $x_{12}R_2x_{11}$ result that $x_{11} \neq x_{12}$, which is in contradiction with (13). Thus the assumption is false.

It is considered that the two relations, R_1 and R_2, are **asymmetric**, therefore, according to (10), not $x_{12}R_1x_{11}$ and not $x_{22}R_2x_{21}$. To demonstrate that relations remain asymmetric after applying the crossover operator must be shown that occur not $x_{22}R_1x_{11}$ and not $x_{12}R_2x_{21}$. It is supposed, using the reductio ad absurdum, that occur $x_{22}R_1x_{11}$ and $x_{12}R_2x_{21}$. It was previously demonstrated, at symmetry property, that in order to have these results must occur $x_{12}R_1x_{11}$ and $x_{22}R_2x_{21}$, that contradict the hypothesis. Thus the made assumption is false.

It is considered that the two relations, R_1 and R_2, are **transitive**, so, according to (11)

$$x_{11}R_1x_{12} \text{ and } x_{12}R_1x_{13} \implies x_{11}R_1x_{13}. \quad (17)$$

$$x_{21}R_2x_{22} \text{ and } x_{22}R_2x_{23} \implies x_{21}R_2x_{23}. \quad (18)$$

To demonstrate that relations remain transitive after applying the crossover operator must be shown that occur

$$x_{11}R_1x_{22} \text{ and } x_{22}R_1x_{13} \implies x_{11}R_1x_{13}. \quad (19)$$

$$x_{21}R_2x_{12} \text{ and } x_{12}R_2x_{23} \implies x_{21}R_2x_{23}. \quad (20)$$

Crossover operator is applied on $x_{11}R_1x_{12}$ and $x_{21}R_2x_{22}$ and are obtained $x_{11}R_1x_{22}$ and $x_{21}R_2x_{12}$. Further, crossover operator is applied on $x_{12}R_1x_{13}$ and $x_{22}R_2x_{23}$ and are obtained $x_{22}R_1x_{13}$ and $x_{12}R_2x_{23}$. Moreover crossover operator is applied on the results $x_{11}R_1x_{22}$ and $x_{22}R_1x_{13}$ and is obtained (19). Further crossover operator is applied on the results $x_{21}R_2x_{12}$ and $x_{12}R_2x_{23}$ and is obtained (20).

It is considered that the two relations, R_1 and R_2, are **coreflexive**, thus, according to (12)

$$x_{11}R_1x_{12} \implies x_{11} = x_{12}. \tag{21}$$

$$x_{21}R_2x_{22} \implies x_{21} = x_{22}. \tag{22}$$

To demonstrate that relations remain coreflexive after applying the crossover operator must be shown that occur

$$x_{11}R_1x_{22} \implies x_{11} = x_{22}. \tag{23}$$

$$x_{21}R_2x_{12} \implies x_{21} = x_{12}. \tag{24}$$

It is supposed that relation R_1 is not coreflexive after applying the crossover operator, meaning that for a new element introduced in X_1 not exists an element in X_1 thereby the two elements are equal and R_1-related. Be x_{22} this new element. Crossover operator is applied on $x_{11}R_1x_{22}$ and $x_{21}R_2x_{22}$ and are obtained $x_{21}R_1x_{22}$ and $x_{11}R_2x_{22}$. According to the previous assumption, $x_{21}R_1x_{22}$ implies $x_{21} \neq x_{22}$, which is in contradiction with (22). Thus the assumption is false.

Following the same reasoning it is supposed that relation R_2 is not coreflexive after applying the crossover operator. This means that for a new element introduced in X_2 not exists an element in X_2 thereby the two elements are equal and R_2-related. Be x_{12} this new element. Crossover operator is applied on $x_{11}R_1x_{12}$ and $x_{21}R_2x_{12}$ and are obtained $x_{21}R_1x_{12}$ and $x_{11}R_2x_{12}$. According to the previous assumption, $x_{11}R_2x_{12}$ implies $x_{12} \neq x_{11}$, which is in contradiction with (21). Thus the assumption is false.

5 Experimental Results

The concept of *evolutionary ontologies* was applied on user centered automatically generated scenes application. This is based on the following methodology: a list of 10 scenes, representing the initial population, were generated randomly. Each scene had a frame. Were used three types of frames: land, water, interior. Every user rated them on a scale from 1 to 10, based on how much he/she liked them. Genetic operators (crossover and mutation) were applied on the scenes selected by a Monte Carlo-based selection operator, while the grade played the role of the fitness function. The evolutionary process reloaded until at least one scene was graded 10 or the maximum number of 20 epochs was met.

To convey the importance of relational crossover operator we make a comparison between it and a conventional crossover operator on a specific case. For this, we consider two scenes which have land as frame. We call them, in the following, Land1 and Land2. Number of elements in a scene may differ from one scene to another. Thus Land1 has 6 elements and Land2 has 4 elements.

The most commonly used crossover operator in genetic algorithms is one-point crossover. We check its effect on chromosomes that have ontological character (as an entity crossover, as defined in [1]). Whether we have a chromosome with n alleles and another chromosome with m alleles, where $n \neq m$, the cutting point is chosen as a random number between 1 and $\min(n, m) - 1$. In our case the cutting point is chosen as shown in Fig. 1a and are obtained two offspring, like those from Fig. 1b.

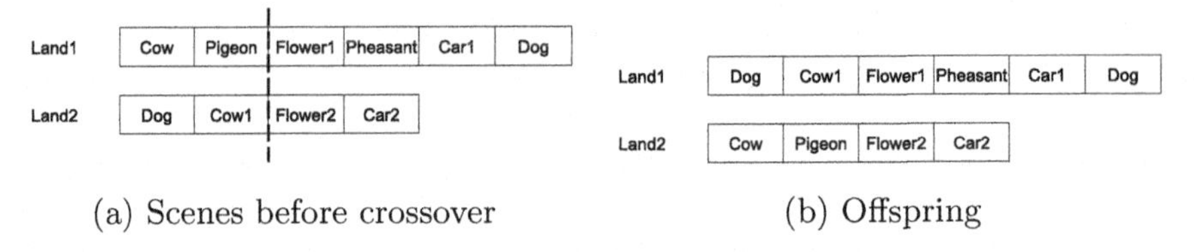

Fig. 1 Entity crossover

As it can be seen in Fig. 1b, in scene Land1 the item Dog appears twice. But visually, this means reducing the diversity as the superposition of two identical objects creates the impression of the existence of a single one. Therefore, we conclude that by applying one-point crossover is possible to reduce the number of elements of the scenes, thus reducing the diversity.

At the opposite pole is the relational crossover, which maintains diversity. As it can be seen in Fig. 1a there are the following relationships: Dog is Land Land2 (and also Dog is Land Land1), Cow is Land Land1, illustrated in Fig. 2.

Applying the relational crossover on the relations from Fig. 2 two offspring are created, as in Fig. 3.

The interesting aspect is the impact of the relational crossover described in this article on the individuals (scenes). For this, we counted the relational differences between two successive generations (for all 10 subjects) in two cases: (a) without applying relational crossover, as defined in [1], respectively (b) applying the relational crossover as shown in this article. In the former case, the different relationships may be introduced by mutation and by the repair operator, whereas in the latter case, the crossover itself induces new relations. Figure 4 depicts the minimum, average and maximum number of different relationships between two successive generations. In the beginning, the number of differences in the two cases ((a) and (b)) is about the same, but it increases when using the relational crossover almost twice than when only the entity crossover is applied.

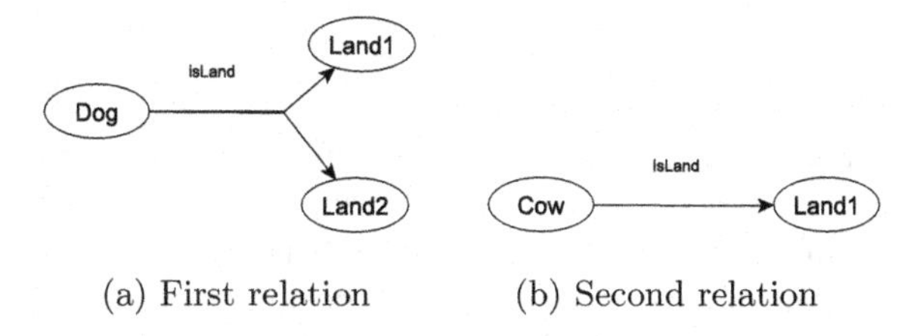

(a) First relation (b) Second relation

Fig. 2 Relations for crossover

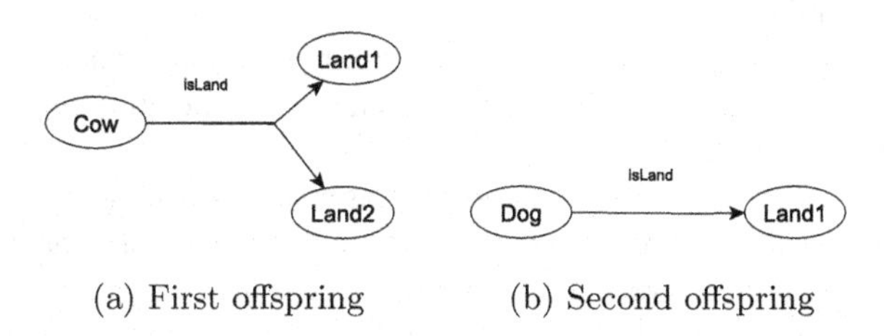

(a) First offspring (b) Second offspring

Fig. 3 The two offspring

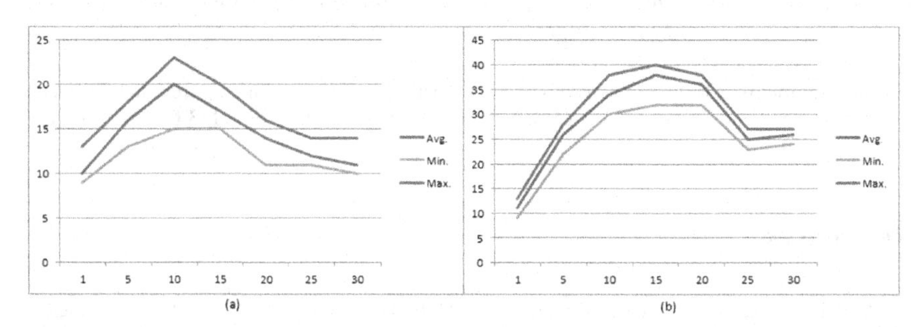

Fig. 4 The average, minimum and maximum number of differences in the two cases: (a) without relational crossover; (b) with relational crossover

From Fig. 4 it is obvious that the changes between two generations are larger when the relational crossover is applied. An interesting remark is that the minimum number of differences in case (b) is higher than the maximum number of differences in case (a). This means that the relational crossover induces a significant variation in the ontological population and thus this leads to a better exploration of the ontological space.

The test cases are based on our own ontologies because as EO is a completely new topic, there are no test ontologies developed yet, as in the case of classical optimization problems, such as (G)TSP or (G)VRP (http://www.iwr.uni-heidelberg.de/groups/comopt/software/). However, the impact of the relational crossover on the diversity of the population is obvious and very efficient.

6 Conclusions

In this article we present the crossover operator applied for recombining ontological relations. We demonstrate that the necessary and sufficient condition as the result of recombination of two binary relations to be valid is that the intersection of the domain and/or of the codomain of the two relations has to be non-empty. In case it can not be reached that result, we demonstrate that a property closure for both relations can be found so that their intersection is non-empty, consequently, the relational crossover can always be applied.

Finally, we prove that relational crossover explores the onto-space in a better way than the entity crossover, so that the individuals in a generation are significantly different than their parents.

Since the fitness function of automatic generator scenes is subjective, we propose as future work, the study of relational crossover in the context of dynamic products, where the costs and the delivery times are objective values.

The experimental results lack of a standardized test set and the ability to compare our results with others. But this is always the case for new topic, as EO is. A future work will focus on developing a test ontological space which can be used for researching evolutionary ontologies.

Acknowledgments The research leading to these results has received funding from the European Communitys Seventh Framework Programme under grant agreement No609143 Project ProSEco.

References

1. Matei O, Contras D, Pop P (2014) Applying evolutionary computation for evolving ontologies. In: 2014 IEEE congress on evolutionary computation (CEC), IEEE (2014), pp 1520–1527
2. Goertzel B (2012) Perception processing for general intelligence: bridging the symbolic/subsymbolic gap. In: Artificial general intelligence. Springer, New York, pp 79–88
3. Lord P (2010) Components of an ontology. Ontogenesis
4. Holland JH (1992) Genetic algorithms. Sci Am 267(1):66–72
5. Malhotra R, Singh N, Singh Y (2011) Genetic algorithms: concepts, design for optimization of process controllers. Comput Inf Sci 4(2):p39
6. Shukla A, Tiwari R, Kala R (2010) Real life applications of soft computing. CRC Press, Boca Raton
7. Simon D (2013) Evolutionary optimization algorithms. Wiley, New York
8. Sivanandam S, Deepa S (2007) Introduction to genetic algorithms. Springer Science & Business Media, Berlin
9. Yu X, Gen M (2010) Introduction to evolutionary algorithms. Springer Science & Business Media, London
10. Yang XS, Gandomi AH, Talatahari S, Alavi AH (2012) Metaheuristics in water, geotechnical and transport engineering. Newnes
11. Király A, Abonyi J (2011) Optimization of multiple traveling salesmen problem by a novel representation based genetic algorithm. In: Intelligent computational optimization in engineering. Springer, Berlin, pp 241–269
12. Nedjah N, de Macedo Mourelle L (2004) Evolvable machines: theory and practice, vol 161. Springer Science & Business Media, Berlin

13. Givens GH, Hoeting JA (2012) Computational statistics, vol 710. Wiley, New York
14. Sels V, Vanhoucke M (2011) A hybrid dual-population genetic algorithm for the single machine maximum lateness problem. In: evolutionary computation in combinatorial optimization. Springer, Berlin, pp 14–25
15. Chen JS, Hou JL (2006) A combination genetic algorithm with applications on portfolio optimization. In: Advances in applied artificial intelligence. Springer, Berlin, pp 197–206
16. Siddique N, Adeli H (2013) Computational intelligence: synergies of fuzzy logic, neural networks and evolutionary computing. Wiley, New York
17. Ju MY, Wang SE, Guo JH (2014) Path planning using a hybrid evolutionary algorithm based on tree structure encoding. Sci World J
18. Gero JS (2002) Computational models of creative designing based on situated cognition. In: Proceedings of the 4th conference on creativity and cognition, ACM (2002), pp 3–10
19. Matei O, Contra D, Pintescu A (2014) Why evolutionary ontologies are a completely different field than genetic algorithms. Carpathian J Electron Comput Eng 7(1):19–24
20. Woronowicz E, Zalewska A (1990) Properties of binary relations. Form Math 1(1):85–89
21. Horridge M, Knublauch H, Rector A, Stevens R, Wroe C (2004) A practical guide to building owl ontologies using the protégé-owl plugin and co-ode tools edition 1.0. University of Manchester
22. Motik B, Patel-Schneider PF, Parsia B, Bock C, Fokoue A, Haase P, Hoekstra R, Horrocks I, Ruttenberg A, Sattler U et al (2009) Owl 2 web ontology language: structural specification and functional-style syntax. W3C Recomm 27(65):159
23. Meyer B (2009) Touch of class: learning to program well with objects and contracts. Springer Science & Business Media, Secaucus
24. Hein JL (2010) Discrete structures, logic, and computability. Jones & Bartlett Publishers, Washington
25. Wang S, Barbosa LS, Oliveira JN (2008) A relational model for confined separation logic. In: 2nd IFIP/IEEE International symposium on theoretical aspects of software engineering, 2008. TASE'08, IEEE (2008), pp 263–270

Intelligent Systems

A Movement Control System Based on Qualitative Reasoning

Przemysław Wałęga and Emilio Muñoz-Velasco

Abstract We present QR_M, a movement control system based on Qualitative Reasoning. The representation of relative movement of an object with respect to another is done by using different components given by qualitative values, such as velocity, orientation, latitude, longitude, etc. These qualitative values are obtained from quantitative data by means of a nonlinear system with hysteresis. We also use composition tables for new data inferring and a table-based control system. The system is implemented in Robotic Operating System ROS and tested with computer simulator STAGE. We show how QR_M works in real applications on the basis of two experiments.

Keywords Qualitative reasoning · Qualitative movement · Movement Control · Collision avoidance

1 Introduction

Space representation and reasoning about its properties interest scientists since antiquity. The field of Spatial Reasoning involves numerous mathematical and philosophical questions that still have no clear answers. Additionally, nowadays, there is also a great interest in Spatial Reasoning systems for practical applications, e.g. mobile robotics or geographic information systems (GIS) [6]. The movement of objects is one of the important aspects that has to be considered in such systems. Reasoning about movement is necessary e.g. in order to control vehicles or robotic manipulators avoiding collisions and other problems.

In many real robotic systems, the information gathered by sensors is noisy or inaccurate due to external factors such as bad weather conditions, low quality of the sensors, among others. Additionally, it happens that Spatial Reasoning needs

P. Wałęga (✉)
Institute of Philosophy, University of Warsaw, Warsaw, Poland

E. Muñoz-Velasco
Department of Applied Mathematics, University of Malaga, Málaga, Spain

Á. Herrero et al. (eds.), *10th International Conference on Soft Computing Models in Industrial and Environmental Applications*, Advances in Intelligent Systems and Computing 368, DOI 10.1007/978-3-319-19719-7_16

to be performed using incomplete data and intuitive information. In general, this information is easy to understand for humans but not for machines. For instance, when a human being is catching a ball, there is a lot of intuitive and incomplete information about the position, velocity and direction of the ball, which is very difficult and slow to process for one machine. One of the approaches that may be used to overcome the above mentioned problems is Qualitative Reasoning (QR) [5] that uses qualitative, human-like representation rather than quantitative, numerical data. Qualitative information is much simpler and may be calculated and transmitted much faster than precise quantitative data. This feature has a significant meaning when an immediate decision needs to be made and movement control in practical applications is surely one of such cases. Another approach to be mentioned is Qualitative Spatial Reasoning [2] - an approach for Qualitative Reasoning about space in a particular formal way that usually uses relational algebras. Both of the above mentioned approaches have numerous applications, e.g. [10, 11].

In this paper, we present a novel approach for Qualitative Reasoning about movement of objects, called QR_M. Our approach is based on the logic PDL_M^F presented in [12], where the movement of an object with respect to another is represented by a tuple whose components include qualitative values representing various aspects, namely absolute velocity, absolute orientation, relative direction of movement, allowed directions of movement and relative position. It is worth to mention that there are several other approaches where the problem of the relative movement of one physical object with respect to another is treated, e.g. [3, 4, 14], however the ideas presented in [12] generalize in some sense several previous approaches. In our approach QR_M we obtain qualitative values for movement description in a novel way, i.e. by means of a nonlinear system with hysteresis on borders of qualitative values. Afterwards we use composition tables - a reasoning method known from Qualitative Spatial Reasoning [2] to infer new qualitative information. Finally, we use a table-based control system. It is worth to mention, that the approach can be modified and adapted to the problem in question with no substantial changes in the general framework.

The QR_M system is implemented in Robotic Operating System ROS [13] - currently one of the most popular open source frameworks for robot software and has been simulated in Stage [8] - a 2D, freeware robotic simulator. We present two examples of QR_M applications, namely for a problem of following another object and for avoiding collisions. We describe computer simulations of above mentioned.

The paper is organized as follows. Section 2 is devoted to introduce the movement representation chosen in our approach. In Sect. 3, the QR_M system is presented: its method for obtaining qualitative values, composition tables for inferring new information and a table-based movement control method. Section 4 provides two real applications of QR_M that have been tested and simulated. Finally, conclusions and prospects of future work are shown in Sect. 5.

2 Movement Representation

In this section, we explain how the movement is represented in our approach. We will use the ideas presented in [12], where the movement of an object with respect to another is given by a tuple of qualitative values which represent, respectively, absolute velocity, absolute orientation, relative direction of movement, allowed directions of movement and relative position. Formally, the movement of one object with respect to another is represented by a tuple $(x_1, \dots, x_7) \in L$, where $L = L_1 \times \cdots \times L_7$ is described in Table 1. It is worth mentioning that the movement representation enables to capture uncertainty of information, e.g. the following velocity description $x_2 = \{\mathsf{v}_1, \mathsf{v}_2\}$ has an intuitive meaning that the velocity is slow or normal. As an example of movement representation let us consider the tuple presented by: [1]

Table 1 Description of the tuple L used in PDL^F_M movement representation

	Description
$L_1 = A \times A$	states that the tuple describes movement of the first object of the L_1 ordered pair with respect to the second one, where $A = \{\mathsf{A}_1, \dots, \mathsf{A}_n\}$ is a set of all objects
$L_2 = 2^{\{\mathsf{v}_0,\mathsf{v}_1,\mathsf{v}_2,\mathsf{v}_3\}}$	first object absolute velocity (zero v_0, slow v_1, normal v_2, quick v_3)
$L_3 = 2^{\{\mathsf{o}_0,\mathsf{o}_1,\mathsf{o}_2,\mathsf{o}_3\}}$	first object absolute orientation (unknown o_0, north o_1, south o_2, east o_3, west o_3)
$L_4 = 2^{\{0,-,+\}} \times 2^{\{0,-,+\}}$	relative movement direction which consists of the first object movement relative to second object (stable 0, moving towards +, moving away from −) and analogously second object movement with respect to the first one
$L_5 = 2^{\{\mathsf{o}_0,\mathsf{o}_1,\mathsf{o}_2,\mathsf{o}_3,\mathsf{o}_4\}}$	allowed absolute orientations of first object movement (unknown o_0, north o_1, south o_2, east o_3, west o_3)
$L_6 = 2^{\{\mathsf{o}_1,\mathsf{o}_2\}} \times 2^{\{\mathsf{d}_0,\mathsf{d}_1,\mathsf{d}_2,\mathsf{d}_3\}}$	latitude position of first object with respect to the second one which consists of orientation (north o_1, south o_2) and distance (zero d_0, close d_1, normal d_2, distant d_3)
$L_7 = 2^{\{\mathsf{o}_3,\mathsf{o}_4\}} \times 2^{\{\mathsf{d}_0,\mathsf{d}_1,\mathsf{d}_2,\mathsf{d}_3\}}$	longitude position of first object with respect to the second one which consists of orientation (east o_3, west o_4) and distance (zero d_0, close d_1, normal d_2, distant d_3)

$$\mathsf{A}_i, \mathsf{A}_j; \mathsf{v}_2\mathsf{v}_3; \mathsf{o}_3; +-; \mathsf{o}_1\mathsf{o}_3; \mathsf{o}_1, \mathsf{d}_1\mathsf{d}_2; \mathsf{o}_3, \mathsf{d}_2$$

[1]Henceforth, for a better reading, we eliminate the curly brackets in the sets and the parenthesis in the tuples, and we use semicolon ";" to separate two consecutive components of the movement.

This tuple represents a movement of an object object A_i with respect to A_j; A_i has a normal or quick velocity, v_2v_3; and east orientation o_3; A_i is moving away from A_j, +, and A_j is moving towards A_i, −; A_i possible orientations are north or east, o_1o_3; A_i is to the north at a close or normal distance with respect to A_j, o_1, d_1d_2 (qualitative latitude); and A_i is to the east at a normal distance with respect to A_j, o_3, d_2 (qualitative longitude).

3 The QR_M System

In this section, we explain the main features of our system QR_M. Firstly, we consider a method using hysteresis, for translating quantitative information (obtained, for instance, by using sensors) into qualitative values, in particular when we are dealing with qualitative values that are changing very close to the borders of a qualitative class. Secondly, we explain how to infer new information from information which is already known, by means of composition tables. Finally in this section, we will use another advantage of the approach given in [12]. In this case, Propositional Dynamic Logic allows us to use some specific relations (called programs) in order to control the movement.

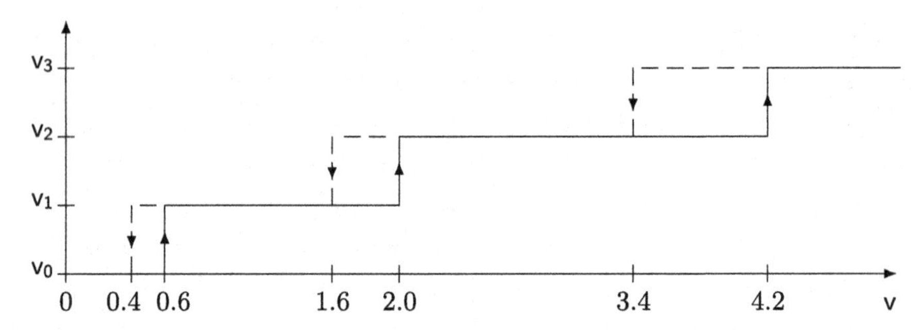

Fig. 1 Hysteresis method for translating quantitative velocity into qualitative data

3.1 *Hysteresis for Qualitative Values*

There are various ways of obtaining qualitative values from quantitative data [7], for instance by means of nonoverlapping subintervals or fuzzy sets, however, most of the methods are affected by small changes of the underlying quantitative values in borderline cases (i.e. when a quantitative value is on the border between two qualitative values). Our approach is useful for obtaining more stable qualitative representation which in general, seems to make the further reasoning system more efficient. Formally speaking, our method uses a nonlinear system with sharp hysteresis loops (for an exhaustive analysis of hysteresis loops and nonlinear systems in general see [9]).

The number of hysteresis in the method is always equal to the number of qualitative values decreased by 1. Hysteresis are located on the borders of qualitative classes. By way of example, let us consider the translation of velocity into 4 qualitative classes v_0, v_1, v_2, v_2, see Fig. 1. In this case, method consists of 3 hysteresis located, close to the borders of each qualitative class, i.e. 0.6, 2.0 and 4.2. Hysteresis prevent from frequent changes of qualitative values, because the borderline values between qualitative values are different depending on whether the quantitative value is increasing or decreasing. For instance, if the value of the velocity is 1.6 and increases to 1.7, then it is still considered in the qualitative class v_2. However, if this value continues increasing to 2.0 then it will be considered in the class v_2 and if it now decreases to 1.7 then it will remain in the same class until it reaches the value 1.6 where it will be translated again into v_1. This feature will be used in order to obtain composition tables in the next section.

3.2 Composition Tables

Composition tables constitute the most common reasoning method in Qualitative Spatial Reasoning [2]. The composition of the movement an object A_i with respect to A_j and a movement of A_j with respect to A_k is a movement of A_i with respect to A_k. The composition tables will infer information of the composition of two movements from the information of both movements. The composition table presented in Table 2 gives information about the composition of two movements in the case of the obtention of qualitative latitude and longitude introduced above. Notice that, in the last row and column of the table, for the composition of two movements with the same orientation o_r in the latitude (or, similarly longitude) and distances d_s and d_u, will be movement with the same orientation o_r, and the distance obtained as the qualitative sum of d_s and d_u. As our approach is focused in controlling moving objects avoiding collisions and other similar situations, it seems to be reasonable to use maximum distance $d_{max\{s,u\}}$ as a representation for the qualitative sum, because, as a matter of fact, this value will be a lower bound of the "real" qualitative sum.

Table 2 Composition table qualitative latitude (longitude), where $r \in \{3,4\}$ and $d, s \in \{0,1,2,3\}$

A_iA_j	A_jA_k	
	o_r, d_0	o_r, d_u
o_r, d_0	o_r, d_0	o_r, d_u
o_r, d_s	o_r, d_s	$o_r, d_{max\{s,u\}}$

3.3 Movement Control

We introduce now a movement control system which is able to modify its velocity and orientation in order to achieve the desired goal. We focus on the modification of velocity, because of the applications presented in the next section. However, the same ideas can be used to control the orientation of the movement. The change of the velocity of object $\mathsf{A_i}$ (denoted by $v_{\mathsf{A_i}}$) is determined by the distance and the velocity difference between the $\mathsf{A_i}$ and the closest obstacle. As stated above, we make use of another feature of the Propositional Dynamic Logic approach presented in [12]: the existence of three relations (called programs) for modifying the velocity. These programs are denoted by *Dec*, *Inc* and *Man*, meaning decreasing, increasing and maintaining the velocity, respectively. We introduce in Table 3 a table-based controlling system that makes use of the above mentioned programs, where *d* denotes qualitative distance between the object and the closest obstacle, whereas *dv* stands for qualitative velocity difference between the object and the closest obstacle. We assume that the velocity difference dv can take seven qualitative values, namely: v_{-3}, v_{-2}, v_{-1}, v_0, v_1, v_2, v_3, meaning zero, and slow, normal and quick, with positive and negative values, where the fact of being positive or negative has a very intuitive interpretation.

Table 3 is interpreted as a set of so called "If-Then" rules which state which program needs to be used, given the values of *v* and *dv*. For instance, the element of the second row and seventh column states that "If $d = \mathsf{d}_1$ and $dv = \mathsf{v}_3$ then use the program *Dec*", if the object and the obstacle are close and the object is moving much faster than the obstacle, then decrease the velocity of the object. The intuitive meaning of all control rules from the Table 3 can be explained in the following general way:

- if the controlled object moves much faster than the obstacle or the distance between them is small, than the controlled object should decrease its velocity (*Dec*),
- if the controlled object moves much slower than the obstacle or the distance between them is big, than the controlled object should increase its velocity (*Inc*),
- otherwise maintain the velocity of the controlled object (*Man*).

Table 3 The movement control rules

d	*dv*						
	v_{-3}	v_{-2}	v_{-1}	v_0	v_1	v_2	v_3
d_0	*Man*	*Man*	***Dec***	***Dec***	***Dec***	***Dec***	***Dec***
d_1	***Inc***	*Man*	*Man*	***Dec***	***Dec***	***Dec***	***Dec***
d_2	***Inc***	***Inc***	*Man*	*Man*	*Man*	***Dec***	***Dec***
d_3	***Inc***	***Inc***	***Inc***	***Inc***	***Inc***	***Inc***	***Inc***

The movement control system enables to maintain a safe distance between the controlled object and the obstacle. Therefore, it may be used in such applications as avoiding collisions with obstacles or following another object. The rules presented in the Table 3 are based on performed experiments, however, they may also be established by means of learning algorithms, e.g. with neural networks. Notice that the change of rules in the table-based system may be easily done, hence the system QR_M may be adapted to various scenarios and controlling strategies.

4 Applications

The system QR_M is implemented in ROS, one of currently most popular open source frameworks for writing robot software [13], capable to control numerous robot platforms, such as: PR2, Robonaut 2, REEM, TurtleBot, iRobot Roomba, Lego Mindstorm and many more. ROS enables us to use QR_M in real robots or perform tests by using STAGE, a 2D freeware robotic simulator [8], which provides a virtual word where different mobile robots together with sensors and objects can be simulated and visualized. The ROS framework for QR_M is divided into several separate modules, called nodes. Every node gathers messages through input topics, processes data and sends messages via output topics to other nodes. The `/stageros` node is the simulator node which publishes information from simulated sensors. The node `/qualitative_values` translate quantitative data into qualitative values as described in the Sect. 3.1. Then, the `/composition` node is used in order to infer new information with the method introduced in Sect. 3.2. Afterwards, `/control` modifies the velocity of the object, depending on its own velocity and the velocity and distance to the obstacle, as described in the Sect. 3.3. All this information is sent to the simulator. The whole process is in a loop which repeats until the end of the simulation. The Stage simulator displays objects and informs about collision whenever it occurs.

In what follows, we present two scenarios, namely following an object and collision avoidance. In both scenarios we will use the foliowing notation: i.e. $\mathsf{A_i}$, $\mathsf{A_j}$, $\mathsf{A_k}$ denote cars, d_{rs} denote the qualitative distance between vehicles $\mathsf{A_r}$ and $\mathsf{A_s}$, where $r, s \in \{i, j, k\}$ and $r < s$. Notice that d_{ij} and d_{jk} are obtained after translation of quantitative data from sensors, as explained in Sect. 3.1. On the other hand, d_{ik} can be inferred by means of the composition table for distances (Table 2). The movement of $\mathsf{A_i}$ is controlled by using the information of the distances and the difference of velocities $v_{\mathsf{A_i}}$ and $v_{\mathsf{A_j}}$, as in Table 3.

4.1 Experiment 1: Following an Object

In the first experiment, we consider only two of objects $\mathsf{A_i}$ and $\mathsf{A_j}$, where $\mathsf{A_j}$ is moving with a constant velocity and orientation, whereas $\mathsf{A_i}$ has to follow $\mathsf{A_j}$. The cars behavior is presented in Fig. 2 by means of 6 screenshots obtained from the simulator in consecutive time points. Notice, that since only qualitative values are used, the exact distance between objects is unknown and as a result, it is impossible to maintain an exactly constant distance between the A_i and the A_j. The distance between them increases and decreases during the simulation, however, it always remains safe. Consequently, A_i follows A_j without having a collision or letting A_j go too far from A_i.

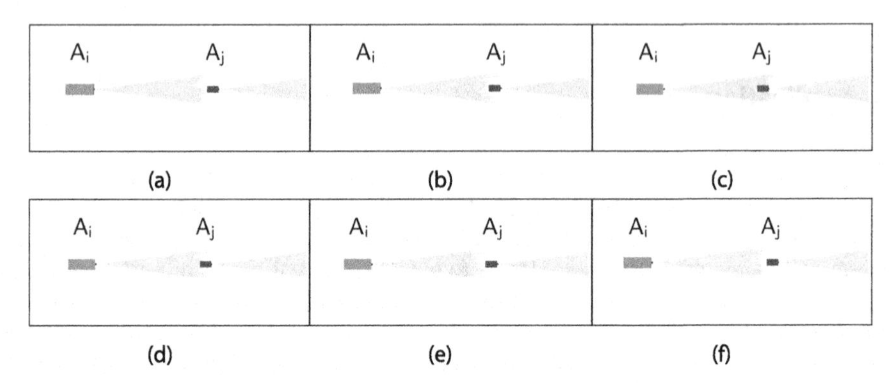

Fig. 2 Experiment 1: Following an object

4.2 Experiment 2: Collision Avoidance

The second experiment models a more complex situation. A number of cars are stuck in a traffic jam, whereas other cars are moving towards the same traffic jam and may not be able to stop in order to avoid a collision. Such a situation is common on highways and leads to dangerous collisions of numerous vehicles. The scenario consists of two moving vehicles, namely A_i and A_j and cars stuck in a traffic jam with A_k being the last car in the traffic jam. To make it more complicated, we assume also that A_i is a truck with long braking distance. We assume that each vehicle has information about its distance to the closest object in front of it. We will study how the information gathered by A_j (namely, the d_{jk} distance) can be used by A_i to avoid the collision. We perform two tests: in the first one, we assume that A_i does not have the information about the distance d_{jk} of its previous car A_j to the traffic jam A_k; while in the second test, this information is available to A_i, so it can infer its distance to the traffic jam d_{jk}, as explained above.

In the first test, we consider what happens if A_j stops suddenly, while A_i cannot infer d_{jk}. In such a case the input values to the movement control of A_i are: the qualitative distance d_{ij} between A_i and A_j and the qualitative velocity difference dv between A_i and A_j. The movement control rules are implemented as presented in the Table 3. In this case, A_j stops suddenly just before the traffic jam, while A_i, having a long braking distance, hits A_j and the collision occurs. The simulation of the Test 1 is presented in the Fig. 3, with collision occurring in frame (f).

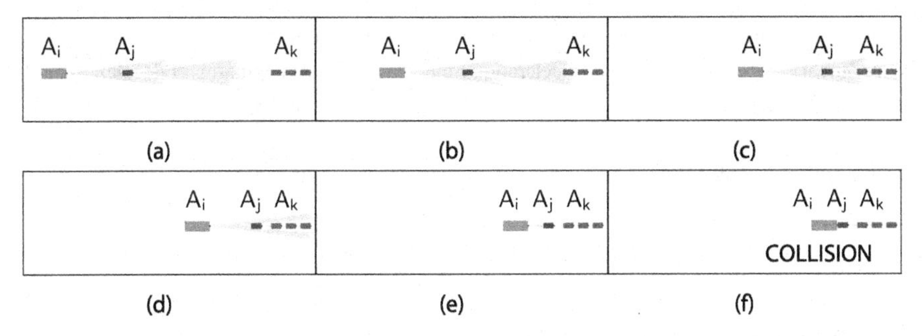

Fig. 3 Experiment 2, Test 1. A_i is unable to infer d_{ik} and, as a result a collision occurs

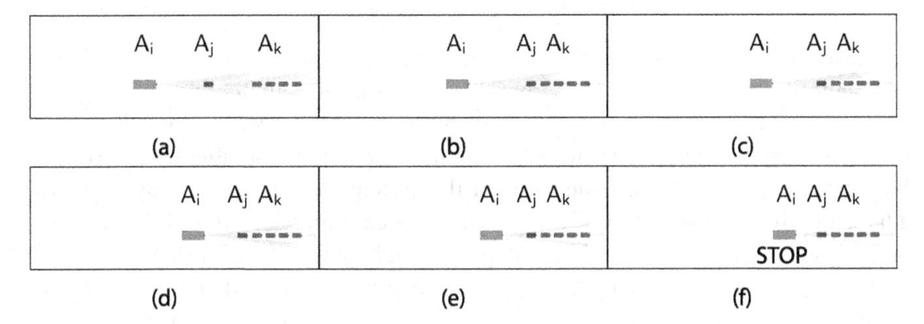

Fig. 4 Experiment 2, Test 2. A_i is able to infer d_{ik} and as a result, it avoids the collision

In the second test, the distance data d_{jk} is sent from A_j to A_i. Since, A_i has an access to d_{ij} and d_{jk}, it can infer d_{ik} with the composition Table 2. Afterwards, the movement control module presented in Fig. 3 is used twice: (a) to maintain a safe distance between A_i and A_j, and (b) to maintain a safe distance between A_i and A_k. Finally, the safer of two programs for movement control is chosen, i.e. the one that leads to smaller velocity of A_i. In other words, A_i movement control performs two tasks, namely (a) tries to avoid the collision with A_j using d_{ij} data and (b) tries to avoid the collision with A_k using the inferred d_{ik} distance. The task (b) may be also considered as a prediction of future A_j movement which enables A_i to start braking earlier and avoiding the collision. The simulation of the second test is presented in Fig. 4.

The comparison of A_i speed control in case of Test 1 and Test 2 is presented in Table 4. In the first case, i.e., without QR_M system, A_i deceleration time (293 ms) is not long enough to stop the vehicle, therefore a crash occur. On the other hand, in the Test 2, A_i begins deceleration much earlier which results in longer deceleration time (472 ms) and collision avoidance.

Table 4 Comparison of deceleration in with and without QR_M system

Experiment	A_i deceleration time	A_i beginning of deceleration
Test 1: collision avoidance without QR_M system	293 ms	945 ms
Test 2: collision avoidance with QR_M system	472 ms	663 ms

5 Conclusions and Future Work

We have presented a system for qualitative movement control called QR_M. This system involves the movement representation method presented in [12] being the first application of this approach. QR_M provides a method for qualitative values representation, namely a translation system with hysteresis. Moreover, composition tables have been introduced in order to infer new information, and a table-based controlling system. One of the main features of the system QR_M is that it may be easily adapted to different scenarios. The system has been implemented in Robotic Operating System ROS and tested with computer simulator STAGE. It has been applied to two experiments: following an object an collision avoidance in a traffic jam. The obtained results are prommising and the method can be applied to more complex scenarios.

Most of methods for robot motion planing in dynamic environment compute Velocity Obstacle (VO) [1, 15], i.e., a set of robot velocities resulting in potential collisions and select robot velocity that is outside VO. As a result, a first order method based on current position and velocity of the robot or a second order method that takes into account the current velocity and path curvature of the moving obstacle [15] are constructed. Even more complex systems are obtained by further modifications, e.g., by mean of Kalman-based observer for estimating the obstacles velocities [1]. Our method with qualitative table-based controlling is much simpler and more human-like than mentioned system for robot motion planing. Therefore, we obtain a fast and universal method that models some aspects of human spatial reasoning.

Since QR_M may be easily extended to representation and reasoning in 3D environment, as a future work, we consider experiments in 3D simulators, e.g., GAZEBO and V-REP, which are already integrated with ROS (in which QR_M is implemented). We are also interested in experiments with real robots, e.g., Lego Mindstorm or TurtleBot that may be controlled by means of ROS. Additionally, we consider extending QR_M system and introducing machine learning algorithms that autonomously determine an optimum number of qualitative values for movement description and rules in table-based controlling system.

Acknowledgments The work presented in this paper is partially supported by the Polish National Science Centre grant 2011/02/A/HS1/00395 and by the Spanish Project TIN12-39353-C04-01.

References

1. Andrea C, Boris G, Fabien S, François C (2013) Avoiding moving obstacles during visual navigation. In: 2013 IEEE international conference on robotics and automation (ICRA), pp 3069–3074. IEEE
2. Cohn AG, Renz J (2008) Qualitative spatial representation and reasoning. Handb Knowl Represent 3:551–596
3. Delafontaine M, Bogaert P, Cohn AG, Witlox F, De Maeyer P, Van de Weghe N (2011) Inferring additional knowledge from qtcn relations. Inf Sci 181(9):1573–1590
4. Escrig MT, Toledo F (2002) Qualitative velocity. In: Topics in artificial intelligence, pp 29–39. Springer
5. Forbus KD (2008) Qualitative modeling. Handb Knowl Represent 3:361–393
6. Frank AU (1992) Qualitative spatial reasoning about distances and directions in geographic space. J Vis Lang Comput 3(4):343–371. Elsevier
7. Gedig M, Stiemer S (2003) Qualitative and semi-quantitative reasoning techniques for engineering projects at conceptual stage. Electron J Struct Eng 3:67–88
8. Gerkey B, Vaughan RT, Howard A (2003) The player/stage project: tools for multi-robot and distributed sensor systems. In: Proceedings of the 11th international conference on advanced robotics, vol 1, pp 317–323
9. Khalil HK, Grizzle J (2002) Nonlinear systems, volume 3. Prentice Hall Upper Saddle River
10. Liu H, Brown DJ, Coghill GM (2008) Fuzzy qualitative robot kinematics. IEEE Trans Fuzzy Syst 16(3):808–822
11. Liu W, Li S, Renz J (2009) Combining RCC-8 with qualitative direction calculi: algorithms and complexity. In; IJCAI, pp 854–859
12. Muñoz-Velasco E, Burrieza A, Ojeda-Aciego M (2014) A logic framework for reasoning with movement based on fuzzy qualitative representation. Fuzzy Sets Syst 242:114–131
13. Quigley M, Conley K, Gerkey B, Faust J, Foote T, Leibs J, Wheeler R, Ng AY (2009) ROS: an open-source robot operating system. In: ICRA workshop on open source software, vol 3, p 5
14. Van de Weghe N, Kuijpers B, Bogaert P, De Maeyer P (2005) A qualitative trajectory calculus and the composition of its relations. In: GeoSpatial Semantics, pp 60–76. Springer
15. Zvi S, Frederic L, Sepanta S (2001) Motion planning in dynamic environments: Obstacles moving along arbitrary trajectories. In: 2001 IEEE international conference on robotics and automation (ICRA), pp 3716–3721. IEEE

A Multi-agent Framework for the Analysis of Users Behavior over Time in On-Line Social Networks

E. del Val, C. Martínez and V. Botti

Abstract The number of people using on-line social networks as a new way of communication is continually increasing. The messages that a user writes in these networks and his/her interactions with other users leave a digital trace that is recorded. In order to understand what is going on in these virtual environments, it is necessary to use systems that collect, process, and analyze the information generated. Currently, there are tools that analyze all the information related to an on-line event once the event finished or on a specific point of time (i.e., without considering the evolution of users' actions during the event). In this article, we present a multi-agent system (MAS) that deals with the analysis of the evolution of users' interactions in events on on-line social networks during a period of time. The system offers a complete vision of what is happening in an event. We evaluated its functionality through the analysis of a set of events on Twitter.

1 Introduction

The way people communicate with each other is changing [9]. Social networks such as Flickr, Linkedin, Facebook and Twitter contain millions of users and are among the most popular sites on the web. Currently, users share their thoughts, preferences, feelings, or political beliefs in on-line social networks. Each user's contribution or interaction with others leaves a digital trace. Therefore, there are vast amounts of data that can be used for research on human behavior. The analysis of this information facilitates the identification of individuals and groups that play central roles in the diffusion of information and it is useful to find opportunities to accelerate knowledge flows and to locate information.

Social networks can be considered to be dynamic processes [12, 13] where, as time passes, individuals join, leave, create, or deactivate social ties thereby altering the structure of the network [7]. However, in general, previous works that analyze on-line social networks focus on the analysis of a specific snapshot and analyze

E. del Val (✉) · C. Martínez · V. Botti
Universitat Politècnica de València, Valencia, Spain

Á. Herrero et al. (eds.), *10th International Conference on Soft Computing Models in Industrial and Environmental Applications*, Advances in Intelligent Systems and Computing 368, DOI 10.1007/978-3-319-19719-7_17

friendship relationships instead of interactions among users. The analysis of the evolution of social networks in real-time is considered a valuable source of information about activities. Monitoring the activity in social networks such as events is however a complex task for humans, due to the well known problem of information overload. The main problem is how to automate the process of monitoring user interactions with the intention of detecting users and/or groups of users behaviour.

Multi-Agent Systems (MAS) is an appropriate technology to deal with the distributed analysis of the evolution of users interactions in on-line social networks. MAS are well-suited to implement complex, distributed software systems. Agents as social computing entities are able to collaborate in the process of collecting, processing, and analyzing information in a distributed way. MAS can adapt to changes in the environment while still achieving overall system goals, dealing with the high dynamicity of Internet. Moreover, MAS provide a foundation to create an architecture that deals with the complexity reduction, flexibility, and scalability needed in the automatic analysis of events in social networks.

In this paper, we propose a MAS for the analysis of the evolution of social behavior in on-line events based on network theory. A set of temporally annotated networks are created by taking as input users interactions collected by the agents. Agents analyze the structural properties and the dynamical patterns of the temporally annotated networks. The analysis of structural properties provides information to answer the following questions: how users interact in social networks depending on the type of event; when the most important structural changes take place; and if there are structural similarities or differences in the evolution of users behavior depending on the type of event. All the information generated during the analysis is provided to the final user through a web interface. Agents that are part of the MAS operate asynchronously and concurrently, which results in computational efficiency. Moreover, new requirements or functionalities can be addressed by adding new modules or by reconfiguring the existing ones.

2 Related Work

The majority of real-world on-line social networks evolve over time and generate an enormous amount of data that is temporally annotated. Currently, a limited number of works have analyzed the evolution of social networks that emerge from users interactions and only a few of them are based on Twitter. For instance, Kumar et al. [8] presented an evolutionary analysis of structural properties of the friendship relationships in *Flickr* and *Yahoo! 360* networks. Borge-Hoelthoefer et al. [2] studied the structural and dynamical patterns of a network made up of Twitter users during the 15M social phenomenon in Spain. The authors detected that network formation was not a gradual process and that the patterns of popularity growth reflected a tendency towards a hierarchical structure. There are other works that focus on political events [3, 10, 11].

In the context of MAS, there are not many approaches that deal with the process of information retrieval from on-line social networks and its analysis. Chau et al. propose a set of *crawler* agents that retrieve user data from social networks such as eBay and a *master* that is responsible for maintaining the system consistency and collecting all the data about users [4]. A similar work is presented in [1] where the authors propose a MAS to monitor users' updates in their profiles in social networks. They consider a *grabber* agent that retrieves information from a user's profile and a *master* agent that organizes the information retrieved in a local repository. Gatti et al. [6] propose the use of MAS in order to simulate the behavior of users in a social network instead of retrieving and analyzing the information of the social network.

Our proposal presents a MAS that is responsible for collecting information from user interactions instead of user's profile or friendship relations as we have seen in previous works based on MAS. The system offers the possibility of analyzing the evolution of interaction patterns of users in events on social networks. This analysis is carried out at *group level* and at *individual level*. Moreover, this approach differs from previous works in the literature since we consider the analysis of on-line social networks as dynamic processes (i.e., an analysis of the evolution of user's interactions).

3 MAS Architecture

The approach that we present in this paper is a MAS with a web interface to facilitate the interaction with users. The general system architecture is composed by the following agents: *Manager*, *Monitor*, *Analyzer*, and *User*. The *User* agent is the intermediary between the system and the final user. The user interacts with this agent via a web interface to indicate which event or events he/she wants to start to follow. The *User* agent receives this information and sends a message to the *Manager* to start tracking an event. The *Manager* is responsible for managing the analysis of each event selected by users in online social networks. The *Manager* takes into account how many users are interested in tracking the social behavior of users in an event and the size of the events in order to create one or more *Monitor* agents. The *Monitor* agents are responsible for collecting user interactions associated to an event that occurs in social networks. These interactions are stored in a database. Periodically, if *Monitor* agents have collected new information, they notify the *Analyzer* agent to conduct an analysis of the event. The *Analyzer* agent can also perform an analysis on user demand. Once the analysis is done, all the data is sent to the *User* agent that is responsible for displaying the results of the analysis associated to the event to the user through the web interface. In order to describe with more detail the proposed MAS, we describe four dimensions (i.e., structural, functional, and dynamical) based on GORMAS methodology [5].

Structural dimension. The structural dimension describes the set of components of the system and their relationships. We have considered the Organizational Unit *SocialNetAnalyzer* (see Fig. 1 Left) where there are defined four roles: *User* (i.e., the role that acts as intermediary between the system and the user), *Manager* (i.e., the role in charge of the management of the retrieval process), *Monitor* (i.e., the role that follows the event and collects the messages of the users that participate on it), and *Analyzer* (i.e., the role that analyzes all user interactions and extract information about the users behavior in the event). In the organizational unit, there is at least one agent per role (i.e., *AnalyzerAg*, *UserAg*, *ManagerAg*, and *MonitorAg*). Depending on the number of users, these events, and the dimension of the events, new agents can be created by the *Manager* in order to deal with new requirements.

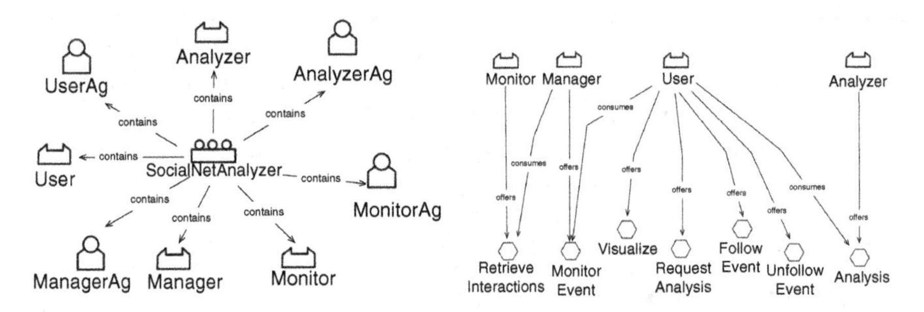

Fig. 1 Structural and functional dimension

Functional dimension. Each role defined along with the organizational unit *SocialNetAnalyzer* offers and/or consumes a set of services that defines the functionality of the system (see Fig. 1 Right). We describe the system functionality as follows:

- *RequestAnalysis*: is a service offered by *User* agents in order to obtain the information related to the analysis of the evolution of an event. This information is processed and shown to the user through a web interface.
- *FollowEvent*: *User* agents offer this service that is in charge of starting the process of analysis of the social interactions associated to an event. *User* agent is responsible of sending a message to the *Manager* agent in order to create or add the necessary *Monitor* agents.
- *UnfollowEvent*: *User* agents provide this service in order to inform the *Monitor agent* that a certain event has one follower less.
- *MonitorEvent*: *Manager* agent provides this service that consists on adding one or a set of *Monitor* agents (i.e., if the event is being followed by several users) to start collecting and storing messages associated to the event. Moreover, the *Manager* adds one or several *Analyzer* agents in order to analyze the events.
- *RetrieveInteractions*: this service is offered by *Monitor* agents and consists on the retrieval of the messages associated to an event on a specific on-line social network.

- *Analysis*: this service is responsible for the analysis of user's interactions in an event. This service is provided by agents that play the role *Analyzer*. *Analyzer* agents retrieve information associated to the event, build temporally annotated networks, and start an analysis at network level and at node level. The information from the analysis is sent to the corresponding *User* agent.
- *Visualize*: this service is provided by *User* agents and provides the visualization of the results of the analysis provided by *Analyzer* agent to the final user.

Dynamical dimension. This dimension defines the interactions and their participant roles. We consider three interaction entities: *FollowEvent* and *MonitorEvent*. Figure 2 (Left) shows the interaction entity *FollowEvent* where the roles *User*, *Monitor*, *Analyzer*, and *Manager* participate. The agent that plays the role *User* initiates the *FollowEvent* interaction unit and the agents that play the roles *Monitor* and *Manager* collaborate initiating the activities *createMonitors*, *createAnalyzers*, *RetrieveInteractions*, and *Analyze*. The execution of these activities depends on the existence of the event in the system. The *Manager* decides to create new *Monitor* and *Analyzer* agents taking into account if the number of users interested in the event, if the event is generating a high number of user interactions, and if the users decide to consider information from different online social networks. Figure 2 (Right) shows the interaction entity *RequestAnalysis* where the roles *User*, *Monitor* and *Analyzer* participate. The agent that plays the role *User* or an agent that plays the role *Monitor* can initiate the *RequestAnalysis* interaction unit. The agent that plays the role *User* can also intiate the activity *ShowResults*. The agents that play the role *Analyzer* collaborates in the interaction unit. The agents with the role *Analyzer* initiates the activities *Network Analysis*, *NodeAnalysis*, and *StatisticsAnalysis*.

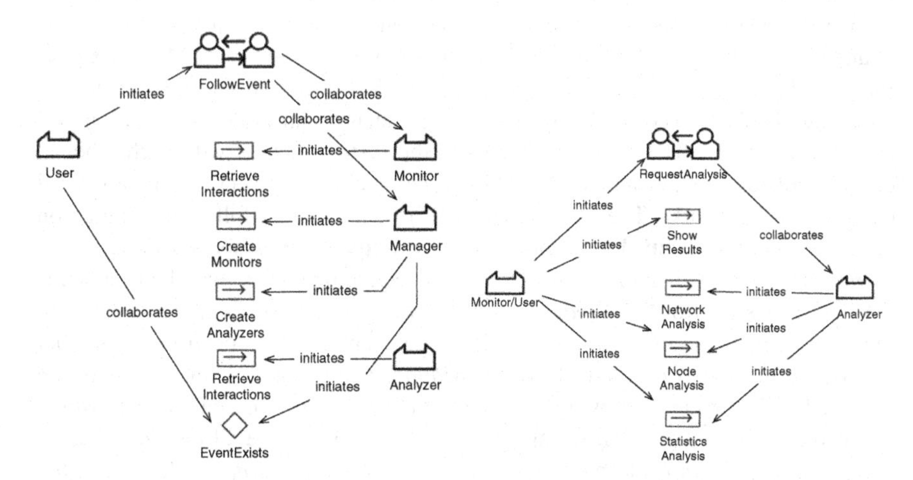

Fig. 2 Dynamical dimension: (Left) *FollowEvent* and (Right) *RequestAnalysis* interaction units

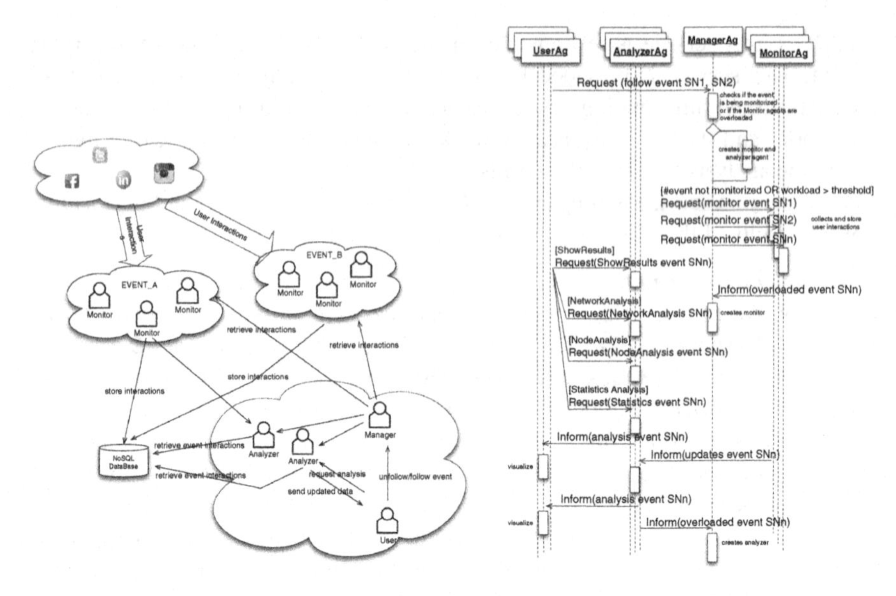

Fig. 3 (Left) MAS framework. (Right) Agent interactions in the framework

4 Framework

The aim of the *SocialNetAnalyzer* framework is the analysis of the evolution of users' interactions in events on-line (see Fig. 3). The framework consists on a set of agents that play the following roles: *User*, *Manager*, *Monitor* and *Analyzer*. The functionality of each agent is implemented as a set of independent services that can be allocated in any host, even in different ones. The agents are implemented using Magentix platform.[1] All the agents of the system are part of the *SocialNetAnalyzer* Organizational Unit. The database used by the agents for the management of the information collected from the events is Neo4J.[2] Agents in the system interact with each other in order to collect, process, and analyze what is happening in a event on one or several on-line social networks. The system dynamics are shown in the following interaction scenarios: **Follow Event**, **Unfollow Event**, and **Request Analysis** (see Fig. 3).

Follow Event. This scenario starts when the agent that represents the user sends a *Request* message to the *Manager* agent in order to start following an event. The message contains a keyword that identifies the event and the social networks that the user is interested in. First of all, the *Manager* agent checks whether there are already *Monitor* agents retrieving user interactions from the required social networks and if there are enough *Monitor* agents to collect and manage all the information. If the event does not exist in the system (i.e., there are no *Monitors* following the event), the *Manager* creates at least one *Monitor* per each social network and event

[1]http://gti-ia.upv.es/sma/tools/magentix2/index.php.

[2]http://neo4j.com/.

requested by the user and one *Analyzer* agent in order to retrieve and analyze user interactions. If the current *Monitor* agents that are in the system are overloaded by the amount of information generated during the event, the *Manager* agent will create more *Monitor* agents in order to deal with the current amount of information. Each *Monitor* agent will store the collected messages in a DB. Something similar happens with the *Analyzer* agents. If the *Analyzer* agent assocaiated to an event receives a number of analysis requestras that it is not able to manage in a reasonble time, it notifies this problem to the *Manager*. Once the agents are no longer overloaded by the number of user interactions in the event, the *Manager* will reduce the number of *Monitor*/*Analyzer* agents associated to the event.

Request Analysis. The network analysis starts for two reasons: (i) the user explicitly requests an analysis of the event he/she is interested in at a certain time; (ii) once all the monitor agents have collected a significant amount of user interactions in the different social networks, they report this to the *Analyzer* in order to update the information about the analysis of the event. In both cases, the *Analyzer* will send the test results to the *User* agent to display them via the web interface. In the case that the user requests the analysis, *Analyzer* agent offers two type of analysis: user's interactions analysis or general statistics about the information generated during the event. The analysis of users interactions, consists on the following process:

- First, the *Analyzer* agent builds a temporally annotated network for the event. This network is updated periodically with the new user interactions collected by the *Monitor* agents in order to see its evolution.
- Once the network is created, the *Analyzer* agent starts an analysis at global and at individual level. At global level, the following metrics are considered: symmetric links, degree distributions, path length, communities, and clustering among others. At individual level, centrality metrics are considered (i.e., betweenness, closeness, in-degree, out-degree, and eigenvector).
- The results associated to the analysis are sent to the *User* agent. *User* agent will show the analysis results in a web interface each time that an analysis is done.

Unfollow Event. This scenario starts when a user that was following an event decides to stop following it. Then, the *Manager* agent checks if there are more users interested in following the event, and, therefore, there are still *Monitor* agents collecting messages about the event. If there are no users interested in the event, the *Manager* agent sends to the *Monitor* and *Analyzer* agents a *Request* message to stop their activity. The *Monitor* agents will inform the *Manager* that they stopped their activity.

5 Analyzing Twitter Events

In order to validate the framework proposed, we considered a scenario where a user decided to start following a set of events in Twitter in order to analyze the evolution of users' behavior in different contexts. Initially, the *User* agent (i.e., the software entity that represents the user) sent a *Request* message to the *Manager* agent per each event the user was interested to follow. Each message contained the hashtag associated to the event and the social network that the user was interested in following. Once the *Manager* received the message, it created at least one *Monitor* agent per each event in order to collect tweets associated to them. These tweets were periodically stored in the database. During the events, the corresponding *Analyzer* studied the event in a set of snapshots. In each snapshot, the *Analyzer* did an analysis that consisted on the following steps: (i) creation of a temporal user's interaction network associated to the event until that point of time; (ii) analysis of the evolution of network properties and centrality metrics; and, (iii) statistics about the information collected from the event.

Temporal Network. The temporal network of users interactions is based on the interactions among users during a time interval. In our scenario, tweets represent user interactions. In the case of Twitter, the *Analyzer* agent classifies tweets into two categories: *global* and *individual*. *Global* tweets are used when the update is meant for anyone that cares to read it. *Individual* tweets are those that involve another user. Individual tweets can be: retweets, mentions, and replies to users. *Retweets* are messages that were previously posted by another user. *Mentions* are messages that are used when a user aims to inform about an update to a specific person. Often, two or more users will have conversations by posting mentions to each other. *Replies to users* are messages where a user mentions another user as a result of a previous message. The *Analyzer* agent builts the temporally annotated network as follows. A user *A* becomes a node of the network when he/she participates by writing a global or individual message (retweet, mention or reply to user) with the hashtag associated to the event or when another user *B* references him/her in an individual message. Each user has an associated label that represents the instant when he/she joined the network. Links of the network are established when a user writes an individual message to an existing or new user.

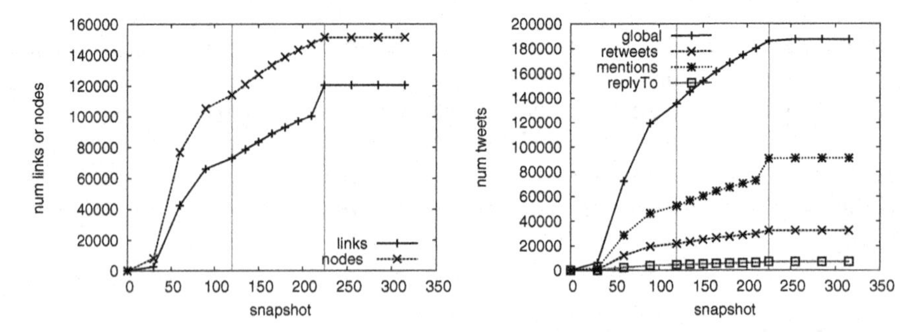

Fig. 4 Structural analysis results in a TV show event (#breakingBad). (Left) Node and edge dynamics; (Right) Tweets dynamics

Network Level Analysis. At the *network level*, the *Analyzer* agent studies the evolution of the following properties over time: (i) types of interactions (i.e., the number of the different types of messages that were generated in an event); (ii) users (the evolution of the participation in the events; (iv) links (how information flows in the network and influences the formation of the giant component); (v) symmetric links (i.e., in which moment there are conversations between users); (vi) distribution of the degree of connection to understand the topology of the network and how it changes or when it remains constant as time passes; (vii) average path length and diameter to understand in which moment there is a change in the network structure; (viii) Clustering (i.e., how the interactions among neighbors evolve).

As an example of the network level analysis, Fig. 4 shows some of the results obtained by the *Analyzer* agent related to the evolution of two structural properties in an event. The Y-axis of the graphs shows the value of the property that the agent analyzed. The X-axis shows the time in minutes before, during, and after the event. It is also possible to compare the evolution of different events. In that case, the *Analyzer* agent normalizes the data in order to facilitate the visualization of the evolution of the structural properties of several events that the user is interested in. Figure 5 shows the comparasion of several events of the same type. The Y-axis of the graphs shows the value of the property that we analyzed. Each event is represented by a set of points of certain color. To compare the results obtained in the different events, *Analyzer* agent normalizes the structural metrics and time snapshots. When the user is interested in comparing several events analyzed, the *Analyzer* agent calculates for each of the properties analyzed, the regression function that best fits the data collected from the events. We do not include all the results due to a lack of space.

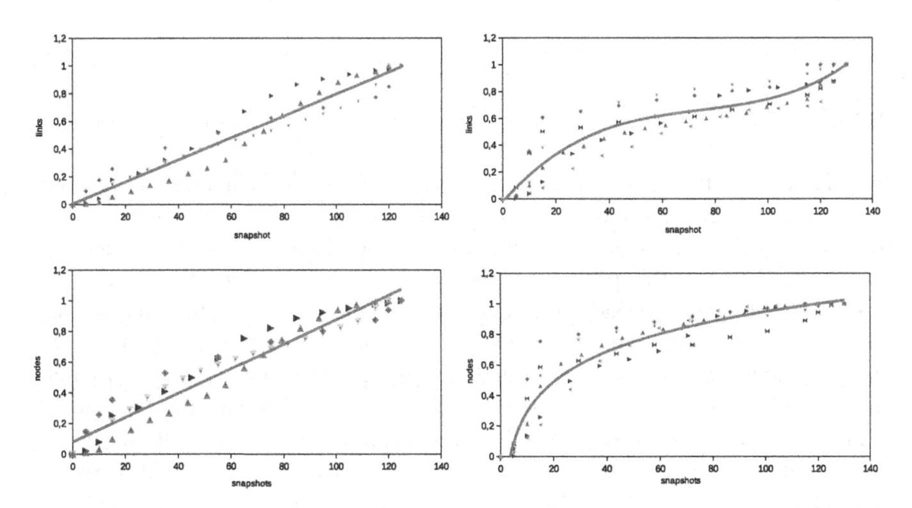

Fig. 5 Structural analysis of several events that can be classified in two types: *Socio-political* (#lomce #24O, #viacatalana, #diada2014, #LoteriadeNavidad) and *TV-shows* events (#lavoz, #topchef12, #OperacionPalace, #breakingbad, #GH, and #GameOfThrones)

Individual Level Analysis. At *individual level*, the *Analyzer* agent studies the evolution of centrality properties of individual users (network nodes) in one or several events. The centrality properties determine which users are the most important ones based on their location in the network. The *Analyzer* agent considers the following centrality metrics: betweenness (i.e., the number of times a user acted as bridge/broker through the shortest path between two other users); in-degree (i.e., indicates whether or not the user was meaningful for other users); out-degree (i.e., indicates the activity of the user); and, eigenvector (i.e., if the users around a node A are influential, it makes A more influential also). We do not include all the results due to lack of space. As an example of the individual level analysis, Fig. 6 shows some of the results obtained by the *Analyzer* agent related to the evolution of betweenness of the users with the highest values.

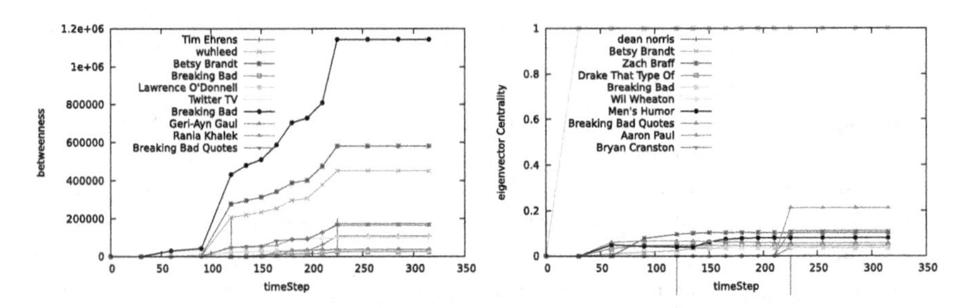

Fig. 6 Structural analysis results in a TV show event (#breakingBad). (Left) Betweenness dynamics; (Right) Eigenvector dynamics

6 Conclusions

In this paper, we have presented a Multi-Agent System for the analysis of the evolution of user's behavior in on-line events. The system is responsible for collecting, processing, and analyzing information retrieved from the activities of the users. The main contribution of this proposal is the consideration of a period of time in the analysis of the events instead of a single point. Moreover, the system offers a complete vision of what is happening taking into account the structural properties of the temporally annotated network generated from users interactions. The system offers an analysis at a global level (i.e., offers a view of the global behavior of users in the event) and at individual level (i.e., which users are playing a key role in the event). The system was tested in several types of events in Twitter in order to observe differences in users behavior.

Acknowledgments This work is supported by the Spanish government grants CONSOLIDER INGENIO 2010 CSD2007-00022, MINECO/FEDER TIN2012-36586-C03-01, TIN2011-27652-C03-01, and SP2014800.

References

1. Abdulrahman R, Neagu D, Holton D (2011) Multi agent system for historical information retrieval from online social networks. Agent Multi-Agent Syst Technol Appl 6682:54–63
2. Borge-Holthoefer J, Rivero A, García I, Cauhé E, Ferrer A, Ferrer D, Francos D, Iñiguez D, Pérez MP, Ruiz G et al (2011) Structural and dynamical patterns on online social networks: the spanish may 15th movement as a case study. PLoS One 6(8)
3. Borondo J, Morales AJ, Losada JC, Benito RM (2013) Characterizing and modeling an electoral campaign in the context of twitter: 2011 spanish presidential election as a case study
4. Chau DH, Pandit S, Wang S, Faloutsos C (2007) Parallel crawling for online social networks. In: Proceedings of the 16th WWW, pp 1283–1284. ACM
5. Esparcia S, Argente E, Julián V, Botti V (2014) Gormas: a methodological guideline for organizational-oriented open mas. In: Handbook on agent-oriented design processes, pp 173–218
6. Gatti M, Cavalin P, Neto SB, Pinhanez C, dos Santos C, Gribel D, Appel AP (2014) Large-scale multi-agent-based modeling and simulation of microblogging-based online social network. In: Multi-agent-based simulation XIV, pp 17–33
7. Kossinets G, Watts D (2006) Empirical analysis of an evolving social network. Science 311(5757):88–90
8. Kumar R, Novak J, Tomkins A (2010) Structure and evolution of online social networks. In: Link mining: models, algorithms, and applications, pp 337–357. Springer
9. Licoppe C, Smoreda Z (2005) Are social networks technologically embedded?: How networks are changing today with changes in communication technology. Soc Netw 27(4):317–335
10. Lotan G, Graeff E, Ananny M, Gaffney D, Pearce I, Boyd D (2011) The revolutions were tweeted: information flows during the 2011 tunisian and egyptian revolutions. Int J Commun 5:1375–1405
11. Morales A, Losada J, Benito R (2011) Structure and dynamics of emerging social networks from twitter's conversation. Int J Complex Syst Sci 1(2):216–220
12. Stockman FN, Doreian P (1997) Evolution of social networks: processes and principles. In: Evolution of social networks, pp 233–250. Routledge
13. Wasserman S, Faust K (1994) Social network analysis: methods and applications. Cambridge University Press

An Emotional-Based Hybrid Application for Human-Agent Societies

J.A. Rincon, V. Julian and C. Carrascosa

Abstract The purpose of this paper is to present an emotional-based application for human-agent societies. This kind of applications are those where virtual agents and humans coexist and interact transparently into a fully integrated environment. Specifically, the paper presents an application where humans are immersed into a system that extracts and analyzes the emotional states of a human group trying to maximize the welfare of that humans by playing the most appropriate music in every moment. This system can be used not only online, calculating the emotional reaction of people in the bar to a new song, but also in simulation, to predict the people's reaction to changes in music or in the bar layout.

1 Introduction

Ubiquitous Computing and Ambient Intelligence (AmI) [1, 2] have changed the concept of smart homes, introducing devices that improve the quality of life of people. Among other applications, we can highlight smart devices that learn our tastes, smart homes that help reducing energy consumption [3], safer homes for elderly [4]. To achieve this, developers may use different Artificial Intelligence (AI) tools, sensor networks, and new and sophisticated embedded devices.

Agent technology allows the development of systems that support the requirements of AmI applications. Concretely it allows the development of systems where

---

J.A. Rincon (✉) · V. Julian · C. Carrascosa
Departamento de Sistemas Informáticos Y Computación (DSIC), Universitat Politècnica de València, Camino de Vera S/n, Valencia, Spain
e-mail: jrincon@dsic.upv.es

V. Julian
e-mail: vinglada@dsic.upv.es

C. Carrascosa
e-mail: carrasco@dsic.upv.es

Á. Herrero et al. (eds.), *10th International Conference on Soft Computing Models in Industrial and Environmental Applications*, Advances in Intelligent Systems and Computing 368, DOI 10.1007/978-3-319-19719-7_18

the main components can be humans along with software agents providing complex services to humans in an environment of whole integration. This kind of applications are what we call a *Human-Agent Society* [5], which can be defined as a new computing paradigm where the traditional notion of application disappears. This new paradigm is based on an immersion of the users in a complex environment that enables computation. The main challenge to achieve real human-agent societies lies in the design and construction of intelligent environments, in which humans interact with autonomous, intelligent entities, through different input and output devices. This means that there are two layers in which humans interact within the environmental and ubiquitous computing intelligence. The first layer is the real world where the human being interacts with other humans and with real objects. The second layer is a virtual layer in which humans interact with virtual entities and objects. This latter layer will be inhabited by intelligent entities (agents), which must be able to perform the different human orders. These virtual environments where agents are involved, are known in the literature as intelligent virtual environments or *IVE*. An *IVE* [6], is a 3D space that provides the user with a collaboration, simulation and interaction with software entities, so he can experience a high immersion level. This immersion is achieved through detailed graphics, realistic physics, AI techniques and a set of devices that obtain information from the real world. As an example, the *JaCalIV E* framework enables the design, programming and deployment of systems of this kind.

As we can think, working with humans is hard and complex. Humans use their emotions in their decision making. Human beings manage themselves in different environments, either in the working place, at home or in public places. At each one of these places we perceive a wide range of stimuli, that interfere in our commodity levels modifying our emotional levels. These variations in our emotional states could be used as information useful for machines. Nevertheless, it is needed that the machines will have the capability of interpreting or recognizing such variations. This is the reason for implementing emotional models that interpret or represent the different emotions. Our proposal is to employ the emotional state of agents (humans or not) in an AmI application. Concretely, we propose in this paper a multi-agent system for controlling automatically the music which is playing in a bar. The main goal of the proposed system is to play music making that all individuals within the bar are mostly as happy as possible. In order to achieve this, an hybrid approach is proposed, where different AI techniques are used. Specifically, we use different algorithms for capturing the environment, for detecting faces and music, for converting inputs into emotional values and for analyzing the decision making of the entities. The application has been developed using *JaCalIV E* framework [7], which is a framework for the design and simulation of intelligent virtual environments (IVEs). This framework differs from other works in the sense that it integrates the concepts of agents, humans, artifacts and physical simulation. The main reason to employ *JaCalIV E* is that allows an easy integration of human beings in the system. This article presents a new way of interaction between humans and intelligent autonomous entities living within intelligent virtual environments using emotions as a form of communication. To do this a number of AI tools were used, thus creating a hybrid application which is composed of multi-agent system, machine learning and statistical classification. All this integration of different tool IA, are tested in an example which is explained in the last section of this article. The rest of the paper is organized as follows. Section 2 presents the main characteristics of the *JaCalIV E* framework. Section 3 explains the proposes AmI application. Finally, some conclusions are presented in Sect. 4.

2 Emotional Models

Over the last few years different approaches have been proposed in order to define the emotional state of an entity. Among them, the best known are the *OCC* and *PAD* emotional models. The *OCC* model designed by Ortony, Clore & Collins is a model frequently used in applications where an emotional state can be simulated. This has allowed to create applications to emulate emotions in virtual humans [8] and to create agents reacting to stress situations [9]. The *OCC* model specifies 22 emotional categories, which are divided into five processes. These processes define the whole system, where the emotional states represent the way of perceiving our environment (objects, persons, places) and, at the same time, influencing in our behavior positively or negatively [10]. However, the *OCC* model utilization presents one complication due mainly to its high dimensionality.

On the other hand, the *PAD* is a simplified model of the *OCC* model. This model allows to represent the different emotional states using three values. These three values are usually normalized in $[-1, 1]$, and correspond to the three components conforming the emotional model (*Pleasure, Arousal, Dominance*). These components can be represented in a $\mathbb{R}^3$ space. Each one of the components conforming the *PAD* model allow to influentiate the emotional state of an individual in a positive or negative way. This influence evaluates the emotional predisposition of such individual, modifying in this way his emotional state. For instance, if the *Pleasure* parameter is modified so that its value is positive (the same can be achieve modifying its opposite *Displeasure* with negative values), then such an individual would tend to prefer pleasant stimuli than unpleasant ones. In the same way, if the *Arousal* parameter is modified so that its value is positive (or *Calm* is modified to have negative values) then the excitation levels and the duration of it will change. Last, the *Dominance* parameter can be modified (or its opposite *Submissiveness*) to indicate the usual inclination to feel in control facing different situations, and the relation existing with feeling controlled in different circumstances.

The existing emotional models are usually employed in applications where multi-agent systems (MAS) are involved, allowing the simulation of human emotions for a lonely entity. Nevertheless, it is not taken into account how these emotions can be detected/captured from humans in a non-invasive way. The need of detecting the emotions of an heterogeneous group of entities can be reflected in the different applications that could be developed. This aspect has an increased interest with the appearance of new smart devices that simplifies that detection. Emotional states turn into valuable information, allowing to develop applications that help to improve the human being life quality.

3 Problem Description

The developed application is thought to be useful for the DJ of a bar or pub along with the managers of such place. The idea behind it is that music may change the emotional state of a person or a group, as it influences in the human mood. But not any kind of music, but music that matches people interests. Concretely the example has been developed in a bar, where there is a DJ in charge of playing music and a specific number of persons listening that music. The main goal of the DJ is to play music making that all the people within the bar are mostly as happy as possible.

In an specific moment, each one of the persons placed in the bar will have an emotional response according to his/her musical taste. That is, depending on the musical genre of the song, people will respond varying their emotional states. Moreover, varying emotions of each person will modify the social emotion of the group. If the DJ could have a way to evaluate the emotional state of the people which is in the bar, he could know the effect that the songs have over the audience. This will help the DJ to decide whether to continue with the same musical genre or not in order to improve the emotional state of the group.

In such a way, the proposed application seeks to identify the different emotional states using them as a tool of communication between humans and agents. To perform this detection we need to use pattern recognition algorithms and image and audio processing techniques in order to detect and classify the different emotional states of humans. Moreover, once we have these models of the people that tend to go to this place, we can be able to make simulations of what would happen in a DJ session with a similar public in the place. It will also be possible to make different kinds of simulations scaling the attending public, changing the layout of the place (where the bars are placed, where are the speakers and the DJ located, or even changing the dimensions of the place).

4 Application Design

In the last years, there have been different approaches for using MAS as a paradigm for modelling and engineering IVEs, but they have some open issues: low generality and then reusability; weak support for handling full open and dynamic environments where objects are dynamically created and destroyed. As a way to tackle these open issues, and based on the MAM5 meta-model [11], the *JaCalIV E* framework was developed [7]. It provides a method to develop this kind of applications along with a supporting platform to execute them.

The presented work has used an extension of both the MAM5 meta-model along with the *JaCalIV E* framework to develop Human-Agent Societies, that is, to include the human in the loop. MAM5 allows to design an IVE as a set of entities that can be virtually situated or not. These entities are grouped inside Workspaces (*IVE Workspaces* in the case of virtually situated). Entities that are divided into Artifacts (*IVE Artifacts* if situated) and Agents (*Inhabitant Agents* if situated). One new type of situated agents, that is, of *Inhabitant Agents* are *Human-Inmersed Agents*, that model the human inside the system.

The application has been developed as a virtual multi-agent system using the *JaCalIV E* framework where there will be different entities. Each one of these entities may represent not only real or simulated human beings, and the DJ agent, but also the furniture or the speakers with their location (Fig. 1). The main tasks of the different type of agents are:

- DJ agent: the main goal of this agent is to achieve a emotional state of happiness for all of the people which are in the bar. When the DJ agent plays a song, it must analyze the emotional state of people. According to this analysis it will select the most appropriated songs in order to improve, if possible, the current emotional state of the audience.
- Human-immersed agent: it is in charge of detecting and calculating the emotional state of an individual which is in the bar, sending this information to the DJ agent. In order to accomplish its tasks, this agent must have access to a variety of input/output information devices as cameras, microphones, ...

In order to facilitate the access to this kind of devices, they have been modeled as artifacts (as can be seen in Fig. 1).[1] Concretely, there has been designed an artifact for managing each camera which allow the face detection; each microphone is managed by an artifact which captures the ambient sound in order to classify the music genre; the music DB has been designed as an artifact employed by the DJ agent (it stores around 1.000 songs classified by genres) and there is an artifact for controlling the multimedia player and the amplifiers for playing songs in the bar with the appropriated volume.

[1]In this figure and in the following ones, we are using data from a simple example with only 3 Human-Immersed Agents.

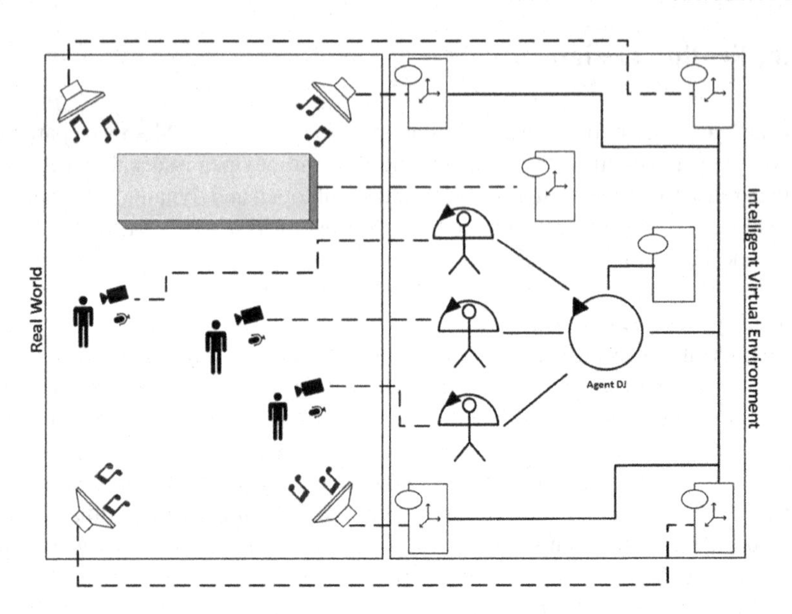

Fig. 1 General scheme of the application

Each one of these entities has been designed using *JaCalIV E* through an XML file describing all its different properties (including physical ones). The XML file allows to describe if you need some kind of sensor to capture information, some type of actuator or simply an agent that does not need real-world information. It is also in this XML where humans are associated with each agent. These XML files are automatically translated into code templates using the *JaCalIV E* framework.

5 Implementation

The proposed application includes two different types of agents, as above commented: the DJ Agent and the Human-Inmersed Agent. Due to space limitations, this paper only details the Human-Immersed Agent. Regarding the DJ Agent, this agent uses the information sent by the Human-Immersed Agent to analyze the group's emotional states, using it to decide which is the following song to play. The main goal of the DJ Agent is that all the humans feel as happy as possible. To achieve this, each of the Human-Inmersed Agent will communicate their emotional state. It is necessary to provide each Human-Inmersed Agent with a series of tools so that the DJ agent would be able to know the emotional state of each human. Those tools will help Human-Immersed agents to perceive the real environment, to interpret human it has associated and be able to classify the different emotional states this human expresses. So, each one of these agents contains both audio processing and image recognition algorithms. In order to achieve the detection of emotional states, each one of the Human-Inmersed Agents needs to perform a serie of 4 processes in order to recognize the human and his/her emotional state.

The first process is responsible for capturing information from the real world. This information is obtained by the Human-Immersed agent using an *Asus Xtion* and microphone. The second process is responsible of extracting the most relevant information, using the different images and ambient sounds. This information allows us to make two steps, the first one is a face detection using the Viola Jones algorithm [12]. The Human-Immersed Agent uses these images to classify the different humans and so identify the human associated to him. The images detected and extracted by the Viola Jones algorithm are resized to $92x112$. Each one of these matrices is transformed to a uni-dimensional vector. This vector is formed by concatenating the rows of the image matrix, so for a $92x112$ image it will have 10304 elements. All the image vectors are grouped forming a matrix (each vector is one row) of NxM, where N represent the number of images and M is the maximum amount of pixels of the initial images. Clearly this new matrix is very big and the number of values is more than is required (Fig. 2).

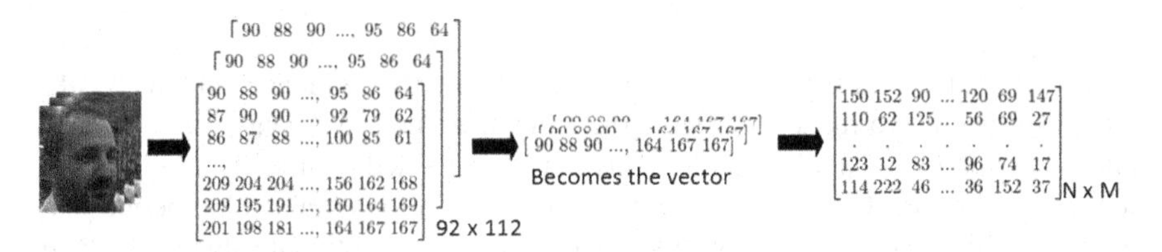

Fig. 2 Processed image

Once the system have the *NxM* matrix with all the images of the data set, it is necessary to reduce dimensionality. The Principal Component Analysis algorithm (*PCA*) allows us to perform this task, reducing the dimensionality as can be checked in Table 1.

Table 1 Comparison between applying or not PCA to image processing

	n-Image (N)	Pixels (M)
Without PCA	30	10304
With PCA	30	2

This new matrix is used with the K-Means algorithm to partition it into k groups (Fig. 3). This is a fast method to determine if the face seen by the agent is the same which has been associated to him. Nevertheless, this does not mean you can not use other more complex face recognition algorithms such as using support vector machines [13], or neural networks [14], among others.

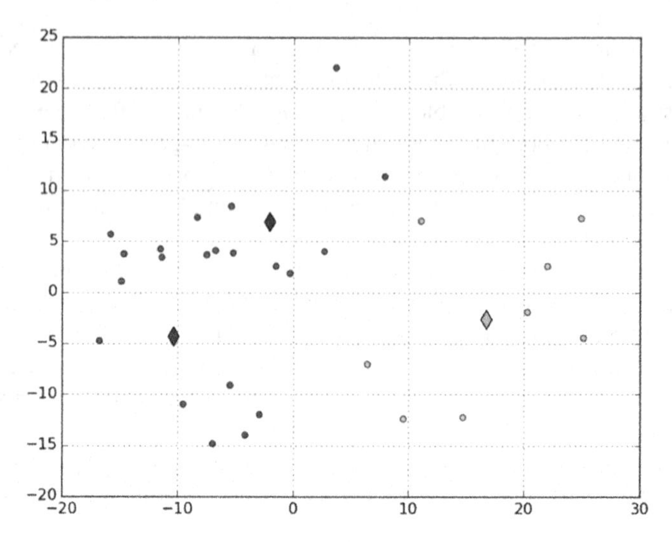

Fig. 3 Face clusters identified in the example of the application

The second step of the second process is to capture the ambient sound to classify the music genre. To make it possible the Human-Immersed Agent uses its microphone to capture the ambient sound. This step is explained in more detail in the following subsection. The third process uses the information obtained by the musical genre classifier to obtain the new emotional value. This emotional value is obtained using a fuzzy logic algorithm, which returns three values corresponding to the *PAD* model [15, 16]. To obtain these values, it is necessary to know how the different musical genre influence on the human. This information allows to modify the membership function of the fuzzy logic algorithm. The variation of this membership function depends on the corresponding human musical preferences, e.g. a human can respond favorably to pop music but not to the blues music. This is the reason why it is needed that each human configure its Human-Immersed Agent before using the system, varying the membership function of the fuzzy logic module. And finally, the fourth process is the responsible of communicating the emotional state to all entities and especially to the DJ agent.

Figure 4 left shows our bar distribution and how is the virtual representation of the different entities representing a human that lives in the real world. To build it we used Unity3D[2] to create the virtual representation, and used free 3D models for the bar and humans representation.[3] Figure 4 right shows a graphical representation in the PAD space of the individual emotions of the agents which are in the bar. In this case, we can see that all the agents have a similar emotional state.

[2]http://unity3d.com/.

[3]http://tf3dm.com/3d-model/vega-strike-starship-bar-economy-class-88446.html,http://tf3dm.com/3d-model/alexia-89488.html,http://tf3dm.com/3d-model/dante-33087.html,http://tf3dm.com/3d-model/girl-44203.html,http://tf3dm.com/3d-model/girl-44203.html.

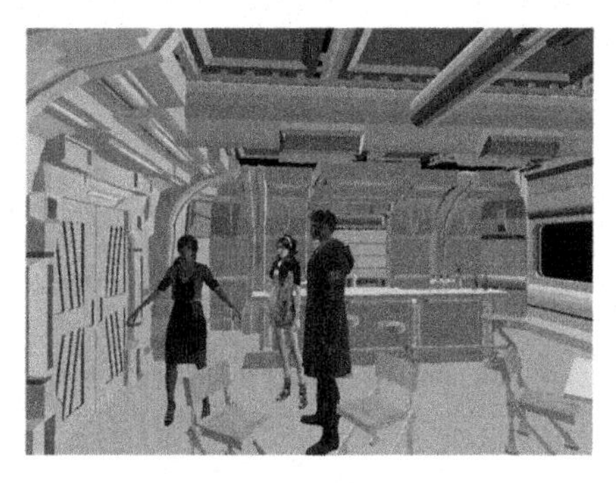

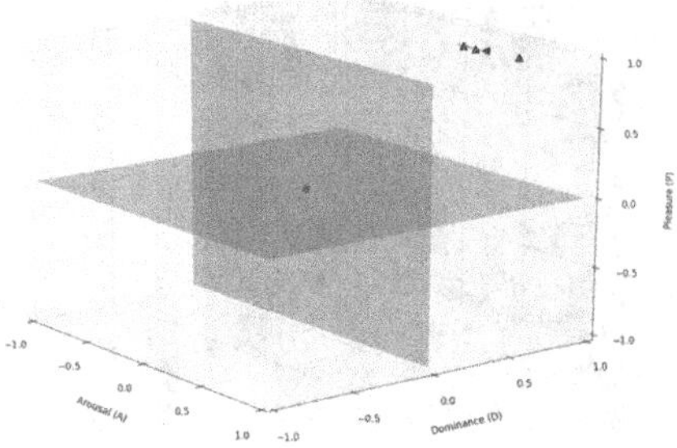

Fig. 4 Virtual bar and emotion representation

5.1 Music Classification

The music classification is made for the Human-Immersed agent, using a statical classifier [17–19]. This classifier allows the agent to recognize the music genre. This is done training the classifier with the musical preferences of the associated human. This learning capability allows the agent to recognize if the song that is playing by the agent DJ, corresponds to his/her musical tastes. At this moment, the classifier only uses six musical classes, corresponding each class with a different genre music (classical music, jazz, country, pop, rock and techno music). In order to classify the music genre, the classifier needs to make two mandatory steps.

In the first step, the Human-Immersed agent captures some information from the environment. Concretely, it is necessary to get the song played by the DJ agent. This can be done using a microphone. This microphone captures the environment sound which is saved as a *.wav* file. Once the file is saved, the spectrum is calculated. In order to obtain the spectrum we use the previously saved *.wav* file applying a band pass butterworth filter. This filter allows us to discriminate some frequencies that people don't listen. Once the sound is filtered, it is necessary to apply a sub-sampling algorithm. This algorithm allows us to reduce the sampling frecuency from 44 KHz (12.769.920 values) to 5.5 KHz (1.596.240 values).

The second step consists of getting the feature vectors for all the songs. This feature vectors allows learning the gaussian model for each genre. This process is mandatory in order to to train our model. Once we have trained the model, the system calculates the unnormalized average of the likelihood. This is done for each test song and genre model. After this, the following process consists of evaluating the model. To do this, it is necessary to create a confusion matrix. This matrix allows the representation of the prediction number for each one of the classes and the calculation of the classification accuracy. Figure 5 shows an example of the confusion matrix, where the diagonal represents the classification. This classification shows that the genre classified by the Human-Immersed agent is, in this example, techno music, as it has the yellow color.

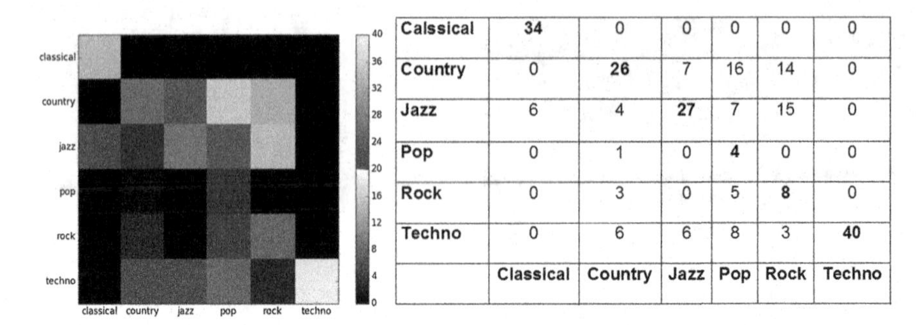

Calssical	**34**	0	0	0	0	0
Country	0	**26**	7	16	14	0
Jazz	6	4	**27**	7	15	0
Pop	0	1	0	**4**	0	0
Rock	0	3	0	5	**8**	0
Techno	0	6	6	8	3	**40**
	Classical	Country	Jazz	Pop	Rock	Techno

Fig. 5 Numerical and graphical confusion matrix

6 Conclusions and Future Work

Multi-agent systems allow the design and implementation of applications where the main components can be humans and software agents interact and communicate with humans in order to help them in their daily activities. In this sense, this paper presents an ambient intelligence application where humans and agents must coexist in a framework of maximum integration. The application has been developed over the *JaCalIV E* framework allowing an easy integration of the human in the multi-agent system and a visualization of the system in a virtual environment. The proposed system is able to extract (in a non-invasive way) and to analyze the social emotion of a group of persons and it can take decisions according to that emotional state. Moreover, it can be used in an online fashion, where the system is reflecting what is happening in the real world at the same time, or it can be used to simulate - predict what would happen in the conditions that are used in the simulation, where this conditions can be about the number of agents and its preferences and also about the layout of the environment. Moreover, as a future work we are applying this system to other application domains, as it can be extracted out of the music domain, and carried out to any other ambient intelligence domain, even an industrial one, where it can monitor and simulate the conditions inside a factory.

References

1. Satyanarayanan M (2002) A catalyst for mobile and ubiquitous computing. IEEE Pervasive Comput 1(1):2–5
2. Mangina E, Carbo J, Molina JM (2009) Agent-based ubiquitous computing. Atlantis Press, World Scientific, Amsterdam; Paris
3. Han D-M, Lim J-H Smart home energy management system using IEEE 802.15. 4 and zigbee. IEEE Trans Consum Electron 56(3):1403–1410
4. Intille SS (2002) Designing a home of the future. IEEE Pervasive comput 1(2):76–82
5. Billhardt H, Julián V, Corchado JM, Fernández A (2014) An architecture proposal for human-agent societies. In: Highlights of practical applications of heterogeneous multi-agent systems,

vol 430, pp 344–357. Springer
6. Hale KS, Stanney KM (2002) Handbook of virtual environments: design, implementation, and applications. Human Factors and Ergonomics. Taylor & Francis
7. Rincon JA, Garcia E, Julian V, Carrascosa C (2014) Developing adaptive agents situated in intelligent virtual environments. In: Hybrid artificial intelligence systems, number 8480 in LNCS, pp 98–109. Springer
8. Becker-Asano C, Wachsmuth I (2010) Affective computing with primary and secondary emotions in a virtual human. Auton Agent Multi Agent Syst 20(1):32–49
9. Jain D, Kobti Z (2011) Simulating the effect of emotional stress on task performance using OCC. In: Advances in Artificial Intelligence, pp 204–209. Springer
10. Ali F, Amin M (2013) The influence of physical environment on emotions, customer satisfaction and behavioural intentions in chinese resort hotel industry. In: KMITL-AGBA Conference Bangkok, pp 15–17
11. Barella A, Ricci A, Boissier O, Carrascosa C (2012) MAM5: Multi-agent model for intelligent virtual environments. In: 10th european workshop on multi-agent systems (EUMAS 2012), pp 16–30
12. Viola P, Jones MJ (2004) Robust real-time face detection. Int J comput vision 57(2):137–154
13. Osuna E, Freund R, Girosi F (1997) Training support vector machines: an application to face detection. In: Computer vision and pattern recognition, 1997. Proceedings., 1997 IEEE computer society conference on, pp 130–136. IEEE
14. Lawrence S, Giles CL, Tsoi AC, Back AD (1997) Face recognition: a convolutional neural-network approach. IEEE Trans. Neural Netw 8(1):98–113
15. Mehrabian A (1997) Analysis of affiliation-related traits in terms of the PAD temperament model. J Psychol 131(1):101–117
16. Nanty A, Gelin R (2013) Fuzzy controlled PAD emotional state of a NAO robot. In: 2013 conference on technologies and applications of artificial intelligence (TAAI), pp 90–96
17. Richert W, Coelho LP (2013) Building Machine Learning Systems with Python. Packt Publishing, Birmingham
18. Holzapfel A, Stylianou Y (2007) A statistical approach to musical genre classification using non-negative matrix factorization. In: Acoustics, speech and signal processing, 2007. ICASSP 2007. IEEE international conference on, vol 2, pp II-693. IEEE
19. Tzanetakis G, Cook P (2002) Musical genre classification of audio signals. IEEE Trans Speech Audio Process 10(5):293–302

Modeling and Analysis of Agent-Based Labor Market

Jae-Min Yu and Sung-Bae Cho

Abstract This paper presents an agent-based model of labor market to investigate the relationship between company and worker. Unlike most of previous studies of labor market we apply a game theoretic approach to defining entities in labor market: companies and workers. A company can choose the level of wages, and workers can select the level of effort to increase the productivity in response to the wages. Company and worker agents are designed to possess the basic attributes in order to reflect the real labor market and their activities are adaptively changed using evolutionary model. Our approach is illustrated with four simulation results: the effect of workers resignation, sick leave, dismissal of companies and productivity growth. Various experiments were conducted, and the interactions between worker and company are analyzed. Especially performance-based reward strategy and non-greedy strategy in job changing are necessary for companies and workers according to the simulation results. The experimental result confirms that the balanced power between worker and company is important in maintenance and extension of labor market. Nash Equilibrium can be maintained in all cases.

1 Introduction

In a real-world labor market, the behavioral characteristics expressed by worksite individuals, diligence or lazy, depend on who is working for whom [1, 2]. A behavior pattern of each individual may heavily affect the state of the labor

J.-M. Yu · S.-B. Cho (✉)
Department of Computer Science, Yonsei University, Seoul, Korea
e-mail: sbcho@yonsei.ac.kr

J.-M. Yu
e-mail: yjam@yonsei.ac.kr

Á. Herrero et al. (eds.), *10th International Conference on Soft Computing Models in Industrial and Environmental Applications*, Advances in Intelligent Systems and Computing 368, DOI 10.1007/978-3-319-19719-7_19

market such as stability or productivity. There have been many studies on the analysis of the behavioral patterns of the individuals and estimating future in economic or social phenomena by using agent-based computational models [3–6].

This paper investigates the possible interactions between workers and companies, and constructs a framework for modeling such strategic interactions. In this framework, there are two types of agent, worker and company. Workers and companies, who are principal members of labor market, act as agent having one's strategies, status, attributes, and actions. The relationship of workers and companies can be simplified as cooperation and defection. Such characteristic behavior of each agent is decided by strategy held by the agent. Base on the evolutionary labor market framework, we discuss the variation of mutual interaction occurred by workers and companies. This paper uses game theory to define formalized interaction between workers and companies. The game theory has been used widely to model multi-agent environment such as social and economic phenomena, where the primary purpose is not to model a dynamic system, but to study how co-evolution can be used in learning strategies for agents [7]. The rest of this paper is organized as follows. Section 2 presents related works for agent-based modeling. Section 3 describes in detail the framework of labor market proposed. Section 4 reports the experiments were designed. The final section presents a conclusion.

2 Related Works

There are various related works using agent-based modeling. Mostly, evolutionary computation focuses on the prediction of future or is used to analyze social phenomena. Table 1 is about previous studies on agent-based modeling. Maniadakis

Table 1 Previous studies on agent-based modeling

Researchers	Topic	Agent modeling	Game theory	Evolutionary computation
H. Quek and K. Tan [5]	Macroscopic behavioral dynamics of civil violence	O	O	O
R. Grass et al. [8]	Behavioral pattern of predator-prey (Ecosystem)	O	X	O
M. Maniadakis et al. [9]	Mechanism of human brain working	O	X	O
H.T. Kim and S.B. Cho [10]	Relationship between companies and workers (Labor market)	O	O	O

introduced a novel agent-based co-evolutionary computational framework for modeling assemblies of brain areas. Quek focused on the design and development of a spatial Evolutionary Multi-Agent Social (EMAS) network to investigate the underlying emergent macroscopic behavioral dynamics of civil violence. Also, Grass investigated the complexity level of agent-based predator/prey ecosystem simulation.

3 Labor Market Model

All agents coexist in the artificial market with its own behavioral strategy determined by the current status and circumstance, and interact with each other repeatedly. Once a worker has been employed in a company, the worker provides labor for wages until resignation from the company or elimination from the system. The amount of wages and earnings is decided according to behavioral tactics of both companies and workers agents. Overall concept of the proposed labor market is depicted in Fig. 1.

Each agent decides its current behavior, cooperation or defection, depending on their own strategies encoded the chromosome, and evolves its worksite strategies over time on the basis of its assets earned by past worksite interactions. The worksite strategies of workers and companies are mutually evolved by means of co-evolutionary system and standard genetic algorithms including reproduction, crossover, mutation, and selection that aims to select more capable agents to survive in the labor market framework [4].

3.1 Structure of Agent Model

In this section, we present some basic features that characterize our simulation model. The model uses discrete-time simulation, and the state of the model is updated only at discrete time intervals. Every agent can perform one behavior at each time. Accurate modeling of agents is crucial for a close-life depiction of human behavior [10]. We adopt multi-agents approach in order to capture the dynamics in the problem. Here, we specify two agents, workers and companies. In each generation, all agents decide their actions based on their evolving strategy and recent cooperation history of opponent worksite partners. We assume that all actions performed by opponent agent can be completely observed.

As generation is repeated, incompetent agents can appear in labor market. By the concept of evolutionary computation, highly capable agent can survive and reproduce child agents who may have high fitness. On the other hand, incompetent agents who possess low fitness should be eliminated from system. In our labor market framework, fitness is calculated based on the only *asset* or *capital* attribute.

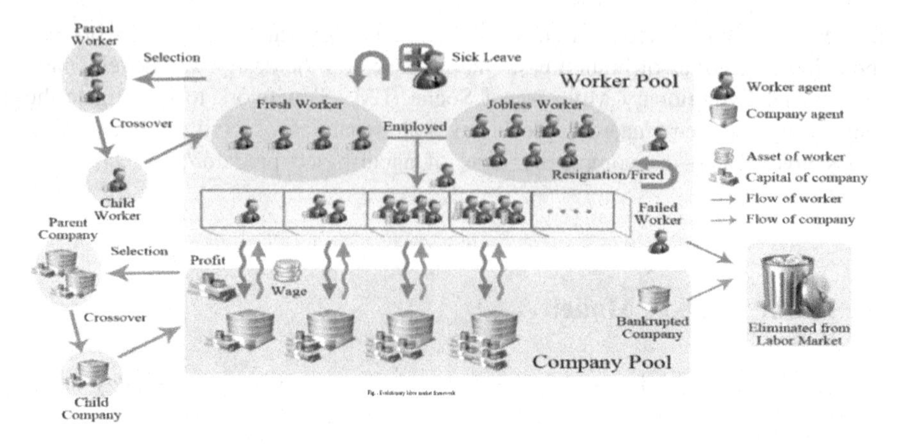

Fig. 1 Evolutioary labor market framework

3.2 Agent Behavior Decision Based on Game Approach

This paper uses pay off table that is slightly different from the traditional IPD game. To satisfy the regularity condition of usual prisoner's dilemma game, Nash equilibrium [11] must be mutual defection, and mutual cooperation dominates alternating cooperation and defection on average, but the proposed payoff table cannot guarantee the requirements [12]. The structure of payoff table represented in Table 2 is constructed by the following steps.

(1) Worker's phase: The worker can decide whether to cooperate or defect with his opponent company. All workers have own productivity attribute and the company gains payoff from the work's productivity. A defector worker produces low productivity ($Prod_L$), whereas a cooperator worker produces high productivity ($Prod_H$).
(2) Company's phase: After a company receives payoff from all of its workers, the company has to pay wage for the workers. A wage is accounted to payoff of each worker. In this situation, if a company cooperates to its opponent worker, the worker will receive high wage ($Wage_H$). Otherwise, the worker will receive low wage ($Wage_L$).

Table 2 Payoff table between worker and company

(Worker, Company)		*Company*	
		Cooperation	Defection
Worker	Cooperation	$(Wage_H, Prod_H - Wage_H)$	$(Wage_L, Prod_H - Wage_L)$
	Defection	$(Wage_H, Prod_L - Wage_H)$	$(Wage_L, Prod_L - Wage_L)$

In case of using the payoff table described in Table 2, 'defection' is selected as dominant strategy by company, whereas worker can select both cooperation and defection as dominant strategy. Hence the Nash equilibrium is (cooperation, defection) or (defection, defection), and each agent's strategy will mostly stay on Nash equilibrium. Although game is infinitely iterated because worker does not need to do cooperative behavior. Worker's payoff is completely determined by company's choice regardless of worker's choice. Obviously, it seems unfair from worker's point of view. Therefore, leading the cooperative behavior of companies, or driving mutual cooperation is an important issue of this paper (Fig. 2).

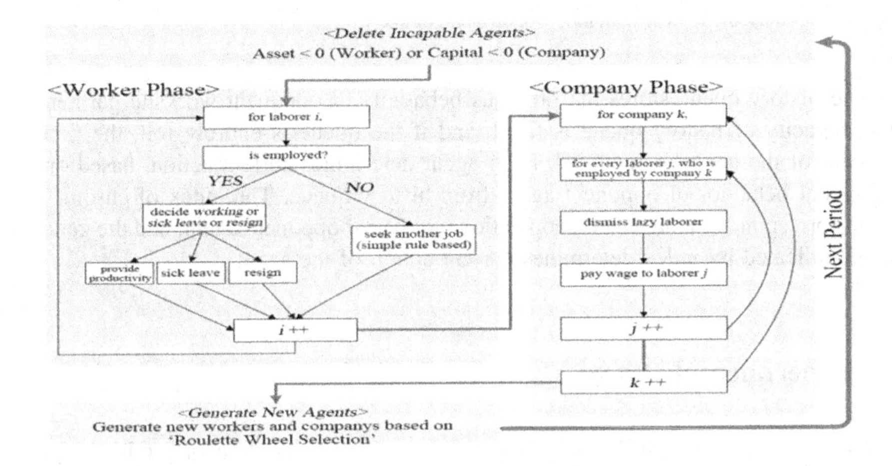

Fig. 2 Simulation process overview

3.3 Representation of Strategy

Several different types of strategy representation schemes have been proposed for the IPD game such as finite state machine, logic tree, If-Skip-Action (ISAc) [13], neural net, Markov chain and look-up table [14]. Ashlock showed that each scheme has different sensitivity on inducing cooperative behavior of agent [14]. In this paper, we use the bit string representation that is generally used for the convenience of computer-based simulation.

The agent's strategy is determined by the chromosome which is an array of 0 or 1 as shown in Fig. 3. The index of the chromosome corresponds to the number of cooperative behaviors of the worksite partner, and the value of a cell in the array determines whether a current action is cooperative or not. The value 0 means defection, and 1 means cooperation. The strategy held by an agent can be maintained until the chromosome becomes extinct during simulation process. The simulation process applied in this paper is described in the next section.

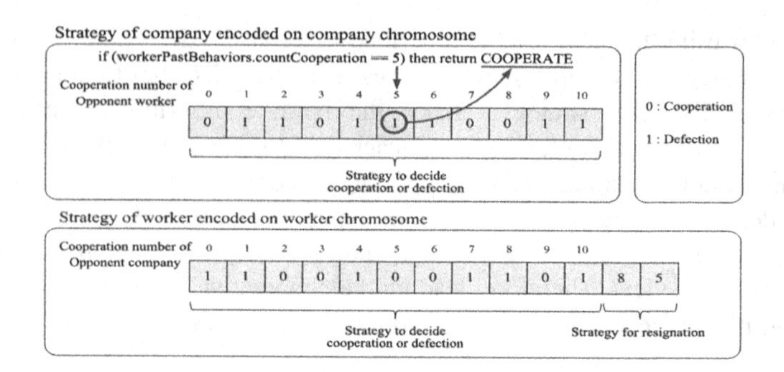

Fig. 3 Chromosome representation of worker and company

The history queue stores the previous behaviors of opponent worksite partner. The capacity of history queue is fixed, and if the queue is entirely full, the first element of the queue is removed. Each agent determines current action, based on the recent behavior of opponent agent from history queue. The index of chromosome corresponds to the recent cooperation number of opponent agent, and the gene value indicated by index determines current action of the agent.

4 Experimental Design

The experimental design focuses results mainly on the variation of productivity, market extent, dominant strategy and cooperative behaviors among the agents. In addition, we add more factors such as labor specialization, and resignation of workers. Other factors such as evolutionary parameters, payoff table and other environment setting are fixed to observe the effect of the factors. Table 3 lists the evolution parameters.

4.1 Productivity Growth of Workers

In the research area of economy, various theories and arguments are investigated on how different characteristics of an individual worker affect his productivity. Although there is a widespread belief that job performance declines with age, several individual-level studies emphasized that it was essential to take into account the effect of worker's experience [15] since older workers can more productive than young workers in tasks that required plentiful experience [16, 17].

The conventional belief is that the productivity growth slows down as age is increasing, which leads to a concave age-productivity profile. In the beginning of working career, a worker's productivity increases fast, but in the later career, productivity growth rate decreases and may turn to fall. If companies need general

Table 3 Parameters of evolutionary engine

Parameter	Value
Initial population of company	50
Maximum population of company	Infinite
Initial population of worker	500
Maximum population of worker	Infinite
Increment rate of worker population (Reproduce rate)	0.02
Sick leave rate	0.01, 0.02
Maximum period of sick leave	5, 20, 100
Mutation rate	0.005
Selection method	Roulette wheel and elitism mixed
Crossover method	1 point crossover

skills rather than company-specific skills from workers, the productivity of a worker should be higher after changing occupation than previous. If skills are job specific, however, new workers should begin with low productivity irrespective of their earlier experience [18]. In this paper, we assume that worker's productivity does not turn to fall, and skills are company specific, so worker's productivity is mathematically expressed as sigmoid function as follows.

$$\mathrm{Prod}_{\mathrm{E,H}} = \mathrm{Prod}_{\mathrm{H}} + \frac{1}{1+e^{\frac{-(continuous-a)}{k}}} : \text{cooperation} \tag{1}$$

$$\mathrm{Prod}_{\mathrm{E,L}} = \frac{Prod_{E,H}}{2} : \text{defection} \tag{2}$$

In Eq. (1), a and k are productivity parameters of determining growth rate of the worker's productivity. The productivity decreases by half when the agent defects to company. By using this policy, we expect that company agents can get more power as compared with before.

4.2 Resignation Strategy of Worker Agents

Basically, worker has two behaviors, cooperation or defection, in the original game model. But in this paper, one more worker action, resign, is added from the company. Originally, if worker is employed in one company at once, the worker must work there until he is removed from the framework owing to financial failure. This means that although company repeats defection, the worker does not have any other defiant way except the defection together.

$$\text{Count}_{cooperation}(companyPastBehaviors) < threshold1 \tag{3}$$

$$\frac{aseet}{livingCost} > threshold2 \tag{4}$$

Resignation can be a way for active resistance whereas defect is passive resistance. As seen in Fig. 3, each worker maintains resignation strategy by using worker chromosome, and two threshold values are represented at the worker's chromosome. The first condition represents that the worker may resign from company if company recently repeats defection until the count of cooperation is below a specific threshold value taken by opponent worker. The second condition implies the worker's financial capability to support his living for some period without any income. If both conditions are satisfied, the worker will certainly resign. We expect that resignation strategy can offer the opportunity to workers to escape from vicious entrepreneur, and exploitative company is hardly survivable in the labor market. This means that this policy can reduce the power of company agents whereas worker power can increase.

4.3 Dismissal of Company Agents

In order to model market as more realistic, we assume that company can dismiss lazy workers. This condition shows that if workers recently repeat defection until percentage of lazy is over specific threshold3 (laziness percentage of half of total continuous service year), the worker may be dismissed from company.

$$\frac{lazyPeriodSum}{age\ of\ worker} \times 100 > threshold3 \tag{5}$$

$$\text{continuous(worker)} > threshold4 \tag{6}$$

$$\text{state(company)}! = \text{ENOUGH\_BLACK} \tag{7}$$

This condition is related with seeking state of workers. If worker is seeking occupation on lazy state, it is applied for accumulation of lazy period. On the other hand, the case of worker on hard state, it is not applied for accumulation. Although a worker leaves the company, Percentage Of Laziness (POL) is preserved. This function causes dismissal of company for frequently lazy worker. In order to fulfill dismissal of company, we need the second and third conditions. The second one means whether continuous service year of workers is over threshold4 (average continuous service year of all workers of company) or not. Last one shows if state of company is not equal to enough black, which means that company has sufficient surplus resources.

4.4 Sick Leave of Worker Agents

If worker is sick, worker can submit sick leave to company. Worker should come back from sick leave after threshold period. Company wants to know worker's state on sick leave whether worker is a lazy person or not. If worker on sick leave is in lazy state, percentage of laziness increases. When worker is return, POL affects on worker's state.

5 Experimental Result and Analysis

We have performed the four experiments with combinations of experiment setting as shown in Table 4. We analyzed the variation of the cooperation rate and the number of agents of the workers and companies. In this section, we present some meaningful experimental result among the various experiments and analyze it.

Table 4 Experiment plans of each eight experiment

	Productivity Growth	Resign	Dismissal	Sick leave (rate, period)
1	Yes	Yes	No	No
2	Yes	Yes	Yes	No
3	No	No	No	Yes (0.02,20)
4	No	No	No	Yes (0.01,20)

Table 4 shows the setting of the experiments. In each experiment, we adjust four effects. Moreover, we change the increase rate of worker population. We will illustrate experimental result by using several graphs and analyze each result from various social and economical view points.

5.1 Result of Dismissal

To provide workers with resistance method, worker resignation is allowed in this experiment. Now workers can get out from company against for defective company. The result is displayed on Fig. 4. Definitely, cooperation rate of companies was increased, but most of companies went into bankruptcy at generation 20 because workers resign from his company excessively. In this case, there are not enough time for worker to learn rational resignation strategy. The number of workers began increasing at generation 25. Because jobless worker is not enough in worker pool, companies are more cooperative to workers. At the generation 205, the number of lazy workers is increasing more than that of hard workers. Because they can submit resignation to their company if company is deceptive, workers do

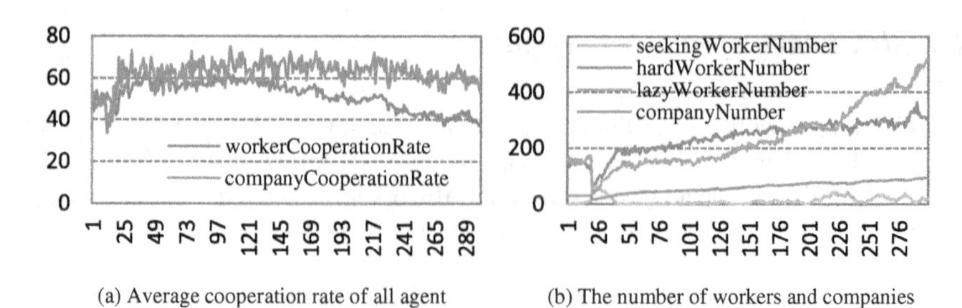

(a) Average cooperation rate of all agent (b) The number of workers and companies

Fig. 4 The result of non-dismissal effect (experiment 1)

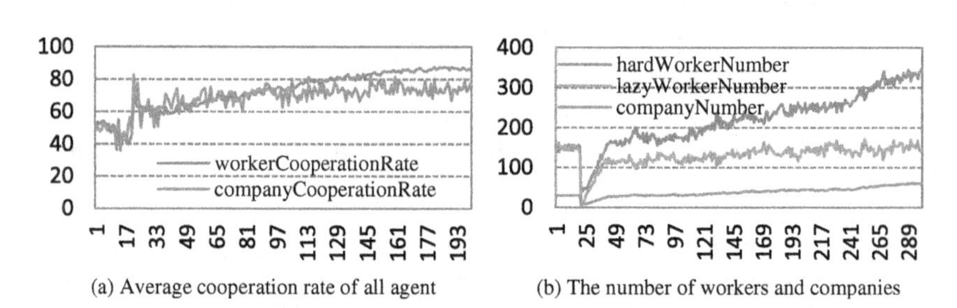

(a) Average cooperation rate of all agent (b) The number of workers and companies

Fig. 5 The result of dismissal effect (experiment 2)

not worry about being fired. High percentage of hard worker is caused by productivity growth and dismissal of company. High of productivity is guaranteed when a worker works hard (Fig. 5).

5.2 Result of Sick Leave

In the experiments 3 and 4, workers are allowed to submit sick leave to company without any resignation, productivity growth, and dismissal. According to these figures, a company is cooperative with defective workers in the beginning, and this tendency has been kept in the overall process. However, in case of rate 0.02, there is higher percentage of lazy worker than 0.01; companies are bankrupted faster than those in the result 3 at generation 200, and workers become more deceptive after several generations. As we can verify sick leave value between 0.01 and 0.02, allowing sick leave and if its value is little high, it is hard for a company to maintain itself. Because they are not fired, the high level percentage of sick leave makes worker lazy (Figs. 6 and 7).

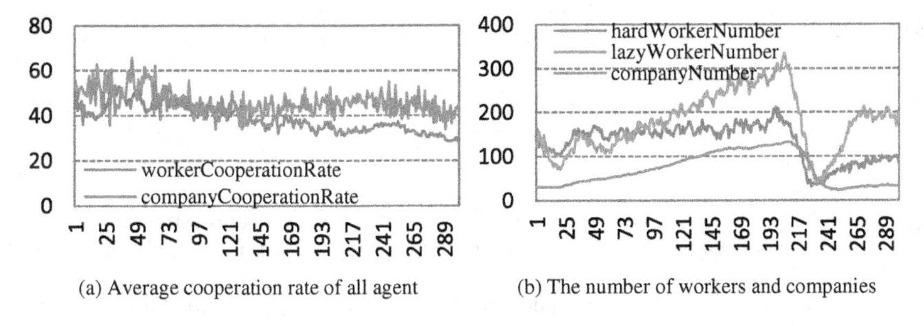

(a) Average cooperation rate of all agent (b) The number of workers and companies

Fig. 6 The result of sick leave in high probability (experiment 3)

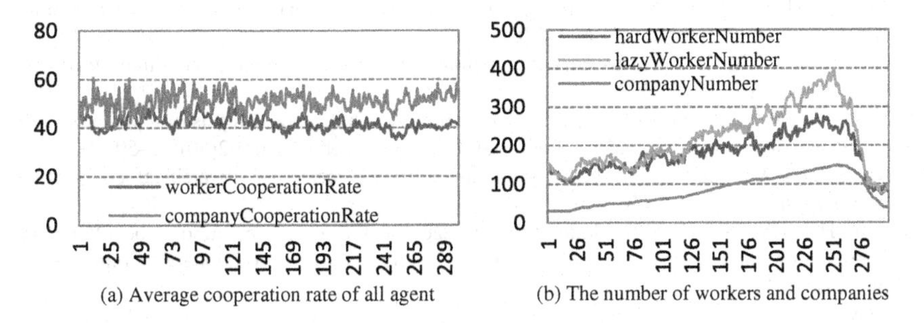

(a) Average cooperation rate of all agent (b) The number of workers and companies

Fig. 7 The result of sick leave in low probability (experiment 4)

6 Conclusions

In this paper, we have focused mainly on constructing evolutionary labor market framework, and many experiments have been performed with various parameters. To construct the relationship rule between worker and company agents, we inducted gift exchange game model which is suitable model for economic situations. In addition, we used co-evolutionary learning to define each agent's strategy and observed the variation of each agent's behavior pattern owing to the evolution of strategy.

The experiments were conducted with the four special policies, growing productivity as worker's experience, resignation strategy of workers, dismissal of company, and sick leave of workers. The first and third policies can contribute for company power, where the second and fourth policies correspond to worker power. Through the various experiments, we have got several meaningful results about variation of cooperation rate and the labor market.

References

1. Tesfatsion L (2001) Structure, behavior, and market power in an evolutionary labor market with adaptive search. J Econ Dyn Control 25:419–457
2. Tesfatsion L (2002) Hysteresis in an evolutionary labor market with adaptive search. Evol Comput Econ Finance 50:189–210
3. Reynolds RG, Kobti Z, Kohler TA, Yap L (2005) Unraveling ancient mysteries: Reimagining the past using evolutionary computation in a complex gaming environment. IEEE Trans Evol Comput 9(6):707–720
4. Yang S-R, Cho S-B (2005) Co-evolutionary learning with strategic coalition for multi-agents. Appl Soft Comput 5(2):193–203
5. Goh CK, Quek HY, Tan KC, Abbass HA (2009) Modeling civil violence: An evolutionary multi-agent, game theoretic approach. IEEE Trans Evol Comput 13(4):780–800
6. Bagnall AJ, Smith GD (2005) A multiagent model of the UK market in electricity generation. IEEE Trans Evol Comput 9(5):522–536
7. Tesfatsion L (2002) Agent-based computational economics: Growing economics from the bottom up. Artif Life 8:55–82
8. Farahani YM, Golestani A, Grass R (2010) Complexity and chaos analysis of a predator-prey ecosystem simulation. Second Int Conf Adv Cognitive Technol Appl 2(40):52–59
9. Maniadakis M (2009) Agent-based brain modeling by means of hierarchical cooperative coevolution. J Artif Life 15(3):293–336
10. Kim HT, Cho S-B (2009) Modeling multi-agent labor market based on coevolutionary computation and game theory. In: IEEE Congress on Evolutionary Computation, pp 2145–2148
11. Mayerson RB (1999) Nash equilibrium and the history of economic theory. J Econ Lit 37 (3):1067–1082
12. Straffin PD (1980) The prisoner's dilemma. UMAP J 1:101–103
13. Ashlock D, Joenks M (1998) ISAc lists, a different representation for program induction. In: Proceeding of the Third Annual Genetic Programming Conference, pp 3–10
14. Ashlock D, Leahy N (2003) A representational sensitivity study of game theoretic Simulations. Soft Comput Ind Appl, pp 67–72
15. Waldman DA, Avolio BJ (1986) A meta-analysis of age differences in job performance. J Appl Psychol 71:33–38
16. Warr P (1995) In what circumstances does job performance vary with age? Eur Work Organ Psychol 3:237–249
17. McEvoy GM, Cascio WF (1989) Cumulative evidence of the relationship between worker age and job performance. J Appl Psychol 74:11–17
18. Ilmakunnas P, Maliranta M, Vainiomaki J (2004) The roles of company and worker characteristics for plant productivity. J Prod Anal 21(3):249–276

SOCO15-SS01 - Soft Computing Methods in Bioinformatics

Medical Edge Detection Combining Fuzzy Mathematical Morphology with Interval-Valued Relations

Agustina Bouchet, Pelayo Quirós, Pedro Alonso, Irene Díaz and Susana Montes

Abstract Image processing represents an important challenge in different fields, especially in biomedical field. Mathematical Morphology uses concepts from set theory, geometry, algebra and topology to analyze the geometrical structure of an image. In addition, it is possible to consider methods where the starting point to analyze an image is a fuzzy relation. This paper studies three methods to image edge detection based on a construction method for interval-valued fuzzy relations which can be understood as a gradient from a morphological point of view. The performance of the proposal in detecting medical image edges is tested, showing the method performing better with regard to a least squared adjust.

Keywords Interval-valued fuzzy sets · Fuzzy sets · Fuzzy mathematical morphology · Gradient · Edge detection

A. Bouchet
Engineering Faculty, Digital Image Processing Group,
National University of Mar del Plata, Mar del Plata, Argentina
e-mail: abouchet@fi.mdp.edu.ar

P. Quirós · P. Alonso
Department of Mathematics, University of Oviedo, Gij ón, Spain
e-mail: uo205956@uniovi.es

P. Alonso
e-mail: palonso@uniovi.es

I. Díaz (✉)
Department of Computer Science, University of Oviedo, Oviedo, Spain
e-mail: sirene@uniovi.es

S. Montes
Department of Statistics and O. R., University of Oviedo, Gij ón, Spain
e-mail: montes@uniovi.es

A. Bouchet
CONICET, Mar del Plata, Argentina

Á. Herrero et al. (eds.), *10th International Conference on Soft Computing Models in Industrial and Environmental Applications*, Advances in Intelligent Systems and Computing 368, DOI 10.1007/978-3-319-19719-7_20

1 Introduction

Edge detection plays an important role in the field of binary, grayscale or colour image processing, primarily for image segmentation. Edge detection can be defined as the process to detect relevant image features. It usually works by detecting discontinuities in brightness [22] as edges are usually the sign of a lack of continuity. In particular, medical images edge detection is one of the most important pre-processing steps in medical image segmentation and 3D reconstruction.

There are many different approaches to edge detection with the aim to extract contour features. Classical edge detectors are based on a discrete differential operators [11], derivatives [14] or wavelet transformations [12]. In addition, there are several approaches to edge detection based on statistical inference. In this case, edge detection is data driven instead of model based [16]. However, the performance of most of these classical methods degrades with noise. To overcome this drawback, Mathematical Morphology (MM) has been introduced. MM provides with an alternative approach to image processing based on concepts of set theory, topology, geometry and algebra. The main idea of this methodology is to compare the objects of interest with a set of predefined and known geometry, called Structuring Element (SE). The use of different shapes and sizes for the SE allows testing and quantifying how the structuring element is, or is not contained in the image.

Fuzzy Mathematical Morphology (FMM) [3] has emerged as an extension of the Mathematical Morphology's binary operators to gray level images, by redefining the set operations as fuzzy set operations, based on the theory of Fuzzy Sets (FS) [24]. It is inspired by the observation that both greyscale images and FS are modeled as mappings from a universe into the unit interval [0,1].

Therefore fuzzy set theory is thus used as a tool here and not to model uncertainty. In particular, the extension of FS so called Interval-Valued Fuzzy Sets (IVFSs) [20] based on generalizing the membership function as a closed interval in [0, 1] is quite useful in detecting edges [8, 18]. In fact, this kind of construction methods are often applied to the detection of edges in gray scale images, which has its most important application in the medical field (see [19]) and other branches of science (see [7]).

The aim of this paper is to define a method to image edge detection based on a construction method for Interval-Valued Fuzzy Relations (IVFR) which can be understood as a gradient from a morphological point of view.

The remainder of the paper is structured as follows: next section revises FMM. Section 3 describes basic concepts regarding IVFSs. In Sect. 4 the weighted construction method for IVFR and its relation with MM are studied. Section 5 draws some experiments and finally in Sect. 6 the main conclusions of this work are highlighted.

2 Fuzzy Mathematical Morphology

The theory of MM is broadly used as a processing tool for enhancement, segmentation, edge detection and filtering, with an important development in biomedical image processing, where objects are characterized by their topology or geometrical structure. MM has been extensively studied and applied to binary and gray level images [21].

The main objective of the MM is to extract information of the geometry and topology of an unknown set in an image. The key of this methodology is the SE, a small set completely defined with a known geometry, which is compared with the whole image. The basic morphological operators of MM are erosion and dilation. From the combination of these operators other more complex arise.

When combined with MM, fuzzy set theory extends the applicability of this model by adding the ability to handle uncertainty. Extension of binary MM to gray level images is obtained via FMM, which combines the power of MM, based on set theory, with the ability of fuzzy logic to handle degrees of membership. FMM has been developed in several directions, until it was combined in a general theoretical framework [1, 2, 5, 10, 13].

Let μ and ν be two fuzzy sets with membership functions $\mu : U \subset R^2 \rightarrow [0, 1]$ and $\nu : U \subset R^2 \rightarrow [0, 1]$. The first one corresponds to a gray scale images and the second one determines the structuring element. Then basic operators are defined as follows.

Definition 1 [5, 6] Let μ be a gray scale image and let ν be an SE. Then the Fuzzy Morphological Dilation and Fuzzy Morphological Erosion of the image μ by the SE ν are defined respectively as:

$$\delta(\mu, \nu)(x) = sup_{y \in U}[t(\mu(y), \nu(y - x))]$$

$$\epsilon(\mu, \nu)(x) = inf_{y \in U}[s(\mu(y), c(\nu(y - x)))],$$

where $t(a, b)$ is a t-norm, $s(a, b)$ is a t-conorm and $c(a) = 1 - a$ is the fuzzy complement operator.

Gradient operators are used in segmentation because they enhance intensity variations in image. These variations are assumed to be edges of objects. This is why gradients are also called "edge detectors". They are presented in the following definition.

Definition 2 [5, 21] Let μ be a gray scale image and let ν be an SE, then

– The basic morphological gradient of the image μ by the SE ν is:

$$Grad_{D,E}(\mu, \nu) = \delta(\mu, \nu) - \epsilon(\mu, \nu).$$

– Gradient by erosion or internal gradient of the image μ by the SE ν is:

$$Grad_E(\mu, \nu) = \mu - \epsilon(\mu, \nu).$$

– Gradient by dilation or external gradient of the image μ by the SE ν is:

$$Grad_D(\mu, \nu) = \delta(\mu, \nu) - \mu.$$

Gradient by erosion detects edges in the positions of higher gray levels in the edges while Gradient by dilation detects edges in the positions of lower gray levels in the edges. Therefore, the morphological gradient grouped the results of these two operators, getting thicker contours.

3 Interval-Valued Fuzzy Sets

IVFSs, introduced by Sambuc in [20], are extensions of classical FS (see [15]). While a FS is defined by a function μ that maps each element $x \in X$ onto a value $\mu(x) \in [0, 1]$. The following definitions introduce basic concepts regarding IVFS. All of them can be found in [20].

Definition 3 A is an IVFS in the finite set X if it is defined by the membership function:

$$A : X \to L([0, 1]), \text{with } L([0, 1]) = \{[a_l, a_u] | (a_l, a_u) \in [0, 1]^2 \text{ and } a_l \leq a_u\},$$

Fuzzy Relations (FR) are a special case of FS which are very useful in the construction of IVFS.

Definition 4 Let X and Y be two finite sets. A FR R in $X \times Y$ is given by the membership function $R : X \times Y \to [0, 1]$.

In the same way as it has been done with the classical Fuzzy Sets, the definition of IVFSs is extended to IVFRs.

Definition 5 Let X and Y be two finite sets. An IVFR R is defined by the membership function $R : X \times Y \to L([0, 1])$.

An IVFR can be denoted by $R = \{((x, y), R(x, y)) | x \in X, y \in Y\}$, where $R(x, y)$ is the degree of strength of the relation between x and y given by an interval.

4 Construction of Interval-Valued Fuzzy Relations for Edge Detection in Images

The following definition states the meaning of a gray scale image in the fuzzy field.

Definition 6 [18] A gray scale image R whose dimensions are $P \times Q$ pixels, is a FR where the finite sets used are $X = \{0, 1, \ldots, P-1\}$ and $Y = \{0, 1, \ldots, Q-1\}$.

This means that gray scale images can be understood as FRs. The edge detection can be solved through morphological operations describing the interaction of the image with a SE. The SE is usually small compared to the image. The notion of a SE is applied to construct an IVFR from the gray scale image represented as FR.

The IVFR construction method is based on two constructors (lower and upper constructors) in order to obtain the limits of each interval. In addition, it involves t-norms and t-conorms [9, 23] which are well known generalizations of conjunction and disjunctions in classical logic. Definitions 7, 8 and 9 establish the basic method to construct an IVFR.

4.1 Non-Weighted Method

Let's describe first the basic construction method.

Definition 7 [4] Let $X = \{0, 1, \ldots, P-1\}$ and $Y = \{0, 1, \ldots, Q-1\}$ two universes, $R \in F(X \times Y)$ a FR, two t-norms T_1, T_2, two t-conorms S_1, S_2, and $n, m \in \mathbb{N}$ such that $n \leq \frac{P-1}{2}$ and $m \leq \frac{Q-1}{2}$,

- $L^{n,m}_{T_1,T_2} : F(X \times Y) \to F(X \times Y)$ is the lower constructor associated to T_1, T_2, n and m. It is defined by

$$L^{n,m}_{T_1,T_2}[R](x,y) = \underset{\substack{i=-n\\j=-m}}{\overset{\substack{m\\n}}{T_1}} (T_2(R(x-i, y-j), R(x,y))),$$

- $U^{n,m}_{S_1,S_2} : F(X \times Y) \to F(X \times Y)$ is the upper constructor associated to S_1, S_2, n and m. It is defined by

$$U^{n,m}_{S_1,S_2}[R](x,y) = \underset{\substack{i=-n\\j=-m}}{\overset{\substack{m\\n}}{S_1}} (S_2(R(x-i, y-j), R(x,y))),$$

$\forall (x, y) \in X \times Y$, where i, j take values such that $0 \leq x-i \leq P-1$ and $0 \leq y-j \leq Q-1$, n and m indicate that the SE is a matrix of dimension $(2n+1) \times (2m+1)$ and $\overset{n}{\underset{i=1}{T}} x_i = T(x_1, \ldots, x_n)$.

Definition 8 [4] Let $R \in F(X \times Y)$ be a FR, $L^{n,m}_{T_1,T_2}[R]$ a lower constructor and $U^{n,m}_{S_1,S_2}[R]$ an upper constructor, then $R^{n,m}$ is defined by:

$$R^{n,m}(x,y) = [L^{n,m}_{T_1,T_2}[R](x,y), U^{n,m}_{S_1,S_2}[R](x,y)],$$

for all $(x,y) \in X \times Y$ is an IVFR from X to Y.

After obtaining both lower and upper constructors from the initial FR (Definition 7), and the IVFR generated by them (Definition 8), the last step of the construction method is to obtain another FR from such IVFR by considering the length of each interval is used.

Definition 9 [4] Given $L^{n,m}_{T_1,T_2}[R]$ a lower constructor and $U^{n,m}_{S_1,S_2}[R]$ an upper constructor, the W-fuzzy relation associated to them is given by:

$$W[R^{n,m}](x,y) = U^{n,m}_{S_1,S_2}[R](x,y) - L^{n,m}_{T_1,T_2}[R](x,y).$$

4.2 Weighted Construction Method

In the method developed by [4], for each element of the relation, the SE is a window centered in that element. Window dimension depends on natural numbers n and m, which are the parameters in both constructors.

However, it can be assumed that the influence of pixels closer to the center of the SE is higher in detecting edges. In this case, the closer the value to the center of the window, the greater the importance that it takes in the definition of the constructors. This fact is modeled in this work by introducing weights in the lower and upper constructors.

In other words, our goal is to obtain the final lower and upper constructors as $L(x,y) = \sum_{i=1}^{k} w_i L^i(x,y)$ and $U(x,y) = \sum_{i=1}^{k} w_i U^i(x,y)$, where w_i are weights that satisfy $w_i \geq w_{i+1}$ and $L^i(x,y)$ and $U^i(x,y)$ the lower and upper constructors of different window sizes respectively, so the smaller windows have more strength in the definition.

Definition 10 [18] Given two finite universes of natural numbers $X = \{0, 1, \dots, P-1\}$ and $Y = \{0, 1, \dots, Q-1\}$, $R \in FR(X,Y)$ a fuzzy relation in $X \times Y$, two t-norms T_1, T_2, two t-conorms S_1, S_2, and $n, m \in \mathbb{N}$ such that $n \leq \frac{P-1}{2}$ and $m \leq \frac{Q-1}{2}$, for any $i = 1, 2 \dots, \max(n,m)$ we can consider the two fuzzy relations $L^i[R]$ and $U^i[R]$ defined by

$$L^i[R](x,y) = L^{\min(i,n),\min(i,m)}_{T_1,T_2}[R](x,y), \;\; U^i[R](x,y) = U^{\min(i,n),\min(i,m)}_{S_1,S_2}[R](x,y).$$

There are several options to define these weights (see [18]). In this work, it is proposed one that assign a different value to each point in the window according to the average of the k windows ($w_i = \frac{1}{k}, \quad i = 1, \dots, k$).

The scheme of the weighted method with the smoothing step is shown in Algorithm 1. In order to obtain the non weighted method point 4 of Algorithm 1 is skipped.

Input: $R \in FR(X, Y)$, $n, m \in \mathbb{N}$, T_1, T_2 t-norms, S_1, S_2 t-conorms, $N \in \mathbb{N}$, α
Output: W-fuzzy relation $W \in FR(X, Y)$

1: Fix $k = max(n, m)$
2: Obtain L^i associated to n, m, T_1 and T_2 $\forall i = 1, \ldots, k$
3: Obtain U^i associated to n, m, S_1 and S_2 $\forall i = 1, \ldots, k$
4: Obtain weights w_i for $i = 1, \ldots, k$
5: Calculate the lower and upper constructors as:

$$L(x, y) = \sum_{i=1}^{k} w_i L^i(x, y) \text{ and } U(x, y) = \sum_{i=1}^{k} w_i U^i(x, y)$$

6: Construct the IVFR $R^{n,m}$ from L and U (Def. 8)
7: Obtain the W-fuzzy relation W_w from $R^{n,m}$ (Def. 9)
8: Calculate W from W_w with the smoothing step defined by the cut point α

Algorithm 1: Weighted construction method algorithm.

4.3 Linking Interval-Valued Fuzzy Relations to Fuzzy Mathematical Morphology

As gray scale images can be represented by FRs, the outputs produced by the construction method are the following:

- **The lower constructor:** it represents a darker version of the original image. Depending on the $t - norms$ chosen, this image can be more or less dark. Note that the lower constructor is a morphological operator (see Sect. 2).
  Depending on the pair of $t - norms$ selected, different lower constructors are obtained (see Fig. 1).
- **The upper constructor:** it represents a brighter version of the original image. Depending on the $t - conorms$ chosen, this image can be more or less bright. Upper constructors are associated to dilatation operators (see Sect. 2). Figure 2 shows the effect of different co-norms in the upper constructors.
- **The W-fuzzy edge image:** it represents the difference of contrast between both constructors. The edges can be identified in this image. Figure 3 represents three W-fuzzy images with different pairs of $t - norms$ and $t - conorms$.

Note that the lower and upper constructors can be seen as fuzzy morphological operators. Thus, there is a parallelism between the definition of gradient and the definition of the constructors. Therefore, assuming μ to be an image and ν to be the SE defined by a $n \times m$-matrix, the gradients defined in terms of constructors are

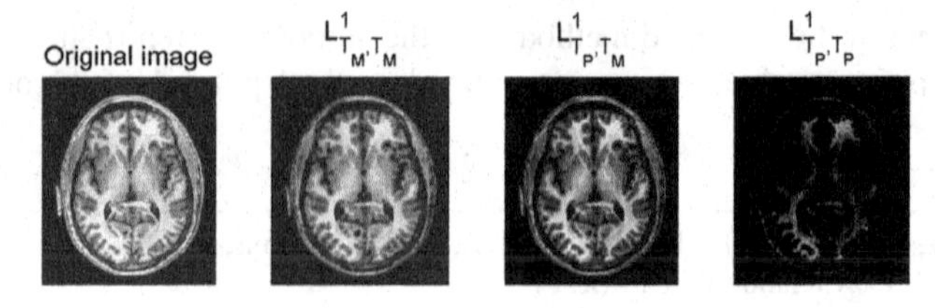

Fig. 1 Examples of lower constructors. T_M and T_P are the minimum and product t-norms respectively

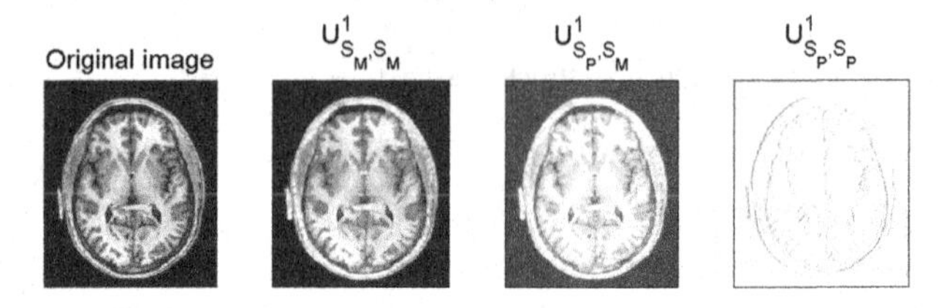

Fig. 2 Examples of upper constructors. S_M and S_P are the maximum and product t-conorms respectively

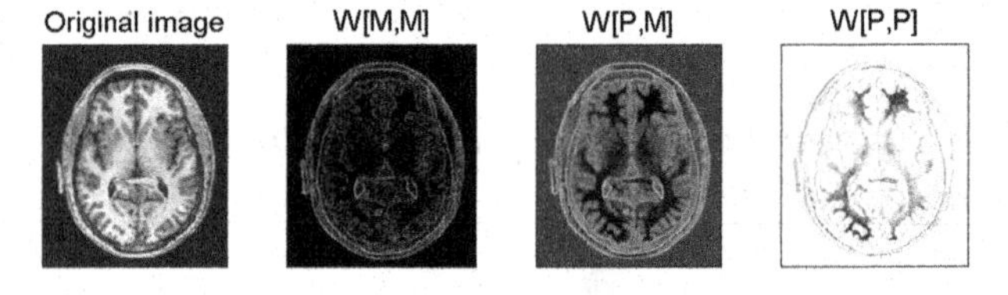

Fig. 3 Examples of W-fuzzy images depending on the used pairs of t-norms and t-conorms, where $W[M,P] = U^{n,m}_{S_M,S_P} - L^{n,m}_{T_M,T_P}$, and analogously for the others

- Gradient: $R^{n,m} = U^{n,m}_{S_M,S_P} - L^{n,m}_{T_M,T_P}$.
- Internal Gradient: $\mu - L^{n,m}_{T_M,T_P}$.
- External Gradient: $U^{n,m}_{S_M,S_P} - \mu$.

5 Experiments

The goal of this section is to compare the performance of the gradients defined in terms of the lower and upper constructors in detecting edges. This comparison is obtained in terms of least squares estimator by comparing respectively the image produced by the lower constructor, the upper constructor and the W-fuzzy image to the benchmark one. In addition it is also checked the effect of weights [18] in the constructors against the original one [4] and different sizes of the SE. To do that, a gray scale images database has been considered. These gray scale images were obtained from the *Berkeley Segmentation Dataset* (see [17]). This database contains original images and its corresponding edge images. The first 25 images from the test set were selected, whose dimensions are 481×321 (or 321×481) pixels. In addition, the t-norms and t-conorms are the standard ones (Minimum-Maximum). The different configurations studied are:

- Non-weighted method with $n = m = 1$ (SEs size: 3×3 matrix) (NW1).
- Non-weighted method with $n = m = 2$ (SEs size: 5×5 matrix) (NW2).
- Weighted method using the average (A) with 2, 3 or 4 terms (2,3,4)(methods noted as A2, A3 and A4).

Figure 4 shows the performance of the morphological gradient, and the internal and external gradients. *X* axis represents each different method: Average with two (A2), three (A3) or four (A4) term; Non weighted method with $n = m = 1$ (NW1) or with $n = m = 2$ (NW2). *Y* axis represents the error (the lower the error, the better the method).

As it can be seen the non-weighted methods with an SE represented by a 3×3-matrix are the one performing better. Regarding the gradients, the external gradient (associated to the upper constructor) is the one performing better independently of the size of the SE as well as of the kind of weighting method selected.

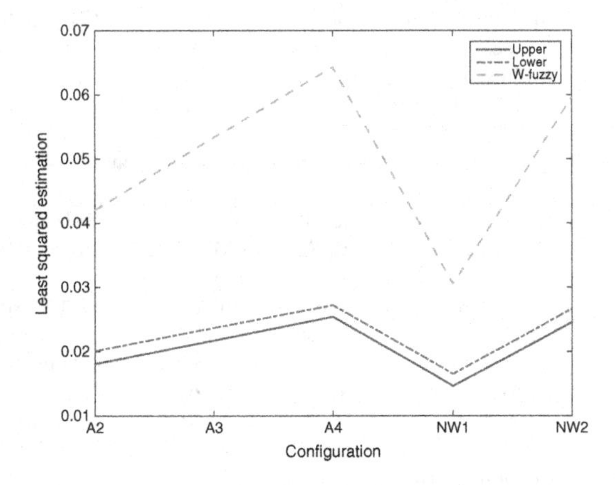

Fig. 4 Comparative of gradients depending on the experimental parameters, where two non-weighted methods (NW1, NW2), and averaging method with 2, 3 and 4 terms

6 Conclusions

This paper presents different methods to construct IVFRs where the starting point is a FR. These methods are understood as gradients, which are well known tools used in Mathematical Morphology to discover edges in images. The behaviour of these gradients have been analyzed depending on several parameters. The results show that the external gradient obtained from the upper constructor in the one performing better.

Acknowledgments This work has been partially supported by MEC and FEDER Grant TEC2012-38142-C04-04 and by ERASMUS Mundus Project EUREKA SD 2013-2591.

References

1. De Baets B, Kerre E, Gupta M (1995) The fundamentals of fuzzy mathematical morphology, Part 1: basic concepts. Int J Gen Syst 23:155–171
2. De Baets B, Kerre E, Gupta M (1995) The fundamentals of fuzzy mathematical morphology. Part 2: idempotence, convexity and decomposition. Int J Gen Syst 23:307–322
3. De Baets B (1997) Fuzzy morphology: a logical approach. In: Ayyub BM, Gupta MM (eds) Uncertainty analysis in engineering and sciences: fuzzy logic, statistics, and neural network approach. Kluwer Academic Publishers, Boston, pp 53–67
4. Barrenechea E, Bustince H, De Baets B, Lopez-Molina C (2011) Construction of interval-valued fuzzy relations with application to the generation of fuzzy edge images. IEEE Trans Fuzzy Syst 19:819–830
5. Bloch I, Maitre H (1995) Fuzzy mathematical morphologies: a comparative study. Pattern Recogn 28:1341–1387
6. Bouchet A, Benalczar Palacios F, Brun M, Ballarin VL (2014) Performance analysis of fuzzy mathematical morphology operators on noisy MRI. Lat Am Appl Res 44(3):231–236
7. Buades A, Coll B, Morel JM (2005) A review of image denoising algorithms, with a new one. Multiscale Model Simul 4(2):490–530
8. Bustince H (2010) Interval-valued fuzzy sets in soft computing. Int J Comput Intell Syst 3(2):215–222
9. Frank MJ (1979) On the simultaneous associativity of $F(x, y)$ and $x + y + F(x, y)$. Aequationes Math 19:194–226
10. Di Gesu V, Maccarone MC, Tripiciano M (1993) Mathematical morphology based on fuzzy operators. In: Lowen R, Roubens M (eds) Fuzzy logic. Kluwer Academic Publishers, pp 477–486
11. Hermosilla T, Bermejo E, Balaguer A, Ruiz LA (2008) Non linear fourth order image interpolation for subpixel edge detection and localization. J Image Vis Comput 26:1240–1248
12. Heric D, Zazula D (2007) Combined edge detection using wavelet transform and signal registration. J Image Vis Comput 25:652–662
13. Kerre E, Nachtegael M (2000) Fuzzy techniques in image processing, vol 52. New York
14. Marr D, Hildreth E (1980) Theory of edge detection. In: Proceedings of the royal society of London, pp 187–217
15. Klir GJ, Wheeler B (1995) Fuzzy sets and fuzzy logic. Prentice Hall, New Jersey
16. Konishi S, Yuille AL, Coughlan JM, Zhu SC (2003) Statistical edge detection: learning and evaluating edge cues. IEEE Trans Pattern Anal Mach Intell 25(1):57–74
17. Martin D, Fowlkes C, Tal D, Malik J (2001) A database of human segmented natural images and its application to evaluating segmentation algorithms and measuring ecological statistics. In: Proceedings of the 8th International Conference on Computer Vision, vol 2, pp 416–423
18. Quirós P, Alonso P, Díaz I, Jurío A, Montes S (2015) An hybrid construction method based on weight functions to obtain interval-valued fuzzy relations. In: Mathematical methods in the applied sciencies (in press)
19. Rulaningtyas R, Ain K (2009) Edge detection for brain tumor pattern recognition, instrumentation, communications, information technology, and biomedical engineering (ICICI-BME), vol 3, pp 1–3

20. Sambuc R (1975) Fonctions Φ-floues. Application l'Aide au Diagnostic en Pathololologie Thyroidienne. Ph. D. Thesis, Univ. Marseille
21. Serra J (1988) Image analysis and mathematical morphology. Academic Press, London
22. Wang R, Gao L, Yang S, Liu Y (2005) An edge detection method by combining fuzzy logic and neural networks. Mach Learn cybern 7:4539–4543
23. Yager RR (1980) On a general class of fuzzy connectives. Fuzzy Sets Syst 4:235–242
24. Zadeh L (1971) Similarity relations and fuzzy orderings. Inf Sci 3:177–200

Shape-Output Gene Clustering for Time Series Microarrays

Camelia Chira, Javier Sedano, José R. Villar, Monica Camara and Carlos Prieto

Abstract The identification of coexpressed genes is a challenging problem in microarray data analysis due to a very high number of genes and low number of samples normally available. This paper presents a shape-output clustering method which is engaged in the analysis of a real-world time series microarray data from the industrial microbiology area. The proposed approach uses the changes in gene expression levels to group genes based on their shape measured over time in several samples. Furthermore, these coexpression patterns are correlated with the measured outputs of production and growth available for each sample. Experiments are performed for time series microarray of a bacteria and an analysis from a biological perspective is carried out. The obtained results confirm the existence of relationships between output variables and gene expressions.

1 Introduction

Time series microarray analysis [2, 4, 8, 11, 12, 16] aims to find the best gene subset that promotes a certain variable or event when subsequent samples are taken from the same biological data at a certain time rate. Finding groups of genes meaningfully

C. Chira (✉)
Technical University of Cluj-Napoca, Cluj-Napoca, Romania
e-mail: camelia.chira@cs.utcluj.ro

J. Sedano · M. Camara
Instituto Tecnológico de Castilla y León, Burgos, Spain
e-mail: javier.sedano@itcl.es

M. Camara
e-mail: monica.camara@itcl.es

J.R. Villar
University of Oviedo, Gijón, Spain
e-mail: villarjose@uniovi.es

C. Prieto
Instituto de Biotecnología de León (INBIOTEC), León, Spain
e-mail: carlos.prieto@inbiotec.es

Á. Herrero et al. (eds.), *10th International Conference on Soft Computing Models in Industrial and Environmental Applications*, Advances in Intelligent Systems and Computing 368, DOI 10.1007/978-3-319-19719-7_21

correlated and extracting knowledge from microarray data represent challenging problems due to the low number of samples usually available and the high dimensionality of the gene space [6, 7, 14].

The current study focuses on the analysis of transcriptomic data in order to infer knowledge about the overall operation of a bacterial specie called *Streptomyces stukubaensis*. In particular, we pay special attention to decipher the production process of a natural product (an immunosuppressant) which is performed in this organism. The objective of this work is to identify genes which are directly implied in the production of the studied natural product and to determine competent genes which are coexpressed with the production ones and are involved in other metabolic processes. For this purpose, microarrays and production values were obtained and a clustering method has been developed as a tool to explore and analyse the microarray data. The method proposed in this paper overcomes some limitations previously identified in [1], where a generic model based on measures from information theory was introduced to identify gene-gene and gene-production correlations for the same microarray data. The methods proposed in [1] include the following steps: grouping of genes by Markov Blanket based on information correlation measures, online validation of groups by checking rate of change similarity and final gene selection based on minimum redundancy and maximum relevance. As shown in [1], gene ranking and selection measures help to identify genes that are involved in the production and growth processes. However, the biological utility of the results is limited due to the big size of resulting groups and the lack of co-expression between the genes of each group [1]. In contrast to these information-based models, the current study avoids using any information theory measure but at the same time aims to capture the correlation in changes between gene expressions and output values. The result should be a set of groups containing genes that follow the same pattern in the gene expression level over time and with the output.

The proposed *Shape-Output Clustering (SOC)* method incorporates knowledge about the shape of gene expression levels and their correlation with the available output values in order to cluster genes from time series microarray data. The technique considers the differences in gene expression levels at consecutive timepoints as well as the difference in the output level for the same samples. This way, the genes in the same group are related with each other and with some production or growth output (corresponding to each sample). Experiments focus on time series microarrays from the Streptomyces stukubaensis bacteria. The results obtained emphasize the ability of the proposed SOC method to identify meaningful genes and groups of genes in order to guide metabolic engineering actions. Furthermore, the biological analysis of results shows the existence of relationships between output variables and gene expressions. Results are also compared to strategies engaging information and statistic based measures from [1] and the findings of this study support a better performance of the SOC method from a biological perspective.

The paper is structured as follows: the specific time series microarray analysis task is detailed, the proposed SOC method is described, experiments and comparative results are presented and the main conclusions of the study are discussed.

2 Time Series Microarray Data Analysis and Problem Specification

This study focuses on deciphering the production process of a natural immunosuppressant product performed by the Streptomyces stukubaensis bacteria. This includes the identification and definition of the transcriptional behavior of main sinthetases and processes related with the metabolic activity. Once the target metabolic process is fully characterized and concurrent pathways which construct other metabolites or bioprocesses are identified, metabolic engineering strategies can be applied. For example, the duplication of genes involved in the target process or the blocking of competing pathways can be performed [13]. Microarray data and production values were obtained and a new methodology which integrates production values in the analysis was implemented to identify the genes that directly enhance the production of the immunosuppressant. Genes for which the expression levels change with the production variable are supposed to be relevant in this metabolic process.

The considered microarray analysis task involves samples measuring the expresssion level of 8848 genes of Streptomyces stukubaensis. The output values available for each sample are as follows: (i) the production - a real value indicating a production level in the studied bacteria, and (ii) the growth - a real value representing the level of growth in the bacteria. Biological samples were extracted as time series, covering 12 key time points for the production and growth. The objective is to select and group those genes which are the most relevant and related with the changes in the production and growth. Coexpressed genes should be identified and a clustering based on time series data needs to be produced.

Solving this problem has a clear potential to generate economical benefits providing a unique resource in order to analyze expression data in combination with an output measure, an objective which is gaining importance due to new omics techniques [3]. Regarding clustering methods, several publications show promising results and different utilities of such analysis [5, 10, 17]. However, not many time series microarrays studies have been developed in the industrial microbiology area. To the best of our knowledge, studies which performed time series microarrays and have measured outputs do not relate both measures and they have been taken into account separately in order to guide the biological interpretation of results in a manual way. Nevertheless, they have not been integrated in the bioinformatics analysis [9, 17].

3 Shape-Output Gene Clustering

The proposed *Shape-Output Clustering (SOC)* method detects gene groups from time series microarray data by taking into account the gene patterns deduced from the distance between consecutive time points values and the gene correlation with the output.

Two genes will have the same index if they follow a similar pattern of changes in gene expression values over time and with regard to the output. The change of expression value from time t to $t + 1$ can be modelled using a shape index. This idea is similar to the first phase (called Algorithm A) of the pattern-based clustering approach presented in [12]. However, the shape index presented in [12] only considers the changes in gene expression levels while our approach further takes into account the correlation with the output.

Let M be the number of genes and N be the number of samples in the microarray data. The rate of change in the gene profile is calculated at each time interval (t_i, t_{i+1}), $i = 1 \ldots N - 1$, as follows:

$$g\_step(t_i, t_{i+1}) = \frac{x_{i+1} - x_i}{t_{i+1} - t_i} \tag{1}$$

where x_i is the gene expression level in sample i (at time point i) and t_i is the value of time point i.

Each gene has a corresponding array of $N - 1$ values of *g_step*, one for each time interval in the dataset. A parameter called ψ is used to decide the significance in the rate of change for each time interval and the corresponding category $g\_change(t_i, t_{i+1})$. The value of ψ represents the level of acceptable difference between two consecutive gene expression levels resulting in $l = 3$ different categories of genes. If $g\_step(t_i, t_{i+1})$ is less than ψ or greater than $-\psi$, the level of change from t_i to t_{i+1} is not significant and the gene category *g_change* assigned for time step (t_i, t_{i+1}) has a 'no change' meaning associated. The other two possible categories correspond to a significant increase - respectively decrease - of the gene expression level from t_i to t_{i+1}.

SOC also requires the computation of the category associated with the rate of change in the output for each time interval. Let *y_step* be the change in the output variable for each time step (computed similarly with *g_step* in Eq. 1, where the gene expression level x is replaced by the output value y). The level of output change *y_change* is computed in the same way and using the same categories with *g_change*.

For each gene, an index called *g_index* is calculated based on the values of $g\_change(t_i, t_{i+1})$ at each time interval and the value *y_change* corresponding the output value rate of change for the same time interval. The index corresponding to a gene is given by Eq. 2.

$$g\_index = \sum_{j=1}^{N-1} l^j * g\_change(t_j, t_{j+1}) * y\_change(t_j, t_{j+1}) \tag{2}$$

where l is the number of different categories that can be assigned.

The main steps of the SOC method are as follows:

(i) for each gene and each time interval, the *g_step* and *g_change* values are computed,
(ii) for each time interval, the output category *y_change* is calculated,

(iii) for each gene, the *g_index* is computed according to Eq. 2, and
(iv) genes having the same *g_index* value are placed in the same group.

The last step of SOC consists of (i) initializing the number of groups $k = 1$ and $G_k = \{g_1\}$; and (ii) for each gene $g_i, i = 2 \ldots M$, if there is a group G_r such that $g\_index(G_r[1]) = g\_index(g_i)$ then $G_r = G_r \cup \{g_i\}$, otherwise set $k = k + 1$ and form new group $G_k = \{g_i\}$. The final result of the SOC method is a set of k groups of genes having the same *g_index* meaning that genes in a group have a similar shape of change and output correlation.

4 Computational Experiments

Experiments focus on the real-world time series microarray data from the considered bacterial specie.[1] This section presents the results obtained by the proposed method, comparing them to the results of information-based methods [1] and analysing them from a biological perspective.

4.1 Dataset Description

The dataset consists of samples collected from 8848 genes. Regarding the microarray platform, a custom gene expression microarray with eight high-definition 15k arrays of Agilent Technologies was used. The microarray design was performed with the software eArray 5.0 and has provided the capacity to measure the expression of 8848 genes.

Biological samples were extracted as a time series, covering the key time points for the production and growth. 12 time points were selected, in which the studied biological processes are performed, and 12 biological replicates were done for each time. From these replicates, 3 cell lines were chosen as samples for the microarray hybridization based on the maximum similarity of their production and growth rates. In this way, we obtained 3 temporal series of 12 time points that belonged to 3 different cell lines.

A normalization process was performed with the limma package [15]. Median and none background correction methods were applied for all results reported in this paper. It should be noted that this is the same normalization used in [1]. Method none computes M and A values without normalization so the corrected intensities are equal to foreground intensities. On the other hand, method median substracts the weighted median of background intensities from the M-values for each microarray.

[1]The microarray dataset obtained and used in this paper is available at request for academic purposes.

4.2 SOC Results

The SOC method has been applied to the considered microarray problem for different values of parameter ψ. The number of groups reported increases with smaller values of ψ. Figure 1 presents the results obtained for ψ ranging from 0.01 to 0.2. The output used for Fig. 1 is the production real value but the number and size of groups follows a similar distribution also with the growth output. As shown in Fig. 1, the number of groups is higher for very small values of ψ but in the same time the size of the largest group is smaller. Additionally, the number of groups having only one gene increases in the same rhythm with the number of groups.

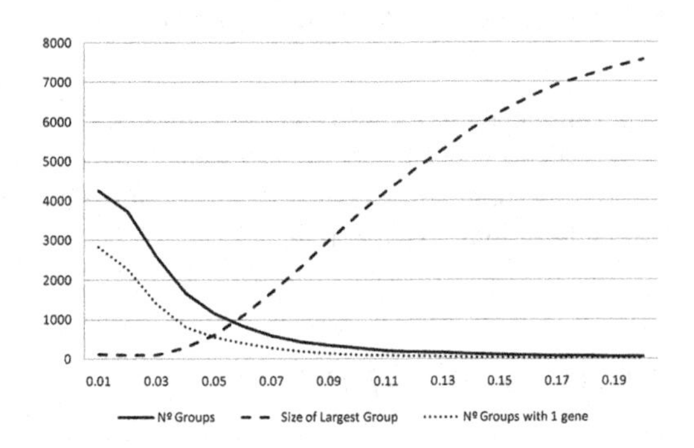

Fig. 1 Number of groups reported by SOC for different ψ values. The size of the largest group and the number of groups containing only 1 gene are also depicted

The size of the groups produced by SOC is illustrated in Fig. 2 for different ψ values. It can be noticed how the size of the largest groups becomes smaller with the ψ value e.g. for $\psi = 0.07$ (solid black line in Fig. 2) the size of groups ranges from 1697 genes down to only 1 gene (however, only the groups larger than 100 genes are depicted in Fig. 2 to allow a better visibility of results).

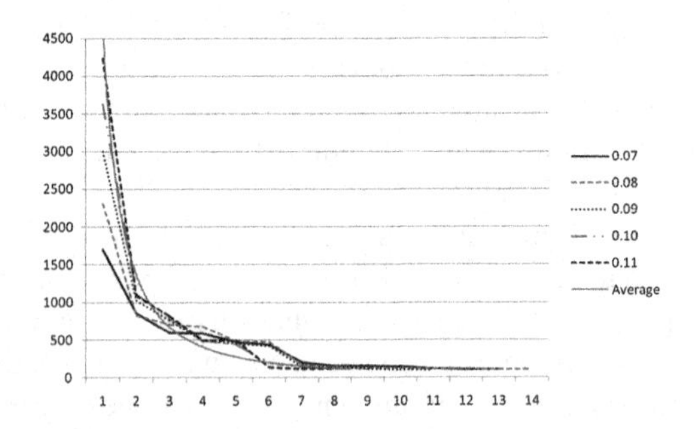

Fig. 2 Size of groups reported by SOC for different ψ values. Only groups with more than 100 genes are depicted

The profiles corresponding to gene clusters are extremely interesting and, as shown in the last subsection, more biologically meaningful compared to the information based clustering. Some examples of gene group profiles are given in Fig. 3 for $\psi = 0.07$.

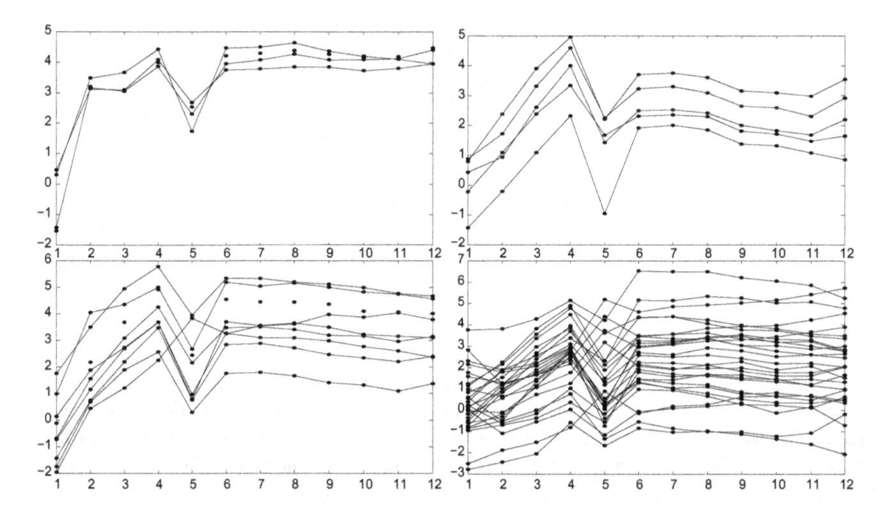

Fig. 3 Different clusters obtained by SOC for $\psi = 0.07$: up left - a group of 4 genes, up right - a group of 5 genes, down left - a group of 9 genes and down right - a group of 28 genes. The x axis represents time succesive points at which samples are taken and the y axis gives the gene expression level at that time

Some groups contain genes with fully similar shape profiles while others allow slight discrepancies between genes in the same group. For example, in Fig. 3, the group down right contains 28 genes from which some present an increase of gene expression level from time step 5–6 (and 6–7) while others actually decrease their expression level at the same time interval (while the shape is the same for the rest of the time points). This is due to the fact that the correlation with the output is considered in computing the *g_index* in the SOC method. Genes are grouped together because they have a similar shape but in a more flexible way in order to include information about the gene-output correlation in the clustering.

Moreover, the SOC method can be applied in connection with the production or growth output (both real-valued). Figure 4 presents the comparison of these two SOC clusterings by showing the pairs of genes which have been grouped in the same way with regard to both outputs (black/white color means the two genes are/are not in the same group and grey means only one method grouped the two genes together). The clustering follows the same general lines in both cases indicating a similar correlation of all genes with both outputs.

4.3 Comparative Results

SOC obtains significantly different results compared to the information based methods [1]. The model proposed in [1] includes several steps: (i) gene sorting according to information measures and correlation with the output, (ii) formation of groups using a Markov Blanket (MB) approach, (iii) validation of groups based on rate of

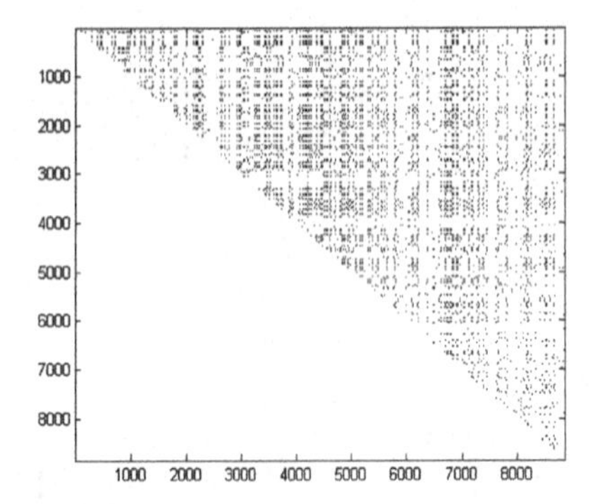

Fig. 4 Matrix comparison of SOC results with same parameters but in connection with two different ouputs: production and growth. Each cell corresponds to a pair of genes and is given one of three colors: white if the two genes are not in the same group in none of the methods, grey if the two genes are in the same group only by one of the two methods and black if both methods group the two genes together

change from one time point to another, and (iv) ranking the genes from each group. Several variants of this model can be specified according to different measures chosen for clustering and validation phases. The measures used in [1] include *Information Correlation Coefficient (ICC)*, *Pearson Correlation Coefficient (PCC)* and *Rate of Change Similarity (RCS)* measure (see [1] for more details). The following variants of the model, also analysed in [1], have been selected for comparisons in the current study: (i) ICC_MB_PCC: genes are sorted and MB clustered based on ICC while the validation of clusters is based on PCC, and (ii) ICC_MB_RCS: genes are sorted and MB clustered based on ICC while the validation of clusters is based on RCS.

Figure 5 presents a direct comparison of groups produced by SOC with $\psi = 0.07$ and the most representative information based methods i.e. ICC_MB_RCS and ICC_MB_PCC. As shown in Fig. 5, the vast majority of gene pairs have a grey

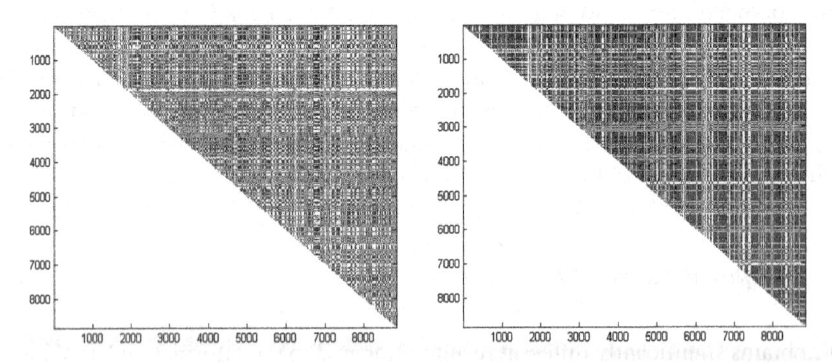

Fig. 5 Comparison of SOC results for production output with most representative information based methods [1]: ICC_MB_RCS (left) and ICC_MB_PCC (right). Colours have the same significance as in Fig. 4

colour associated in the matrix indicating that one of the methods grouped them together while the other method did not. An unsignificant number of genes have been grouped together by both SOC and ICC_MB_RCS, respectively by both SOC and ICC_MB_PCC.

4.4 Biological Perspective

The functional analysis of genes has revealed a biological consistency of each resulting group. It showed that the genes involved in the same biological process tend to be in the same group. This analysis was developed with the predictive annotation of genes to functional categories and the graphical representation of gene expression profiles for each group. Moreover, this analysis has showed that the biological function is coherent with time points in which an expression change is produced. In this way, the co-expression has achieved the initial objectives of reducing the size of groups and improving the biological significance of results.

Regarding the production process, the genes which belong to the production cluster were mainly inserted in two groups that have similar shape index. This observation proves that the combination of co-expression and ranking is working as desired and it is useful to identify the key genes involved in production. Furthermore, this approach can predict the function of novel genes and defines the molecular signatures of main cellular processes.

5 Conclusions and Future Work

A gene clustering method for time series microarray data with production outputs has been presented and analysed in this paper. The objective was to provide meaningful gene profiles based on their expression levels over time and the correlation with different outputs available for each sample. The gene profiles obtained by the proposed method are shown to be more biologically meaningful compared to information based methods.

Future work focuses on extending the SOC method in several ways. The inclusion of a ranking phase based on information theory measures for the groups produced by SOC might provide more relevant results compared to directly applying information based grouping on time series microarray data. Another future direction is to allow different level of importance to each time point in the dataset so that the resulted groups are more faithful to some time intervals considered to be more significant than others.

Acknowledgments This research has been supported through Junta de Castilla y Len projects BIO/BU09/14, CCTT/10/BU/0002 and Fundacin Universidad de Oviedo project FUO-EM-340-13.

References

1. Chira C, Sedano J, Villar JR, Prieto C, Corchado E (2013) Gene clustering in time series microarray analysis. In: Proceedings of International joint conference SOCO'13-CISIS'13-ICEUTE'13 - Salamanca, Spain, pp 289–298, 11-13 Sept 2013
2. Coffey N, Hinde J (2011) Analyzing time-course microarray data using functional data analysis - a review. Stat Appl Genet Mol Biol 10: Article 23
3. Dharmadi Y, Gonzalez R (2004) DNA microarrays: experimental issues, data analysis, and application to bacterial systems. Biotechnol Prog 20(5):1309–1324
4. Ernst J, Bar-Joseph Z (2006) Stem: a tool for the analysis of short time series gene expression data. BMC Bioinformatics 7(1):191
5. Kang A, Chang M (2012) Identification and reconstitution of genetic regulatory networks for improved microbial tolerance to isooctane. Mol BioSyst 8:1350–1358
6. Larrañaga P, Calvo B, Santana R, Bielza C, Galdiano J, Inza I, Lozano JA, Armañanzas R, Santafé G, Pérez A, Robles V (2006) Machine learning in bioinformatics. Briefings Bioinf 7(1):86–112
7. Lee C-P, Leu Y (2011) A novel hybrid feature selection method for microarray data analysis. Appl Soft Comput 11:208–213
8. Liu T, Lin N, Shi N, Zhang B (2009) Information criterion-based clustering with order-restricted candidate profiles in short time-course microarray experiments. BMC Bioinformatics 10(1):146
9. Nieselt K, Battke F, Herbig A, Bruheim P, Wentzel A, Jakobsen O, Sletta H, Alam M, Merlo M, Moore J, Omara W, Morrissey E, Juarez-Hermosillo M, Rodriguez-Garcia A, Nentwich M, Thomas L, Iqbal M, Legaie R, Gaze W, Challis G, Jansen R, Dijkhuizen L, Rand D, Wild D, Bonin M, Reuther J, Wohlleben W, Smith M, Burroughs N, Martin J (2010) The dynamic architecture of the metabolic switch in streptomyces coelicolor. BMC Genomics 11(1):10
10. Pandey G, Yoshikawa K, Hirasawa T, Nagahisa K, Katakura Y, Furusawa C, Shimizu H, Shioya S (2007) Extracting the hidden features in saline osmotic tolerance in saccharomyces cerevisiae from dna microarray data using the self-organizing map: biosynthesis of amino acids. Appl Microbiol Biotechnol 75:415–426
11. Peddada SD, Lobenhofer EK, Li L, Afshari CA, Weinberg CR, Umbach DM (2003) Gene selection and clustering for time-course and doseresponse microarray experiments using order-restricted inference. Bioinformatics 19(7):834–841
12. Phan S, Famili F, Tang Z, Pan Y, Liu Z, Ouyang J, Lenferink A, O'connor MM-C (2007) A novel pattern based clustering methodology for time-series microarray data. Int J Comput Math 84:585–597
13. Pickens L, Tang Y, Chooi Y-H (2011) Metabolic engineering for the production of natural products. Annu Rev Chem Biomol Eng 2(1):211–236
14. Saeys Y, Inza I, Larrañaga P (2007) A review of feature selection techniques in bioinformatics. Bioinformatics 23(19):2507–2517
15. Smyth G, Speed T (2003) Normalization of cdna microarray data. Methods 31(4):265–273
16. Storey JD, Xiao W, Leek JT, Tompkins RG, Davis RW (2005) Significance analysis of time course microarray experiments. Proc Nat Acad Sci U S A 102(36):12837–12842
17. Tummala S, Junne S, Paredes C, Papoutsakis E (2003) Transcriptional analysis of product-concentration driven changes in cellular programs of recombinant clostridium acetobutylicum-strains. Biotechnol Bioeng 84(7):842–854

Heuristics for Apnea Episodes Recognition

Silvia González, José Ramón Villar, Javier Sedano, Joaquín Terán, María Luz Alonso Álvarez and Jerónimo González

Abstract The Sleep Apnea is a respiratory disorder that affects a very significant number of patients, with different ages. One of the main consequences of suffering from apneas is the increase in the risk of stroke onsets. This study is concerned with an automatic identification of apnea episodes using a single triaxial accelerometer placed on the center of the chest. The relevance of this approach is that the devices for home recording and the analysis of the data can be highly reduced, increasing the patient comfort during the data gathering and reducing the time needed for the data analysis. A very simple heuristic has been found useful for identifying this type of episodes. For this study, normal subjects have been evaluated with this approach; it is expected that data from patients that might suffer apneas will be available soon, so the performance of this approach on real scenarios can be reported.

J. Ramón Villar
Instituto Tecnológico de Castilla y León. C/López Bravo 70, Pol. Ind. Villalonquejar, 09001 Burgos, Spain
e-mail: villarjose@uniovi.es

S. González (✉) · J. Sedano
Computer Science Department, University of Oviedo, ETSIMO, C/Independencia 13, 33004 Oviedo, Spain
e-mail: silvia.gonzalez@itcl.es

J. Sedano
e-mail: javier.sedano@itcl.es

J. Terán · M. Luz Alonso Álvarez
Hospital Universitario de Burgos, Unidad de Sueño y Unidad de Investigación, 09006 Burgos, Spain
e-mail: joaquinteransantos@yahoo.es

J. González
Faculty of Humanities, University of Burgos, C/Villadiego, s/n, 09001 Burgos, Spain
e-mail: jejavier@ubu.es

Á. Herrero et al. (eds.), *10th International Conference on Soft Computing Models in Industrial and Environmental Applications*, Advances in Intelligent Systems and Computing 368, DOI 10.1007/978-3-319-19719-7_22

1 Introduction

The Respiratory Rate (RR) is a vital sign that provides important information about the patient's health; it is widely accepted that an abnormal RR could indicate a variety of pathological conditions including respiratory, cardiovascular and metabolic disorders [1], the apnea among them. Apneas, which are a sleep disorders, not only affect adults but some children as well. The Sleep Apnea/Hypopnea Syndrome (SAHS) is characterized by drowsiness, cardiorespiratory neuropsychiatric disorders, functional abnormalities of the upper airway that leads to repeated obstruction during sleep and cardiorespiratory neuropsychiatric disorders. SAHS causes reductions in blood Oxygen Saturation (SaO_2) and produces non-restorative sleep. SAHS is associated with increased cardiovascular risk as well as psychosocial complications such as daytime somnolence, depression, and fatigue [2]. There are three forms of apnea: Central Sleep Apnea (CSA), Obstructive Sleep Apnea (OSA), and complex or mixed sleep apnea (CSA followed by a OSA) [3]. In all them exists an absence or reduction of >90 % respiratory signal in the nasal cannula for >10 seconds in the presence or absence of respiratory effort, respectively. The most striking feature of obstructive respiratory events is that they are at their most severe and frequent in the supine sleeping position: indeed, more than half of all OSA patients can be classified as supine related OSA. Moderate to severe OSA occurs in 10–17 % of middle aged men and 3–9 % of middle-aged women with a higher prevalence among obese subjects [4]. It is noteworthy that patients with OSA have been reported to have more obstructive events in the supine position than in the lateral position, and when this is the case, forcing a change to the nonsupine position during sleep can be an effective treatment [5].

There exists supporting evidence of supine OSA to be due to unfavorable airway geometry, to a reduced lung volume, or to an inability of the airway dilator muscles to adequately compensate the airway collapses. The mechanisms of supine related OSA can be overcome by the use of Continuous Positive Airway Pressure (CPAP) [6]. CPAP uses mild air pressure to keep the airways open, and it is mainly used by people who have breathing problems, such as sleep apnea. Thus, the challenge is to identify the OSA episodes for correctly diagnose the patient. This medical study, known as Polysomnography (PSG), is typically carried out at the sleep laboratories or, depending on the case, in the patient's home using portable devices [7]. The PSG is a continuous recording of different biometrics, including the air flow, the electrocardiogram, etc. However, the PSG devices are expensive, and the patients have limited or constrained movements, making the sleep experience uncomfortable and different as usual. Alternatively, OSA has been also studied based in electrocardiograms [8], in exhaled breath condensate [9], or in accelerometers and other sensors for apneas detection [10]. The accelerometers have been also reported for identifying the volume of airflow [11]. Interesting enough to mention, the studies reported by Lapi et colleagues proposed the identification of the breath rate based on two triaxial accelerometers (3DACC) placed on both sides of the chest even if the movements are limited by the patient's sleeping position [1].

This study is focused on identification of OSA episodes based on the measurements gathered from a single 3DACC located in the center of the chest. The advantages of using this approach include reducing the interference in the patient's normal sleeping and a decreasing in the cost of the equipment and data analysis. This research proposed some algorithms based on simple heuristics to identify the patient's posture and breath rate, as well as a threshold method for identifying the OSA episodes. Data gathered from normal patients have been obtained and the algorithms have been tested.

Our study aims to detect respiratory diseases, as a recent study [12], that integrates textile electrodes on beds, as studies that used more than one accelerometer [1, 11], or a combination with other sensors [10], while we use only one sensor in a localized area. Our approach is wearable and comfortable, decreasing the final cost of the system.

The organization of this contribution is as follows. The next section deals with the details and the algorithms proposed by this study. Section 3 includes the description of the experimentation and the obtained results. Finally, the extracted conclusions are presented.

2 A Solution for OSA Recognition

The hypothesis we propose is that the absence or reduction of the respiratory signal can be detected by means of placing a single 3DACC on the center of the chest wall, just at the end of the breastbone. This location would allow to identify the movements made during inspiration and expiration produced during respiration according to the current position of the patient, while might reduce the disturbance of sensory system in the patient's sleeping process. Eventually, the apnea episodes can be identifying by measuring the time between consecutive respiration cycles.

Thus, the problem is reduced to identification of the breath cycles and then measurement of the time elapsed between breaths. Using a 3DACC means that according to the posture of the patient, there will be one axis -either x, y or z- that performs better, that is, there is an axis which is more sensitive to the movements for each posture. Therefore, this study includes the following steps: (1st) data preprocessing, (2nd) posture identification, (3rd) breath cycle identification, and (4th) apnea identification. From the input Acceleration (A) vector, the raw acceleration, the Body Acceleration (BA) is obtained by filtering the raw acceleration with a high band pass filter. Similarly, the Gravity (G) components can be extracted using a low band pass filter [13]. Let us call the three components of the A, BA and G accelerations as $a_{i,j}$, $b_{i,j}$ and $g_{i,j}$, respectively, with i the correlative sample number and $j \in \{x, y, z\}$ being the axis.

2.1 The Posture Identification Stage

For this study only five well-known postures are to be considered: (1) lying with the face up -aka Decubito Supino, DS-, (2) lying on his left side -Decubitus Left Lateral, DLL-, (3) lying with the face down -Decubito Prono, DP-, (4) lying on his right side -Decubitus Right Lateral, DRL-, and (5) partially lying on his right side -Decubitus Partially Lateral, DPL-. To determine the posture of a subject the angle $\delta_{j\in\{x,y,z\}}$ of the gravity components with respect to the gravity vector direction (refer to Table 1), that is, $\delta_{j\in\{x,y,z\}}(g_{i,x}, g_{i,y}, g_{i,z}) = arcos(\frac{g_{i,j}}{\sqrt{(g_{i,x}^2 + g_{i,y}^2 + g_{i,z}^2)}})$.

2.2 Marking the Breath Cycles

The breath cycle includes inhalation and exhalation activities. The diaphragm extends the chest wall -increasing the chest's volume- during the inhalation, and returns to its initial state -compressing the chest- during the exhalation [14]. During the sleeping, the rate and depth of breathing are reduced unconsciously; the duration of breathing cycle takes about 3 seconds, 1s spent in inhalation and 2 s in exhalation [15].

Table 1 Posture identification based on the δ_z and δ_y values

Postures	Considering 4 postures	Considering 5 postures
Decubito supino	$\delta_z \in [0, \frac{7}{36}\pi]$ and $\delta_y \in [\frac{7}{36}\pi, \frac{2}{3}\pi]$	$\delta_z \in [0, \frac{7}{45}\pi]$ and $\delta_y \in [\frac{16}{45}\pi, \frac{2}{3}\pi]$
Decubito prono	$\delta_z \in [\frac{2}{3}\pi, \pi]$ and $\delta_y \in [\frac{7}{36}\pi, \frac{2}{3}\pi]$	$\delta_z \in [\frac{26}{45}\pi, \pi]$ and $\delta_y \in [\frac{16}{45}\pi, \frac{2}{3}\pi]$
Decubito lateral left	$\delta_y \in [0, \frac{7}{36}\pi]$ and $\delta_z \in [\frac{7}{36}\pi, \frac{2}{3}\pi]$	$\delta_y \in [0, \frac{7}{45}\pi]$ and $\delta_z \in [\frac{16}{45}\pi, \frac{2}{3}\pi]$
Decubito lateral right	$\delta_y \in [\frac{2}{3}\pi, \pi]$ and $\delta_z \in [\frac{7}{36}\pi, \frac{2}{3}\pi]$	$\delta_y \in [\frac{26}{45}\pi, \pi]$ and $\delta_z \in [\frac{16}{45}\pi, \frac{2}{3}\pi]$
Decubito prono partial		$\delta_z \in [\frac{2}{3}\pi, \frac{26}{45}\pi]$ and δ_y $[\frac{7}{45}\pi, \frac{16}{45}\pi]$

To record the breath rate it is only necessary to measure the cyclical movements, for which a 3DACC is provided. More specifically, the BA should be enough to measure these movements. However, the breath's movements when sleeping are soft movement, that is, there are no high acceleration values (refer to the left part in Fig. 1.

For instance, if the subject is in supine position, the BA will be slightly higher than 0 when inhaling, while when exhaling the BA will be slightly negative. In order to magnify these effects, a sliding window is proposed. Within the sliding window, the positive and the negative values of the BA are added independently. These semi integrals allow to scale up the effects of the diaphragm movements (right part of Fig. 1). Finding the maximum peaks in each semi-cycle, and taking into account that an inhalation precedes an exhalation, the peaks for the breath cycle can be determined.

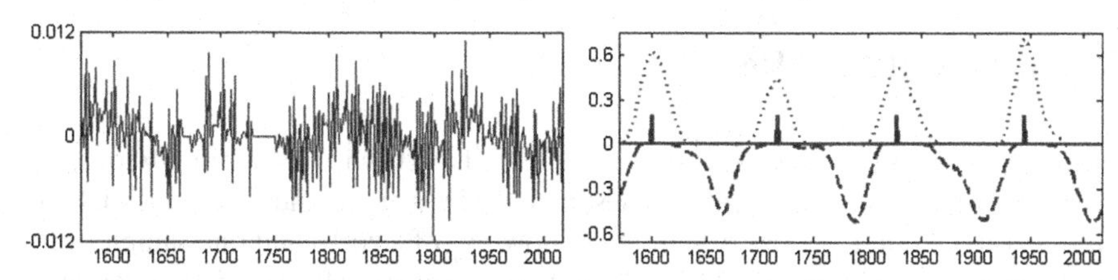

Fig. 1 Left part, the evolution of b_i^x while sleeping. Right part, the aggregation of both the positive values (dotted) and negative values (dashed). Also, the peaks that marks the breath cycles

Nevertheless, a better result can be afforded if the most remarkable axis is considered for aggregation instead of the BA modulus. In Fig. 2 the evolution of the components of the BA for each posture are depicted. more specifically, if we focus on the supine position, the most remarkable x axis, so the $b_{i,x}$ is used for determining the breath cycles while the subject rests in that posture. Lets denote $w_{i,j}$, $s^+(w_{i,j})$ and $s^-(w_{i,j})$ the sliding window at sample i for the axis j, the sum of the positive and the sum of the negative values in $w_{i,j}$, correspondingly. Then, the pseudo-code shown in Algorithm 1 describes the identification of the breath cycles and the estimation of the time between peaks.

In this algorithm the current patient's posture is calculated; then the main component of the BA is determined. This means that, although the calculations are carried out for the three components, only those corresponding to the current main component are used for determining the peaks and the time between peaks. However, every time the activity level of the subject is highly modified due to a change in the position, the calculations are reset. The output of the algorithm is the set of peaks, the corresponding time stamps and the time between consecutive peaks. The apnea events are detected when the time between breath cycles is higher than certain value. This value can be obtained from training and can be related with the breath rate. This algorithm is run in batch mode after gathering the data; nevertheless, it can be adapted for on-line deployment. This issue is rather interesting, as far as this can lead us to develop a device for helping patients suffering of apneas to detect them on the fly and alarm the patient suggesting a change in the posture.

Moreover, there are two thresholds: γ^+ and γ^-. Setting these parameters, as well as the apnea time interval, require further analysis. There are several ways of determining these values: extracting them from the data as general values for all the population [1], adaptive thresholds [7], a mixture of both methods [16], etc. In this study, the thresholds depicted in Table 2 were considered after analyzing the data from several patients. The same values were applied for each axis.

Table 2 Main threshold used during this study. $\beta = 0.15$ for the smoking patients

	Postures				
Threshold	DS	DLL	DP	DRL	DPL
γ^+	0.26	0.1	0.12	0.15	0.08
γ^-	0.26-β	0.1	0.12-(β/2)	0.1	0.08

3 Experiments and Results

The data gathering has been carried out using a 3DACC with a sampling frequency of 16 Hz, placed in the middle of the chest onto the breastbone and fixed with sticking plasters directly on the skin. A sliding window $w_{i,j}$ of size 16 and one sample shift has been used for computing the of $s^+(w_{i,j})$ and $s^-(w_{i,j})$, a sliding window for each axis $j, j \in \{x, y, z\}$. For this study, four female subjects ranging from 26 to 53 years old were asked to lay on a bed, remaining in each posture during a period of time (see Table 3). Evidence of differences among the subjects were found, specially for the smoker subject. Therefore, the time for determining an apnea occurrence was settle to 11.875 s for the non-smoker subjects and 11.25 for the smoker one. For each posture, the subjects should run a breath sequence consisting on 20 normal breaths followed by 5 apnea simulations; the first three simulations were characterized with high respiratory effort, the two latter ones were low-intensity simulations. To simulate a high-effort (low-effort) apnea, the subject was asked to perform normal breaths, then to hold their breath during 20 (10) seconds, then exhaling and inhaling. At least, the subject should hold the breath during 10 (5) seconds to be considered a valid high-effort (low-effort) apnea simulation.

The posture identification was performed satisfactorily. For the sake of space, the results are not shown. However, it can be said that the percentages of success were: DS 100 %, DLL 100 %, DP 100 %, DRL 100 % and DPL ranging from 72 % for the smoker subject to 100 % of two healthy subjects.

With respect to the breath cycle identification, the results are displayed in Table 4. It is worth noting that the almost all the breath cycles were identified using this simple approach of thresholds. Once these breath cycles are found correctly, identifying the periods of time that exceeds the settled limits for apnea recognition also works almost perfectly. There appeared some problems with the data gathered from some subjects in certain postures -shown as – (–) in the above referred table-. For these cases, the amount of signal was not enough to properly work with the given thresholds.

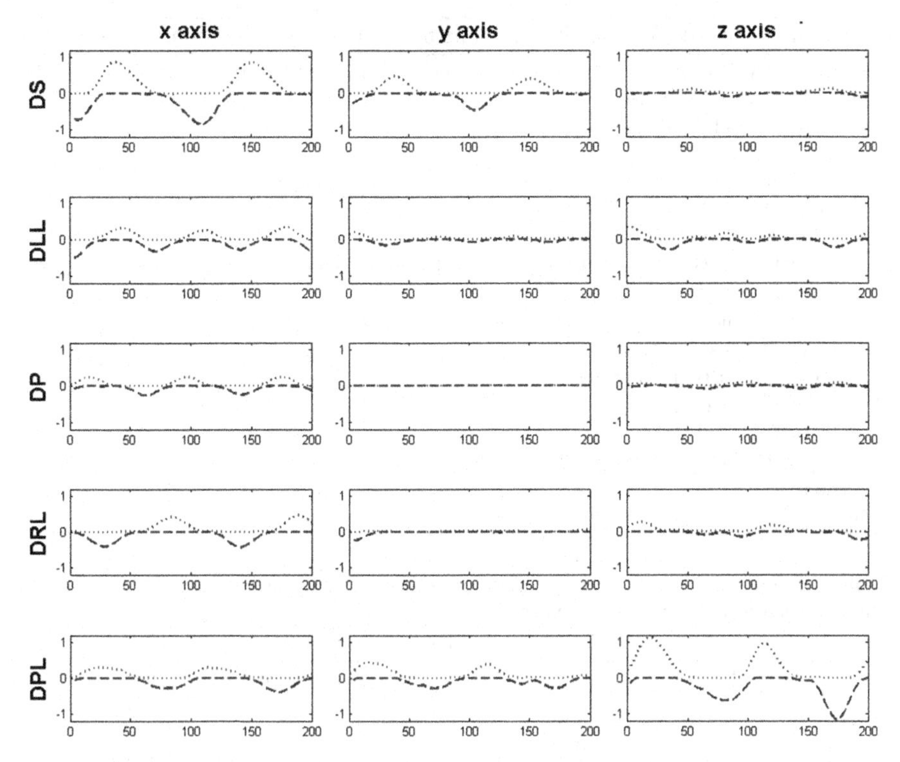

Fig. 2 Evolution of the $s^{+}(w_{i,j})$ and $s^{-}(w_{i,j})$ for each posture

Table 3 Data and facts from the subjects

Ind	Age	Smoke	Problems with apnea simulation	Number of samples according to the posture				
				DS	DLL	DP	DRL	DPL
1	27	No	No	4503	3134	3567	4194	4202
2	53	No	No	4863	4714	5041	4005	4711
3	26	Yes	Yes	4196	3671	3426	3481	3179
4	27	Yes	No	2513	2619	2539	2582	2641

4 Conclusions

A proposal for identifying breath cycles and time between breaths is proposed using a single 3DACC and an threshold based algorithm. Results from the experiments showed promising results, though there are several drawbacks. The main one is concerned with the thresholds: they should be adapted to the subject's breath movements. The time for apnea recognition is also a parameter that should be set for each

Algorithm 1 Breath identification and time between peaks estimation.

state is INHALING and $S_j^+ = \gamma_j^+$, *and* t_j^+=0 and $S_j^- = \gamma_j^-$, *and* $t_j^- = 0$
for each $b_{i,j}, j \in \{x, y, z\}$ **do**
  compute δ_j and posture$_j$
  determine the main $b_{i,j}$ component
  update the sliding window $w_{i,j}$ and compute $s^+(w_{i,j})$ and $s^-(w_{i,j})$
  if $s^+(w_{i,j}) > S_j^+$ **then**
    if state is INHALING **then**
      state is EXHALING and $tbp_i = t_j^+ - t_j^-$ *time between peaks*
    end if
    $S_j^- = \gamma_j^-$ and $S_j^+ = s^+(w_{i,j})$ and $t_j^+ = t_i$ *time at sample i*
  else if $s^-(w_{i,j}) < S_j^-$ **then**
    if state is EXHALING **then**
      state in INHALING
    end if
    $S_j^+ = \gamma_j^+$ and $S_j^- = s^-(w_{i,j})$ and $t_j^- = t_i$ *time at sample i*
  end if
end for

Table 4 Results for identifying breath cycles and apnea events for each subject. The number of apnea events is included between parenthesis

Ind	DS	DLL	DP	DRL	DPL
1	20 (5)	18 (5)	20 (5)	20 (5)	20 (5)
2	20 (4)	20 (5)	20 (5)	17 (5)	20 (5)
3	18 (5)	20 (5)	– (–)	20 (5)	– (–)
4	20 (5)	19 (5)	– (–)	19 (5)	19 (5)

patient. So the challenge is to extract the correct relationships between the patient data in order to properly set these parameters and thresholds.

Although we are unable to detect breath in DP position in two of the patients, the approach continues to be valid because supine OSA is the dominant phenotype of the OSA syndrome [6]. Hence, time between breaths is a valid method, due the high percentages of apneas detection.

Future work envisages the implementation of the system in real patients, easing the contrast of the presented approach. Due of the great difficulty implied by this, we have not been able to add the results in this study.

Acknowledgments This research has been funded by the Spanish Ministry of Science and Innovation, under projects TIN2011-24302 and TIN2014-56967-R, Fundaci6n Universidad de Oviedo project FUO-EM-340-13, Junta de Castilla y Le6n projects BIO/BU09/14 and SACYL 2013 GRS/822/A/13.

References

1. Lapi S, Lavorini F, Borgioli G, Calzolai M, Masotti L, Pistolesi M, Fontana GA (2014) Respiratory rate assessments using a dual-accelerometer device. Respir Physiol Neurobiol 191:60–66
2. Stepnowsky CJ, Palau JJ, Zamora T, Ancoli-Israel S, Loredo JS (2011) Fatigue in sleep apnea: the role of depressive symptoms and self-reported sleep quality. Sleep Med 12:832–837
3. Lloberes P, Durán-Cantolla J, Martínez-García MÁ, Marín JM, Ferrer A, Corral J, Masa JF, Parra O, Alonso-Álvarez ML, Terán-Santos J (2011) Diagnosis and treatment of sleep apnea-hypopnea syndrome. Arch Bronconeumol 47(3):143–156
4. Badran M, Ayas N, Laher I (2014) Insights into obstructive sleep apnea research. Sleep Med 15:485–495
5. Duran-Cantolla J, Barbe F, Rigau J, Oreja D, Martinez-Null C, Santaolalla CE (2013) A new vibratory postural device for the treatment of positional obstructive sleep apnea (OSA). A pilot study Sleep Med 14:e111
6. Joosten SA, O'Driscoll DM, Berger PJ, Hamilton GS (2014) Supine position related obstructive sleep apnea in adults: pathogenesis and treatment. Sleep Med Rev 18:7–17
7. Doukas C, Petsatodis T, Boukis C, Maglogiannis I (2012) Automated sleep breath disorders detection utilizing patient sound analysis. Biomed Signal Process Control 7:256–264
8. Sannino G, Falco ID, Pietro GD (2014) Monitoring obstructive sleep apnea by means of a real-time mobile system based on the automatic extraction of sets of rules through differential evolution. J Biomed Inf 49:84–100
9. Antonopoulou S, Loukides S, Papatheodorou G, Roussos C, Alchanatis M (2008) Airway inflammation in obstructive sleep apnea: Is leptin the missing link? Respir Med 102:1399–1405
10. Al-Mardini M, Aloul F, Sagahyroon A, Al-Husseini L (2014) Classifying obstructive sleep apnea using smartphones. J Biomed Inf 52:251–259
11. Rodes C, Chillrud S, Haskell W, Intille S, Albinali F, Rosenberger M (2012) Predicting adult pulmonary ventilation volume and wearing complianceby on-board accelerometry during personal level exposure assessments. Atmos Environ 57:126–137
12. Peltokangas M, Verho J, Vehkaoja A (2012) Night-time ekg and hrv monitoring with bed sheet integrated textile electrodes. IEEE Trans Inf Technol Biomed 16:935–942
13. Villar JR, Gonzlez S, Sedano J, Chira C, Trejo-Gabriel-Galan JM Improving human activity recognition and its application in early stroke diagnosis. Int J Neural Syst. DOI: 10.1142/S0129065714500361
14. Patiño J, Restrepo J, Rodríguez E (2005) Gases sanguíneos, fisiología de la respiración e insuficiencia respiratoria aguda. Editorial Medica Panamericana Sa de
15. Myrrha MAC, Vieira DSR, Moraes KS, Lage SM, Parreira VF, Britto RR (2013) Chest wall volumes during inspiratory loaded breathing in COPD patients. Respir Physiol Neurobiol 188:15–20
16. Bsoul M, Minn H, Member S, Tamil L, Member S (2011) Apnea medassist: real-time sleep apnea monitor using single-lead ecg. In: IEEE Trans Inf Technol Biomed 416–427

Energy-Efficient Sound Environment Classifier for Hearing Aids Based on Multi-objective Simulated Annealing Programming

Alberto Cocaña-Fernández, Luciano Sánchez, José Ranilla, Roberto Gil-Pita and David Ayllón

Abstract A methodology for designing classifiers through multicriteria metaheuristics is introduced. The purpose of the new method is to jointly optimize the expected total classification cost and the energy consumption of the device the classifier is implemented on. A numerical study is provided, where different alternatives are implemented on a hearing aid. This aid is capable of automatically classifying the acoustic environment that surrounds the user and choosing the parameters of the amplification that are best adapted to the user's comfort. The proposed method attains relevant improvements in energy consumption with small to negligible loss in classification accuracy with respect to a selection of algorithms.

1 Introduction

The overall performance of machine learning classification algorithms is usually assessed in terms of the expected total classification cost in the target population. In the most frequent case, this cost is the proportion of misclassified instances. In other cases, different penalties and rewards are assigned to incorrectly and correctly classified instances [1]. Notwithstanding this, there are cases where the desired cost scheme is not related to the confusion matrix of the classifier. In particular, costs derived from energy consumption and computational efficiency do not fit into the cost-sensitive classification framework and are seldom addressed by regular classification schemes.

Despite this, consumptions and efficiencies play a key role when classifiers are implemented in energy-constrained devices that must work at low clock rates to minimize power consumption and maximize battery life. An example of this are state-of-

A. Cocaña-Fernández · L. Sánchez (✉) · J. Ranilla
Computer Science Department, University of Oviedo, Gijón, Spain
e-mail: luciano@uniovi.es

R. Gil-Pita · D. Ayllón
Department of Signal Theory and Communications, University of Alcalá, Madrid, Spain

Á. Herrero et al. (eds.), *10th International Conference on Soft Computing Models in Industrial and Environmental Applications*, Advances in Intelligent Systems and Computing 368, DOI 10.1007/978-3-319-19719-7_23

the-art hearing aids that are capable of automatically classifying the acoustic environment that surrounds the user, and selecting the parameters of the amplification that are best adapted to the user's comfort [2]. This self-adaptation of hearing aids is done by means of Sound Environment Classifiers (SEC). User comfort is improved as the user is not required to manually tune the device but the most suitable program is automatically chosen [3]. They also work as a support for speech enhancement, noise reduction [4, 5], or voice activity detection [6].

In previous approaches to sound environment classification in hearing aids, the feature extraction of sound signals was based on Mel Frequency Cepstral Coefficients (MFCCs) [7]. MFCCs render good results in terms of error rate, but require high computational resources [8], which further increases power consumption and reduces battery life. Since the problem here consists in finding a sound classifier that minimizes both error rates and power consumption, and these are conflicting objectives, the traditional approach would be to calculate a set of Pareto-optimal solutions. This can be done by means of a multi-objective optimization algorithm, addressing both the problem of selecting the best subset of MFCC-based features and the selection of the most suitable classification scheme learnt upon a provided training set with the previously chosen features.

In this paper it is shown that this approach is fundamentally limited because, as mentioned before, the learning algorithms for regular classification schemes, such as Support Vector Machines, decision trees or rule-based systems, are designed solely to minimize the expected error rate. This is expected to yield suboptimal results in terms of energy efficiency. A new approach is proposed for that reason where energy-efficient fuzzy classification systems are found by means of multi-objective simulated annealing programming, addressing the feature selection and classification problems altogether in order to undercut energy consumption without a significant loss in classification power.

The remainder of the paper is as follows. Section 2 explains the MFCC-based sound feature extraction. Section 3 details the cost evaluation of a classifier. In Sect. 4 the learning algorithm is introduced. Experimental results are given in Sect. 5. The paper concludes in Sect. 6.

2 MFCC-Based Sound Feature Extraction

Prior to the classification of the environment, sound signals are processed in order to transform them into the Mel Frequency Cepstral Coefficients, which provide a compact representation of the spectral envelope, having most of the energy concentrated in the first coefficients [9]. The steps that the MFCC calculation requires are:

1. Apply the Short Time Fourier Transform (STFT) to the input sound signals.
2. Transform the energy of the STFT onto the Mel scale by multiplying its values by a triangular filter bank. These filters are determined by the behaviour of the human psychoacoustic system, approximated through the frequency in Mel scale.

3. Apply the Discrete Cosine Transform (DCT) to the logarithm of the spectral coefficients, converting the Mel frequency spectral coefficients to a cepstral domain.

Once the MFCCs are calculated, features are determined from temporal statistics of each MFCC, where the most common statistics are the mean and the standard deviation, as these have been successfully used in sound environment classification problems for hearing aids [10].

In this paper, instances in the classification problem comprise features that are defined as follows. Let F be the number of MFCCs, let M_{mt} be the value of the m-th MFCC of the temporal frame t, and let T be the number of frames between classifications. Then, every classification instance can be expressed as $(c_1 \dots c_F \dots c_{2F})$ where the i-th value $(i = 1 \dots F)$ represents the average value of the i-th MFCC calculated as:

$$c_i = \frac{1}{T} \sum_{t=1}^{T} M_{mt} \tag{1}$$

and where the j-th value $(j = F + 1 \dots 2F)$ represents the standard deviation value of the $(j - F)$-th MFCC calculated as:

$$c_j = \sqrt{\frac{1}{T} \sum_{t=1}^{T} M_{mt}^2 - \left(\sum_{t=1}^{T} M_{mt} \right)^2} \tag{2}$$

3 Multiobjective Feature and Classification Scheme Selection

Standard classification schemes do not take into account the computational efficiency of the classifier or the energy consumption of the physical device the classifier is implemented on. At most, limits can be imposed on the length of the algorithm implementing the decision system, i.e. the height of a decision tree or the number of hidden nodes of a neural network, to name some. In this respect, the search of the best balance between energy consumption and classification performance is divided into two stages: (a) many different classifiers are independently designed on the basis of their classification performances and (b) a search is conducted on the results of the first stage that finds the best compromise between computational complexity and accuracy.

Standard learning algorithms are blind to energy restrictions, that are not present in the feature selection stage neither in the learning stage. As a consequence of this, the most suitable classifier is found by searching among the results of multiple blind learnings. In this paper a non-standard method will be described (see Sect. 4) that jointly optimizes the classification performance and the energy consumption during feature selection (which is implicit in the new algorithm) and also during the rule learning stage. In any case, the evaluation of the computational cost of an algorithm that will be used in both approaches is described in this section thus a numerical comparison between standard and non-standard classifiers can be performed later (see Sect. 5).

Given a set comprising the results of different classification algorithms that have been applied to the same problem –stage (a)– it makes sense to find a set of non-dominated solutions comprised of a subset of features plus a classification scheme –stage (b)–. A solution to this last problem can be expressed as $(b_1, b_2 \ldots, b_F \ldots b_{2F}, s)$, where b_i $(i = 1 \ldots 2F)$ is a bivaluated value indicating whether the i-th feature is chosen or not, and where s takes integer values, each one representing a classification scheme, such as SVM, k-NN, C4.5, PART, etc. Each solution has a fitness that consists of two values: classification error rate and the computational cost measured as the number of instructions per second required for the hearing aid to do the feature extraction and the classification. This computational cost can be calculated as

$$C_T = \frac{2F_s}{N_s}\left(\sum_{m=1}^{F'} C_M(m)\right) + \frac{2F_sF'}{N_sT}C_S + \frac{2F_s}{N_sT}C_C \tag{3}$$

where $C_M(m)$ is the computational cost associated to the computation of the m-th MFCC, C_S is the computational cost associated to the evaluation of the statistics of each MFCC, C_C is the computational cost associated to the classifier, F_s is the sampling frequency, F' is the number of selected MFCCs, and N_s is the number of samples obtained dividing the input sound signals into segments at the first step of the MFCC calculation.

In particular, the cost associated to the computation of each MFCC can be approximated by 27 instructions per frame, and the cost associated to the evaluation of the statistics of every MFCC can be approximated by 8 instructions if only the average is computed, and 40 instructions if both average and standard deviation are computed.

The classification cost depends on the classification scheme chosen. Let N be the total number of features extracted from the MFCCs and N_c the number of sound environment classes, then the computational cost of using a SVM as classifier is

$$C_C^{SVM} = \binom{N_c}{2} \times 2 \times N \tag{4}$$

In the case of either ruled-based systems or decision trees, all costs rely on the path followed to get the output of the classifier, since its cost is not deterministic, and not in every classification would require the same features. Therefore, let C_T^{min} and C_T^{max} be the minimum and maximum possible computational costs, respectively. Then, the computational cost of using a decision tree or a rule-based system as classifier is

$$C_T = \frac{C_T^{max} + C_T^{min}}{2} \tag{5}$$

4 Multicriteria Design of Fuzzy Systems with Simulated Annealing

In this section a new design procedure is described that has been successfully used for obtaining a fuzzy rule-based classification system [11] which balances performance and energy efficiency. The learning algorithm used is the MOSA-P [12], a multi-objective extension of the SA-P algorithm defined in [13].

In short, the multi-objective simulated annealing (MOSA) algorithm is based on a variable sized population whose individuals are mutated at each iteration. The fitness of the new mutated individuals is calculated and compared to their corresponding predecessor search point. If the new individual dominates the current search point, it replaces its parent in the population. If neither the new individual nor its parents dominates the other, both are included in the population. If the new individual is dominated, it still might be included in the intermediate instead of its parent. This decision is made on a random basis, with a probability that depends on a cooling pattern, and decreases with both the distance and the time.

The population is regularly sampled by means of a selection operator that prevents it from growing past a certain size. Also, elitism is applied to the algorithm by keeping a set of non-dominated solutions aside from the population. This set will be the output of the algorithm.

Further information on the MOSA-P algorithm, as well as regarding mutation and selection operators can be found in reference [12].

4.1 *Representation of an Individual*

Fuzzy rule-based classifiers are individuals of the search population. Each of them consists in turn in a valid chain in the context free grammar defined by the following production rules:

CLASSIFIER $\rightarrow$ if CONDITION then class is e_1
if CONDITION then class is e_2
...
if CONDITION then class is e_{N_c}
LOGICAL-CONS

CONDITION $\rightarrow$ ASSERT$_1$ | ASSERT$_2$ | ... | ASSERT$_{2F}$
| (CONDITION $\vee$ CONDITION)
| (CONDITION $\wedge$ CONDITION)
| K_1 | K_2 | ... | K_{N_l}

$$\text{ASSERT}_1 \rightarrow c_1 \; is \; V_{11} \mid c_1 \; is \; V_{12} \mid \ldots \; c_1 \; is \; V_{1L_1}$$
$$\ldots$$
$$\text{ASSERT}_N \rightarrow c_N \; is \; V_{N1} \mid c_N \; is \; V_{N2} \mid \ldots \; c_N \; is \; V_{NL_N}$$

$$\text{LOGICAL-CONS} \rightarrow K_1 \; K_2 \; \ldots \; K_{N_l}$$

where N_c is the number of sound environment classes, $c_1 \ldots c_N$ are the aforementioned features and L_i the number of linguistic terms in feature i. Also this grammar includes a chain of N_l logical constants taking values between 0 and 1.

The output class given by a classifier following this grammar is the one at the consequent of the rule with the highest compatibility grade in its antecedent. This compatibility grade is computed using these semantic actions (Fig. 1):

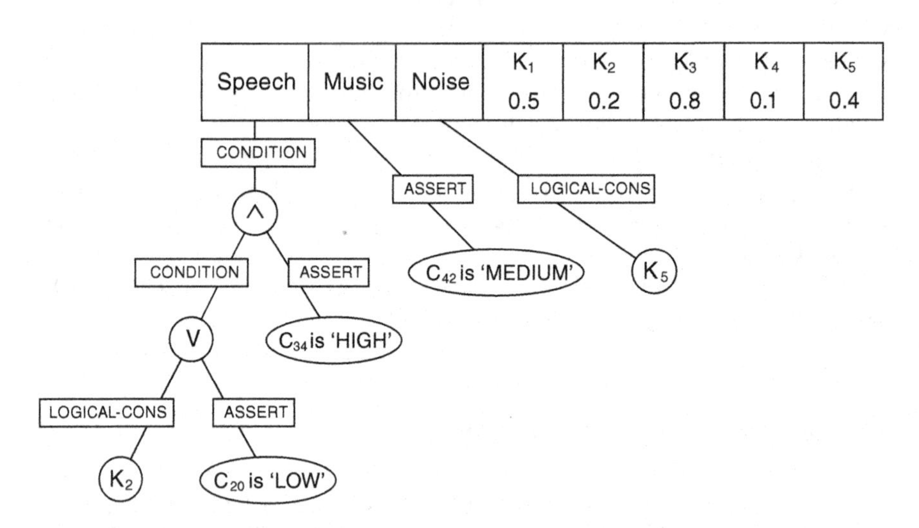

Fig. 1 Genotype of an individual in the MOSA algorithm. An individual comprises a set of rules and a chain of logical constants

Production	Action
COND → $C_1 \wedge C_2$	COND.value = C_1.value × C_2.value
COND → $C_1 \vee C_2$	COND.value = C_1.value + C_2.value − (C_1.value × C_2.value)
COND → CONSTANT	COND.value = CONSTANT.value
COND → ASSERT	COND.value = ASSERT.value
ASSERT → $c_N \; is \; V_{NN_L}$	ASSERT.value = $\widetilde{T}_{N_L}(c_N)$

where $\widetilde{T}_1, \ldots, \widetilde{T}_{N_L}$ are triangular fuzzy sets forming a uniform fuzzy partition [14] of the domain of the variables c_N.

Lastly, the fitness of each MOSA-P individual is a pair of numbers, comprising both the training error and a cost calculated as described in the previous section.

5 Experimental Results

The experimental setup used to evaluate standard and non-standard approaches is based on a database comprised of 1776 seconds of audio, including samples of the three considered classes ($N_c = 3$): speech (including speech in a quiet environment, speech in a noisy ambient, and speech combined with music), music (including vocal music and instrumental music) and noise. The database files were sampled with a sampling frequency of $F_s = 16$ KHz and 16 bit per sample, a frame size of 8 ms, and decisions were made with time slots of 16 ms ($T = 4$).

The results obtained using the approach described in Sect. 2 were calculated using the Non-dominated Sorting Genetic Algorithm II (NSGA-II), through its implementation in the MOEA Framework,[1] and WEKA implementations of SVM, C4.5, PART and k-NN (with $k = 1$ and $K = 5$) [15]. In Table 1 the results of these classifiers in both training and test are shown. Observe that the expected cost of the decision trees is better but SVMs are the most accurate classifiers in the test sets. Other approaches with comparable accuracy but higher computational cost (k-NN, kernel classifiers, etc.) were evaluated but they are not included in the results as they were dominated by either SVM or C4.5.

Table 1 Sample of the set of non-dominated solutions obtained in the experiment by means of NSGA-II based feature and classifier selection

Training error (%)	Test error (%)	Cost (MIPS)	Classifier
23.838	23.872	134625	SVM
24.250	24.718	125000	SVM
24.307	23.872	113000	SVM
24.888	24.051	95375	SVM
25.750	24.051	81000	SVM
26.406	25.744	75750	SVM
26.762	25.744	73000	SVM
27.905	28.205	62500	SVM
28.861	29.385	41843	C4.5
29.273	28.974	32218	C4.5
30.304	29.564	28500	C4.5
31.878	29.410	27218	C4.5
31.934	29.000	26312	C4.5

The application of the new method results in a wide range of classifiers with different balances between cost and accuracy, as shown in Table 2. Observe that MOSA-P solutions undercut the cost of standard classifiers but the best accuracy found so far

[1]MOEA Framework, a Java library for multiobjective evolutionary algorithms, http://www.moeaframework.org/.

was not better in test than that of the most complex SVM. This is an expected result, as the implementation of MOSA-P in this paper is guided to optimize the train error of the classifier, which is simpler than needed for competing with SVMs (see [16] for a discussion about the subject).

Figure 2 combine the results of the preceding tables. Standard classifiers are depicted by red and green colors. MOSA-P solutions are marked in blue color. Observe in Fig. 2 that the Pareto front of the MOSA-P classifiers is under the combined Pareto front of the standard classifiers for error rates higher than or equal to 25 %. The most interesting classifiers from the point of view of the energy balance are positioned around 26–28 % in this case. However, these are preliminary results as it is expected that the use of more complex definitions of the fitness improves the behaviour of the new type of classifier in the left area of the Pareto front, as mentioned.

Table 2 Sample of the set of MOSA-P non-dominated solutions obtained in the experiment

Training error (%)	Test error (%)	Cost (MIPS)
24.756	25.609	130937
24.794	25.634	92437
25.094	25.763	82562
25.412	25.686	70250
25.937	26.173	48750
26.181	26.250	41375
27.061	28.993	40625
27.118	27.454	28875
27.455	29.223	25937
27.942	30.172	25562
28.298	29.198	25187
28.861	30.813	25062
29.067	29.890	23062
30.491	29.890	17687
31.859	34.504	10937
34.389	37.093	10437
34.464	36.734	9812
37.725	39.272	8187
66.660	66.675	125

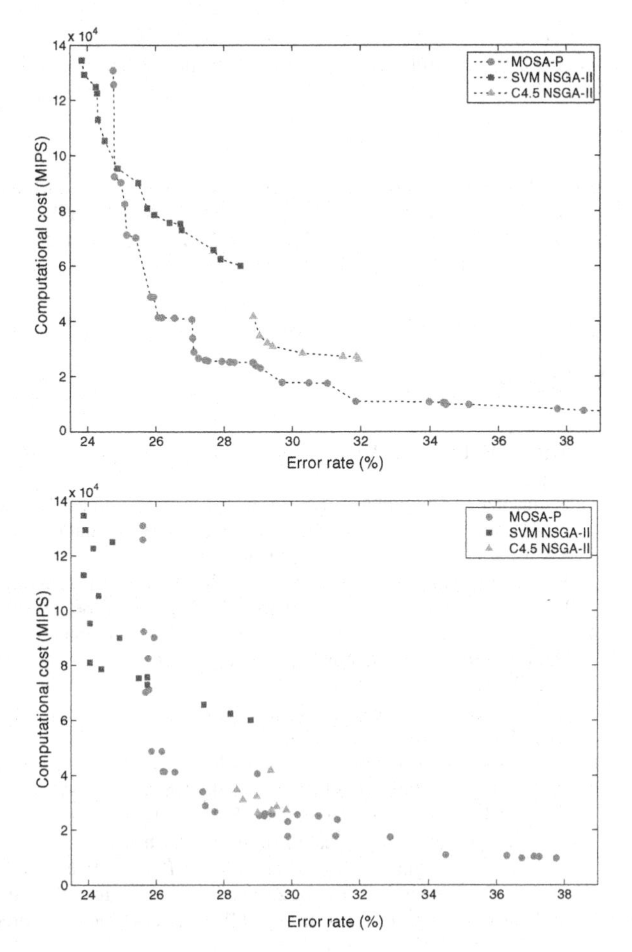

Fig. 2 Upper part: Results obtained in the experiment using the training set. Lower part: Pareto Efficient Frontier obtained in the experiment using the test set

6 Concluding Remarks and Future Work

A method has been introduced where the energy efficiency of a classifier is jointly optimized along with its classification accuracy. The definition of the new problem generalizes that of cost-sensitive classifiers as the cost function is not related to the confusion matrix but to the number and properties of the selected features and the average computational complexity of the decision algorithm implementing the classifier. The final purpose of the new methodology is to find efficient classifiers that can be implemented in hearing aids. These devices require simple algorithms with a low energy profile that are not easy to find via standard algorithms.

There is still room for improvement in the new method. The cost component that defines the accuracy of the classifier was defined solely in terms of the training error. While there is an implicit regularization given by both the intrinsic feature selection and the limit in the length of the decision system that prevents overfitting, it is expected that better results will be obtained when the training error is replaced by a different loss function which is based on an upper bound of the expected population error.

Acknowledgments This work has been partially supported by "Ministerio de Economía y Competitividad" from Spain/FEDER under grants TIN2011-24302, TIN2014-56967-R, TEC2012-38142-C04-02 and TEC2012-38142-C04-04, and the Regional Ministry of the Principality of Asturias under grant FC-15-GRUPIN14-073.

References

1. Witten IH, Frank E, Hall MA (2011) Data mining: practical machine learning tools and techniques. Elsevier
2. Hamacher V, Chalupper J, Eggers J, Fischer E, Kornagel U, Puder H, Rass U (2005) Signal processing in high-end hearing aids: state of the art, challenges, and future trends. EURASIP J Adv Signal Process 2005(18):2915–2929
3. Büchler M, Allegro S, Launer S, Dillier N (2005) Sound classification in hearing aids inspired by auditory scene analysis. EURASIP J Adv Signal Process 2005(18):2991–3002
4. Li M, McAllister H, Black N, De Perez T (2001) Perceptual time-frequency subtraction algorithm for noise reduction in hearing aids. IEEE Transa Biomed Eng 48(9):979–988
5. Maj JB, Royackers L, Moonen M, Wouters J (2005) SVD-based optimal filtering for noise reduction in dual microphone hearing aids: a real time implementation and perceptual evaluation. IEEE Trans Biomed Eng 52(9):1563–1573
6. Marzinzik M (2000) Noise reduction schemes for digital hearing aids and their use for hearing impaired. Ph.D. thesis, Carl von Ossietzky University Oldenburg
7. Dong R, Hermann D, Cornu E, Chau E (2007) Low-power implementation of an HMM-based sound environment classification algorithm for hearing aid application. In: Proceedings of the 15th european signal processing conference, Poznan, Poland
8. Xiang J, McKinney MF, Fitz K, Zhang T (2010) Evaluation of sound classification algorithms for hearing aid applications. In: 2010 IEEE international conference on acoustics, speech and signal processing, pp 185–188. IEEE
9. Hunt M, Lennig M, Mermelstein P (980) Experiments in syllable-based recognition of continuous speech. In: IEEE international conference on ICASSP '80 acoustics, speech, and signal processing, vol 5, pp 880–883
10. Gil-Pita R, Alexandre E, Cuadra L, Vicen R, Rosa-Zurera M (2009) Analysis of the effects of finite precision in neural network-based sound classifiers for digital hearing aids. EURASIP J Adv Signal Process 2009(1):456945
11. Jang JSR, Sun CT (1996) Neuro-fuzzy and soft computing: a computational approach to learning and machine intelligence
12. Sanchez L, Villar J (2008) Obtaining transparent models of chaotic systems with multi-objective simulated annealing algorithms. Inf Sci 178(4):952–970
13. Sánchez L, Couso I, Corrales J (2001) Combining GP operators with SA search to evolve fuzzy rule based classifiers. Inf Sci 136(1–4):175–191
14. Ishibuchi H, Nakashima T, Nii M (2004) Classification and modeling with linguistic information granules: advanced approaches to linguistic data mining. Adv Inf Process
15. Hall M, Frank E, Holmes G, Pfahringer B, Reutemann P, Witten IH (2009) The WEKA data mining software. ACM SIGKDD Explor Newsl 11(1):10
16. Trawinski K, Cordón O, Sanchez L, Quirin A A Genetic Fuzzy Linguistic Combination Method for Fuzzy Rule-Based Multiclassifiers. IEEE Trans Fuzzy Syst 21(5):950–965

SOCO15-SS02 - Optimization, Modeling and Control Systems (OMCS)

Modeling the Electromyogram (EMG) of Patients Undergoing Anesthesia During Surgery

José Luis Casteleiro-Roca, Juan Albino Méndez Pérez, Andrés José Piñón-Pazos, José Luis Calvo-Rolle and Emilio Corchado

Abstract All fields of science have advanced and still advance significantly. One of the facts that contributes positively is the synergy between areas. In this case, the present research shows the Electromyogram (EMG) modeling of patients undergoing to anesthesia during surgery. With the aim of predicting the patient EMG signal, a model that allows to know its performance from the Bispectral Index (BIS) and the Propofol infusion rate has been developed. The proposal has been achieved by using clustering combined with regression techniques and using a real dataset obtained from patients undergoing to anesthesia during surgeries. Finally, the created model has been tested with very satisfactory results.

Keywords EMG · BIS · Clustering · SOM · MLP · SVM

1 Introduction

One of the facts that has contributed to the outstanding development in some scientific fields is related to the existing synergy with other disciplines. Probably, the most representative examples are the sciences and the techniques around the medicine topics. Typical examples are: robotics, instrumentation, materials and so on.

The present research shows a new advance in this sense, specifically with the Electromyogram signal (EMG) when a patient is undergoing surgery with general

J.L. Casteleiro-Roca · A.J. Piñón-Pazos · J.L. Calvo-Rolle (✉)
Departamento de Ingeniería Industrial, University of A Coruña,
Avda. 19 de Febrero s/n, 15495, Ferrol, A Coruña, Spain
e-mail: jose.luis.casteleiro@udc.es

J.A.M. Pérez
Dpto. de Ingeniería de Sistemas y Automática y Arquitectura y
Tecnología de Computadores, University of La Laguna, Avda. Astrof.
Francisco Sánchez s/n, 38200 Santa Cruz de Tenerife, Spain

E. Corchado
Departamento de Informática y Automática, Universidad de Salamanca,
Plaza de la Merced s/n, 37008 Salamanca, Spain

Á. Herrero et al. (eds.), *10th International Conference on Soft Computing Models in Industrial and Environmental Applications*, Advances in Intelligent Systems and Computing 368, DOI 10.1007/978-3-319-19719-7_24

anesthesia. This research is focused on general anesthesia with Propofol. The measurement of the hypnotic state is done with the Bispectral Index (BIS). This index varies between 0 (no electrical activity) and 100 (awake state). The target for general anesthesia is normally established in 50. The hypothesis in this work is that EMG is correlated to BIS and infusion rate. Then, the objective is to predict the EMG value in terms of BIS and Propofol rate. The EMG signal is used by the clinician to assess the muscular relaxation of the patient [1–3].

For the EMG signal prediction, many different methods can be considered. The common regression methods are typically based on Multiple Regression Analysis (MRA) techniques, that are very common in applications in different fields [4–6]. However, these methods have limitations and do not provide a good performance [5]. In order to increase this feature, many new proposals have been developed. These proposals are based on Soft Computing techniques, both simple or hybrid. As it is shown in [4, 7–20] these techniques improve the first ones mentioned above.

This study implements a hybrid model to predict the EMG signal from the BIS signal and the Propofol infusion rate. To develop the model, K-means clustering algorithm is used to create groups of data with similar behavior. Previously, a Self-Organization Map (SOM) is employed with the aim of obtaining a first estimation of the optimal clusters quantity. Then, several regression methods were verified for each group to select the best one based on the lowest Mean Squared Error (MSE) reached.

This paper is structured in the following way. After the present section, the case of study is described, the Bispectral index. Then, the model approach and the tested algorithms taken into account in the research are shown. The results section shows the best configuration achieved by the hybrid model. After the results, the conclusions and future works are presented.

2 Case of Study

When a patient is undergoing surgery with general anesthesia, a proper dose of Propofol should be administrated to achieve an adequate hypnosis level [21]. To monitor the anesthesia level, the Bispectral Index Signal (BIS) is measured [22, 23]. As a consequence of the anesthesia level and the BIS signal, the ElectroMyoGram signal (EMG) varies according them [23]. The studied problem could be represented as shown in Fig. 1.

The patient is under anesthesia while the right dose of drug is administrated. When the surgery starts the patient is wake up, and then, with the Propofol drug help, is achieved the adequate hypnosis level. During the procedure, the provide level of drug ($mg/Kg/h$) is controlled to achieve the BIS desire level (50). Also, the EMG signal (*Volts*) depends on the drug quantity. At the end of the surgery the administration of the Propofol is stopped, consequently, the patient wakes up.

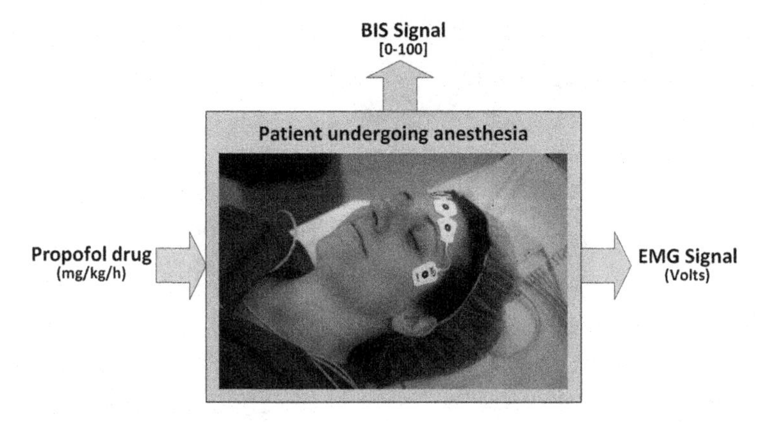

Fig. 1 Case of study. Input/Output representation

3 Model Approach

The scheme defined for the model approach is shown in Fig. 2. Taking into account the system behavior and the test accomplished, it is possible to divide the dataset in several operation ranges. Consequently, some clusters are created, and for each one, a regression model is implemented for the single output. As shown in Fig. 2, the global model has two inputs (the *–Propofol–* drug and the Bispectral index *–BIS–*) and one output (Electromyogram signal *–EMG–*). The cluster selector block connects the chosen models with the output. On each figure cluster block, only the best model is implemented; the cluster for a specific input is selected based on the Euclidean distance between the input and the centroids on each cluster.

The modeling process is shown in Fig. 3. Despite that the figure only shows the data division for training and testing, the dataset has been processed by using cross-validation (holdout) to ensure the best results for the achieved model.

3.1 The Dataset Obtaining and Description

The dataset has been obtained from several patients undergoing general anesthesia with Propofol drug during surgery. The three variables employed on this research (BIS, EMG and Propofol infusion rate) have been monitored during surgeries. A preconditioning stage was considered for BIS and EMG. The dataset is composed with the data of a total of xx patient, recording new set of values every 5 seconds. Due to the signals vary slowly at the acquisition phase, a low pass filter was implemented to avoid the measurement noise. The induction phase and the recovery phase were not considered in this study. Thus, the results obtained are only valid to predict EMG in the maintenance phase of surgery. With the conditions exposed above, the employed dataset contains 2788 samples.

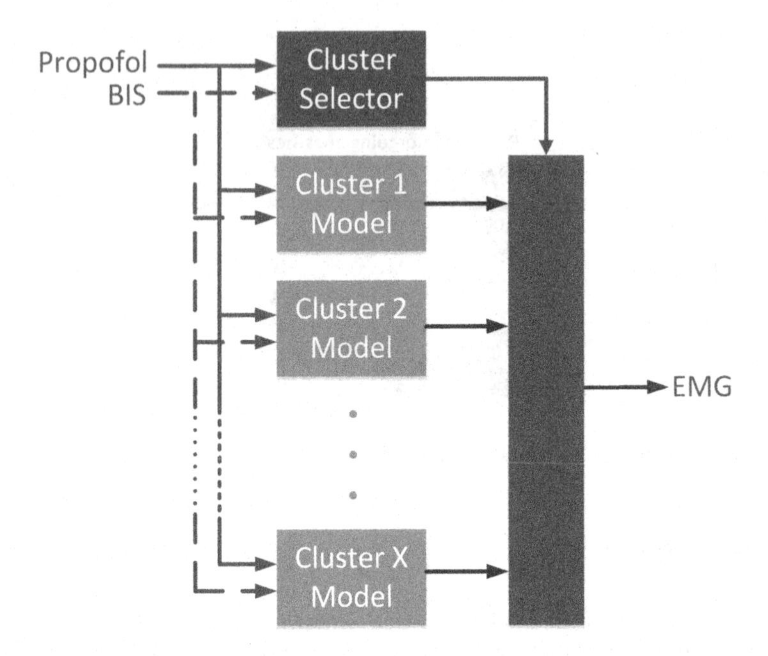

Fig. 2 Model approach

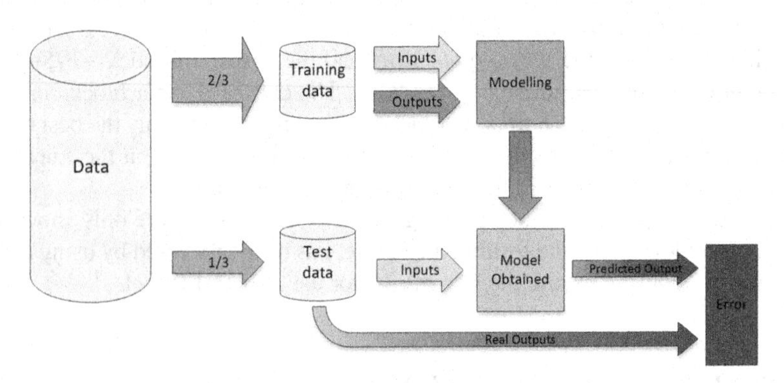

Fig. 3 Modeling process

3.2 Used Techniques

The techniques tested in the present study, with the aim of achieving the best model, are described below.

The procedure to obtain the hybrid model is through clustering by using the SOM and K-means algorithms. Below, different intelligent regression techniques are contemplated to accomplish the models. Then, only the best one is chosen according with the MSE criteria.

Self-Organizing Map (SOM) Self-Organizing Map (SOM) [24] is a type of Artificial Neural Network (ANN) that maps a high dimensionality vector onto a low dimensional one. This type of ANN uses unsupervised competitive learning [25, 26]. The learning process is recurrent until the network reaches a precise result or a maximum number of iterations have been finished. Its configuration is based on an array tied to N inputs by an N-dimensional weight vector. The process provides the geometry of the data. As a result, a $2D$ graphical representation is possible to show. Then, possible relationships are discovered by comparing different component maps with another one. The method provides an idea of the number of the required clusters.

Data Clustering - the K-means algorithm Clustering techniques are procedures of data grouping where similarity is measured [27, 28]. These algorithms try to organize unlabeled feature vectors into groups, such samples within a cluster are similar to each other [28]. The K-means method is a frequently used clustering algorithm with square-error criterion, which minimizes error function.

The clustering will depend on the initial cluster centroids and on the K value (number of groups). The K value choice is the most critical election because it needs certain previous knowledge of the quantity of clusters present in the data, which is extremely uncertain. The K-means clustering algorithm is computationally effective, it works well if the data are close to its cluster, the cluster is hyperspherical in shape and well-separated in the hyperspace.

Artificial Neural Networks (ANN). Multi-Layer Perceptron (MLP) A Multi-Layer Perceptron (MLP) is a feedforward Artificial Neural Network (ANN) [29, 30]. It is one of the most used ANNs due to its simple configuration and its robustness. In spite of this fact, the ANN architecture must be carefully chosen in order to achieve satisfactory results. MLP is made up by one input layer, one or more hidden layers and one output layer. Each layer has neurons, with an activation function. In a typical configuration, all layer neurons have the same activation function. This function could be a step, linear, log-sigmoid or tan-sigmoid.

Support Vector Regression (SVR), Least Square Support Vector Regression (LS-SVR) Support Vector Regression (SVR) is based on the algorithm of the Support Vector Machines (SVM) for classification. In SVR the main aim is mapping the data into a high-dimensional feature space F through a nonlinear plotting and doing linear regression in this space [31].

The Least Square algorithm of SVM is called LS-SVM. The solution estimation is obtained by solving a system of linear equations, and it is similar to SVM in terms of performance generalization [28]. The use of LS-SVM algorithm to regression is well-known as LS-SVR (Least Square Support Vector Regression) [32]. In LS-SVR, the insensitive loss function is replace by a classical squared loss function, which makes the Lagrangian by solving a linear Karush-Kuhn-Tucker (KKT).

Polynomial Regression Usually, a polynomial regression model [33–36] could also be defined as a linear summation of basis functions. The quantity of basis functions depends on the number of inputs of the system, and the degree of the employed polynomial. The model becomes more complex when the degree rises.

4 Results

The model was obtained using the current value of BIS signal, and the Propofol infusion rate quantity of drug (Propofol). To include the dynamic of the modeled system, the last three previous values of the inputs were included to train the models; also, the previous values of the desired output (the EMG signal) were included.

Figure 4 shows one of the results of the SOM analysis of the dataset. In this figure, different regions in the dataset structure should be appreciated. It is necessary to remind that the SOM neighbor weight distances represents the border of different clusters in the dataset with dark lines.

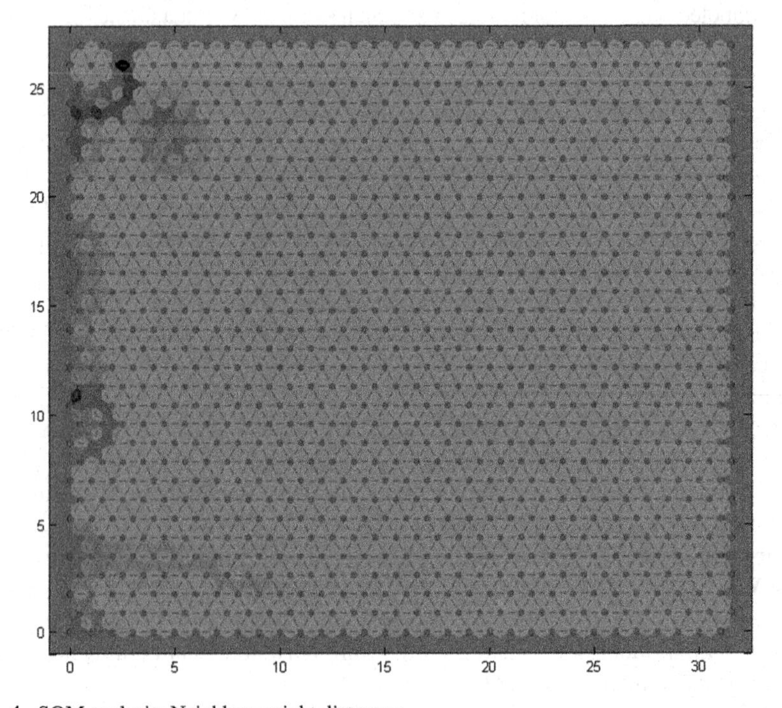

Fig. 4 SOM analysis: Neighbor weight distances

The SOM analysis is very helpful to decided the number of clusters in a dataset when this clusters are well defined. However, in this dataset the clusters are not valid to use the method. As the number of clusters is not previously known, the model was trained with different configuration of clusters, created by using the K-means algorithm.

In each cluster, using cross-validation, 2/3 of the samples were used to train the models, and the other samples were used to calculate the MSE to select the best one. The algorithm performance depends on the initial state, the process was performed 20 times with a random initialization, and finally the best result was stored. The

K-means configuration was ranged from 2 to 10 clusters with the aim to create 9 different topologies. The global model was taking into account too.

The MLP-ANN regression algorithm was trained for different configurations; always with one hidden layer, but the number of neurons in the hidden layer varies from 2 to 15. The activation function of this neurons was tan-sigmoid for all tests, and the output layer neuron had a linear activation function. The training algorithm used was Levenberg-Marquardt; gradient descent was used as learning algorithm, and the performance function was set to mean squared error.

The LS-SVR was trained with the self-tuning implemented with the toolbox for MatLab developed by KULeuven-ESAT-SCD. The kernel of the model was set to Radial Basis Function (RBF), and the type was 'Function Estimation' to perform regression. The optimization function is 'simplex' and the cost-criterion is 'leaveoneoutlssvm' with 'mse' as a performance function.

For Polynomial regression, the order of the polynomial trained varies from 1st to 20th order.

Table 1 shows the best MSE achieved for each cluster with the corresponding test data(with all the different configurations tested). Moreover, in the last column is shown the worst MSE achieved for each configuration. The best hybrid model was chosen taking into account the worst cases, and then selecting the best.

Table 2 shows the best regression technique used and its configuration for each cluster.

The best configuration achieved for the model was the one that divides the data in 2 different clusters, as is shown in Table 3. In this table, the best configuration for each algorithm is presented. Each cluster named was based on the quantity of the samples (2763 samples in the big cluster, and only 25 in the small cluster).

5 Conclusions

This study provides a precise way of modeling the Electromyogram (EMG). The accomplished model predicts the EMG from the Bispectral index (BIS) signal and the Propofol drug quantity provided to the patient

This model was obtained from a real dataset. The approach is based on a hybrid intelligent system, by combining different regression techniques on local models. After some tests, the analysis of the results shows that the best model configuration has 2 clusters. The regression techniques employed on the clusters were ANN with different configurations (5 and 3 neurons in the hidden layer). The best average MSE obtained with this configuration was 1.1083.

Table 1 Best MSE for each cluster

N° of Clusters	Cluster 1	Cluster 2	Cluster 3	Cluster 4	Cluster 5	Clusters MSE
Global Model	1.4702	-	-	-	-	1.4702
2 Clusters	1.1083	0.0118	-	-	-	1.1083
3 Clusters	3.0819	1.1470	1.2154	-	-	3.0819
4 Clusters	0.8627	0.9678	12.1164	17.3731	-	17.3731
5 Clusters	0.6532	0.2730	38.1696	0.5758	0.8678	38.1696
6 Clusters	0.6307	1.3774	0.6544	1.8068	17.6927	17.6927
7 Clusters	2.8374	0.5853	0.8183	0.8000	17.0372	17.0372
8 Clusters	1.0466	1.2601	1.8061	0.5447	1.0011	24.9626
9 Clusters	2.3795	53.3880	1.2450	22.6021	0.0092	53.3880
10 Clusters	0.4696	18.1680	0.3254	0.9348	0.0173	18.1680
N° of Clusters	Cluster 6	Cluster 7	Cluster 8	Cluster 9	Cluster 10	Clusters MSE
Global Model	-	-	-	-	-	1.4702
2 Clusters	-	-	-	-	-	1.1083
3 Clusters	-	-	-	-	-	3.0819
4 Clusters	-	-	-	-	-	17.3731
5 Clusters	-	-	-	-	-	38.1696
6 Clusters	0.4948	-	-	-	-	17.6927
7 Clusters	0.0055	0.8909	-	-	-	17.0372
8 Clusters	24.9626	0.8936	0.0020	-	-	24.9626
9 Clusters	1.1125	0.4152	1.2255	1.9976	-	53.3880
10 Clusters	0.8402	1.1774	1.5591	0.0177	0.0966	18.1680

This analysis could be applied to several different systems with the aim of improving other specifications like: efficiency, performance, features of the obtained material. It is important to emphasize that quite satisfactory results have been obtained with the approach proposed in this research.

Table 2 Best regression technique for each cluster

N° of Clusters	Cluster 1	Cluster 2	Cluster 3	Cluster 4	Cluster 5
Global Model	Poly-04	-	-	-	-
2 Clusters	ANN-05	ANN-03	-	-	-
3 Clusters	ANN-03	Poly-02	LS-SVR	-	-
4 Clusters	LS-SVR	LS-SVR	ANN-04	LS-SVR	-
5 Clusters	LS-SVR	ANN-06	Poly-01	LS-SVR	ANN-04
6 Clusters	ANN-08	Poly-01	LS-SVR	ANN-02	Poly-01
7 Clusters	ANN-02	ANN-07	LS-SVR	ANN-05	Poly-09
8 Clusters	Poly-03	ANN-07	LS-SVR	Poly-06	ANN-06
9 Clusters	Poly-01	ANN-15	ANN-11	Poly-01	ANN-05
10 Clusters	LS-SVR	Poly-01	Poly-01	LS-SVR	ANN-05
N° of Clusters	Cluster 6	Cluster 7	Cluster 8	Cluster 9	Cluster 10
Global Model	-	-	-	-	-
2 Clusters	-	-	-	-	-
3 Clusters	-	-	-	-	-
4 Clusters	-	-	-	-	-
5 Clusters	-	-	-	-	-
6 Clusters	LS-SVR	-	-	-	-
7 Clusters	ANN-15	LS-SVR	-	-	-
8 Clusters	ANN-02	LS-SVR	ANN-03	-	-
9 Clusters	LS-SVR	Poly-02	ANN-13	Poly-01	-
10 Clusters	ANN-03	LS-SVR	Poly-08	ANN-13	ANN-14

Table 3 Best result for the final configuration of the proposal

	Big Cluster	Small Cluster
Train samples	1842	17
Test samples	921	8
Best ANN	5 neurons	3 neurons
Best Polynomial	First order	First order
ANN MSE	1.1083	0.0118
LSSVM MSE	1.6241	6.9461
Poly MSE	1.3546	58.5147

Acknowledgments This study was conducted under the auspices of Research Project *DPI*2010 − 18278, supported by the Spanish Ministry of Innovation and Science.

References

1. Coelho AL, Lima CA (2014) Assessing fractal dimension methods as feature extractors for EMG signal classification. Eng Appl Artif Intell 36:81–98
2. Xing K, Yang P, Huang J, Wang Y, Zhu Q (2014) A real-time EMG pattern recognition method for virtual myoelectric hand control. Neurocomputing 136:345–355
3. Phinyomark A, Quaine F, Charbonnier S, Serviere C, Tarpin-Bernard F, Laurillau Y (2013) A feasibility study on the use of anthropometric variables to make musclecomputer interface more practical. Eng App Artif Intell 26(7):1681–1688
4. Ghanghermeh A, Roshan G, Orosa JA, Calvo-Rolle JL, Costa ÁM (2013) New climatic indicators for improving urban sprawl: a case study of tehran city. Entropy 15(3):999–1013
5. Calvo-Rolle JL, Quintian-Pardo H, Corchado E, Del Carmen Meizoso-López M, García RF (2015) Simplified method based on an intelligent model to obtain the extinction angle of the current for a single-phase half wave controlled rectifier with resistive and inductive load. J Appl Logic 13(1):37–47
6. Calvo-Rolle JL, Fontenla-Romero O, Pérez-Sánchez B, Guijarro-Berdinas B (2014) Adaptive inverse control using an online learning algorithm for neural networks. Informatica 25(3):401–414
7. García RF, Rolle JLC, Gomez MR, Catoira AD (2013) Expert condition monitoring on hydrostatic self-levitating bearings. Expert Syst Appl 40(8):2975–2984
8. Calvo-Rolle JL, Casteleiro-Roca JL, Quintián H, Del Carmen Meizoso-Lopez M (2013) A hybrid intelligent system for PID controller using in a steel rolling process. Expert Syst Appl 40(13):5188–5196
9. García RF, Rolle JLC, Castelo JP, Gomez MR (2014) On the monitoring task of solar thermal fluid transfer systems using NN based models and rule based techniques. Eng Appl Artif Intell 27:129–136
10. Quintián H, Calvo-Rolle JL, Corchado E (2014) A hybrid regression system based on local models for solar energy prediction. Informatica 25(2):265–282
11. Nieto PG, Torres JM, de Cos Juez F, Lasheras FS (2012) Using multivariate adaptive regression splines and multilayer perceptron networks to evaluate paper manufactured using eucalyptus globulus. Appl Math Comput 219(2):755–763
12. de Cos Juez F, Nieto PG, Torres JM, Castro JT (2010) Analysis of lead times of metallic components in the aerospace industry through a supported vector machine model. Math Comput Model 52(78):1177–1184 (Mathematical Models in Medicine, Business and Engineering 2009)
13. de Cos J, Sanchez F, Ortega F, Montequin V (2008) Rapid cost estimation of metallic components for the aerospace industry. Int J Prod Econ 112(1):470–482 (Special Section on Recent Developments in the Design, Control, Planning and Scheduling of Productive Systems)
14. Casteleiro-Roca J, Calvo-Rolle J, Meizoso-López M, Piñón-Pazos A, Rodríguez-Gómez B (2015) Bio-inspired model of ground temperature behavior on the horizontal geothermal exchanger of an installation based on a heat pump. Neurocomputing 150 Part A(0):90–98
15. Calvo-Rolle JL, Machón-González I, López-García H (2011) Neuro-robust controller for non-linear systems. DYNA 86(3):308–317
16. Quintián-Pardo H, Calvo-Rolle JL, Fontenla-Romero O (2012) Application of a low cost commercial robot in tasks of tracking of objects. Dyna 175:24–33
17. Vilar-Martínez XM, Montero-Sousa JA, Calvo-Rolle JL, Casteleiro-Roca JL (2014) Expert system development to assist on the verification of TACAN system performance. DYNA 89(1):112–121

18. Garcia RF, Calvo-Rolle JL, Gomez MR, Catoira AD (2013) Expert condition monitoring on hydrostatic self-levitating bearings. Expert Syst Appl 40(8):2975–2984
19. Garcia RF, Calvo-Rolle JL, Castelo JP, Gomez MR (2014) On the monitoring task of solar thermal fluid transfer systems using NN based models and rule based techniques. Eng Appl Artif Intell 27:129–136
20. Machon-Gonzalez I, Lopez-Garcia H, Calvo-Rolle J (2010) A hybrid batch som-ng algorithm. In: The 2010 International Joint Conference on Neural Networks (IJCNN), pp 1–5
21. Litvan H, Jensen EW, Galan J, Lund J, Rodriguez BE, Henneberg SW, Caminal P, Villar Landeira JM (2002) Comparison of conventional averaged and rapid averaged, autoregressive-based extracted auditory evoked potentials for monitoring the hypnotic level during propofol induction. J Am Soc Anesthesiologists 97(2):351–358
22. Sánchez SS, Vivas AM, Obregón JS, Ortega MR, Jambrina CC, Marco ILT, Jorge EC (2009) Monitorización de la sedación profunda. el monitor BIS. Enfermería Intensiva 20(4):159–166
23. Pérez JAM, Torres S, Reboso JA, Reboso H (2011) Estrategias de control en la práctica de anestesia. Rev Iberoamericana de Automática e Informática Ind RIAI 8(3):241–249
24. Kohonen T (1990) The self-organizing map. Proc IEEE 78(9):1464–1480
25. Kohonen T, Oja E, Simula O, Visa A, Kangas J (1996) Engineering applications of the self-organizing map. Proc IEEE 84(10):1358–1384
26. Kohonen T (1997) Exploration of very large databases by self-organizing maps. In: International Conference on Neural Networks, vol 1, pp PL1-PL6
27. Qin A, Suganthan P (2005) Enhanced neural gas network for prototype-based clustering. Pattern Recogn 38(8):1275–1288
28. Kaski S, Sinkkonen J, Klami A (2005) Discriminative clustering. Neurocomputing 69(13):18–41
29. Wasserman P (1993) Advanced Methods in Neural Computing, 1st edn. Wiley, New York
30. Zeng Z, Wang J (2010) Advances in Neural Network Research and Applications, 1st edn. Springer Publishing Company, New York
31. Vapnik V (1995) The Nature of Statistical Learning Theory. Springer, Berlin (1995)
32. Li Y, Shao X, Cai W (2007) A consensus least squares support vector regression (LS-SVR) for analysis of near-infrared spectra of plant samples. Talanta 72(1):217–222
33. Heiberger R, Neuwirth E (2009) Polynomial regression. In: R Through Excel. Use R, pp 269–284. Springer New York
34. Casteleiro-Roca J, Calvo-Rolle J, Meizoso-López M, Piñón-Pazos A, Rodríguez-Gómez B (2014) New approach for the QCM sensors characterization. Sens Actuators A: Physical 207: 1–9
35. Wu X (2007) Optimal designs for segmented polynomial regression models and web-based implementation of optimal design software, pp AAI3337619. State University of New York at Stony Brook, Stony Brook, New York
36. Zhang Z, Chan SC (2011) On kernel selection of multivariate local polynomial modelling and its application to image smoothing and reconstruction. J Signal Process Syst 64(3):361–374

Real Time Parallel Robot Direct Kinematic Problem Computation Using Neural Networks

Asier Zubizarreta, Mikel Larrea, Eloy Irigoyen and Itziar Cabanes

Abstract The calculation of the Direct Kinematic Problem (DKP) is one of the main issues in real-world applications of Parallel Robots, as iterative procedures have to be applied to compute the pose of the robot. Being this issue critical to robot Real-Time control, in this work a methodology to use Artificial Neural Networks to approximate the DKP is proposed and a comprehensive study is carried out to demonstrate experimentally the Real-Time performance benefits of the approach in a 3PRS parallel robot.

Keywords Parallel robots · Direct kinematic problem · Artificial neural network

1 Introduction

Parallel robots have become the focus of many researchers and manufacturers in the last decade. Their structure is far different from the one of serial robots, where a single chain connects the base of the robot to the end effector. Parallel robots [1], however, are composed by several kinematic chains that connect a fixed base to a mobile platform with end effector. This mechanical configuration provides many advantages over traditional serial robots, such as higher stiffness, precision and the possibility

This work was supported in part by the Government of Spain under project DPI2012-32882, the Government of the Basque Country (Project IT719-13) and UPV/EHU under grant UFI11/28.

A. Zubizarreta (✉) · M. Larrea · E. Irigoyen · I. Cabanes
University of the Basque Country (UPV/EHU), Bilbao, Spain
e-mail: asier.zubizarreta@ehu.eus

M. Larrea
e-mail: m.larrea@ehu.eus

E. Irigoyen
e-mail: eloy.irigoyen@ehu.eus

I. Cabanes
e-mail: itziar.cabanes@ehu.eus

Á. Herrero et al. (eds.), *10th International Conference on Soft Computing Models in Industrial and Environmental Applications*, Advances in Intelligent Systems and Computing 368, DOI 10.1007/978-3-319-19719-7_25

of reducing the overall moving mass of the robot. Due to this advantages, commercial parallel robots have been developed to execute high precision, high speed and acceleration or high load handling tasks.

In order to control the robot, it is required to calculate the Direct Kinematic Problem (DKP), which is obtained by estimating the location of the end effector based on the measurable position of the actuators of the robot. However, due to the mechanical complexity of parallel robots, no analytical solution can be found to the DKP in the general case. Hence, iterative numerical procedures should be used to calculate the position of the end effector each control time step.

One of the most used numerical procedures is the one proposed by Newton-Raphson (N-R). This requires an initial guess of the solution, i.e. the end effector location, and a indefinite number of iterations, which will vary depending on the distance from the initial guest to the closest root of the equation system required to solve the DKP. Hence, the computational cost is usually high, which represents a key issue when implementing advanced controllers for parallel robots.

Being high speed tasks one of the main areas of application of parallel robots, the time optimization of the DKP computation has become of great interest in both industry and academia. Some authors, such as Yang et al. [2], have proposed to modify the classical Newton-Raphson algorithm in order to increase its efficiency. Other works [3–5] focus on reducing the nonlinear equation set of the DKP to an univariate polynomial that can be solved more efficiently. These approaches, however, are very sensitive to measurement errors in practice, as an small amount of data is used to estimate the rest of the variables.

The introduction of extra sensors has also been discussed [6, 7]. The measure of some passive or nonactuated joint variables using a limited amount of extra sensors can be used to efficiently solve the DKP analytically. However, this approach increases the cost of the robot, and adds complexity to its calibration.

The need for bounded but not exact accuracy in the solution has also led some authors to apply universal approximators to solve the DKP. For instance, Wang [8] applied Taylor approximations to solve the DKP. Among the solutions proposed in this set, Artificial Neural Networks (ANN) have also been used due to their nature as universal approximators [9]. ANNs are composed by several layers of interconnected artificial neurons, which simulate the behaviour of biological ones by means of a nonlinear activation function that computes an output value based on the inputs of each neuron. Based on this structure, an ANN is able to *learn* high complex input-output data relations such as the DKP and even extrapolate outputs that have not been learned by the network.

In the literature, although architectures such as Radial Basis Function (RBF) [10], polynomial neural networks [11] or Adaptive-network-based fuzzy inference system [12] have been applied, most of the works are based on the use of the widely studied Multi Layer Perceptron (MLP) architecture to approximate the DKP of complex robots such as the Hexa [13], the Hexapod [14, 15] or other architectures [10].

Hence, Artificial Neural Networks have the potential to become an effective solution to the computation of the DKP problem in parallel robots. However, most of the works proposed in this area do not focus on two key issues that have to be considered in this application: the definition of a bounded error and the computational time benefit of the approach.

In this work, an experimental study is carried out for the 3PRS parallel robot in which the methodology to solve the DKP using ANN is detailed and a real time computational cost comparison is carried out. For that purpose, the 3PRS parallel robot and its kinematic relations are detailed in Sect. 2. The DKP estimation procedure using Artificial Neural Networks is detailed in Sect. 3, in which the proposed methodology is explained. Section 4 analyzes the error distribution and time efficiency of the selected ANN. Finally, the most important ideas are summarized in the conclusions.

2 The 3PRS Parallel Robot

The 3PRS (Fig. 1a) is a three degrees-of-freedom parallel robot that provides displacement in its z axis and two independent orientation angles θ_x and θ_y. The robot is composed by a fixed platform where three guides $\mathbf{A_iB_i}$ are attached. The end effector of the robot is located in $\mathbf{P}$ point of the mobile platform. Three spherical joints $\mathbf{C_i}$ connect the mobile platform to each of the three limbs $\mathbf{B_iC_i}$. A slider is located in point $\mathbf{B_i}$, allowing a prismatical motion of the limb along the fixed guide. In addition, a rotational joint is attached to the slider in point $\mathbf{B_i}$. This restricts the motion of the limb to the plane π_i that contains the points $\mathbf{A_i}$, $\mathbf{B_i}$, $\mathbf{C_i}$.

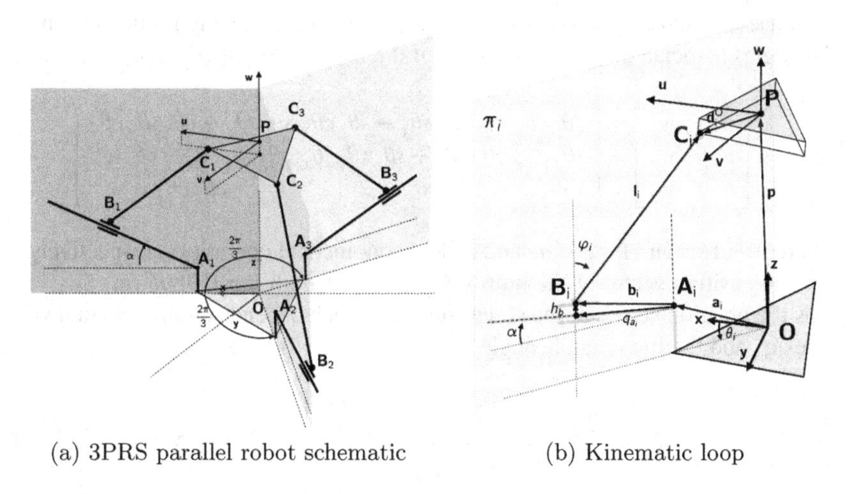

(a) 3PRS parallel robot schematic (b) Kinematic loop

Fig. 1 3PRS parallel robot

This robot is designed for positioning (i.e. telescopes) or machine tool applications where the tracking of the location of the end-effector associated to the mobile frame $\mathbf{P}(u, v, w)$ with respect to the fixed frame $\mathbf{O}(x, y, z)$ is required. Due to the difficulty measuring directly the end-effector location (position and orientation) $\mathbf{x} = \left[x\ y\ z\ \theta_x\ \theta_y\ \theta_z\right]^T$, it is usually estimated using the positioning data of the actuators $\mathbf{q_a}$, which measure the position of the slider $\mathbf{B_i}$ along the fixed linear guide. The geometric relations between the active joint variables $\mathbf{q_a} = \left[q_{a_1}\ q_{a_2}\ q_{a_3}\right]^T$ associated to the actuators and the output variables (z, θ_x, θ_y) that define the location of the end-effector define the kinematic model of the robot.

The vectorial closure loop equations of each of the three limbs of the 3PRS robot (Fig. 1b) can be calculated in the fixed frame $\mathbf{O}(x, y, z)$ as follows,

$$\mathbf{p} + \mathbf{d_i^O} - \mathbf{l_i} - \mathbf{b_i} - \mathbf{a_i} = \mathbf{0}_{3\times1}, \quad i = 1, 2, 3 \tag{1}$$

where each term of this equation is defined as,

- $\mathbf{p} = \left[x\ y\ z\right]^T$ defines the position of the end-effector in the fixed frame.
- $\mathbf{d_i^O} = \mathbf{R}\,\mathbf{d_i}$ is the projection in the fixed frame of the constant vector $\mathbf{d_i} = \mathbf{PC_i}$ defined in the moving frame,

$$\mathbf{d_i} = \left[r\ -r\,sin\,(2\pi, (i-1)/3)\ -h_p\right]^T \quad \mathrm{i} = 1,2,3 \tag{2}$$

where $r = 0.3638$ (m) is the radius of the mobile platform and $h_p = 0.04$ (m) its height.
In order to project $\mathbf{d_i}$ a rotation matrix $\mathbf{R}$ that defines the orientation of the moving frame with respect to the fixed one is required. This matrix can be defined by selecting a set of Euler angles. In this paper, the Roll-Pitch-Yaw convention has been used, which allows to define the orientation of the moving frame in terms of the angles associated to x, y and z axes of the fixed frame $(\theta_x, \theta_y, \theta_z)$,

$$\mathbf{R} = \begin{bmatrix} u_x\ v_x\ w_x \\ u_y\ v_y\ w_y \\ u_z\ v_z\ w_z \end{bmatrix} = \begin{bmatrix} c\theta_z\, c\theta_y & c\theta_z\, s\theta_y\, s\theta_x - s\theta_z\, c\theta_x & s\theta_z\, s\theta_x + c\theta_z\, s\theta_y\, c\theta_x \\ s\theta_z\, c\theta_y & c\theta_z\, c\theta_x + s\theta_z\, s\theta_y\, s\theta_x & s\theta_z\, s\theta_y\, c\theta_x - c\theta_z s\theta_x \\ -s\theta_y & c\theta_y\, s\theta_x & c\theta_y\, c\theta_x \end{bmatrix} \tag{3}$$

where c and s stand for *cosine* and *sine* trigonometical operations, respectively.

- $\mathbf{l_i}$ is the position vector of the limb $\mathbf{B_iC_i}$ of fixed length $l_i = 0.9805$ (m).
- $\mathbf{b_i}$ is the position vector of the slider joint $\mathbf{B_i}$, which depends on the actuated variables $\mathbf{q_a}$ and the linear guide angle $\alpha = -0.7854$ (rad).

$$\mathbf{b_i} = \begin{bmatrix} q_{a_i}\ \cos\alpha\ \cos(2\pi\,(i-1)/3) \\ q_{a_i}\ \cos\alpha\ \sin(2\pi\,(i-1)/3) \\ q_{a_i}\ \sin\alpha \end{bmatrix} \quad i = 1, 2, 3 \tag{4}$$

- $\mathbf{a_i}$ is the constant position vector of each of the three linear guides with respect to the fixed frame,

$$\mathbf{a_i} = a_i\left[r_a \cos\,(2\pi\,(i-1)/3)\ -r_a\,sin\,(2\pi, (i-1)/3)\ 0\right]^T \quad \mathrm{i}=1,2,3 \tag{5}$$

where $r_a = 0.425$ (m).

The 3PRS presents only 3 degrees of freedom, being the independent variables the displacement along the z axis and the angles θ_x and θ_y. However, due to the configuration of the robot, movement in the x and y axes and the θ_z angle is not null. This effect is known as parasitic motions [16] and increases the complexity of the calculation of the kinematic relations of the 3PRS robot. Hence, three constraint equations have to be introduced to relate the independent and parasitic variables,

$$\begin{aligned} \Gamma_1 &= h_p\, w_y - r\, u_y - y = 0 \\ \Gamma_2 &= h_p\, w_x + \tfrac{r}{2}\left(u_x - v_y\right) - x = 0 \\ \Gamma_3 &= v_x - u_y = 0 \end{aligned} \tag{6}$$

where as seen in Eq. 3, w_x, w_y, v_x, v_y, u_x and u_y depend on the RPY Euler angles $(\theta_x, \theta_y, \theta_z)$.

If a distance contraint $l_i^2 = ||\mathbf{l_i}||$ is imposed to each of the vectorial closure loop Eq. 1,

$$\Gamma_{i+3} = q_{a_i}^2 + B_i\, q_{a_i} + C_i = 0 \quad i = 1, 2, 3 \tag{7}$$

where

$$\begin{aligned} B_i &= \left(-2\, CA_{ix} \cos\alpha\right) / \cos(2\pi\,(i-1)/3) \\ &\quad -2\, CA_{iz} \sin\alpha - 2\, CA_{iy} \cos\alpha\, \sin(2\pi\,(i-1)/3) \\ C_i &= ||\mathbf{CA_i}|| + h_b^2 - l_i^2 + \left(2\, h_b\, CA_{ix} \sin\alpha\right) / \cos(2\pi\,(i-1)/3) \\ &\quad +2\, h_b\, CA_{iy} \sin\alpha\, \sin(2\pi\,(i-1)/3) - 2\, h_b\, CA_{iz} \cos\alpha \end{aligned}$$

where $\mathbf{CA_i} = \mathbf{p_x} + \mathbf{R}\,\mathbf{d_i} - \mathbf{a_i} = \left[CA_{ix}\ CA_{iy}\ CA_{iz}\right]^T$, for $i = 1, 2, 3$.

Combining Eqs. 6 and 7 the Direct Kinematic Problem is defined. In order to solve it, the values of the actuated joint variables q_{a_i} are considered to be known and the six output variables (independent z, θ_x, θ_y and parasitic x, y, θ_z) are to be calculated using the six equation system by means of an iterative numerical procedure such as Newton-Raphson,

$$\mathbf{x_{k+1}} = \mathbf{x_k} - \mathbf{J}^{-1}\, \mathbf{\Gamma}(\mathbf{x_k}, \mathbf{q_a}) \tag{8}$$

where $\mathbf{\Gamma}(\mathbf{x_k}, \mathbf{q_a}) = 0$ is the equation system defined by Eqs. 6 and 7 and $\mathbf{J} = \partial\mathbf{\Gamma}/\partial x$ is the jacobian of the parallel robot.

3 Kinematic Model Based on ANNs

The DKP of parallel robots can be defined as an input-output relationship $\mathbf{x} = f\left(\mathbf{q_a}\right)$ in which the output coordinates $\mathbf{x}$ (cartesian position and orientation of the platform) are to be calculated in terms of the active joint variables, $\mathbf{q_a}$, which can be measured. However, as in the general case no explicit expression can be obtained, approximators can be used to avoid iterative calculation approaches such as the one detailed in the previous section.

ANNs have been widely used as function approximators. Both MLP [9] and RBF [17] topologies, for instance, have been demonstrated to provide universal approximator properties. In this work, MLP architecture is selected. As the main goal of the model is to reduce the computational cost of the DKP, the number of hidden neurons in the MLP will be required to be as small as possible in order to propagate the signals as fast as possible.

The number of inputs and outputs of the network is defined by the DKP problem, being the inputs the three joint variables q_{a_i} associated to the position of the three linear motors (Fig. 1) and the output the z, θ_x and θ_y independent coordinates that define the location of the mobile platform. A single hidden layer is selected, whose optimum number of neurons is to be estimated experimentally.

For that purpose, an iterative batch training method has been selected in which network architecture vary from 40 to 90 neurons with steps of 5 neurons in the hidden layer. Nguyen-Widrow approach has been selected to initialize the network parameters (weights and biases) and a set of 20 nets are trained for each architecture (modifying the number of neurons in the hidden layer) so that the effect of random weight initialization can be reduced. The net with less approximation error is selected as the representative for each network architecture. According to the universal approximator property as stated in [9], any type of activation function can be implemented in the MLP. Due to the existence of both positive and negative input-output data, the hyperbolic tangent and linear activation functions have been selected for the hidden and output layer respectively [18] for all networks trained.

A set of training and test examples are required for the networks to learn the DKP mapping. For that purpose, the effective workspace of the 3PRS parallel robot has been extracted from the mathematical model with an offline uniform sweep of the workspace. The parameters used for the distribution of the examples are summarized in Table 1. This way, a set of 192.238 examples have been created for training the set of ANNs. The test set has a more dense grid, being composed by of 2.836.219 examples. It should be noted that both sets contain different examples.

Table 1 Workspace dimensions for training and validation

	z			θ_x			θ_y		
	min (m)	max (m)	points	min (°)	max (°)	points	min (°)	max (°)	points
Training	0.5500	1.0500	200	−30	30	50	−30	30	50
Test	0.5500	1.0500	200	−30	30	200	−30	30	200

The training set is used to train each of the different network architectures and its different iterations. For that purpose Levenberg-Marquardt training algorithm has been selected with initial $\mu = 0.001$. All training set has been used to train each network, and no validation set has been defined. The training has been configured to stop by gradient level criteria, establishing the limit in 10^{-10}. A total of 220 nets have been trained.

Once the nets have been trained, their performance has been tested. As solving the DKP is critical to robot control the *maximum approximation error* in each of the output coordinates (z, θ_x, θ_y) associated to the test set inputs has been considered as the performance index of each network.

4 Real-Time Performance Analysis

This section shows and discuses the different aspects of the simulation and the results obtained with the ANN model. In order to analyze the effect of the number of hidden layer neurons and their performance, the best ANN for each architecture is defined first. As 20 nets of each architecture with N neurons in the hidden layer have been trained, the one with lesser absolute maximum error using the Test set has been selected as the representative one.

A Matlab based analysis tool has been developed to calculate the maximum absolute error and analyze in simulation the error distribution of each network in the workspace of the 3PRS robot (Fig. 2). From this analysis, several conclusions have been extracted. First, due to the particular mechanical structure of the 3PRS robot, errors in the orientation outputs θ_x and θ_y are an order of magnitude lesser than those of the z displacement, i.e., the ANN based approximation is able to approximate more accurately orientation parameters. Second, the distribution of errors shows that the maximum error deviation is usually located in the limits of the workspace. Hence, the Training set used should emphasize examples of these areas in order to reduce the approximation error.

The Real Time performance of the best networks has been evaluated in Labview RT environment. Each network has been implemented and optimized for RT execution and their error and timing performance has been analyzed. The Newton-Raphson approach has been also tested in RT (Eq. 8). Results are shown in Fig. 3, where the maximum error produced by the ANN in contrast to the mathematical model is presented. These subfigures are related to the error in the three outputs, namely z, θ_x and θ_y.

As the ANNs are to be used for estimating robot poses, a maximum admissible error has been defined by design for each coordinate estimation. This way a maximum of 0.1 mm in the z coordinate, and a maximum of 0.1° in the θ_x and θ_y orientation angles. These limits correspond to a medium precision commercial robot.

As it can be seen in Fig. 3, all the ANNs comply with the required error in case of the θ_x and θ_y coordinates. However, as stated in the previous analysis, the error in the z coordinate is the limiting factor. Of the trained networks, only the ones with 55, 70, 75 and 90 neurons on the hidden layer comply with the defined limits. The best ANN regarding to the overall error is the network with 70 neurons in the hidden layer. Hence, it is possible to obtain bounded error approximations of the DKP using ANNs.

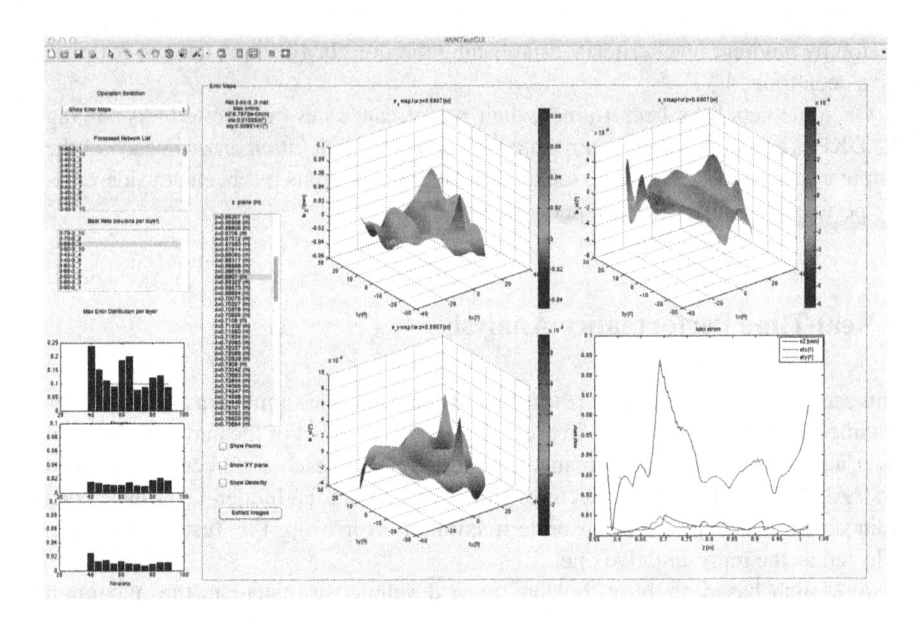

Fig. 2 Validation tool GUI

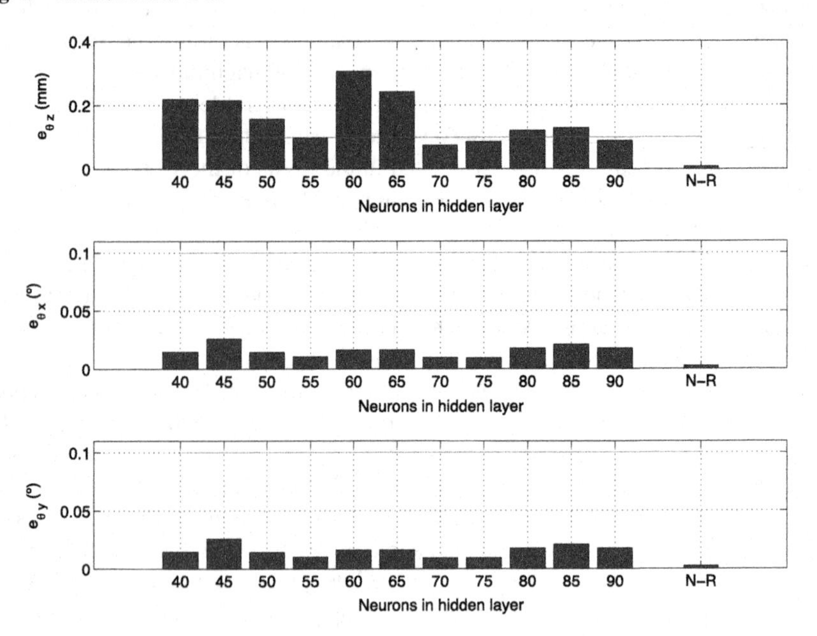

Fig. 3 Maximum error vs hidden layer size

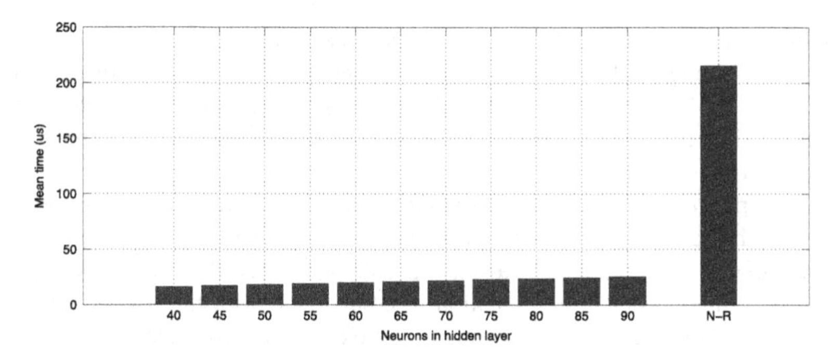

Fig. 4 Hidden layer size vs. execution time

The time performance of the best ANNs is analyzed next. The Newton-Raphson (N-R) approach has also been implemented in Real Time and in order to obtain equivalent precision results, the number of iterations has been limited by the relative maximum error defined also for ANNs. In a similar way, these limits are used to compute randomly the initial iteration point used for the N-R approach. The mean computational time for the Test set of each ANN and N-R is shown in Fig. 4. The execution platform is a National Instruments PXI-8102 1.9 Ghz industrial PC.

As it can be seen, the computational time (in μs) required by the traditional N-R iterative approach is approximately six times higher than the ANN based approach, in which no iterations are required. Note that the mean number of iterations required to compute each of the examples of the Test set in the N-R approach is 3.9. Hence, even the time performance of a single iteration in the N-R approach requires higher computational time than the ANN solution. This is mainly produced due to the need of inverting the 6×6 Jacobian matrix (Eq. 8), which has also to be computed using numerical methods.

In addition, it can be clearly seen in Fig. 4 that the number of neurons in the hidden layer is directly related to the computational cost of the approach. However, an increased number of neurons does not always guarantee better approximation as seen in Fig. 3 and an experimental study has to be carried out in order to determine the best computational cost/precision ratio. For the particular case analyzed in this work, the network with 55 neurons in the hidden layer will be selected to implement the DKP.

5 Conclusions

Real Time pose estimation is critical in parallel robotic control applications. The Direct Kinematic Problem is required to estimate the position of the end effector of the robot in terms of the actuated joints position. Traditional approaches use iterative procedures that although precise, have high computational cost.

In this work the use of ANN has been proposed to reduce the computational time of the DKP solving while maintaining the required accuracy in the estimation. For that purpose a ANN definition methodology has been proposed, in which a set of network architectures with different number of neurons in the hidden layer have been defined and trained.

A proper performance index, the absolute maximum overall error has been defined to establish the best networks for the particular case of DKP estimation and their time performance has been analyzed. Results have demonstrated that the ANN approximation is 6 times faster in computation in comparison with the traditional iterative approach while maintaining acceptable approximation errors. Hence, ANN have demonstrated to be one of the best alternatives to the Real Time computation of DKP in Parallel Robotics.

References

1. Merlet JP (2006) Parallel Robots. Springer
2. Yang C, He J, Han J, Liu X (2009) Real-time state estimation for spatial six-degree-of-freedom linearly actuated parallel robots. Mechatronics 19(6):1026–1033
3. Innocenti C (2001) Forward kinematics in polynomial form of the general stewart platform. J Mech Des 123(2):254–260
4. Lee TY, Shim JK (2001) Algebraic elimination-based real-time forward kinematics of the 6–6 stewart platform with planar base and platform. In: Proceedings of the 2001 IEEE international conference on robotics and automation, vol 2, pp. 1301–1306
5. Huang X, Liao Q, Wei S (2012) Closed-form forward kinematics for a symmetrical 6–6 stewart platform using algebraic elimination. Mech Mach Theory 45(2):327–334
6. Baron L, Angeles J (2000) The direct kinematics of parallel manipulators under joint-sensor redundancy. IEEE Trans Robot Autom 16(1):12–19
7. Bonev I, Ryu J, Kim SG, Lee SK (2001) A closed-form solution to the direct kinematics of nearly general parallel manipulators with optimally located three linear extra sensors. IEEE Trans Robot Autom 17(2):148–156
8. Wang Y (2007) A direct numerical solution to forward kinematics of general stewart-gough platforms. Robotica 25:121–128
9. Hornik K, Stinchcombe M, White H (1989) Multilayer feedforward networks are universal approximators. Neural Netw 2(5):359–366
10. Zhang D, Lei J (2011) Kinematic analysis of a novel 3-dof actuation redundant parallel manipulator using artificial intelligence approach. Rob Comput Integr Manuf 27(1):157–163
11. Boudreau R, Darenfed S, Gosselin C (1998) On the computation of the direct kinematics of parallel manipulators using polynomial networks. IEEE Trans Syst Man Cybern 28(2):213–220
12. Sadjadian H, Taghirad HD (2006) Comparison of different methods for computing the forward kinematics of a redundant parallel manipulator. J Intell Rob Syst 44(3):225–246
13. Dehghani M, Ahmadi M, Khayatian A, Eghtesad M, Farid M (2008) Neural network solution for forward kinematics problem of hexa parallel robot. In: Proceedings of the 2008 American control conference, pp. 4214–4219
14. Seng Yee C, Bin Lim K (1997) Forward kinematics solution of stewart platform using neural networks. Neurocomputing 16(4):333–349
15. Sang LH, Han MC (1999) The estimation for forward kinematic solution of stewart platform using the neural network. In: Proceedings of the 1999 IEEE/RSJ international conference on intelligent robots and systems, vol 1, pp. 501–506

16. Carretero JA, Podhorodeski RP, Nahon MA, Gosselin C (2000) Kinematic analysis and optimization of a new three degree-of-freedom spatial parallel manipulator. Trans-Am Soc Mech Eng J Mech Des 122(1):17–24
17. Park J, Sandberg IW (1991) Universal approximation using radial-basis-function networks. Neural Comput 3(2):246–257
18. Hagan M, Menhaj M (1994) Training feed-forward networks with the marquardt algorithm. IEEE Trans Neural Netw 5:989–993

Reinforcement Learning in Single Robot Hose Transport Task: A Physical Proof of Concept

Jose Manuel Lopez-Guede, Julián Estévez and Manuel Graña

Abstract In this paper we address the physical realization of proof of concept experiments demonstrating the suitability of the controllers learned by means of Reinforcement Learning (RL) techniques to accomplish tasks involving Linked Multi-Component Robotic System (LMCRS). In this paper, we deal with the task of transporting a hose by a single robot as a prototypical example of LMCRS, which can be extended to much more complex tasks. We describe how the complete system has been designed and built, explaining its different main components: the RL controller, the communications, and finally, the monitoring system. A previously learned RL controller has been tested solving a concrete problem with a determined state space modeling and discretization step. This physical realization validates our previous published works carried out through computer simulations, giving a strong argument in favor of the suitability of RL techniques to deal with real LMCRS systems.

Keywords Reinforcement learning · Linked multicomponent robotic systems · LMCRS · Hose transport · Proof of concept

1 Introduction

Linked Multi-Component Robotic Systems (LMCRS) [1] are composed of a collection of autonomous robots linked by a flexible one-dimensional link introducing additional non-linearities and uncertainties when designing the control of the robots to accomplish a given task. A paradigmatic task example is the transportation of a hose-like object by the robots (or only one robot in its simplest form). There are no relevant references in the literature about this line of research apart from the work done by our group. The first attempts to deal with this problem, modeled the task as a cooperative control problem [2], but this resulted in a too low level approach lacking the intended autonomous learning. The work reported in [3, 4] represented

J.M. Lopez-Guede (✉) · J. Estévez · M. Graña
Computational Intelligence Group of the Basque
Country University (UPV/EHU), San Sebastian, Spain
e-mail: jm.lopez@ehu.es

Á. Herrero et al. (eds.), *10th International Conference on Soft Computing Models in Industrial and Environmental Applications*, Advances in Intelligent Systems and Computing 368, DOI 10.1007/978-3-319-19719-7_26

a great breakthrough: a powerful tool was developed based on Geometrically Exact Dynamic Splines (GEDS) [5–7] to execute accurate simulations of LMCRS, allowing to assess the dynamical effects of the linking element of the LMCRS [8]. Using that tool, the work focused on the autonomous learning of optimal policies by Reinforcement Learning (RL) [9] reformulating the task as a Markov Decision Process (MDP) [9–11]. Within the RL paradigm, the Q-Learning algorithm [9, 12, 13] was implemented because it allows the agent to learn from its experience with the environment, without previous knowledge. Several works have been reported [14–19] using that algorithm improving the optimality of the reached results. Following the philosophical approach of using RL techniques that learn only from the experience, the TRQ-Learning algorithm was introduced in [20] reaching better results with boosted convergence. However, these results were always the fruit of computer simulations.

The main objective of the paper is to execute a proof of concept physical experiment of the task in the simplest instance of a LMCRS (with only one robot) to demonstrate that the computational simulation results are transferable to real physical world systems. To achieve this objective first it is necessary to build a complete physical system composed of the hose to transport, the robot which transports the hose, the RL controller that controls the execution of the task, the communications interface connecting the RL controller and the robot and, finally, a perception system to monitor the evolution of the task execution.

The paper is structured as follows. Section 2 details the design of the different parts of the system and their construction to carry out the proof of concept experiments. The concrete experimental design is given in Sect. 3. Section 4 discusses the results obtained in the experimental realization. Finally, Sect. 5 summarizes the obtained results and addresses further work.

2 System Design and Construction

2.1 Global Scheme Control

Figure 1 shows the main specific modules of which the system is composed in this proof of concept, and whose boxes are highlighted with a thicker trace:

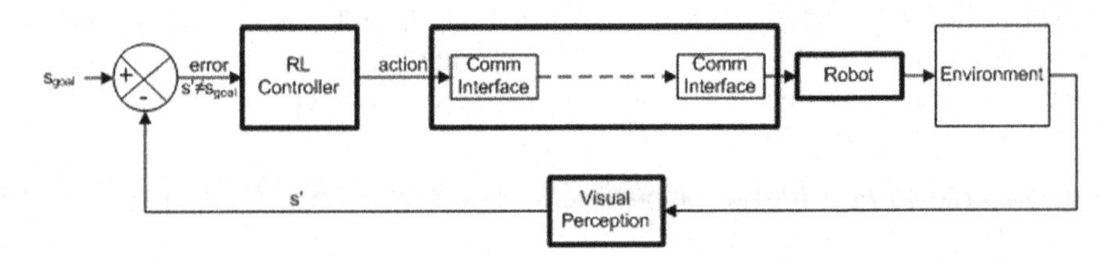

Fig. 1 Specific closed loop of the general system

- A control module, built according to the optimal policy learned by means of a RL algorithm.
- A communications module in charge of the transmission of the action to be executed by the robot by means of a wireless interface built for the occasion.
- The actuator that exerts the action on the environment, i.e., the robot.
- The perception module sensing and monitoring the environment after the actions has been executed to build the new system's state representation.

2.2 RL Control Module

This module is the responsible for determining the optimal action to be carried out by the agent to reach the goal state (s_{goal}) given an actual state (s'). We have to clarify that the RL controller has been previously learned by means a RL algorithm, and at this point it is executed without any retraining. Therefore, at this stage the controller is in the *exploitation* phase, so that it is neither able to learn new knowledge nor to improve its performance. The previous required learning phase has been carried out using TRQ-Learning algorithm, reaching a performance of 92 % successful goals. Anyway, that training phase is carefully described in previous papers [14–20] and it is beyond the scope of this paper.

2.3 Robot Manager

The actions that the agent can execute in the environment are movements of the SR1 robot. That robot is quite simple, so we have adapted it to our purpose. We have divided the software functions which we have built into two groups according to their abstraction level.

Low level functions This set of low abstraction level functions include all the settings and functions that are necessary to carry out our purpose, but that are not in the interface that the robot offers at RL level. The first operation that we had to do was the calibration of the two servo motors to determine the width of pulses necessary to reach a given velocity. This operation was performed only once, but there are other supporting functions executed on demand by the high abstraction level functions.

High level functions This set of high abstraction level functions includes the functions that the interface of the robot offers at RL level.

The functions that the robot interface offers to the controller are the following:

- **move_straight(time, velocity)**: it is used to move straight the SR1 robot at specified velocity (between −200 mm/s and +200 mm/s) during the specified time (between 0 and 9999 s). The product of these magnitudes (ignoring the sign) gives us the absolute distance of the movement.
- **turn(angle)**: it turns the SR1 robot the specified angle at fixed angular velocity.

- **move(time, velocity, angle)**: it first calls the **turn(.)** function to turn the robot in a desired orientation, and then calls the **move_straight(.)** function to move the robot straight during the specified time at the desired velocity.

2.4 Communications Module

Wireless communications The platform that has been used in this implementation does not have wireless capabilities even in its most complete embodiment. It can only receive commands through a wired RS-232 interface, but it is not suitable to the environment of our proof of concept because it would introduce a disturbing element (another wire) in the problem which was not considered when the learning agent was trained. So, we have extended the basic SR1 platform with wireless capabilities with the 19200 baud Easy-Radio ER400TRS Data Transceiver at 433 MHz. This is a transparent radio modem that allows us sending and receiving data frames up to 192 bytes, limitation that we take into account while designing the communications interface. On the host PC side, the complementary wireless device is the Devantech RF04 radio module, which allows transparent plug and play connection to the host PC. This device also includes another Easy-Radio ER400TRS Data radio modem that can send and receive data of 192 bytes, but the USB-serial data converter which includes has a buffer of only 128 bytes.

Communication protocol A complete specification of the communication protocol design is given by means of the syntax of the messages, their semantics and the procedure used to coordinate them. In this case, the protocol provides a connectionless service.

- **Syntax**

Syntax specifies the form of the messages and the size and type of each of the fields of which they are composed. In this case there are only two types of messages. The first is the *command*, sent from the RL controller located in the computer to the SR1 robot. The second is the *response* sent from the SR1 robot to the computer. It is not specified in more detail due to space issues.

- **Semantics**

Semantics specifies the meaning of the messages and fields of which they are composed.

1. The meaning of the *command* message is to allow the RL controller to send to the SR1 robot the specification of the action to be executed. By means the **move(time, velocity, angle)** interface, the RL controller tells the robot to turn **angle** degrees, and to move straight during **time** seconds at **velocity** m/s.
2. The meaning of the *response* message is to inform the RL controller by the SR1 robot manager about the reception of the *command* message. Each time that a *command* message is correctly received, the SR1 robot manager sends an 'OK' message to indicate that reception.

– **Procedures**

The procedures specify the order in which the different messages are sent between the two interlocutors. In this case, the procedures of the protocol are quite simple because it is a connectionless protocol: first the RL controller send to the SR1 robot manager the action (movement) to carry out through the *command* message, and the SR1 robot manager answers with a *response* message if the *command* message has been correctly received. If this does not happen, the RL controller will not receive that 'OK' *response* message and a timeout will be triggered. In that circumstance, the RL controller will be able to decide if the same *command* message must be sent again or not if it suspects that a permanent problem is happening (i.e., the same *command* message has been sent several times and the *response* message does not come back).

2.5 Visual Perception Module

In our proof of concept implementation the sensing is carried out by means of a visual perception module which monitors the relevant objects of the system, i.e., the robot that performs the task and the hose to transport, and translates the situation of the system into the state space model used in the experiment.

The complete description and design of the visual perception system and several experiments are reported in [21].

3 Experimental Design

3.1 Transportation Task Specification

The transportation task can be described as follows: given a flexible hose with one end attached to a fixed position (which could be a source of some fluid or electrical energy), transport its tip from a given position P_r to a goal position P_g by means of a mobile robot attached to it. The robot is able to execute a reduced set of actions (movements) to achieve that objective, avoiding collision with the hose itself which may interfere with reaching the goal position while ensuring that there is not any segment of the hose going outside of the predefined working area.

3.2 Experimental Settings

Working area and hose The concrete environment in which the previously described task is carried out is a perfect square of 2 m of side. The hose length at rest is 1 m, and one end of it is fixed in the center of the square, which will be the center of the coordinate reference system.

Visual perception The complete system monitoring is done through an overhead web cam, placed just in the center of the working area, 3.5 m over the coordinate reference and deliberately misaligned with the working area. The ground on which the experiments have been performed has been chosen on purpose because it is realistic enough for the difficulty of the task of identifying objects in the images. It is not specified in more detail due to space issues.

Actions, state models, spatial discretization steps Several elements regarding the Markovian formulation of the problem must be formalized. First, the set of actions (movements) that the SR1 robot can execute is specified in Eq. 1:

$$A = \{North, South, East, West\} \tag{1}$$

On the other hand, we have to define the state modeling, i.e., the environment perception by the agent and the discretization step, because the visual perception system must describe the real working environment according to them. The experiments has used a spatial discretization step of 0,5 m with the state model $X = \{P_r, P_g, i, c\}$, where P_r and P_g are the discrete two-dimensional coordinates of the robot and the goal position respectively, i is set to 1 if the line intersects the hose and c is set to 1 if the box with corners and intersects the hose

4 Physical Realization Results

Figure 2 shows the realization of the system. In the left column there is the evolution of the system according to the simulator, while in the right column there are the steps of the real world physical realization. The episode of the experiment starts from the initial situation of the system shown in Fig. 2(a, b), with the SR1 robot and the tip of the hose in the position P_r. The goal position P_g is indicated through the bold cross that is in the 4-th quadrant. Along the five necessary steps, all intermediate states of the system are shown in Figs. 2(c, e, g, i, k) for the simulated system and in Figs. 2(d, f, h, j, l) for the real world physical realization until the goal position P_g is reached. Each of these steps is of a length of 0.5 m. We can notice that the path that the SR1 has followed is optimal, because it is impossible to reach the goal position P_g with a lower number of intermediate steps starting from the position P_r. Comparing the simulator implementation versus the real physical world implementation realizations, we may notice that the shape of the hose (the linking element) differs slightly in some steps because the linking element implemented in the simulator is an ideal hose, where ideal twisting and bending forces have been implemented. However, the real world hose that has been used in the experiment is conditioned by folds, biases and flaws derived from the position in which it has been placed from the manufacturing moment.

These minimal differences are not an issue for the realization of the proof of concept experiment which can be considered successful because the real world realization shows that the linking element and the global system follow the behaviour

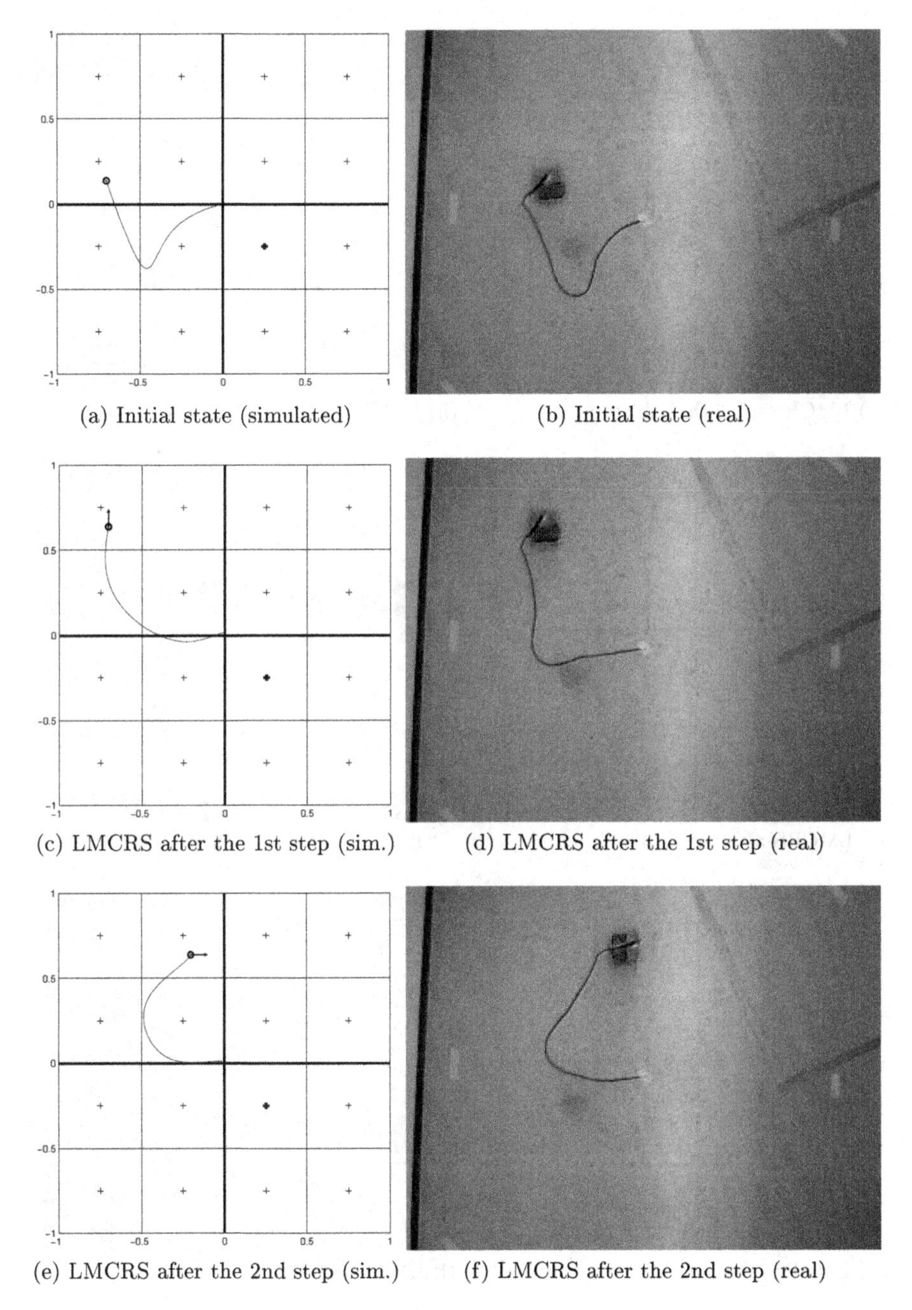

(a) Initial state (simulated) (b) Initial state (real)

(c) LMCRS after the 1st step (sim.) (d) LMCRS after the 1st step (real)

(e) LMCRS after the 2nd step (sim.) (f) LMCRS after the 2nd step (real)

Fig. 2 A realization of the system with state model $X = \{P_r, P_g, i, c\}$ and spatial discretization step = 0.5 m

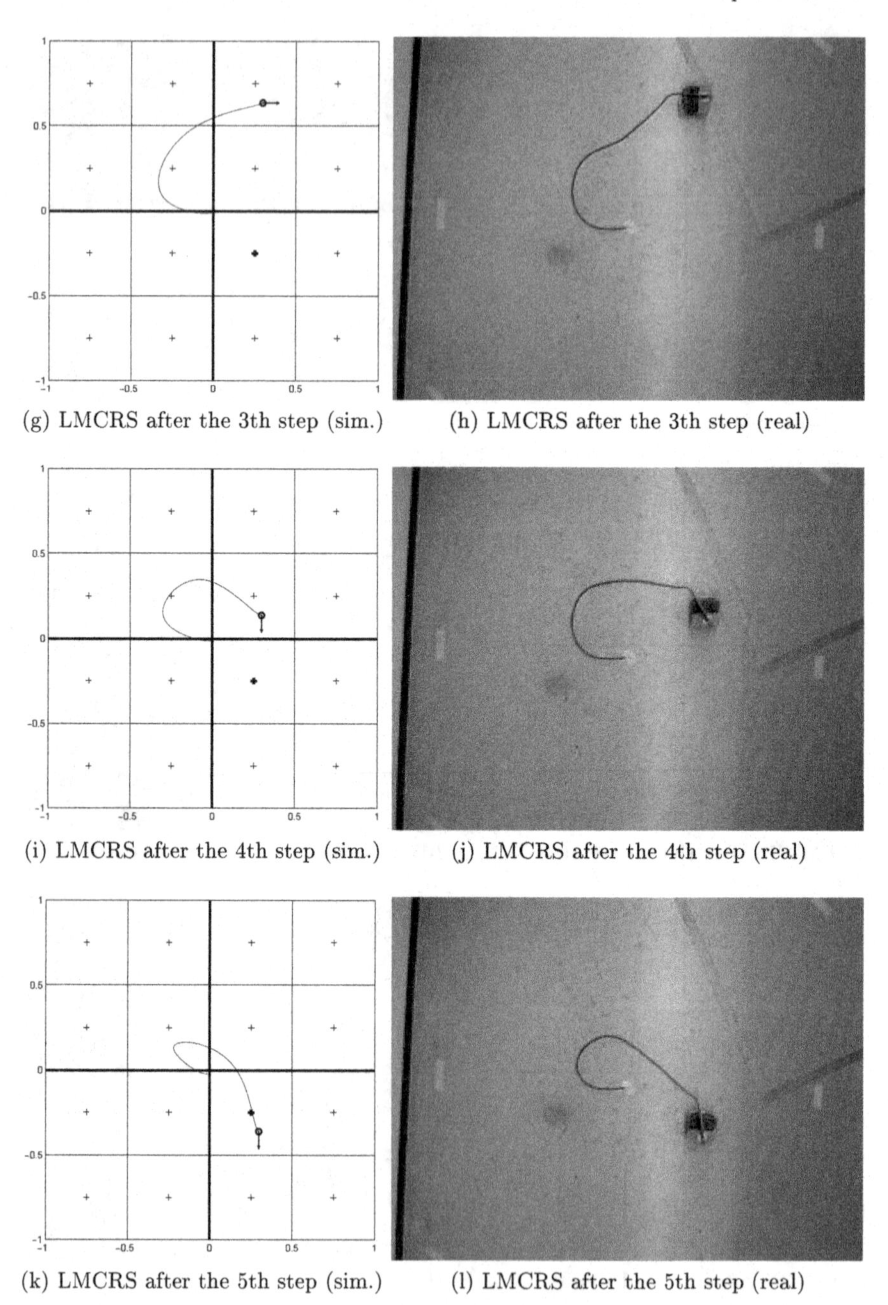

(g) LMCRS after the 3th step (sim.) (h) LMCRS after the 3th step (real)

(i) LMCRS after the 4th step (sim.) (j) LMCRS after the 4th step (real)

(k) LMCRS after the 5th step (sim.) (l) LMCRS after the 5th step (real)

Fig. 2 (continued)

predicted by the simulated system. We can conclude that the real physical implementation validates our previous simulations, and that the Q matrix learned previously in an autonomous way by means of RL techniques to solve this is concrete system is adequate.

5 Conclusions

This paper carries out the first proof of concept experiment of a single-robot hose transport task driven by a RL controller, which serves as a prototypical LMCRS task. We started referencing several of our previous work where the task was approached from a computational point of view, since there we used tools as a simulator based on GEDS to allow the agent to learn autonomously from its experience with the environment. Once the learner agent has reached a specified level of knowledge and could exploit it, we demonstrated through simulations that the single-robot case of hose transport task was successfully done. However, proof of concept experiments were still necessary to demonstrate that these results are transferable to real physical world. In this paper we have reported a successfully real world proof of concept experiment, validating our computational tool (simulator) and demonstrating that RL techniques are well suited to deal with real LMCRS tasks.

Acknowledgments The research was supported by the Computational Intelligence Group of the Basque Country University (UPV/EHU) through Grant IT874-13 of Research Groups Call 2013-2017 (Basque Country Government).

References

1. Duro R, Graña M, de Lope J (2010) On the potential contributions of hybrid intelligent approaches to multicomponen robotic system development. Inf Sci 180(14):2635–2648
2. Lopez-Guede JM, Graña M, Zulueta E (2008) On distributed cooperative control for the manipulation of a hose by a multirobot system. In: Corchado E, Abraham A, Pedrycz W (eds) Hybrid artificial intelligence systems. Lecture notes in artificial intelligence, vol 5271, pp 673–679. 3rd international workshop on hybrid artificial intelligence systems, pp 24–26. University of Burgos, Burgos, Spain
3. Echegoyen Z (2009) Contributions to visual servoing for legged and linked multicomponent robots. Ph.D. dissertation, UPV/EHU
4. Echegoyen Z, Villaverde I, Moreno R, Graña M, d'Anjou A (2010) Linked multi-component mobile robots: modeling, simulation and control. Rob Auton Syst 58(12, SI):1292–1305
5. Boor CD (1994) A practical guide to splines. Springer
6. Rubin M (2000) Cosserat theories: shells. Kluwer, Rods and Points
7. Theetten A, Grisoni L, Andriot C, Barsky B (2008) Geometrically exact dynamic splines. Comput Aided Des 40(1):35–48
8. Fernandez-Gauna B, Lopez-Guede J, Zulueta E (2010) Linked multicomponent robotic systems: basic assessment of linking element dynamical effect. In: Manuel Grana Romay MGS, Corchado ES (eds.) Hybrid artificial intelligence systems, Part I, vol 6076. Springer, pp 73–79
9. Sutton R, Barto A (1998) Reinforcement learning: an introduction. MIT Press

10. Bellman R (1957) A markovian decision process. Indiana Univ Math J 6:679–684
11. Tijms HC (2004) Discrete-time Markov decision processes. John Wiley & Sons Ltd, pp 233–277. http://dx.doi.org/10.1002/047001363X.ch6
12. Watkins C (1989) Learning from delayed rewards. Ph.D. dissertation, University of Cambridge, England
13. Watkins C, Dayan P (1992) Technical note: Q-learning. Mach Learn 8:279–292. doi:10.1023/A:1022676722315. http://dx.doi.org/10.1023/A:1022676722315
14. Fernandez-Gauna B, Lopez-Guede J, Zulueta E, Graña M (2010) Learning hose transport control with q-learning. Neural Netw World 20(7):913–923
15. Graña M, Fernandez-Gauna B, Lopez-Guede J (2011) Cooperative multi-agent reinforcement learning for multi-component robotic systems: guidelines for future research. Paladyn. J Behav Rob 2:71–81.doi:10.2478/s13230-011-0017-5. http://dx.doi.org/10.2478/s13230-011-0017-5
16. Fernandez-Gauna B, Lopez-Guede JM, Zulueta E, Echegoyen Z, Graña M (2011) Basic results and experiments on robotic multi-agent system for hose deployment and transportation. Int J Artif Intell 6(S11):183–202
17. Fernandez-Gauna B, Lopez-Guede J, Graña M (2011) Towards concurrent q-learning on linked multi-component robotic systems. In: Corchado E, Kurzynski M, Wozniak M (eds) Hybrid artificial intelligent systems. Lecture notes in computer science, vol 6679. Springer, Berlin/Heidelberg, pp 463–470
18. Fernandez-Gauna B, Lopez-Guede J, Graña M (2011) Concurrent modular q-learning with local rewards on linked multi-component robotic systems. In: Ferrández J, Alvarez Sánchez J, de la Paz F, Toledo F (eds) Foundations on natural and artificial computation. Lecture notes in computer science, vol 6686. Springer, Berlin/Heidelberg, pp 148–155
19. Lopez-Guede JM, Fernandez-Gauna B, Graña M, Zulueta E (2011) Empirical study of q-learning based elemental hose transport control. In: Corchado E, Kurzynski M, Wozniak M (eds) Hybrid artificial intelligent systems. Lecture notes in computer science, vol 6679. Springer, Berlin/Heidelberg, pp 455–462
20. Lopez-Guede J, Fernandez-Gauna B, Graña M, Zulueta E (2012) Improving the control of single robot hose transport. Cybern Syst 43(4):261–275
21. Lopez-Guede JM, Fernandez-Gauna B, Moreno R, Graña M (2012) Robotic vision: technologies for machine learning and vision applications. In: José García-Rodríguez MC (ed.) IGI Global

Bridging Classical Control with Nature Inspired Computation Through PID Robust Design

P.B. de Moura Oliveira, Hélio Freire, E.J. Solteiro Pires and J. Boaventura Cunha

Abstract Nature and biological inspired search and optimization methods are simple and powerful tools that can be used to design classical industrial controllers. In this paper a particle swarm optimization (PSO) algorithm based technique is deployed to design proportional integrative and derivative controllers to fulfill minimum robustness constraints. PID robustness design using maximum sensitivity and complementary sensitivity values is re-addressed and formulated within a constrained PSO. Results are presented and analyzed regarding the control objective of load disturbance rejection and compared with other techniques.

Keywords PID control · Particle swarm optimization · Robust control

1 Introduction

The use of proportional integrative and derivative (PID) controllers within industrial feedback control loops is quite representative of its practical relevance. Indeed, despite the development of more sophisticated control techniques, such as model predictive control (MPC), in some process control applications PID control is dominant. This relevance of use is one of the major reasons justifying PID research over the last decades [1], and makes the topic inclusion mandatory in both introductory and advanced feedback control courses. Moreover, its practical digital implementation has been promoted by the inclusion of dedicated software in major programmable logic controllers in the market.

PID controllers can be designed using a myriad of techniques [2]. Given the variety of techniques, their classification and appropriate selection for particular system dynamics can be a complicated task. Many classical design techniques are

P.B. de Moura Oliveira (✉) · H. Freire · E.J. Solteiro Pires · J. Boaventura Cunha
Department of Engineering, School of Sciences and Technology, INESC TEC – INESC Technology and Science, UTAD Pole, 5001–801 Vila Real, Portugal
e-mail: oliveira@utad.pt

Á. Herrero et al. (eds.), *10th International Conference on Soft Computing Models in Industrial and Environmental Applications*, Advances in Intelligent Systems and Computing 368, DOI 10.1007/978-3-319-19719-7_27

based on the use of tuning rules, consisting in mathematical functions of time-domain time and/or frequency domain system response characteristics. As far as authors are aware, there is not a single PID tuning technique which can be universally applied to all types of system dynamics. Optimization techniques can be used alternatively and complementary to classical controller design, making the design procedure more flexible. The use of search and optimization techniques depends of using appropriate cost functions and involves a computational burden which has been gradually attenuated with the ever increasing processing power. As it will be explained in the sequel, PID control design requires achieving system performance in terms of tracking and regulation, all well as fulfilling minimum robustness criteria. PID design techniques involving constrained optimization formulations, such as convex-concave have been proposed [3]. Nature and biological inspired meta-heuristics constitute a powerful alternative to design PID controllers considering several design criteria and robustness criteria. In the last 20 years well established bio-inspired meta-heuristics such as: genetic algorithms [4], differential evolution [5], particle swarm optimization PSO [6] have been proposed to tackle this problem considering single and multi-criteria formulations. If the number of simultaneous design criteria is higher than 3, many-objective optimization techniques can be applied [7].

Classical control system robustness measures are the well-known frequency-domain gain and phase margins. These margins should not be used individually, as achieving a good gain margin does not guaranty a good phase margin and vice versa. Even if these two measures are used simultaneously resulting in good design values, this not guaranties the system robustness [8]. Thus, a measure introduced by Smith [10] termed vector margin, representing the minimum distance vector from point $(-1 + j0)$ to the system loop Nyquist plot. Vector margin has been used as a convenient robustness measure as it guaranties minimum values for both gain and phase margins. This paper addresses the problem of optimizing PI-PID controllers using transient response time-domain error based criteria, subject to robustness constraints. The bio-inspired metaheuristic deployed here is a classical PSO algorithm. The motives for selecting the PSO algorithm over other more recently proposed algorithms (e.g. gravitational search optimization, firefly optimization etc.) lies both in its broad and wide acceptance and simplicity of implementation. Thus, the paper aims in contributing to clarify, some important aspects bridging classical PID control robustness design and particle swarm optimization. A tutorial approach is adopted, for concepts exposition sake, which may be advantageous for different types of readers: students, researchers and industrial practitioners.

The remainder of the paper is structured as follows. Section 2 presents some introductory PID control notions and concepts. Section 3 revises the robustness measures based on Nyquist plots circle interpretations. In Sect. 4, the design methodology based on the PSO algorithm is presented outlining the key implementation aspects. Section 5 presents some simulation experiments results obtained with the proposed method and analysis discussion. Finally, Sect. 6 presents some concluding remarks and outlines further research work.

2 PI-PID Controller Design Issues

Consider the single-input single-output control system illustrated by the following block diagram (Fig. 1):

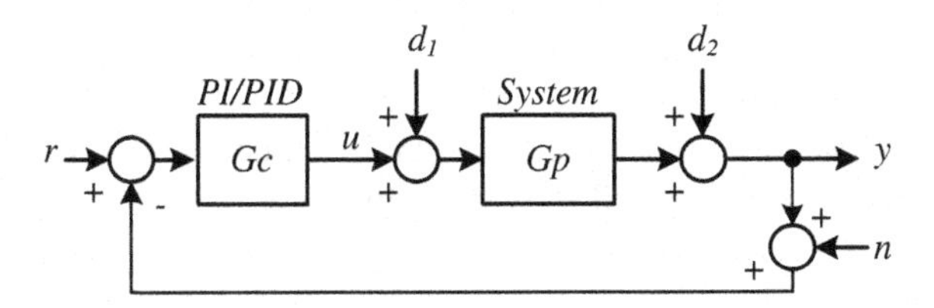

Fig. 1 General feedback loop

where: r represents the reference input (set-point), y the controlled output, d_1 a load disturbance, d_2 an output disturbance, n a noise signal, u the controller output, G_c the controller and G_p the system or process to control. PID controllers can be governed by different equations depending on several approaches and implementation issues. A classical representation, known as parallel form, can be represented in the Laplace domain for PID and PI controllers, by the transfer function Eqs. (1) and (2) respectively:

$$G_{cPID}(s) = K_p\left[1 + \frac{1}{sT_i} + sT_d\right] = K_p + \frac{K_i}{s} + sK_d \tag{1}$$

$$G_{cPI}(s) = K_p\left[1 + \frac{1}{sT_i}\right] = K_p + \frac{K_i}{s} \tag{2}$$

where: K_p, T_i, T_d, K_i and K_d represents, respectively, the proportional gain, integrative time constant, derivative time constant, proportional gain and derivative gain. PID control design requires selecting the control modes and tuning the controller parameters. In the derivative action case, the PID controller is usually implemented incorporating a first order filter which is also beneficial to turn the non-proper Eq. (1) into a proper and physically realizable function. While selecting 2 or 3 parameters may not appear to be much of an optimization problem, it is important to state that in large process control plants there may be hundreds of such controllers to adjust, which in many cases are used with other common techniques involving additional adjustable control parameters, such as: cascade control, feedforward control, etc. Studies [12] reveal that in some cases instead of finding PID controllers in operation PI controllers are selected by plant operators, for simplicity of use in detriment of performance.

The controller can be designed to achieve optimal performance by minimizing common error based criteria obtained from applying a step change in either the reference, load disturbance or output disturbance inputs. Set-point tracking and load disturbance are the most common design objectives. Two of the most used criteria are the integral of absolute error (IAE) and integral of time weighted absolute error (ITAE), represented respectively in (3):

$$IAE = \int_0^{\infty} |e(t)|dt, \quad ITAE = \int_0^{\infty} t|e(t)|dt \tag{3}$$

Controller design should be accomplished taking into account the controller effort. A form to evaluate the control effort (CE), is by using Eq. (4):

$$CE = \int_0^{\infty} |u(t)|dt \tag{4}$$

A topic which has been receiving significant research efforts recently is how to design PID controllers to reject noise [10, 11]. The design considering noise rejection is out of the scope of this study. As the core of this work is based on the robustness design the respective analysis and constraint formulation is detailed in the next section.

3 Robustness Controller Design Criteria

Consider the Nyquist loop transfer function represented in Fig. 2a). In this case $L(j\omega) = Gc(j\omega)Gp(j\omega)$, α_{min} represents the vector margin whose magnitude corresponds to the minimum distance between the point $-1 + j0$ and the loop polar plot, GM and PM are the gain and phase margins, respectively. The vector margin amplitude corresponds to the circle radius with center in $-1 + j0$. From Fig. 2a) it is possible to establish the correspondences between GM and α_{min} (5) and PM and α_{min} (6) [9]:

$$(1-\alpha_1) = \frac{1}{GM} \Leftrightarrow \alpha_1 = 1 - \frac{1}{GM}, \quad \alpha_{min} \leq \alpha_1 \Rightarrow GM \geq \frac{1}{1-\alpha_{min}} \tag{5}$$

$$PM = 2\arcsin\left(\frac{\alpha_2}{2}\right), \quad \alpha_{min} \leq \alpha_2 \Rightarrow PM \geq 2\arcsin\left(\frac{\alpha_{min}}{2}\right) \tag{6}$$

As the sensitivity function, S, defined in (7) represents the inverse of the distance between point $-1 + j0$ and the loop frequency response. L, the maximum sensitivity value, M_s, can be used to determine the vector margin, as follows:

$$S(j\omega) = \frac{1}{1+L(j\omega)}, \quad M_s = \max_{\omega}|S(j\omega)| = \frac{1}{\min_{\omega}|1+L(j\omega)|} = \frac{1}{\alpha_{min}} \Rightarrow \alpha_{min} = \frac{1}{M_s} \tag{7}$$

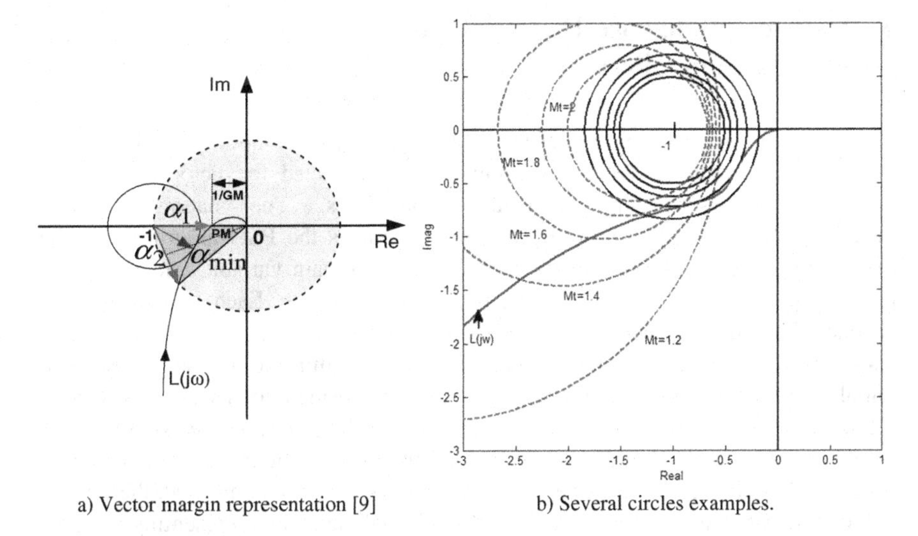

a) Vector margin representation [9]

b) Several circles examples.

Fig. 2 Loop Nyquist plots with circles constraints

An advantage of using the vector margin is that by determining the maximum sensitivity value two estimates for both gain and phase margins are obtained using Eqs. (5–7). However, these estimates are conservative and the degree of conservativeness depends of the type of system Nyquist plot and the vector margin value. While in terms of robustness the conservative values are good, in the other hand it may degrade performance values achieved. Another classic robustness measure is the maximum complementary sensitivity value, M_t, represented by expression (8), with T representing the complementary sensitivity function:

$$T(j\omega) = \frac{L(j\omega)}{1 + L(j\omega)}, \quad M_t = \max_{\omega} |T(j\omega)| \tag{8}$$

The maximum complementary sensitivity value is related with another circle in the Nyquist plot defined by, center, c_t, and radius, r_t, evaluated respectively with (9) [3]:

$$c_t = -\frac{M_t^2}{M_t^2 - 1}, \quad r_t = \frac{M_t}{M_t^2 - 1} \tag{9}$$

In Fig. 2b) several circles are presented for the M_s values (solid line) and using the same values for M_t circles (dashed line). As it can be observed from this figure, increasing both M_s and M_t implies decreasing the respective circle radius. An illustrative example of a system loop transfer function Nyquist plot (L) is presented in Fig. 2b). The overall objective is to design a PI/PID controller for which the corresponding loop Nyquist plot is outside both circles, which usually are formulated as constraints within the optimization problem [3].

4 Swarm Intelligence Design of PI-PID Robust Controllers

The approach proposed here consists in a single objective optimization using as main criteria the ITAE (3) index minimization subjected to vector margin and complementary sensitivity robustness constraints. The natural inspired meta-heuristic selected to perform the optimization is the PSO algorithm for the reasons presented in the introduction. PSO uses two main variables characterizing particles position, x, and velocity, v, in the search space. Each swarm member encodes the controller parameters: *{Kp,Ki}* or *{Kp,Ki,Kd}* for the PI or PID case, respectively. As in any population based search algorithm the first step consists in initializing the population. This is usually performed using a totally blind procedure filling the swarm using a random procedure. Depending on the given problem this may be advantageous or disadvantageous. In this case, if a totally random procedure is used to initialize the swarm, depending of the system to be controlled, there is the risk of a significant number (if not all) of the initial elements representing unstable controllers. This may compromise the search effectiveness. Thus, to avoid this scenario, an interview stage is performed to every randomly generated element, within a predefined gain parameter interval, to evaluate is the system is stable. Only stable elements are allowed in the first swarm.

After the swarm is initialized the stable controllers are evaluated both in terms of time-domain performance and frequency-domain robustness conditions. The velocity of each particle is updated using Eq. (10):

$$v_i(t+1) = \omega\, v_i(t) + c_1\varphi_1 [p_i(t) - x_i(t)] + c_2\varphi_2 [p_g(t) - x_i(t)] \tag{10}$$

with: i representing the element number, t representing the iteration number, ω is an inertia weighting factor, c_1 and c_2 the cognitive and social constants, respectively, φ_1 and φ_2 represent randomly generated numbers with uniform distribution in the range [0,1]. In this case $c_1 = c_2 = 2$. In (10) p_i represent particles individual best value and p_g the global best particle of the entire swarm (fully connected model) or specified neighborhood. All position and velocity vectors are d-dimensional. The inertia weight is usually decreased linearly (or otherwise) between a maximum starting value and a minimum ending value. The velocity value evaluated with (10) is used in the particle incremental position Eq. (11). The PSO here is executed until a predefined maximum number of iterations is achieved.

$$x_i(t+1) = x_i(t) + v_i(t+1) \tag{11}$$

A very important aspect in the proposed technique concerns the robustness constraints handling by the PSO algorithm. The vector margin defined by (7) and maximum complementary function (8) are used as constraints. In order to evolve the PSO to fulfill predefined values, a penalty function is used to penalize swarm elements not satisfying these robustness margin values.

5 Simulation Results and Discussion

Two models used in [3] are considered in this paper representing an open-loop unstable system, Gp_1, and a triple pole system, Gp_2, both represented in (12):

$$G_{p1}(s) = \frac{1}{(s-1)(0.1s+1)}, \quad G_{p2}(s) = \frac{1}{(s+1)^3} \tag{12}$$

The control objective considering is to optimize the system load rejection response considering a unit step input, but the same procedure is equally applicable to set-point tracking. The PSO conditions used are: swarm size $m = 30$, 60 iterations, and ω was linearly decreased between 0.9 and 0.4. The PI controller parameter range used was [0.1, 20] and [0.1, 20] for K_p and K_i, respectively. In the case of the PID controller the range for K_d is [0.1, 20]. All simulations were carried out in Matlab/Simulink environment assuming the simulation time to be shown in the time-response plots, with a fixed solver step interval of 0.01 s. PI control was applied to system Gp_1 and PID control to Gp_2. The results obtained are presented in Fig. 3 and Table 1, for the 4 different cases (I-IV). In Fig. 3 part a) presents the load unit step response and part b) the Nyquist plots with the corresponding vector margin (*VM*) and M_t circles.

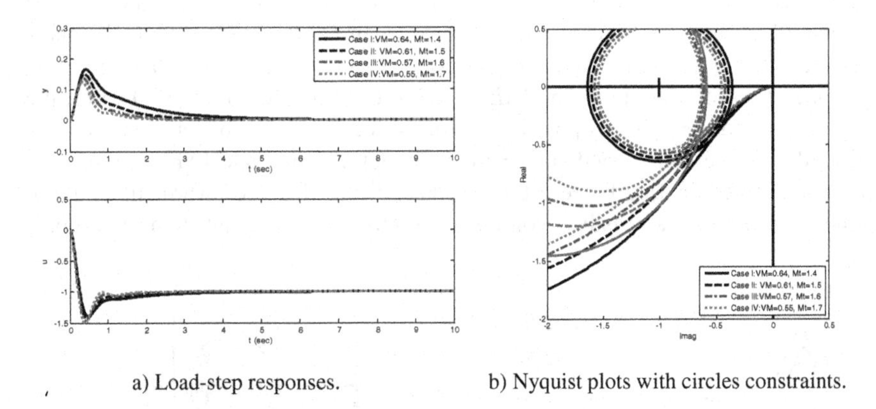

a) Load-step responses.

b) Nyquist plots with circles constraints.

Fig. 3 PI control results for system Gp1

As for system Gp_1, the most influent constraint in the design is the M_t constraint, a minimum value of $VM = 0.5$ was used for all the 4 cases. The M_t values considered were *{1.4, 1.5, 1.6, 1.7}* corresponding respectively to cases I to IV. As it can be observed from Fig. 3, the *VM* value obtained is always superior to 0.5 and the different minimum Mt values were achieved. As the M_t is decreased the load-response improves and the control effort increases.

The results achieved for the PID control are presented in Fig. 4 and Table 1, in the same format used for PI control. For this system the most influent constraint is

Table 1 PI-PID gains obtained

System	Case	Kp	Ki	ITAE	System	Case	Kp	Ki	Kd	ITAE
Gp_1	I	7.23	4.35	36.6	Gp_2	I	3.89	2.59	3.85	175.5
	II	7.87	7.03	13.9		II	5.86	4.38	5.32	88.8
	III	8.58	9.94	7.7		III	8.24	7.25	7.11	50.0
	IV	9.34	13.06	4.9		IV	10.96	11.44	9.02	30.4

the vector margin. Thus a minimum value of $M_t = 1.1$ was used for all cases. The VM values considered were {0.71, 0.63, 0.56, 0.5} corresponding respectively to cases I to IV. The correspondence between these values and M_s values is {*1.4, 1.6, 1.8, 2*}, respectively. The results presented in Fig. 4, show that the PSO can exactly match the specified VM constraint, while the values for the Mt constraint are superior to the defined minimum, 1.1, in all cases.

A comparison between the results presented with the proposed PSO technique and the ones reported in [3] using concave-convex optimization and an IAE Minimization method is presented in Fig. 4. A PI controller was used for Gp_1 (case a)) and a PID controller for Gp_2 (case b)). The gains obtained for the PI controller designed with the PSO, with two constraints $VM = 0.71$ and $M_t = 1.4$ are $\{K_p = 4.84,\ K_i = 2.00\}$ corresponding to an $ITAE = 93.9$, while the concave-convex gains presented in [3] are $\{K_p = 0.71, K_i = 0.63\}$ corresponding to an $ITAE = 115.3$. The PID controller gains obtained with the PSO, to achieve $VM = 0.71$ are the ones presented for case I in Table I, corresponding to an $ITAE = 175.5$, while the gains presented in [3] using a IAE minimization method are $\{K_p = 3.81, K_i = 3.32, Kd = 4.25\}$ corresponding to an $ITAE = 202.3$. Both for the PI and PID cases, the results obtained with the proposed PSO robust technique are better than the ones obtained with the constrained concave-convex optimization and IAE minimization results presented in [3]. Indeed, the results indicate that the swarm intelligence technique proposed can easily accommodate robustness constraints to design robust

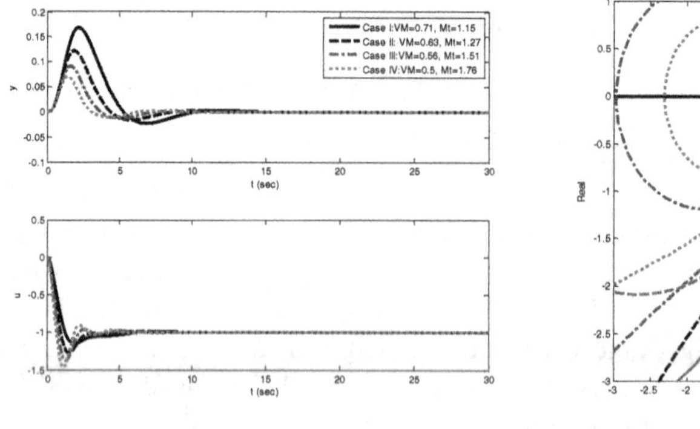

a) Load-step responses. b) Nyquist plots with circles constraints.

Fig. 4 PID control results for system Gp_2

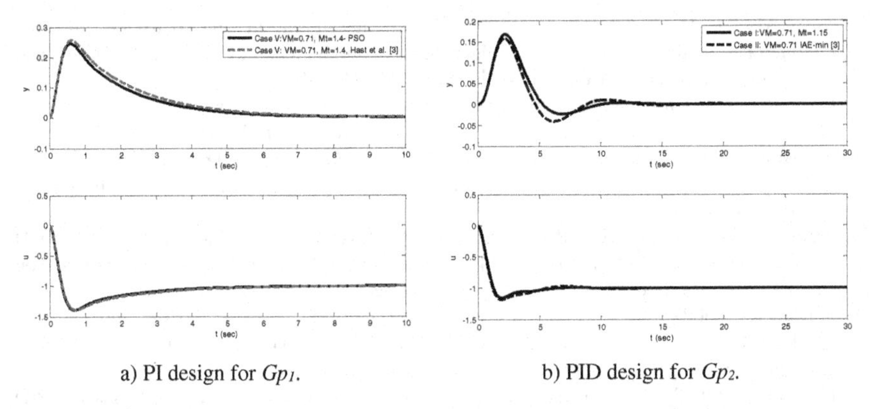

a) PI design for Gp_1. b) PID design for Gp_2.

Fig. 5 Comparison between the proposed PSO based technique and the results reported in [3]

PI-PID controllers, and constitute a simple and powerful alternative to other more complex optimization techniques (Fig. 5).

6 Conclusion

In this paper a technique based in a swarm algorithm was proposed to design PI-PID controllers such as to fulfill minimum robustness constraints. Robustness design specifications were presented, based on vector margin and maximum complementary sensitivity circles using Nyquist plots. A comparison was made relative to the results presented in [3] showing that the proposed PSO based method provides a better performance, keeping the same robustness design constraints. So far, the results achieved with the proposed technique clearly indicate that it can cope very well with the robustness constraints and be used as an alternative to more complex optimization techniques. Further research is undergoing to fully evaluate the proposed technique addressing the following issues: (i) set-point tracking objective design; (ii) multi-criteria design using many-objective techniques, (iii) swarm dynamics analysis.

References

1. Aström KJ, Hägglund T (2001) The future of PID control. Control Eng Pract 9:1163–1175
2. Ang KH, Chong G, Li Y (2005) PID control system analysis, design, and technology. IEEE Trans Control Syst Technol 13(4):559–576
3. Hast M, Aström KJ, Bernhardsson B, Boyd S (2013) PID design by convex-concave optimization. In: 2013 European control conference (EEC). Switzerland, pp 4460–4465

4. Jones AH, De Moura Oliveira PB (1995) Genetic auto-tuning of PID controllers. In: First international conference on genetic algorithms in engineering systems: innovations and applications, GALESIA. (Conf. Publ. No. 414), pp 141–145
5. Dong R (2009) Differential evolution versus particle swarm optimization for PID controller design. In: Fifth international conference on natural computation, IEEE, pp 236–240
6. De Moura Oliveira PB, Cunha JB, Coelho JP (2002) Design of PID controllers using the particle swarm algorithm. Proceedings of the IASTED MIC'2002, February. Innsbruck, Austria, pp 263–26
7. Freire HF, De Moura Oliveira PB, Solteiro Pires E, Bessa M (2014) Many-objective PSO PID controller tuning. In: CONTROLO'2014 – Proceedings of the 11th Portuguese conference on automatic control. Lecture notes in electrical engineering, vol 321, pp 183–192
8. Aström KJ (2002) Control system design, Chapter 7, p 255
9. Franklin GF, Powell JD, Naeiini AE (1991) Feedback control of dynamic systems. Addison-Wesley, pp 324–325
10. Hägglund T (2012) Signal filtering in PID control. In: IFAC conference on advances in PID control PID'12, Brescia (Italy), March, WePl.1
11. Segovia VR, Aström KJ, Hägglund T (2013) Noise filtering in PI and PID Control. In: 2013 American control conference (ACC), Washington, DC, USA, June, pp 1763–1770
12. Ender DB (1993) Process control performance: not as good as you think. Control Eng 180–190

A First Intelligent Approach to Path Following Algorithm for Quadcopters

Pablo García Auñón, Matilde Santos Peñas and Jesus Manuel de la Cruz García

Abstract In this paper, an algorithm to follow a given trajectory with a quadcopter is presented. This algorithm depends on a group of parameters whose selection strongly influences its performance. At the same time, those parameters depend on the velocity of the aerial vehicle and the curvature of the reference path. The main contribution of this paper is to show how the original algorithm can be enhanced if these parameters are chosen using a fuzzy-logic decision system based on a dimensionless variable, rather than taking them as constants. Simulation results are encouraging.

Keywords Computational intelligence · Fuzzy logic · Path following · Quadcopter

1 Introduction

The quadcopter, an unmanned aerial vehicle (UAV), and its control have been studied in the last two decades by many research groups all around the world. However, it was only when the size, power consumption and weight of computers and sensors have decreased, when these autonomous vehicles became popular. Nowadays, they are used in a wide range of applications, such as searching and rescuing activities, close and far inspection of facilities, aerial photography, and for researching purpose, among others. Even some companies, such as the German DHL or Amazon, are trying to use them as a regular way of transporting goods to isolated places. Their regularization in the European frame would boost their use for more civil proposes.

P.G. Auñón (✉)
Superior Technical School of Computer Engineering, UNED, Madrid, Spain
e-mail: pablogarciaaunon@gmail.com

M.S. Peñas
Computer Sciences, University Complutense of Madrid, Madrid, Spain
e-mail: msantos@ucm.es

J.M. de la Cruz García
Faculty of Physics, University Complutense of Madrid, Madrid, Spain
e-mail: jmcruz@ucm.es

Á. Herrero et al. (eds.), *10th International Conference on Soft Computing Models in Industrial and Environmental Applications*, Advances in Intelligent Systems and Computing 368, DOI 10.1007/978-3-319-19719-7_28

Quadcopters are unstable systems that, when properly controlled, present a maneuverability much higher than other aircrafts. Moreover, the lack of complex mechanical systems, its ability to hover and takeoff vertically, and its relative low price, are other advantages, see [6]. On the other hand, they cannot carry big payloads and their endurance and range are limited.

For some applications, following a given path accurately is needed. Take as example the searching of an object on the earth surface with a limited vision system, in which a specified path must be followed to cover the entire surface. In case of landing on a moving surface, such a ship, the importance of tracking properly a trajectory becomes crucial. There are many ways of generating the path, and soft computing techniques have resulted very useful, as in [3], where genetic algorithms were used.

Once the trajectory is generated, different control approaches can be applied to this type of UAVs for path following. In [5], for instance, a simple linear controller was used to follow a path generated by way-points. For a high maneuverability quadcopter, a more complex predictive control is presented in [4], keeping the yaw angle fixed and being the trajectories recalculated online. A similar technique was also used in [7]. In [10], Lyapunov functions were applied, firstly to control the dynamics of the quadcopter, and then to track a given path. An improvement on path following strategy is described in [2], where a robust compensator was added to a classical PD controller. Intelligent control techniques (see [9]) have also provided useful tools to improve the results of guiding autonomous systems.

In this paper, an adaptation of the control law presented in [11] is developed. The original work was designed for a fixed-wing aircraft; it is based on a Lyapunov function which ensures the convergence of the aircraft to a frame placed at a virtual point, which moves along the desired path. The new proposed algorithm not only gives good results following given trajectories, as it will be shown, but it can also be adapted to following targets whose movement is not previously known. This algorithm aims to be implemented in a real quadcopter, in the frame of the SALACOM[1] project, supported by the Spanish Ministry of Industry. Before carrying out tests on the real system, simulations of the algorithm must be performed.

The proposed algorithm depends on some parameters that must be well chosen. These parameters might take different values depending on the conditions of the mission. They were considered constants in the original work. On the contrary, we have used a fuzzy rule-base system to select the best parameters of the control algorithm and to analyze their influence on the system response. This is the main contribution of this paper.

The rest of the paper is organized as follows. In Sect. 2, the UAV system and its reference frames are described. In Sect. 3, the control algorithm is presented in detail. Section 4 shows its application to the aerial vehicle; control parameter selection by means of a fuzzy rule-base is discussed. The results are explained. Finally, the conclusions and future works are exposed.

[1]"Sistema Autnomo para la Localizacin y Actuacin ante Contaminantes en el Mar", Autonomous Systems for the Location and Actions against Contaminants on the Sea, ISCAR, UCM.

2 System Description

The system under consideration consists of a real quadcopter with a low level controller implemented. This controller has as inputs the desired altitude, and the magnitude and direction of the desired velocity in the xy plane, in the $\{\boldsymbol{I}\}$ frame. The quadcopter provides its global position, current altitude, absolute velocity, attitude, and roll, pitch and yaw angular velocities.

It is assumed that the controller is able to stabilize the quadcopter when the magnitude and direction of the reference speed is given. Independently of the flight direction, the controller is also able to take the aerial vehicle to a given altitude. The algorithm uses, in addition to the commonly applied North-East-Down (NED) reference system, here named as $\{\boldsymbol{I}\}$, the next frames:

Transport frame $\{\boldsymbol{F}\}$

- Origin P: virtual point on the path.
- $\boldsymbol{t}$: tangent to the path at P.
- $\boldsymbol{n}_1$: first normal vector.
- $\boldsymbol{n}_2$: second normal vector.

The relationship between the basis vectors of the Bishop and the Frenet reference frames is (see [1]):

$$\begin{bmatrix} \boldsymbol{t} \\ \boldsymbol{n} \\ \boldsymbol{b} \end{bmatrix} = \begin{bmatrix} 1 & 0 & 0 \\ 0 & \cos\beta & \sin\beta \\ 0 & -\sin\beta & \cos\beta \end{bmatrix} \begin{bmatrix} \boldsymbol{t} \\ \boldsymbol{n}_1 \\ \boldsymbol{n}_2 \end{bmatrix} \tag{1}$$

where

$$\frac{d\beta(l)}{dl} = \tau(l) \tag{2a}$$

$$k_1 = \lambda\cos\beta \tag{2b}$$

$$k_2 = \lambda\sin\beta \tag{2c}$$

being τ the curvature of the path, l the length, λ its torsion, and k_1 and k_2 the Bishop curvatures, which will be used afterwards. The transformation matrix can be calculated as follows:

$$\boldsymbol{R}_I^F = \begin{bmatrix} \boldsymbol{t} \\ \boldsymbol{n}_1 \\ \boldsymbol{n}_2 \end{bmatrix}_{\{I\}} \tag{3}$$

if the basis vectors are row vectors expressed in the $\{\boldsymbol{I}\}$ frame.

Velocity frame $\{\boldsymbol{W}\}$

- Origin Q: center of gravity of the quadcopter.
- $\boldsymbol{w}_1$: pointing to the velocity direction.
- $\boldsymbol{w}_2$: right-hand rule.
- $\boldsymbol{w}_3$: perpendicular to $\boldsymbol{w}_1$, in the vertical plane of symmetry of the quadcopter.

Auxiliary frame $\{D\}$

- Origin Q: center of gravity of the quadcopter.
- $\boldsymbol{b_{1D}}$: pointing d meters ahead of P in the $\boldsymbol{t}$ direction.
- $\boldsymbol{b_{2D}}$: perpendicular to $\boldsymbol{b_1}$ in the plane determined by $\boldsymbol{t}$ and $\boldsymbol{n_1}$.
- $\boldsymbol{b_{3D}}$: right-hand rule.

These basis vectors are then calculated as:

$$\boldsymbol{b_{1D}} = \frac{d \cdot \boldsymbol{t} - y_F \cdot \boldsymbol{n_1} - z_F \cdot \boldsymbol{n_2}}{\sqrt{d^2 + y_F^2 + z_F^2}} \tag{4a}$$

$$\boldsymbol{b_{2D}} = \frac{y_F \cdot \boldsymbol{t} + d \cdot \boldsymbol{n_1}}{\sqrt{y_F^2 + z_F^2}} \tag{4b}$$

$$\boldsymbol{b_{3D}} = \boldsymbol{b_1} \wedge \boldsymbol{b_2} \tag{4c}$$

where d is a constant parameter (a distance), and y_F and z_F are the coordinates of the point Q in the $\boldsymbol{n_1}$ and $\boldsymbol{n_2}$ plane, $\{\boldsymbol{F}\}$ frame.

In Fig. 1 the frames $\{\boldsymbol{F}\}$, $\{\boldsymbol{W}\}$ and $\{\boldsymbol{D}\}$ have been represented. Recall that P is the virtual point, to be followed, and Q is the center of gravity of the quadcopter. Frame $\{\boldsymbol{I}\}$ has not been represented.

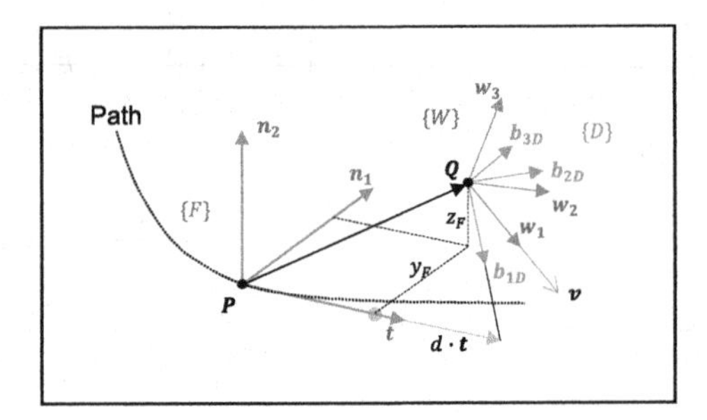

Fig. 1 Reference frames used

For the sake of the test, the quadcopter yaw rate will be limited to ±0.5 rad/s, and its velocity up to 3 m/s. Such limitations are always taken into account for unmanned aerial vehicles but probably higher values might be achievable.

The variable used to select the parameters of the algorithm is dimensionless so that the results can be applied to any system with different characteristics.

3 Control Algorithm

The original path following algorithm is described in [11] which, although developed for fixed-wing UAVs, can be also used for quadcopters with some modifications. The algorithm mainly consists of orienting the velocity vector of the UAV to a virtual point, P, which moves along the desired trajectory with a controlled pace. A one-dimensional error function Ψ is defined as the misalignment between the vectors $\boldsymbol{w_1}$ and $\boldsymbol{b_{1D}}$, see Fig. 1. The altitude is controlled by the embedded quadcopter regulator.

The position of the center of gravity of the UAV, Q, and its derivative, in the frame $\{\boldsymbol{F}\}$, are expressed as:

$$\boldsymbol{r_Q}\Big|_{\{F\}} = \begin{bmatrix} x_F \\ y_F \\ z_F \end{bmatrix} \tag{5}$$

$$\dot{\boldsymbol{r}}_{\boldsymbol{Q}}\Big|_{\{F\}} = \begin{bmatrix} \dot{x}_F \\ \dot{y}_F \\ \dot{z}_F \end{bmatrix} = -\dot{l} \cdot \boldsymbol{t} - \boldsymbol{w_{F/I}} \wedge \boldsymbol{r_Q} + v \cdot \boldsymbol{R^F_W} \cdot \boldsymbol{w_1} \tag{6}$$

where

$$\boldsymbol{w_{F/I}}\Big|_{\{F\}} = [0, -k_2(l) \cdot \dot{l}, k_1(l) \cdot \dot{l}] \tag{7}$$

is the angular velocity of the frame $\{\boldsymbol{F}\}$ with respect to $\{\boldsymbol{I}\}$ expressed in $\{\boldsymbol{F}\}$, and $\boldsymbol{R^F_W}$ is the rotation matrix from frame $\{\boldsymbol{W}\}$ to $\{\boldsymbol{F}\}$. The velocity of the virtual point P along the path is chosen using the following expression:

$$\dot{l} = \left(v \cdot \boldsymbol{w_1} + k_l \cdot \boldsymbol{r_Q}\Big|_{\{F\}} \right) \cdot \boldsymbol{t} \tag{8}$$

where k_l is the coefficient which controls the convergence of the virtual point P proportionally to the distance between this point and the quadcopter (vector $\boldsymbol{r_Q}\Big|_{\{F\}}$). With Eqs. 6, 7 and 8, it is possible to obtain the dynamics of the point Q in the $\{\boldsymbol{F}\}$ frame. The error function Ψ is defined:

$$\Psi(\hat{R}) = \frac{1}{2}\text{tr}\left[(\boldsymbol{I_3} - \boldsymbol{\Pi^T_R}\boldsymbol{\Pi_R})(\boldsymbol{I_3} - \hat{\boldsymbol{R}})\right] = \frac{1}{2}(1 - \hat{R}_{11}) \tag{9}$$

which is a positive-defined function and takes into account the misalignment between the velocity direction vector $\boldsymbol{w_1}$ and $\boldsymbol{b_{1D}}$. In the previous equation, $\hat{\boldsymbol{R}} = \boldsymbol{R^D_W}$ and $\boldsymbol{\Pi_R} = \begin{bmatrix} 0\,1\,0 \\ 0\,0\,1 \end{bmatrix}$. By deriving it we finally get:

$$\dot{\Psi}(\hat{R}) = \boldsymbol{e_{\hat{R}}} \cdot \left(\begin{bmatrix} q \\ r \end{bmatrix} - \boldsymbol{\Pi_R}\hat{\boldsymbol{R}}^T \left(\boldsymbol{R^D_F} \cdot \boldsymbol{w_{F/I}}\Big|_F + \boldsymbol{w_{D/F}}\Big|_D \right) \right) \tag{10a}$$

$$\boldsymbol{e_{\hat{R}}} = \frac{1}{2}\boldsymbol{\Pi_R}\left((\boldsymbol{I_3} - \boldsymbol{\Pi^T_R}\boldsymbol{\Pi_R})\hat{\boldsymbol{R}} - \hat{\boldsymbol{R}}^T(\boldsymbol{I_3} - \boldsymbol{\Pi^T_R}\boldsymbol{\Pi_R}) \right)^{\vee} = \frac{1}{2}\left[\hat{R}_{13}, \hat{R}_{12}\right]^T \tag{10b}$$

$$\left(\boldsymbol{w_{D/F}}\Big|_D \right)^{\vee} = \boldsymbol{R^D_F}\dot{\boldsymbol{R}}^F_D \tag{10c}$$

where q and r are the pitch and yaw velocities, respectively, and $(\cdot)^{\vee}$ denotes *vee map*, see [8]. Equations 6, 8, and 10 define the overall path-following error dynamics. In order to ensure that the error function Ψ converges to zero, its derivative must be negative-defined and therefore, the next control law is proposed:

$$\begin{bmatrix} q_c \\ r_c \end{bmatrix} = \Pi_R \hat{R}^T \left(\boldsymbol{R}_F^D \cdot w_{F/I}\Big|_F + w_{D/F}\Big|_D \right) - 2 \cdot k_{\hat{R}} \cdot \boldsymbol{e}_{\hat{R}} \tag{11}$$

being $k_{\hat{R}}$ a positive convergence constant. The error derivative function is finally $\dot{\Psi}(\hat{R}) = -2 \cdot k_{\hat{R}} \cdot \boldsymbol{e}_{\hat{R}}{}^2$, which is negative-defined, i.e., ensures the convergence of the system.

The above pitch and yaw variables are expressed in the $\{\boldsymbol{W}\}$ frame and must be transformed to the body frame since the low level controller of the quadcopter has as input the yaw rate.

3.1 Application of the Control Algorithm to the UAV

To test the path following control algorithm, a sinusoidal trajectory has been defined (Fig. 2, red line) in the $X_{\{I\}}$ and $Y_{\{I\}}$ plane. It begins at point [0,10]. At the same time, the UAV should be linearly going up in the $Z_{\{I\}}$ direction. The quadcopter starts at position [0,0]; it converges to the reference path quickly and follows the trajectory successfully (Fig. 2, blue line). In Fig. 3 the same trajectory is shown (red line) in 3D, and it is possible to see how the aerial vehicle also reaches the desired height (blue line).

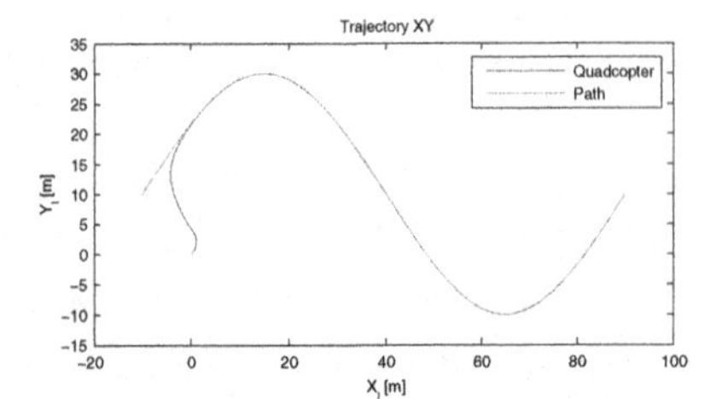

Fig. 2 Path and trajectory of the quadcopter in the $XY|_{\{I\}}$ plane (Colour figure online)

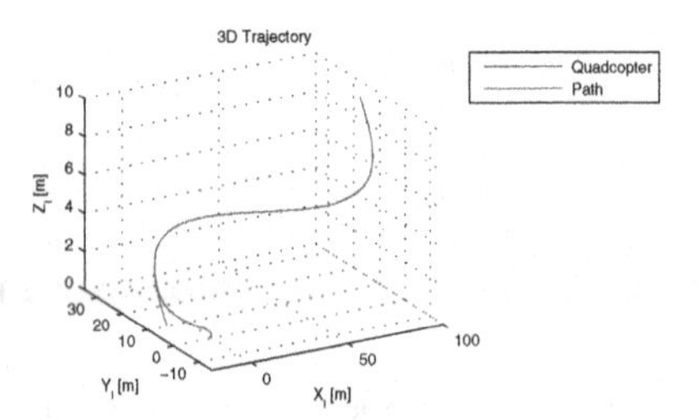

Fig. 3 Path and trajectory of the quadcopter (Colour figure online)

4 Parameters Determination by Fuzzy Logic

As it has been shown, there are three parameters involved in the algorithm: d (Eq. 4), k_l (Eq. 8) and $k_{\hat{R}}$ (Eq. 11). Those parameters have an important influence on the trajectory, mainly depending on two factors: the curvature of the path and the velocity of the quadcopter. If the desired velocity and the curvature of the path are too high, the quadcopter will not be able to follow it due to the limitation of the yaw rate.

Two fuzzy variables will be used, the velocity of the quadcopter, v, and Ω, whose definition is:

$$\Omega = \frac{\bar{\tau}}{q_{max}} \tag{12}$$

where q_{max} is the maximum yaw rate (constant) and $\bar{\tau}$ is the mean curvature of the reference path. Note that $1/\Omega$ is the maximum velocity with which the quadcopter could follow the trajectory. In Fig. 4 the membership functions for both variables have been represented.

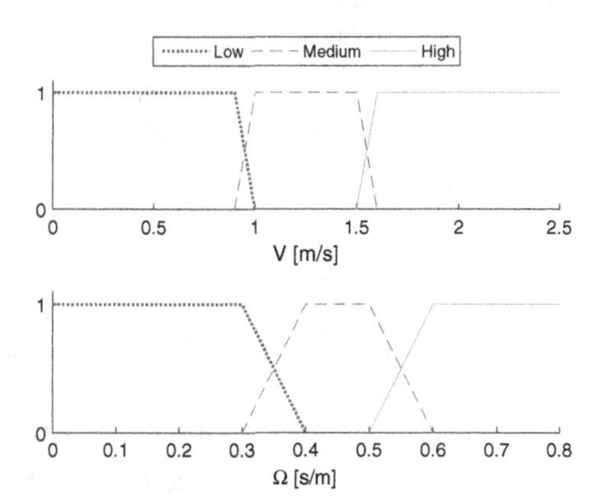

Fig. 4 Membership functions of the input variables

A dimensionless variable, κ, is proposed to obtain the parameters of the algorithm. Its value is defined by the fuzzy rules described in Table 1, based on the fuzzy variables v and Ω. They have been defined using expert knowledge. For instance, low values of κ imply that the path can be followed easily; and the higher κ is, the more difficult following the trajectory is, due to the limitation of the yaw rate.

Table 1 Fuzzy rules to estimate κ, based on the fuzzy variables v and Ω

			v	
		low	medium	high
$\boldsymbol{\Omega}$	**low**	low	low	medium
	medium	medium	medium	high
	high	medium	high	high

For each of the three fuzzy values of κ, a set of values of the algorithm parameters are defined. The knowledge regarding how these parameters influence the trajectories can be summarized as follows:

- For medium values of the distance d, if $k_{\hat{R}}$ is low, results give medium errors, independently of κ.
- For high values of d, better results are achieved by decreasing $k_{\hat{R}}$.
- For low values of κ, d must be low.
- The higher κ is, the lower k_l and the higher d must be.
- For high values of κ, $k_{\hat{R}}$ must decrease
- For low values of κ, $k_{\hat{R}}$ must increase.
- Despite the fact that k_l has smaller influence on the results than the other two parameters, high values of this parameter provide better results when κ is low.

Having said that, the set of values, depending on κ, have been chosen, see Table 2. To calculate the final values of the parameters, the weighted average of the fuzzy variables will be used. For instance, if $v = 1.3$ m/s and $\Omega = 0.55$ s/m, we have:

Table 2 Set of values chosen for each value of κ

κ	d	k_l	$k_{\hat{R}}$
high	5	0.05	2
medium	3	0.15	6
low	1	0.50	4

$$\kappa = 0.5 \cdot medium + 0.5 \cdot high$$

The values of the parameter are finally calculated as follows in this example:

$$\begin{aligned} d &= 0.5 \cdot 3 + 0.5 \cdot 5 = 4 \\ k_l &= 0.5 \cdot 0.15 + 0.5 \cdot 0.05 = 0.10 \\ k_{\hat{R}} &= 0.5 \cdot 6 + 0.5 \cdot 2 = 4 \end{aligned}$$

4.1 Discussion of the Results

Once the mean curvature of the path has been calculated, and knowing the velocity, κ is then determined and the best values of the parameter are selected. For instance, in Fig. 5 a path with $\kappa = high$ has been represented as well as three trajectories with different configurations. The best configuration of the system is given by the fuzzy-logic system, i.e. $d = 5$, $k_l = 0.05$ and $k_{\hat{R}} = 2$ (blue line). The other two configurations correspond to $\kappa = medium$ (green line, $d = 3$, $k_l = 0.15$ and $k_{\hat{R}} = 6$), and $\kappa = low$ (yellow line, $d = 1$, $k_l = 0.50$ and $k_{\hat{R}} = 4$). These two last configurations converge faster to the desired path but they present oscillations, specially the configuration for $\kappa = low$, in yellow. Even though the configuration for $\kappa = medium$ provides good results as well, it is slightly worse.

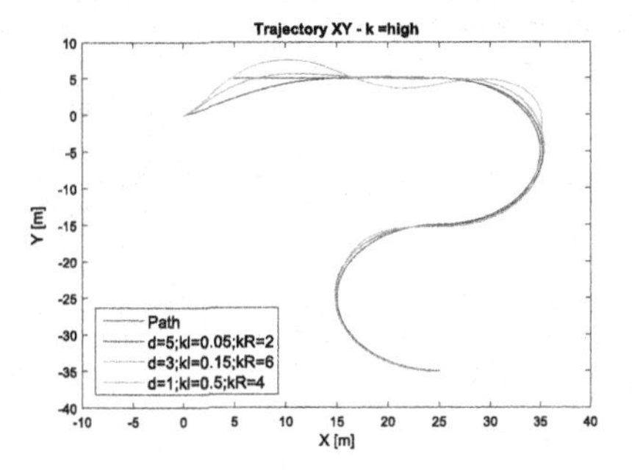

Fig. 5 Trajectories for $\kappa = high$ with different configurations

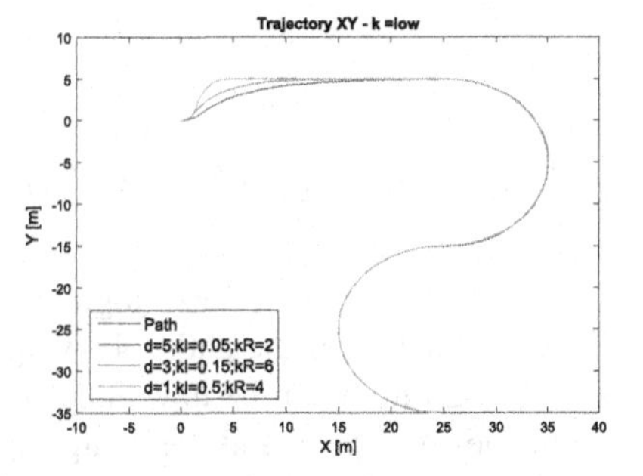

Fig. 6 Trajectories for $\kappa = low$ with different configurations

Another example of the behavior depending on the configuration has been represented in Fig. 6, this time with $\kappa = low$. As it has been already said, the lower the value of κ is, the easier the path can be followed, and therefore all three configurations accomplish the path-following task successfully. However, the values of the parameters proposed by the fuzzy system provide better results, being able to stay on the track from the very beginning.

The results have been measured by integrating the perpendicular distance between the reference path and the actual quadcopter trajectory. In Fig. 7 this error, in $[m^2]$, has been represented for a wide range of κ values, which is now numerically defined as follows for the sake of this comparison:

$$\kappa = \nu \cdot \Omega \tag{14}$$

The three configurations proposed by the fuzzy system and the trajectories of Figs. 5 and 6 were used to measure the error. It can be seen that the parameters have been properly chosen; for every range of values of κ, the best configuration is the one given by the fuzzy decision system. Note that for $\kappa > 0.8$ the error increases quickly, since the necessary yaw rate to follow the path approaches to the maximum achievable by the quadcopter.

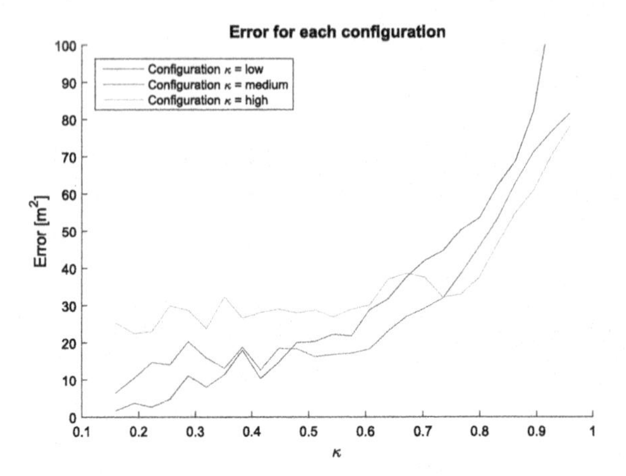

Fig. 7 Path following error, depending on the configuration and the value of κ

5 Conclusions and Future Works

As it has been shown, fuzzy-logic can be applied to simplify the selection of the three parameters of the path-following algorithm, based on a new defined variable κ. The algorithm can be now adapted in order to achieve better results depending on the desired speed and the complexity (curvature) of the path. The analysis here exposed is valid for any other unmanned aerial vehicle, since it was carried out using a dimensionless variable.

As future work, in order to improve the algorithm, the virtual point position expression can be modified so that the system can cope with higher curvature of the path. On the other hand, long paths can be divided into segments with different values of κ to better adapt the algorithm to them.

Acknowledgments The authors would like to thank the Spanish Ministry of Science and Innovation (MICINN) for support under project DPI2013-46665-C2-1-R, and the ISCAR team (UCM) for the quadrotor model.

References

1. Bishop RL (1975) There is more than one way to frame a curve. Am Math Monthly 246–251
2. Dongxuan L, Ting Z, Weijie L, Jinku L, Yina R, Di M (2014) Robust trajectory control of a four-axis dual rotor aircraft. In: Guidance, navigation and control conference, pp 927–931

3. Garca Aun P, Santos M (2014) Use of genetic algorithms for unmanned aerial systems path planning. Decis Making Soft Comput 9:430–435
4. Hehn M, DAndrea R (2011) Quadrocopter trajectory generation and control. In: IFAC world congress, vol 18, pp 1485–1491
5. Hoffmann GM, Waslander SL, Tomlin CJ (2008) Quadrotor helicopter trajectory tracking control. In: Guidance, navigation and control conference, pp 1–14
6. Martnez SE, Toms-Rodrguez M (2014) Three-dimensional trajectory tracking of a quadrotor through PVA control. RIAI 11:54–67
7. Mejari M, Gupta A, Singh NM, Kazi F (2013) Trajectory tracking of quadrotor with bounded thrust using model predictive control. In: Advances in robotics conference, pp 1–6
8. Lee T, Leok M, McClamroch NH (2010) Control of complex maneuvers for a quadrotor UAV using geometric methods on SE(3)
9. Santos M (2011) Intelligent control: a practical approach. Rev Iberoamericana de Automtica e Informtica Ind 8:283–296
10. Salazar-Cruz S, Palomino A, Lozano R (2005) Trajectory tracking for a four rotor mini-aircraft. In: DC and ECC conference, pp 2505–2510
11. Xargay Mata E (2013) Time-critical cooperative path-following control of multiple unmanned aerial vehicles. Doctoral dissertation

Adaptive Neural Control-Oriented Models of Unmanned Aerial Vehicles

J. Enrique Sierra and Matilde Santos

Abstract From real input/output data, different models of an unmanned aerial vehicle are obtained by applying adaptive neural networks. These models are control-oriented; their main objective is to help us to design, implement and simulate different intelligent controllers and to test them on real systems. The influence of the selected training data on the final model is also discussed. They have been compared to off-line learning neural models with satisfactory results in terms of accuracy and computational cost.

Keywords Adaptive neural networks · Soft computing · Modeling · Identification · Unmanned aerial vehicles (UAV)

1 Introduction

One of the main steps in the design of a controller is to obtain a good model of the system dynamic. In control engineering, system identification is one of the most important approaches in order to obtain reliable and useful mathematical models of complex systems. Basically there are two ways to approach this target: the first one is to apply the well-known physical equations that govern the dynamic behavior of the real system, and use this knowledge to generate the differential equations that

J. Enrique Sierra (✉)
Languages and Computer Systems, University of Burgos, Burgos, Spain
e-mail: jesierra@ubu.es

M. Santos
Computer Science Faculty, Complutense University of Madrid, Madrid, Spain
e-mail: msantos@ucm.es

Á. Herrero et al. (eds.), *10th International Conference on Soft Computing Models in Industrial and Environmental Applications*, Advances in Intelligent Systems and Computing 368, DOI 10.1007/978-3-319-19719-7_29

represent it. This methodology is straightforward when all variables that describe the behavior of the system are known, and the relations between them are not too complicated. But in most of the cases, the equations may be either too complex or impossible to obtain. The second way is a set of identification techniques. The system is considered as a black-box. There is not physical knowledge about the system but it is possible to have access to data that are obtained from experiments. Analyzing this information by the application of different identification strategies, it is possible to estimate the internal relationship between inputs and outputs [1].

Among identification methods we may highlight soft computing techniques. These strategies provide a set of tools which makes easy to generate models of complex non-linear systems that can be used for different purposes. Neural networks have been proved as an efficient and useful tool to obtain black-box models from input/output data [3, 13]. Nevertheless, off-line learning neural networks have the disadvantage of being very sensitive to the training dataset. The results may vary in function of the dataset selected to train the learning system. On the contrary, on-line learning methods, like adaptive neural networks, are much more robust because they are continuously learning from the real data of the system [2]. It is not necessary to select any dataset to train the net. In addition, they can be used to identify models of systems whose dynamic changes over time, as for example, systems that work in outdoor environments subject to changing weather conditions.

Unmanned aerial vehicles (UAVs) provide many possibilities for researching and other useful task [4]. In this paper, using real input/output data, different models of an unmanned aerial vehicle (UAV) have been obtained by applying system identification based on adaptive neural networks. They have been compared to previous model obtained by neural networks [9].

Other works have also applied these neural strategies, such as [6], but they have not used adaptive neural networks. Other examples of the application of soft computing techniques in order to model complex non-linear systems can be found in [5, 7, 8].

Given the importance of the model of a process in order to design an efficient control system, in this paper we propose the application of adaptive neural networks to the identification of the UAV. This model will be integrated in the architecture of a control system (Fig. 1) and will be tested on a real prototype.

The paper is organized as follows. Section 2 describes UAV we are working with. In Sect. 3, the neural networks configuration that has been applied to identify the model from real data is presented. Differences between off-line and adaptive neural identification are also explained in this section. The results are compared and discussed in Sect. 4. Conclusions end the paper.

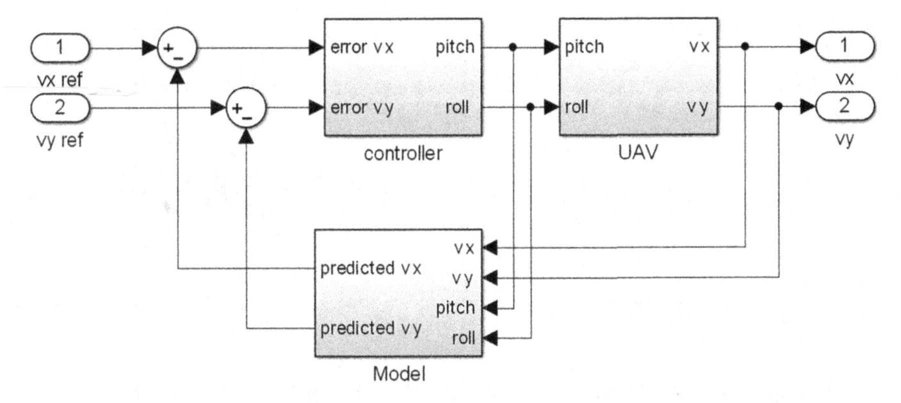

Fig. 1 Integration of the neural model in a control scheme

2 UAV Dynamic Behaviour

We work with the unmanned aerial vehicle ARDRONE 2.0, by Parrot (Fig. 2). This system is carbon fiber and plastic PA66. It is equipped with four high efficiency brushless motors, powered by one 11 V 1000 mAh lithium battery, which gives up to 12 min flight autonomy. It weighs 380 g without casing and 420 g with casing. It has a three-axis accelerometer, a two-axis gyroscope (pitch and roll), and a high precision gyroscope for the yaw angle [10].

Figure 3 depicts the coordinate system of the UAV. The pitch is controlled setting the speed of the longitudinal motors. If the speed of the rotors 1 and 2 increases regarding rotors 3 and 4, the vehicle pitches down and the lift forces cause the vehicle to move along the x-axis direction (forward movement). If the speed of the rotors 1 and 2 decrease regarding rotor 3 and 4, the vehicle pitches up and the lift forces cause the vehicle to move backward. On the other hand, the roll control is carried out by changing the speed of the lateral motors. If the speed of rotors 1 and 4 increases respect to the rotors 2 and 3 the vehicle moves along the direction of the y-axis (left-side movement). If the speed of rotors 1 and 4 decreases regarding rotors 2 and 3, the vehicle rolls and the lift forces cause the vehicle to move toward right.

Fig. 2 UAV (Unmanned aerial vehicle), Parrot [10]

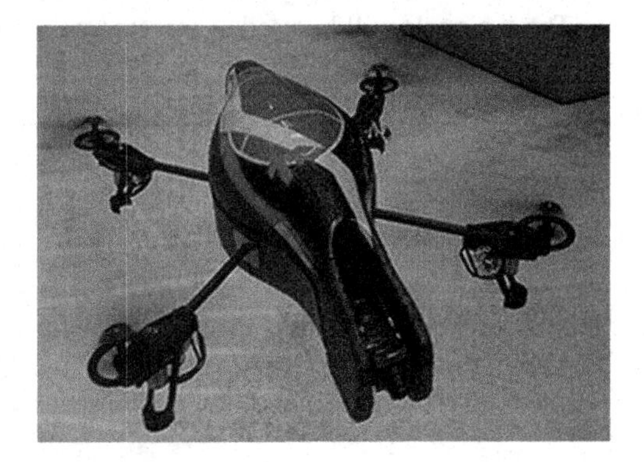

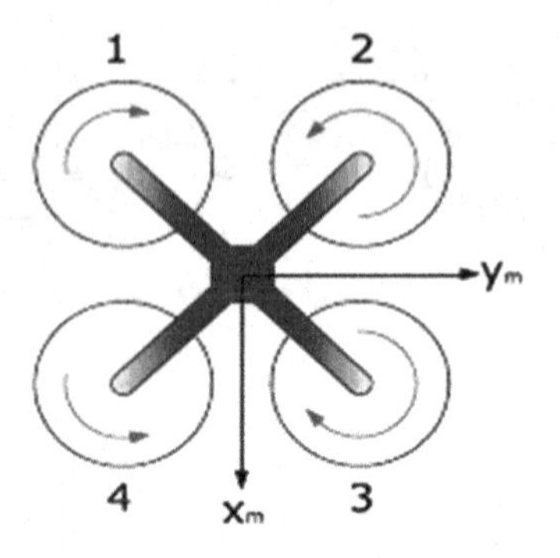

Fig. 3 Coordinate system of the UAV

The input and output signals of the UAV are shown in Fig. 4, provided by the Control Engineering Group of the Spanish Committee of Automatic (CEA) [11]. The pitch signal and velocity in the x-axis are shown in blue meanwhile the roll signal and the velocity in the y-axis are represented in green. As it may be seen in the figure, the input signals consist of a train of steps with different amplitude. This kind of signals is traditionally used in identification to try to extract the maximum possible information from the system.

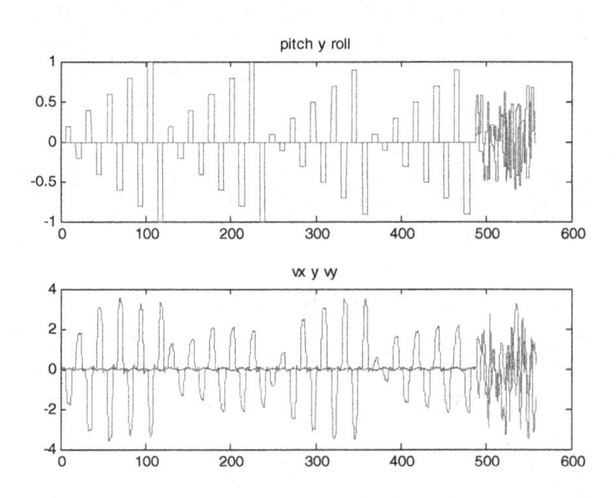

Fig. 4 Pitch and roll (inputs) and velocity (output) [11]

These signals will be used as targets by the identification techniques. They have been obtained from the real system, during 9 min of flight, sampled with a time period of 60 ms [11].

3 Neural Networks Identification

Supervised neural networks can be successfully applied to model complex systems [6, 12]. In this study, back-propagation multilayer feed-forward networks have been selected due to their well-known properties as universal approximators. These networks consist of several layers of neurons connected feedforward; a different activation function can be assigned to each layer. Each neuron in a particular layer is connected to all neurons in the next layer (full connection). The connection between the i-th and j-th neurons (two consecutive layers) is characterized by the weight coefficient, w_{ij}, and the i-th neuron by the threshold coefficient b_i and so on. The output of the i-th neuron is given by:

$$x_i = \mathrm{f}(b_i + \sum_{j=1}^{N} w_{ij}\mathrm{x}_j) \tag{1}$$

where f is the activation function and N is the number of neurons in the previous layer. In order to reduce the computational cost, the number of hidden layers of the network has been fixed to one. In this hidden layer, the activation function was fixed to the hyperbolic tangent sigmoid; the linear transfer function was assigned to the input and output layers. The training algorithm used was Levenberg-Marquardt; gradient descent was used as learning algorithm, and the performance function was set to mean squared error.

3.1 Model Identification by Applying off-Line Learning

The first step when modeling a real system is to analyze the input and output signals, in order to figure out the best way to relate those data. The UAV, as any real system, is causal; therefore, the value of the outputs at any moment depends on the value of the previous inputs. That is, the next value of the state variables of the system will be a function of the present state of the system and the inputs. The system state can be estimated using the past values of the output.

Thus, the discrete output variables, velocity in the x-axis and in the y-axis, are given by the following expressions,

$$v_x[k+1] = f_{v_x}\begin{pmatrix} v_x[k] & u_\theta[k] & u_\phi[k] \\ v_x[k-1] & u_\theta[k-1] & u_\phi[k-1] \\ \cdots & \cdots & \cdots \\ v_x[k-N_{v_x}] & u_\theta[k-N_{u_\theta}] & u_\phi[k-N_{u_\phi}] \end{pmatrix} \tag{2}$$

$$v_y[k+1] = f_{v_y}\begin{pmatrix} v_y[k] & u_\theta[k] & u_\phi[k] \\ v_y[k-1] & u_\theta[k-1] & u_\phi[k-1] \\ \cdots & \cdots & \cdots \\ v_y[k-M_{v_y}] & u_\theta[k-M_{u_\theta}] & u_\phi[k-M_{u_\phi}] \end{pmatrix} \tag{3}$$

with sampling time t_s; u_θ refers to pitch signal and u_ϕ to roll signal.

Once the network has learnt, that is, it converges to a stationary value, the function f (Eqs. 2 and 3) is kept constant.

Using off-line learning methods, the first step is to generate a dataset with all the input/output pairs of data. Then, this set is split in order to use part of it for training and the other for testing the model. But the selection of the data for training is a delicate task. As it will be shown, the accuracy of the model depends on the data used to train the net. Results prove that off-line methods may be too sensitive to the partition of the dataset. Another disadvantage of these methods is that they do not capture the dynamics when it is changing over time.

3.2 Model Identification by Applying on-Line Learning

In order to perform a fair comparison between the results provided by adaptive networks (on-line learning) and off-line learning neural networks, the same configuration of the neural network is used.

In this case, due to adaptive learning, f is different each sampling time. This evolution of f leads to reformulate Eqs. 2 and 3. The new expressions are given by Eqs. 4 and 5. In this case f_{v_x} and f_{v_y} are changing over time, and this evolution is represented by the Eqs. 6 and 7.

$$v_x[k+1] = f_{(k)_{v_x}} \begin{pmatrix} v_x[k] & u_\theta[k] & u_\phi[k] \\ v_x[k-1] & u_\theta[k-1] & u_\phi[k-1] \\ \cdots & \cdots & \cdots \\ v_x[k-N_{v_x}] & u_\theta[k-N_{u_\theta}] & u_\phi[k-N_{u_\phi}] \end{pmatrix} \tag{4}$$

$$v_y[k+1] = f_{(k)_{v_y}} \begin{pmatrix} v_y[k] & u_\theta[k] & u_\phi[k] \\ v_y[k-1] & u_\theta[k-1] & u_\phi[k-1] \\ \cdots & \cdots & \cdots \\ v_y[k-M_{v_y}] & u_\theta[k-M_{u_\theta}] & u_\phi[k-M_{u_\phi}] \end{pmatrix} \tag{5}$$

$$f_{(k+1)_{v_x}} = g_{v_x} \left(f_{(k)_{v_x}} \quad \begin{matrix} v_x[k] & u_\theta[k] & u_\phi[k] \\ v_x[k-1] & u_\theta[k-1] & u_\phi[k-1] \\ \cdots & \cdots & \cdots \\ v_x[k-N_{v_x}] & u_\theta[k-N_{u_\theta}] & u_\phi[k-N_{u_\phi}] \end{matrix} \right) \tag{6}$$

$$f_{(k+1)_{v_y}} = g_{v_y} \left(f_{(k)_{v_y}} \quad \begin{matrix} v_x[k] & u_\theta[k] & u_\phi[k] \\ v_x[k-1] & u_\theta[k-1] & u_\phi[k-1] \\ \cdots & \cdots & \cdots \\ v_x[k-M_{v_y}] & u_\theta[k-M_{u_\theta}] & u_\phi[k-M_{u_\phi}] \end{matrix} \right) \tag{7}$$

Therefore, $f_{(k)}$ and $f_{(k+1)}$ do not represent values obtained by the evaluation of the function, but the shape of the function at k and $k+1$, respectively. Using these expressions, it will be shown how the shape of the function changes over time.

4 Discussion of the Results

Figure 5 shows the evolution of the MSE of the models obtained by different configurations of the adaptive neural networks. Velocity in the x-axis is represented on the left part of this figure, and velocity in the y-axis on the right.

The configuration of the network is indicated by the number linked to the curve in the legend. For example, "err5" (blue line) means that the parameters $(N_{v_x}, N_{u_\theta}, N_{u_\phi})$ of Eqs. 2 and 4 are set to 5, in the case of v_x. This value also corresponds to the parameters $(M_{v_y}, M_{u_\theta}, M_{u_\phi})$ of Eqs. 3 and 5 for the velocity in the y-axis. The MSE decreases as the neural network learns. Where N_i and M_j are the number of consecutive samples for each system variable i and j used to train the net. That is, the learning improves the identified model continuously.

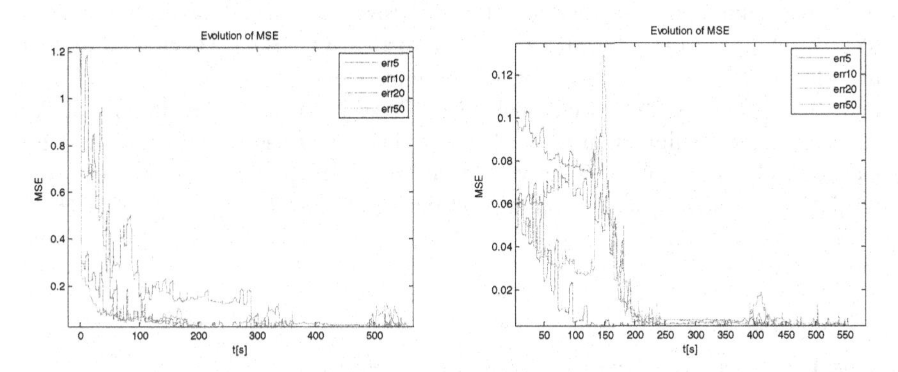

Fig. 5 MSE model evolution of velocity in the x-axis (left), and in the y-axis (right)

Figure 6 shows how the selection of the training data influences the final result. The parameters for the velocity in both axes, $(N_{v_x}, N_{u_\theta}, N_{u_\phi})$ and $(M_{v_y}, M_{u_\theta}, M_{u_\phi})$, were set to (5, 5, 5) in the on-line case. The number associated to each curve means the number of samples used to train the network. The samples were taken from a random point of the dataset, consecutively. This experiment was randomly executed over 9000 times. The MSE error was calculated during each execution and its value is indicated on the y-axis. The MSE error obtained with the adaptative network (on-line learning) is shown in dark blue; the MSE curves obtained by the application of the off-line methods are represented by different colors depending on the number of samples taken (for example, 50 samples, green line).

As it is possible to see, on-line MSE is almost always smaller than the error of the off-line models. In addition, in these figures it may be noticed how the MSE is very dependent on the data selected for training.

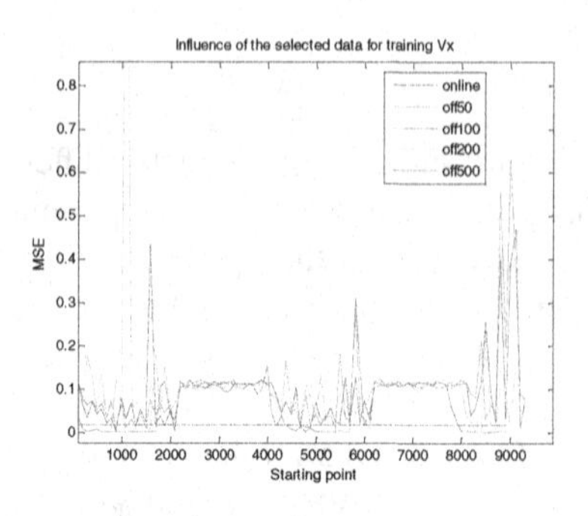

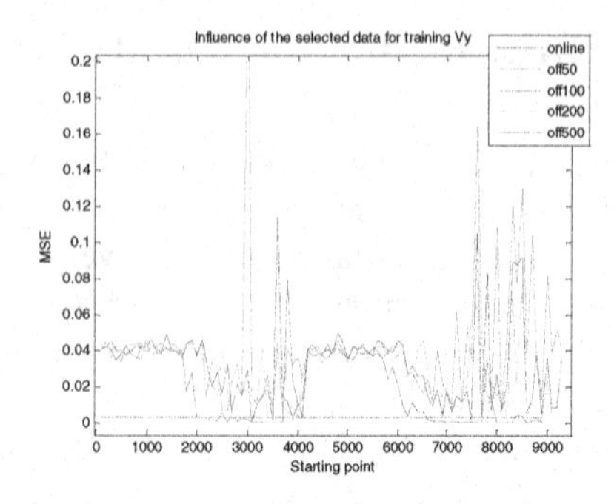

Fig. 6 Influence of selected training data on v_x (left) and on v_y (right)

Table 1 shows the comparison of the MSE obtained by applying on-line and off-line training. Every column of each row has the same configuration, $(N_{v_x}, N_{u_\theta}, N_{u_\phi})$ and $(M_{v_y}, M_{u_\theta}, M_{u_\phi})$. In this test, 200 samples have been used to train the off-line model. For each type of training, the best configuration has been underlined. Besides, the best global results have been boldfaced. In two of the cases, the best configuration is (8, 8, 8); in the rest of cases configurations with values (2, 2, 2) and (6, 6, 6) have showed less MSE. In any case, the best total results have been obtained by applying on-line learning with the configuration (8, 8, 8). To summarize, in this experiment on-line models provide less MSE than off-line methods.

Table 1 Comparison of MSE for on-line and off-line techniques

Configuration	v_x		v_y	
$N_{v_x}, N_{u_\theta}, N_{u_\phi}$	On-line	Off-line	On-line	Off-line
2, 2, 2	0.01552	0.09038	0.004223	0.01015
4, 4, 4	0.01101	0.11556	0.002758	0.04318
6, 6, 6	**0.00320**	0.06491	0.002689	0.02205
8, 8, 8	0.01478	0.05856	**0.001584**	0.02863
10, 10, 10	0.02060	0.05908	0.002710	0.02115

5 Conclusions and Future Works

In this work, models of an unmanned aerial vehicle have been obtained by applying neural networks, both with on-line and off-line learning strategies. Results prove the better performance of adaptive neural networks.

Off-line and on-line learning strategies have been analyzed, and their advantages and disadvantages have been presented. In addition, it has been shown the influence of the partition of the dataset for training and validation on the model error.

The results are satisfactory: the curves of the MSE indicate that the network is able to learn on-line and to improve the identified model continuously. Therefore, it has been proved how soft computing techniques are quite useful to identify and model these complex systems.

These models, obtained from real data, are control-oriented; the main objective is to use them to design, simulate and implement different controllers and to test them on real systems, that is proposed as future works.

Acknowledgments The authors would like to acknowledge the data provided by the Control Engineering Group of the Spanish Committee of Automatic [10]. Author Matilde Santos would like to thank the Spanish Ministry of Science and Innovation (MICINN) for support under project DPI2013-46665-C2-1-R. Authors would also like to thank the reviewers for their useful comments.

References

1. Billings SA (2013) Non-linear system identification: NARMAX methods in the time, frequency, and spatio-temporal domains. Wiley
2. Deng J (2013) Dynamic neural networks with hybrid structures for nonlinear system identification. Eng Appl Artif Intell 26(1):281–292
3. Han HG, Qiao JF (2012) Adaptive computation algorithm for RBF neural network. IEEE Trans Neural Netw Learn Syst 23:2
4. Hoffer NV, Coopmans C, Jensen AM, Chen Y (2014) A survey and categorization of small low-cost unmanned aerial vehicle system identification. J Intell Rob Syst 74(1–2):129–145
5. Nemes A (2015) Synopsis of soft computing techniques used in quadrotor UAV modelling and control. Interdiscipl Descr Complex Syst 13(1):15–25
6. San Martin R, Barrientos A, Gutierrez P, del Cerro J (2006) Unmanned aerial vehicle (UAV) modelling based on supervised neural networks. In: Proceedings of the international conf robotics and automation, IEEE, pp 2497–2502
7. Santos M, López R, de la Cruz JM (2006) A neuro-fuzzy approach to fast ferry vertical motion modelling. Eng Appl Artif Intell 19:313–321
8. Santos M (2011) Un enfoque aplicado del control inteligente. RIAI 8(4):283–296
9. Sierra JE, Santos M (2013) Estudio comparativo de modelos de un vehículo aéreo obtenidos mediante técnicas analíticas y basadas en redes neuronales. In: CEDI, pp 1270–1279
10. Parrot. http://www.parrot.com
11. Control Engineering Group, Spanish Committee of Automatic (CEA). http://www.ceautomatica.es/og/ingenieria-de-control
12. Sayama H, Pestov I, Schmidt J, Bush BJ, Wong C, Yamanoi J, Gross T (2013) Modeling complex systems with adaptive networks. Comput Math Appl 65(10):1645–1664
13. Ugalde HMR, Carmona JC, Alvarado VM, Reyes-Reyes J (2013) Neural network design and model reduction approach for black box nonlinear system identification with reduced number of parameters. Neurocomputing 101:170–180

SOCO15-SS03 - Soft Computing Approaches for Knowledge Extraction

Data Mining for Predicting Traffic Congestion and Its Application to Spanish Data

E. Florido, O. Castaño, A. Troncoso and F. Martínez-Álvarez

Abstract The purpose of this paper is the development and application of patterns and behavioral models of time series data collected by sensors belonging to the Spanish Directorate General for Traffic. The extraction of these patterns will be used to predict the behavior and effects on the system as accurately as possible to facilitate early notifications of traffic congestions, therefore minimizing the response time and providing alternatives to the circulation of vehicles. Decision trees, artificial neural networks and nearest neighbors algorithms have been successfully applied to a particular location in Sevilla, Spain.

Keywords Traffic congestion · Prediction of retentions · Data mining · Time series

1 Introduction

The Southwest Division of the Spanish Directorate General for Traffic (DGT) manages a total of 220 filling stations and 214 data roadside cameras amongst other resources and equipment. One of the most important goal of this organization is to prevent traffic congestion and reduce their duration and length once they have occurred. To achieve this, the capacity of the existing infrastructure could be increased, but this incurs in high logistics costs and the effective and flexible management of resources.

E. Florido · O. Castaño · A. Troncoso · F. Martínez-Álvarez (✉)
Department of Computer Science, Pablo de Olavide University of Seville, Seville, Spain
e-mail: fmaralv@upo.es

E. Florido
e-mail: eflonav@alu.upo.es

O. Castaño
e-mail: ocascal@alu.upo.es

A. Troncoso
e-mail: ali@upo.es

Á. Herrero et al. (eds.), *10th International Conference on Soft Computing Models in Industrial and Environmental Applications*, Advances in Intelligent Systems and Computing 368, DOI 10.1007/978-3-319-19719-7_30

Traffic congestion on the bypass SE-30, specifically on the bridge of the Fifth Centenary (kilometer 12), is a latent problem in the city of Seville. The volume that transits in either direction varies for different slots of time. Northbound and southbound traffic are separated by a moveable barrier, which permits the number of lanes allocated to north or southbound traffic to be changed to accommodate heavy demand in either direction, to respond to unusual traffic conditions, and to balance lane availability with real-time traffic demand. It is impossible to leave the reversible lane from its beginning to its end by any alternative path. Therefore, the proper management of this lane and even the possibility of providing alternatives to traffic at peak times becomes an essential task.

This paper presents an approach of general-purpose methodology devoted to predict congestion of vehicles on roads. As a case of study, it is evaluated at the kilometric point that was noted above. For the preparation of this study, traffic data collected by detectors on the road through the stations of data collection (SDC) have been used, which are transferred to the Southwest Division. It has been carried out with the support of the Spanish DGT. After applying different algorithms, four different scenarios of possible retention were predicted, those in which the anticipation of traffic retention was of 15, 30, 45 and 60 min. In these scenarios, on average, the yielding has an accuracy over 85 %.

A traffic jam can be caused by two reasons. First because of a high volume of vehicles circulating simultaneously. When this volume exceeds the design capacity of the roadway, the route becomes saturated and bottlenecks occur. Typically, this volume is 2000 vehicles per lane and hour, two cars per second, and is calculated for a road lane in optimum condition (width of two to three meters and the pavement in good condition). Such retentions are the most common. They take place periodically, and when people travel to and from work, as well as in holiday outings.

Secondly, traffic jams can be caused by incidents such as vehicle breakdowns, accidents, spills of objects on the road, rain, snow, and other environmental factors. These facts lead to a mobility reduction at a certain point of the road. The accumulation of leads to overtaking to dodge obstacles and this ends up forming a traffic jam.

Once this forms, it grows up quickly and those vehicles that arrive at the retention circulating at high speed have to suddenly brake up and force those circulating behind them to do the same. The result is a retention that propagates backwards at high speed, at a rate of five meters for each vehicle that reaches the retention [9].

There are a number of factors contributing to an increasing bottleneck. One is the very composition of traffic [14]. A lane along with four trucks in a row does not have the same capacity as one that is only occupied by cars. In general, heavy vehicles travel slower and generate longer lines of vehicles. Another factor that negatively affects traffic is the interaction between cars. When traffic density is above 1700 vehicles per hour any overtaking or maneuvering causes a reduction in the speed of the road [7]. Branch-road or slip road may also complicate matters in jam, as they bring an extra supply of vehicles to a road which, in itself, is already saturated.

Decreasing speed helps to prevent traffic jams. It allows the flow to move more evenly. If all vehicles driving down a road are at a low speed, the capacity of track circulation would not saturate so quickly. The mass of cars would move slowly but evenly. It is a very complex problem and possibly the most feasible solution would be to create more and better infrastructures. The problem is that as more roads and highways are built the number of cars also increases, with eventually the same traffic issues arising. Different solutions have been attempted in cities around the world, with varying degrees of success [3].

Some highways have added new lanes to extend them in a general way, using the hard shoulder or decreasing the size of the existing lanes if feasible. These adjustments are both expensive and time consuming. As it depends on the population growth and on the sales of vehicles, this is a medium-term solution [1].

A less costly solution would be to study how traffic behaves using information obtained from roadside sensors [5] based at particular segments of the road network, referencing these variables: [8] congestion, intensity, occupancy, average speed of vehicles, etc. Obtaining a model for traffic control varies depending on the area of study.

The rest of the paper is divided as follows. Section 2 focuses on the methodology used for the prediction of congestions at the kilometric point under study. The results are presented and discussed in Sect. 3. Finally, the conclusions of this paper are summarized in Sect. 4.

2 Methodology

To be able to correctly model the traffic behavior, the data received by the SDC must be interpreted in a consistent manner, to enable its subsequent filtering. The system communicates directly with the DGT, where the information servers and data are located. These servers are the first level of processing to remove erroneous data, values out of range, and they are in charge of keeping communication between the data collection equipments and the control center. These devices are Universal Remote Stations (URS) and allow the management of the field equipment as if they were their own peripherals of the URS.

2.1 Information Provided by the SDC

The SDC uses a set of sensors installed in the roadway to detect the passage of vehicles. An SDC is capable of controlling up to a total of 48 sensing points which may be inductive (coil), piezoelectric sensors (axes of sensors), virtual infrared ties, microwave or laser (each with their corresponding additional equipment). Depending on the characteristics of the sensors installed on the road (for example single or double detectors) a set of variables of traffic or another is obtained.

The data on which this paper is based are electromagnetic coils; coils are formed by a single conductor embedded into the pavement making several turns and forming a polygon in the driveway, located in the passageway of vehicles. The operation is as follows: the SDC is constantly waiting for pulses from the digital inputs produced by the electromagnetic coils placed in the driveway. Once an input is detected,the system stores and analyzes the entire sequence. The SDC has got defined some series of sequences that match the passage of vehicles. If the sequence matches one of the predefined patterns this sequence is analyzed and the data of the detected vehicle are verified and counting occurs. Occasionally (integration period), data is accumulated and remain ready to be uploaded to servers.

These data refer to the following traffic variables obtained from all vehicles registered by the lower body system, detector (consisting of double coils) or lane during the integration period (IP), 1 min:

1. Intensity (INTENSITY). Refers to the number of vehicles that have been registered by the detector D during the IP integration period of 1 min. It is expressed in vehicles per hour number.
2. Occupation (TOCUP). This refers to the arithmetic mean of the occupation time of the vehicles that have been registered by the detector D in the period of integration. It is expressed in percentage % compared to a IP of 1 min.
3. Average speed (AVESP). Refers to the arithmetic mean of the speed of the vehicles that have been registered by the detector D in the integration period; measured in Km/h.
4. Average distance (GAP). Arithmetic mean of the distance between two consecutive vehicles registered by the detector D in the integration period; measured in meters.
5. Average length (AVELEN). Arithmetic mean of the length of the vehicles that have been registered by the detector D in the integration period; measured in decimeters.
6. Composition length (CLASLONG). Number of vehicles classified by length, according to the categories specified under the configured thresholds that have been recorded by the detector D in the period of integration. It is expressed in % of the corresponding category (light or heavy).
7. Composition speed (CLASSP). Number of vehicles classified by speed, according to the categories specified under the configured thresholds that have been recorded by the detector D in the period of integration. The system allows to obtain up to 3 different classifications for speed. It is expressed in % of the corresponding category (light or heavy).
8. Kamikaze alarm (VEH_INV). Refers to the boolean variable indicating the existence of any vehicle circulating in wrong way during the period of integration. It is expressed as YES = 1 or NO = 0.
9. Reverse direction (SENS_CIR). Refers to the boolean variable indicating that the circulation in the period of integration has suffered a modification from the usual sense. This happens in reversible lanes, in a particular instant the sense of circulation may vary and this could occur in the middle of a period of integration. It is expressed as YES = 1 or NO = 0.

10. Error (ERR_REL). The percentage of time within the period of integration in which if vehicles have passed the detector D has not registered them. This parameter represents the reliability of the measurement obtained in the period of integration. It is expressed in %.
11. Congestion (CONGESTION). Refers to the boolean variable indicating the existence of activation of an alarm of congestion determined by the HIOCC algorithm, developed by the Transport and Road Research Lab (TRRL) [4] and which comes implemented in the SDC in the period of integration. It is expressed as YES = 1 or NO = 0. This variable will be the class label and therefore the one wanted to be correctly predicted.

Finally, information about future occurrence of traffic congestion is included in the class. This class, which is binary and only accepts *0* and *1* as possible values, is generated in a similar way as in [10]. That is, if congestion is to happen within the following preset minutes (in this study, those minutes are 15, 30, 45 or 60, as described in Results), then an *1* is assigned to the class. If not, a *0* is assigned.

2.2 Data Mining Algorithms Used

Spanish DGT asks for results easily interpretable, therefore the C4.5 algorithm [15] has been selected as control algorithm. The visualization of its results leads to rules quickly understandable. Additionally, for comparative purposes, two other algorithms have been chosen due to their different underlying nature: artificial neural networks [11] and nearest neighbors [6]. The first one is the paradigmatic soft computing algorithms and the second one uses simple Euclidean distance as fitness function. The implementations used are those implemented in KEEL [2], a free mining software tool.

C4.5 algorithm is a method for generating a decision tree. Generating a decision tree is to select an attribute as root of the tree and create a branch for each of the possible values for that attribute. With each resulting branch (new node of the tree), the same process is done, another attribute is selected and a new branch for each possible attribute value is generated. This procedure continues until the samples are classified through one of the paths of the tree. The end node of each path is a leaf node, to which the corresponding class is assigned.

Therefore, the objective of the decision trees is obtaining rules or relationships that allow classification based on the attributes. This algorithm allows the use of the concept of information gain, build decision trees when some of the samples show unknown values for some of the attributes, working with attributes that have continuous values, decision tree pruning and obtaining classification rules.

Artificial neural networks (ANN) are a family of statistical learning algorithms inspired by biological neural networks and are used to estimate or approximate functions that may depend on a large number of inputs and are generally unknown. ANNs are generally presented as systems of interconnected neurons that can calculate values from inputs, and are able of machine learning and pattern recognition, thanks to its adaptive nature.

Finally, the algorithm of the nearest neighbors (NN) is based on the proximity between the sample to be classified and the samples that form the training set. Therefore, it is a deterministic algorithm whose critical point is to determine the number of neighbors to use. The most common strategies simply choose the closest (the sample belonging to the training set that gives the shortest distance from the sample that you want to be labeled) or take an odd number of neighbors to ensure no tie between their classes, whether it is a binary classification.

3 Experiments

First, the parameters used to evaluate the quality of the results are presented in this section. Then, the preprocessing applied to the input data and the generation of the class labels in the four scenarios analyzed are described. Finally, the results of applying decision trees, artificial neural networks and nearest neighbors algorithms are detailed.

3.1 Quality Parameters

The quality parameters used to evaluate the different methods are now presented. It was decided to use the most common ones in the literature. On the one hand, a set of indicators that quantify the successes and mistakes of classifiers is calculated:

1. True positive, TP. It is defined as the number of times the classifier assigns a 1 to the instance that is being classified (predicts the occurrence of a retention), and this indeed happens during the next few minutes.
2. True negative, TN. It is the number of times it has been predicted that retention does not occur during the next minutes, and in fact it does not.
3. False positive, FP. The number of times it has been erroneously detected a retention in the next minutes, that is, the number of times the classifier assigns a label with value 1 when really ought to assign a 0.
4. False negative, FN. The number of times a retention has not been detected but indeed there was a retention within the next several minutes.

On the other hand, from these four indicators, the quality parameters are properly calculated. In particular:

1. Sensitivity, S. It is the proportion of correctly identified retentions, regardless the value of the FP. Mathematically it is expressed as: $S = TP/(TP + FN)$.
2. Specificity, E. It is the ratio of correctly identified negatives. Mathematically it is calculated as: $E = TN/(TN + FP)$.
3. Positive Predictive Value, PPV. This value measures the reliability of the TP, that is, the certainty associated with each TP. In other words, the relationship between TP and FP. Mathematically formulated as: $PPV = TP/(TP + FP)$.

4. Negative Predictive Value, NPV. Measuring the reliability of the TN, that is, the certainty associated with each TN. In other words, it is the relationship between TN and FN. Mathematically: $NPV = TN/(TN + FN)$.

3.2 Selection and Transformation of Data

For this study, data were obtained from the detector located at kilometer 12 descending on the SE-30 during the last quarter of 2011 using a total of 8926 records for testing, and 27321 for training. The attributes VEH_INV, SENS_CIR, AVELEN, CLASLONG, CLASSP and ERR_REL were discarded due of its low relevance to the study.

Moreover, within the entire set of records were also discarded those whose ERR_REL exceeded a 5 %, about 0.056 % of the data. There have also been ruled out those records in which less than 15 samples were obtained per each quarter of an hour and those with zero values in the samples in the night intervals; although these values are correct for the fact that the early morning traffic is intermittent and often dispersed.

To be able to see the importance of the attributes through point clouds, it has been observed that there are relationships of type:

1. A larger number of vehicles implies a less distance between them.
2. The distance between vehicles decreases with increasing occupancy rate.
3. A higher intensity produces a lower occupancy of the road.

On the other hand, the *Congestion* has been taken as class label attribute, this attribute takes values YES = 1 or NO = 0. As data mining techniques have been applied to various intervals of time, the class has been moved to analyze the behavior at 15, 30, 45 and 60 min. That is, for each of the four scenarios evaluated, the class label corresponds to whether there will be or not congestion in the next 15, 30, 45 or 60 min.

3.3 Results

In this section are shown the results obtained after the applying the selected algorithms in the four scenarios analyzed, as well as a discussion of themselves. Since Spanish DGT collects data every fifteen minutes, prediction of traffic congestion has been made for 15, 30, 45 min and one hour ahead with C4.5, ANN and NN algorithms. It has been proven that traffic congestion keeps no relation with events happened more than one hour earlier [13].

Table 1 shows the results for the prediction of traffic congestion in four possible scenarios: 15, 30, 45 min and one hour of anticipation when the C4.5 algorithm was applied, which was the one that reported the best result. It is also important to note that the choice of this algorithm was done according to the instructions received from the DGT because an easy interpretation of the results were wanted. In this sense, the structure shown by the trees are of immediate interpretation for any professional who is not related to data mining.

It can be seen that the results obtained were in general very competitive, since none of the four scenarios showed under the 85 % on average. But particular emphasis must be done to the values obtained in the scenario with one hour in advance, these results had, on average, the 99 % of accuracy.

It is noted that C4.5 reported the best result. It is also important to note that the choice of this algorithm was done according to the instructions received from the DGT because an easy interpretation of the results were wanted. In this sense, the structure shown by the trees are of immediate interpretation for any professional who is not related to data mining.

In particular, it can be appreciated that after applying the method C4.5 to 60 min a total of 654 (TP) alarms were correctly generated, which means that a total of 654 retentions had been correctly predicted. Particularly noteworthy is the fact that only 14 false alarms (FP) were generated against more than 9000 time intervals evaluated. As in many other fields [12], this a particularly desirable situation. This makes that out of the 668 registered retentions, 654 were achieved to predict.

On the other hand, it is shown how 8543 (TN) non-alarm states were classified and that there had only been 13 not alerted alarms (FN), making that of 8556 traffic flow situations registered, 8543 were correct.

The rest of the scenarios evaluated also showed favorable results on average. However, it should be noted that false alarms were significantly higher for the other time slots (228 for 15 min, compared to the 14 ones recorded for one hour of anticipation). This fact must be taken in mind very carefully because activating an erroneous policy can cause even more troubles than the failure to implement it.

Figure 1 shows the tree constructed by C4.5 for the prediction of retentions with one hour of anticipation. It can be appreciated the simplicity of the rules obtained: one retention will occur in one hour if the time of occupancy is greater than 27 and the gap is less than or equal to 7. In all other cases, no retention will occur. The rest of the trees generated are not here shown for reasons of space. However, note that the relationships found were not as direct nor as simple as the one shown because on average the depth of the three trees was four and they had a total of twelve leaves in mean.

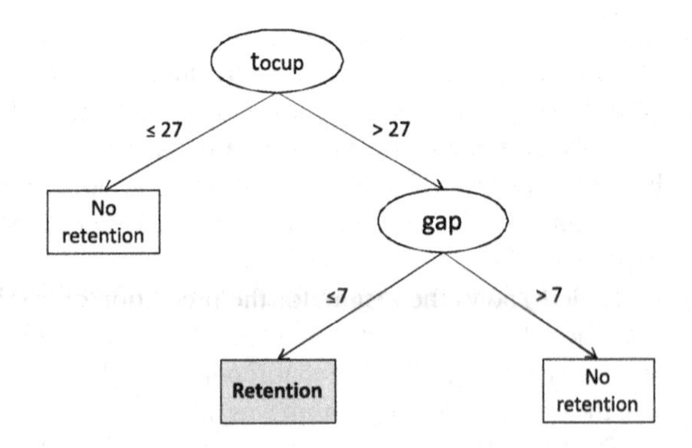

Fig. 1 Tree generated by the C4.5 algorithm for the case of 60 min of anticipation

Table 1 Results of applying C4.5, ANN and NN

	C4.5				ANN				NN			
Parameter	15 min	30 min	45 min	60 min	15 min	30 min	45 min	60 min	15 min	30 min	45 min	60 min
TP	440	511	595	654	196	912	340	397	278	220	175	134
TN	8443	8453	8502	8543	7780	7645	7667	7720	8169	8109	8085	8006
FP	228	157	73	14	472	401	328	271	390	448	471	549
FN	115	104	54	13	778	267	889	835	389	448	493	534
Sensitivity	79.3 %	83.1 %	91.7 %	98.1 %	20.1 %	77.4 %	27.7 %	32.2 %	41.7 %	32.9 %	26.2 %	20.1 %
Specificity	97.4 %	98.2 %	99.1 %	99.8 %	94.3 %	95.0 %	95.9 %	96.6 %	95.4 %	94.8 %	94.5 %	93.6 %
PPV	65.9 %	76.5 %	89.1 %	97.9 %	29.3 %	69.5 %	50.9 %	59.4 %	95.5 %	94.8 %	94.3 %	96.8 %
NPV	98.7 %	98.8 %	99.4 %	99.8 %	90.9 %	96.6 %	89.6 %	90.2 %	41.6 %	32.9 %	27.1 %	19.7 %
Mean	85.3 %	89.2 %	94.8 %	98.9 %	55.4 %	84.6 %	66.0 %	69.6 %	68.6 %	63.9 %	60.5 %	56.8 %

To assess the quality of the values obtained by C4.5, results for ANN and NN applied to the same four scenarios described above and for the same training and testing periods are also reported in Table 1. The ANN was composed of as many inputs as features, one hidden layer with seven neurons, with a feedforward topology, a sigmoid activation function and a backpropagation-based learning paradigm.

It can be seen that the best results for ANN (30 min of anticipation) is worse than the worst results for the C4.5 (one hour of anticipation). Furthermore, if in some cases the false alarms generated by C4.5 were considered too high, with this algorithm the number rises, reaching the value of 472 for the case of 15 min of anticipation.

In the case of NN, unlike what happened to C4.5 and ANN, the quality of the prediction is diminishing as the temporal horizon is enlarged. The results of NN are the poorest of all, because its best result (prediction to 15 min) is just three points percentage worse than the worst of ANN. It is worth noting the large number of FP obtained in the four scenarios evaluated.

From the traffic management point of view, several conclusions can be drawn. First, that only two features are apparently responsible for traffic congestion: time of occupation and gap between cars. This result is of utmost relevance since simple policies can be deployed to reduce both features' impact. Moreover, the simplicity of the model proposed turns it into a powerful tool able to be used by almost any person. It is also worth noting the low false positive rate achieved by C4.5. In this sense, economical losses associated with useless congestion policies are totally undesirable.

Additionally, data mining has been proven to generate results with easy interpretation in this application field. Further research will be conducted directed towards the discovery of new relevant features, the application of feature selection techniques, the development of ad hoc methods, and the exploration of new spots across Spain in order to generalize the results here introduced.

4 Conclusions

In this paper patterns and models of behavior have been searched using data collected by the detectors of the Spanish DGT, especially at the kilometer 12 descending the bypass SE-30 in Seville, Spain. The extraction of these patterns has been used to predict traffic jams early enough to avoid retention at that point, minimizing response time to this event and giving alternatives to the circulation. To achieve this, it has be successfully applied the well known C4.5 algorithm and obtained results with an error ranging from 85 to 95 %, depending on the anticipation time of the prediction made. Moreover, another algorithms such as ANN or NN have been successfully applied but with less accuracy.

Acknowledgments The authors wish to thank the Dirección General de Tráfico (DGT) for providing the data and to the Junta de Andalucía through research project P12-TIC-1728.

References

1. Ahmane M, Abbas-Turki A, Perronnet F, Wu J, El-Moudni A, Buisson J, Zeo R (2013) Modeling and controlling an isolated urban intersection based on cooperative vehicles. Transp Res Part C Emerg Technol 28:44–62
2. Alcalá-Fdez J, Sánchez L García S, del Jesús MJ, Ventura S, Garrell JM, Otero J, Romero C, Bacardit J, Rivas VM, Fernández JC, Herrera F (2008) KEEL: a software tool to assess evolutionary algorithms for data mining problems. Soft Comput 13(3):307–318
3. Carlson RC, Papamichail I, Papageorgiou M, Messmer A (2010) Optimal mainstream traffic flow control of large scale motorway networks. Transp Res Part C Emerg Technol 18:193–212
4. Collins JF (1993) Automatic incident detection: experience with TRRL algorithm HIOCC. TRRL Suppl Rep 775:1–6
5. Dunkel J, Fernandez A, Ortiz R, Ossowski S (2011) Event-driven architecture for decision support in traffic management systems. Expert Syst Appl 38:6530–6539
6. Hart PE (1968) The condensed nearest neighbor rule. IEEE Trans Inf Theory 18:515–516
7. Hernández JZ, Ossowski S, García-Serrano A (2002) Multiagent architectures for intelligent traffic management system. Transp Res Part C Emerg Technol 10(5):473–506
8. Lee WH, Tseng SS, Shieh WY (2010) Collaborative real-time traffic information generation and sharing framework for the intelligent transportation system. Inf Sci 180:62–702
9. Mahmud K, Gope K, Chowdhury SMR (2012) Possible causes and solutions of traffic jam and their impact on the economy of Dhaka city. J Manage Sustain 2(2):p112
10. Martínez-Álvarez F, Reyes J, Morales-Esteban A, Rubio-Escudero C (2013) Determining the best set of seismicity indicators to predict earthquakes. Two case studies: Chile and the Iberian Peninsula. Knowl-Based Syst 50:198–210
11. McCulloch WS, Pitts W (1943) A logical calulus of ideas immanent in nervous activity. Bull Math Biophys 5:115–133
12. Morales-Esteban A, Martínez-Álvarez F, Reyes J (2013) Earthquake prediction in seismogenic areas of the Iberian Peninsula based on computational intelligence. Tectonophysics 593:121–134
13. Portugais B, Khanal M (2014) Adaptive traffic speed estimation. Proc Comput Sci 32:356–363
14. Wei Qu Z, Xing Y, Song XM, Duan YZ, Wei F (2012) A study on the coordination of urban traffic control and traffic assignment. Discr Dyn Nat Soc 2012(12):367–368
15. Quinlan JR (1993) C4.5: programs for machine learning. Morgan Kaufmann Publishers

Summarizing Information by Means of Causal Sentences Through Causal Questions

C. Puente, A. Sobrino, E. Garrido and J.A. Olivas

Abstract The aim of this paper is to introduce a system able to configure an automatic answer from a proposed question and summarize information from a causal graph. This procedure has three main steps. The first one is focused in the extraction, filtering and selection of those causal sentences that could have relevant information for the system. The second one is focused in the composition of a suitable causal graph, removing redundant information and solving ambiguity problems. The third step is a procedure able to read the causal graph to compose a suitable answer to a proposed causal question by summarizing the information contained in it.

Keywords Causal questions · Causality · Causal sentences · Causal representation · Causal summarization

1 Introduction

Providing causal knowledge is a main objective of the scientific practice, and consubstantial to disciplines as Physic or Engineering.

C. Puente (✉) · E. Garrido
Advanced Technical Faculty of Engineering ICAI, Pontifical Comillas University, Madrid, Spain
e-mail: Cristina.Puente@upcomillas.edu

E. Garrido
e-mail: Eduardo.Garrido@upcomillas.edu

A. Sobrino
Faculty of Philosophy, University of Santiago de Compostela, Santiago, Spain
e-mail: alejandro.sobrino@usc.es

J.A. Olivas
Information Technologies and Systems Dept, University of Castilla-La Mancha, Ciudad Real, Spain
e-mail: Joseangel.olivas@uclm.es

Á. Herrero et al. (eds.), *10th International Conference on Soft Computing Models in Industrial and Environmental Applications*, Advances in Intelligent Systems and Computing 368, DOI 10.1007/978-3-319-19719-7_31

In effect, Physic pursues explain and predict natural phenomena and in both tasks causality plays a relevant role. Explanation involves a general statement, usually a physical law and a singular sentence. Physical laws are paradigmatically causal statements. Thus, causality is a main part of the explanation activity. Prediction demands anticipating the future. To predict the effect of any change, naturally or artificially implemented, involves controlling causal dependencies. In this task, inductive or probabilistic logic plays a mayor role: conditional probabilities and the Markov Principle permits to anticipate the behavior of a causal net performing interventions on it, analyzing the dependence or independence of the involved variables and performing a causal inference according to that [1].

Regarding engineering, causality is usually related to the man-machine interaction and the qualities that the description of a system should have. A system should be (a) clear for customers, (b) clear for the system developers and (c) clearly expressed and analyzed itself if required. The language of causation seems to be a natural way for reaching clarity in the specification of the requirement of complex systems [2]. Causal logic offers an appropriate vocabulary and rules for explaining and predicting convoluted processes in terms of cause-effect connections, validating the Popper dictum that the most important fact in science is not precision, but clarity. Usually, both in Physics or Engineering, causal processes are referred to single links between cause and effect.

But sometimes, causal influence is not deployed as a contiguous cause-effect step, but as a kind of flux browsing from a prior cause, crossing intermediate causes, to a final effect. Fields like medicine shows paradigmatically long causal dependencies, usually called mechanisms. A mechanism is a complex of parts, interacting causally in a process leading to one or more effects. Causal mechanisms are used to explain diseases. F. ex., the possible causal mechanism(s) for explaining the association between alcohol consumption and the risk of developing breast and colorectal cancers involves the following factors: increased hormonal receptor levels, increased cell proliferation, a direct stimulatory effect, DNA adduct formation, increase cyclic adenosine monophosphate, change in potassium channels and modulation of gene expression. In medicine causes are complex and diseases are frequently provoked by a set of factors, conveniently connected. Mechanisms illustrate the interrelation and dependence of causes in a network that shows how causality flows from a prior cause to the effect [3]. Then, causality is not only a matter of single causal connections, but also, as Medicine shows, of long causal dependencies articulated in mechanisms.

From the beginning, IA systems attempted to automatically extract causal knowledge from texts in order to simulate causal systems and, so, automatically answer causal questions. Extracting causal knowledge from texts always was a sensible and rewarding task. If the aim was to retrieve isolated causal sentences, Garcia [4], Khoo et al. [5] or Chan and Choi [6] systems achieved, with good efficiency, the aim. But as we said previously, causal knowledge is frequently articulated as causal mechanisms. In that case, up to now, no computer system deals with the extraction of causal mechanisms. And nor with the generation of linguistic summaries from mechanisms. Usually, summaries are made from single sentences, syntactically arranging them in an appropriate order. Our approach uses causal mechanisms to focus on the content -not the structure- of the texts.

In sum, the aim of our paper is twofold: to extract causal mechanisms from texts and to get from them a linguistic summary based on the essential information contained in them (causal information), not on any kind of formal or structural property as usual. We think that, in both tasks, our initiative presents innovations.

2 Causal Questions

Causal explanation is commonly seen as a static process: if we do not understand an event or fact, an explanation of it should be furnished. But causal explanation is alternatively seen as a dynamic act: something provides an explanation to someone about something. In this view, causal explanation adopts the form of some kind of specialized conversation. The conversational model of causal explanation radically diverges from the more usual one called 'attributional model', which describes causal processes as out of context explanations [7].

Despite current searchers do not include true interrogative facilities in natural language, performing web queries a rudimentary dialogue is established between the computer and the human being. Having a search engine able to discriminate interrogative particles and able to direct the answer to what is pointed at by the particle seems to be a primary goal. In this paper we focus on the relevance of interrogative lexicons in order to provide appropriate causal descriptions.

The way we ask a question is relevant to broad or narrow the range of potential answers. Comparing a *yes/no* question with a *when* or a *how* question; the required answer to the first seems to be less complex than the response to the second ones. Interrogative particles involved in interrogative sentences are, among others, which, who, when, where, what, how and why. As seen in Fig. 1, we can classify these particles depending on the potential complexity of their answers [8]:

Fig. 1 Pyramid of questions' complexity

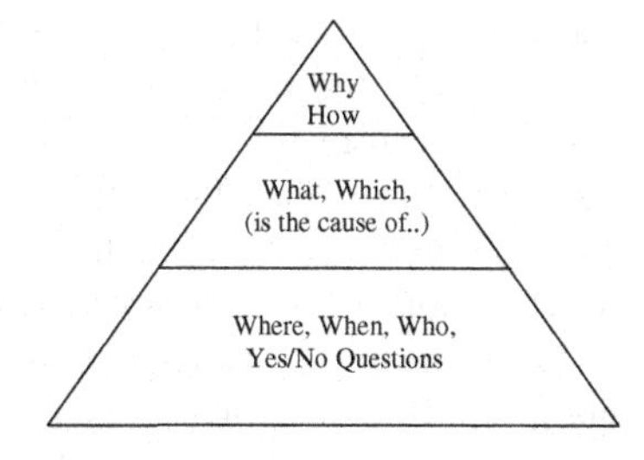

Ascending in the pyramid using interrogative particles there is more and more demand for complex answers to questions, stimulating reflective thinking and a deeper level of conversation. The identification of causal patterns -that we will address in a subsequent point of this paper- will be a useful tool to isolate and match causal slots containing causal content. Finally, questions including interrogative particles above the upper line are related to deepest answers; evoking rather than an answer the justification of the response. How questions frequently refer to a process or mechanism that show the way the answer is reached. In turn, What…is the cause, refers to the cause or causes that are asked for. Last, why questions usually presuppose some external knowledge about the query in order to answer it and are related to the prior cause or to the minimum path in the mechanism that must be followed to get the answer.

Taking these premises into account, in this paper we present a system to extract causal knowledge from texts, create a causal graph and compose a summary of information using that graph as an answer to a causal question. The first part of the system is focused on the extraction of causal sentences from texts belonging to different genres or disciplines, using them as a database of knowledge about a given topic. Once the information has been selected, a question is proposed to choose those sentences where this concept is included. These statements are treated automatically in order to achieve a graphical representation in form of causal graph. The second part is in charge of the generation of an answer by reading the information represented by the causal graph obtained in the previous step. Redundant information is removed, and the most relevant information is classified using several algorithms such as collocation algorithms like SALSA or classical approaches like keywords depending on the context, TF-IDF algorithm. The part of the system generates an answer in natural language thanks to another procedure able to build phrases using a generative grammar.

3 Algorithm to Extract and Represent Causal Sentences

In [9], Puente, Sobrino, Olivas and Merlo described a procedure to automatically display a causal graph from medical knowledge included in several medical texts.

A Flex and C program was designed to analyze causal phrases denoted by words like ‘cause’, ‘effect’ or their synonyms, highlighting vague words that qualify the causal nodes or the links between them. Another C program received as input a set of tags from the previous parser and generated a template with a starting node (cause), a causal relation (denoted by lexical words), possibly qualified by fuzzy quantification, and a final node (effect), possibly modified by a linguistic hedge showing its intensity. Finally, a Java program automated this task. A general overview of the extraction of causal sentences procedure is the following:

Once the system was developed, an experiment was performed to answer the question *What provokes lung cancer? (question introduced by the user about a given topic)*, obtaining a set of 15 causal sentences related to this topic which served as input for a causal graph representation. The whole system (presented in Fig. 2), was unable to answer the question directly, but was capable of generating a causal graph with the topics involved in the proposed question as shown in Fig. 3.

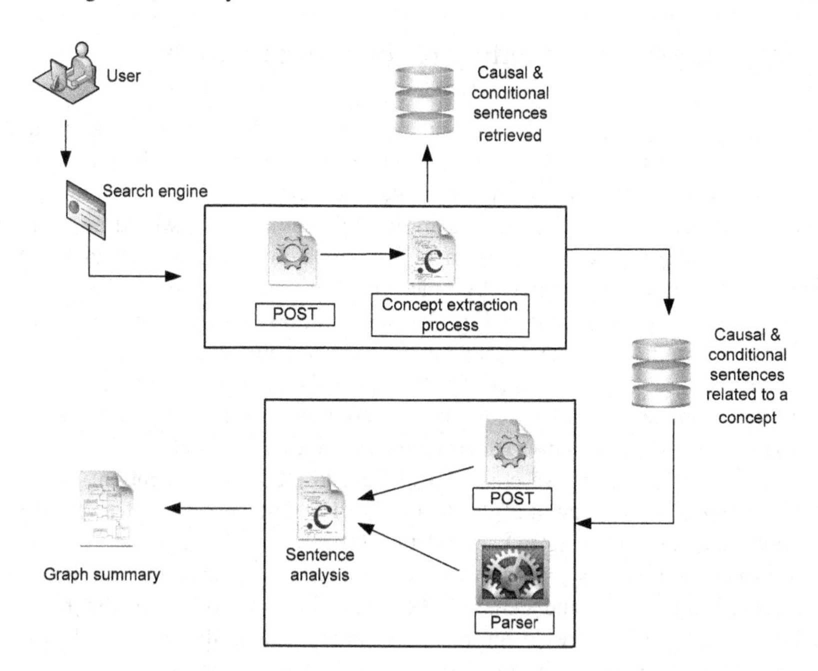

Fig. 2 Extraction and representation of causal sentences

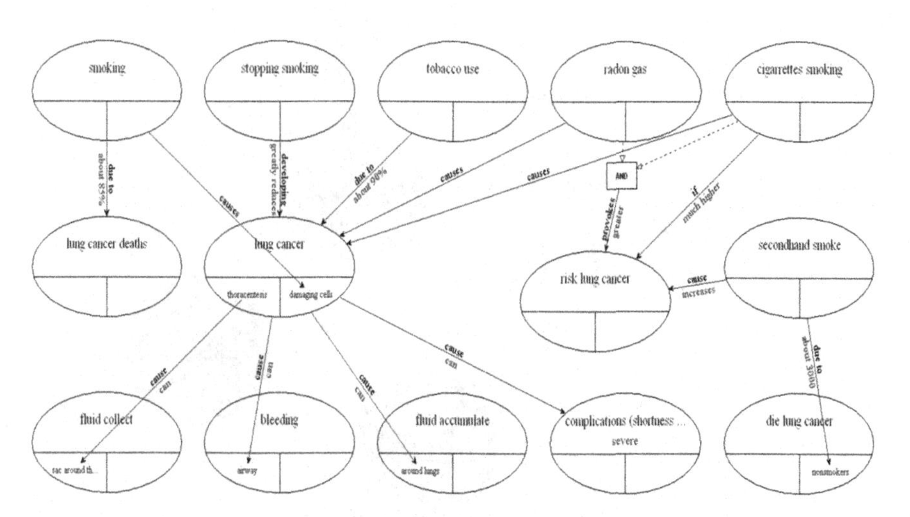

Fig. 3 Causal representation related to the question *What provokes lung cancer?*

Using this causal graph, and the analysis of causal questions we go a further step in this paper to generate the answer to a proposed question by means of a summary, by processing the information contained in the causal nodes and the relationships among them.

4 Summarizing the Content of the Causal Graph

The ideal representation of the concepts presented in Fig. 3 would be a natural language text. This part of the article presents the design of a possible approach to do so. The size of the graph could be bigger than the presented one as not all the causal sentences are critical to appear in the final summary. It would be useful to create a summary of the information of the graph in order to be readable by a human as if it was a text created by other human.

The causal graph depicted in Fig. 3 has the following deficiency: represented words in the graph could have a similar meaning in comparison to other concepts. For example, "*smoking*" and "*tobacco use*" have a similar meaning in the graph so one of these concepts could be redundant. In addition to synonymy, other semantic relations such as hyperonymy or meronymy are relevant as well.

To solve this problem, we created a process to read the concepts of the graph sending them to a ontology like Wordnet or UMLS. The aim is to retrieve different similarity degrees according to each relation [10].

To produce a summary, we need several computational steps to read the graph, reduce the redundancy, and generate the summary. The following diagram presented in Fig. 4, shows the design of the summary system that is created to solve this issue, including the main processes and the main tools needed.

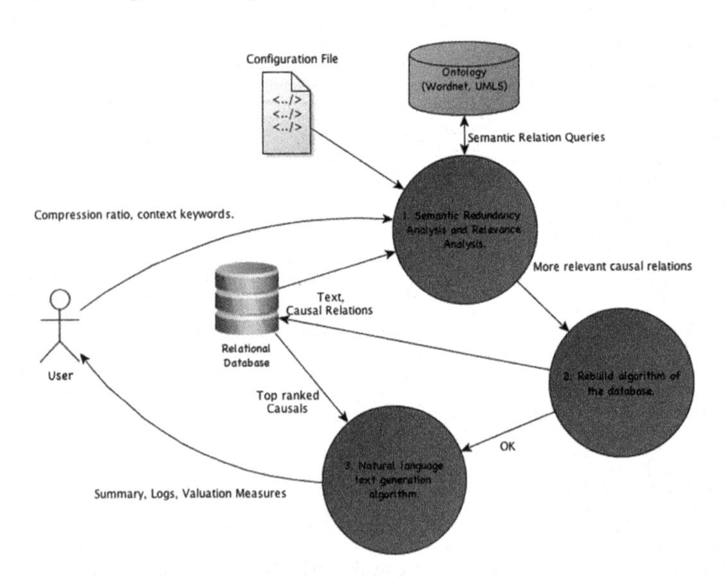

Fig. 4 Basic design of the summarization process

The first problem to solve is the redundancy among nodes. A redundancy analysis process is created to solve this problem taking into account the multiple synsets of every word of the concepts that is been analyzed. It is also taken into account the context of the text having keywords of every context and other measures.

To do so, Wordnet synsets are queried from Java using *Jwnl* and *RiWordnet* tools to find out the meaning of these terms. The output of the process consists of possible relations between all pairs of compared entities, declaring the type and intensity of this relationship.

The degree of their similarity with other concepts is computed as well, being a measure used in the relevance analysis. Different algorithms of similarity between concepts such as Path Length, Leacock and Chodorow [11] or Wu Palmer [12], are executed through platforms like Wordnet::Similarities.

A comparison matrix (Fig. 5), is then generated with all this information, showing the similarity between terms according to different semantic relations. Those concepts with higher similarity degrees with others are considered as the redundant ones.

$$M = \begin{pmatrix} 0 & m_{12} & m_{13} & . & . & . & m_{1n} \\ 0 & 0 & m_{23} & . & . & . & m_{2n} \\ . & . & . & . & . & . & m_{3n} \\ . & . & . & . & m_{ij} & . & m_{in} \\ . & . & . & . & . & . & . \\ . & . & . & . & . & 0 & m_{n-1n} \\ 0 & 0 & 0 & . & . & 0 & 0 \end{pmatrix}$$

Fig. 5 Comparison matrix built by the semantic redundancy algorithm

After running the whole process, a list of semantic relations between entities is obtained, as seen in the results of Fig. 6, providing the information about the relations and the sequence of entities which are going to be deleted. This is the entry for the graph reconstruction algorithm. Additionally, a report on this first version is obtained:

Fig. 6 Final results

```
Final results
==================
Synonyms: 6
Hypernymy/Hyponymy: 13
Meronymy/Holonymy: 0
Entailment: 0
Verb groups: 0
Non related: 72
Total compared concepts: 91
Percentage of reduction of the graph: 79.12088 %
=====================================
Concepts to review:
-> lung cancer deaths
-> risk lung cancer
-> die lung cancer
-> stopping smoking
-> tobacco use
-> cigarettes smoking
-> secondhand smoke
-> fluid collect
-> fluid accumulate
=====================================
```

Once a relation has been found, the next challenge is choosing which term is the most relevant. In the example mentioned above, the question would be what is the most important concept, “smoking” or “tobacco use”. In [13] we proposed a mechanism that gives the answer to that question performing an analysis of the relevance of each concept. To do so, classical measures that analyses the frequency of concepts in the text like TF-Algorithm are used. Connective algorithms that analyses the graph, as SALSA or HITS [14], are also used.

When a causal relation is going to be moved to other concept because it is erased according to the semantic redundancy or relevance ranking algorithm, if the causal relation also exists in the preserved concept two measures are provided. In order to see which implication degree is the resultant one an expression is proposed:

$$\mathrm{NGa} = (1 - \mathrm{s})^*\mathrm{Ga} + \mathrm{s}^*(\mathrm{relA}/(\mathrm{relA} + \mathrm{relB})^*\mathrm{Ga} + \mathrm{relB}/(\mathrm{relA} + \mathrm{relB})^*\mathrm{Gb});$$
$$/ \ \mathrm{NGa}\ [0, 1] \forall \{\mathrm{s}, \mathrm{relA}, \mathrm{relB}, \mathrm{Ga}, \mathrm{Gb}\}\ [0, 1]$$

Denoting NGa the new degree of the concept A, B the one which is going to be erased, s [0,1] the semantic similarity between the two concepts, Ga and Gb [0,1] the implication degree of the concepts, and relA and relB [0,1] the relevance of both terms according to the relevance ranking algorithm. Using this expression the new implication degree is calculated in function of all the parameters of both nodes.

If the implication does not exist in the node which is not going to be erased, then the expression of the new degree is the following one:

$$\mathrm{NGa} = \mathrm{s}^*(\mathrm{relB}/(\mathrm{relA} + \mathrm{relB})^*\mathrm{Gb});$$
$$/ \ \mathrm{NGa}\ [0, 1] \forall \{\mathrm{s}, \mathrm{relA}, \mathrm{relB}, \mathrm{Gb}\}\ [0, 1]$$

After these analyses, the information of the graph has been summarized obtaining this new graph (Fig. 7):

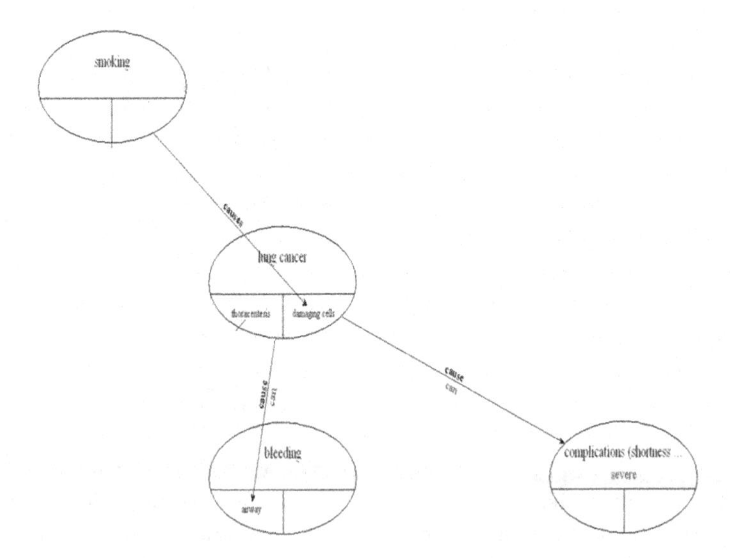

Fig. 7 Causal graph summarized

The summary process has a configuration module depending on the user's preferences and the nature and context of the text to be analyzed. All the modules and measures can be parameterized by means of a weight-value algorithm. In order to have a readable graph, the information should be expressed using natural language sentences. Then, the last process consists of an algorithm that generates natural language from the top ranked causal sentences by the semantic redundancy and relevance analysis. We have performed two experiments varying the compression rate to evaluate the obtained results and check the configuration of the algorithm. In the first experiment, we used a compression rate of 0.3, obtaining as a result the following summary:

"*Cigarettes smoking causes die lung cancer occasionally and lung cancer normally. Tobacco use causes lung cancer constantly and die lung cancer infrequently. Lung cancer causes die lung cancer seldom and fluid collect sometimes. It is important to end knowing that lung cancer sometimes causes severe complication.*"

The original text length is 1497 characters and the summary length is 311, so the system has been able to achieve the compression rate, being the summary less than the 30 % of the original text. In this case, the main information has been included, removing redundant information. The system has chained sentences with the same causes to compose coordinate sentences and reduce the length of the final summary. As seen, the grammatical and semantic meaning is quite precise and accurate, without losing relevant information. In the second experiment, the compression rate was the lowest, removing all the redundant and irrelevant information. This new summary represents a 10 % of the original text, obtaining the following result:

"*Lung cancer is frequently caused by tobacco use. In conclusion severe complication is sometimes caused by lung cancer.*"

In this case, the system just takes the information of the three most relevant nodes, one cause, one intermediate node, and an effect node, creating a summary with the most relevant nodes included in the graph. The length of the summary is of 118 characters, what represents less than a 10 % of the length amount of the original text. Therefore the system is able to modify its behavior considering different configurations of the weights of redundancy and relevance algorithms and the compression rate.

5 Conclusions and Future Works

There are quite IR suggestions to retrieve causal sentences from texts. Extracted causal knowledge varies from single causal sentences to causative rules linking, in a general way, principles and effects. Templates are used to match syntactic and semantic evidences of causality. But as we largely evidence is this paper, causality is not a single process prior_cause-to-final_effect. Areas as engineering, biology or medicine, show causality as a complex and evolutionary process with multiple causes evolving over time. In most disciplines, instead of causality, it is frequently speak of causal mechanisms, denoting how a phenomenon comes about or how some meaningful processes work. This is particularly true in the case of medicine. Our proposal deals not preferentially with the extraction of single causal sentences from texts, but with the organization of them in a casual mechanism, showing how a prior cause is transformed in effects that, in turn, become intermediate causes attempting the final effect.

Summarizing texts is another main task of the A.I. business. There are a lot of systems shortening the contents from the most varied sources. Medicine, law or email threads are some of the investigated areas. Most proposals are focused more on the text structure than on the displayed content. For example, it is frequently conjectured that main content usually appears in conclusions, or that an underlined text must be considered as relevant. Our approach puts forward content as the main source to obtain good summaries. Causal content is largely considered as relevant, as it answers to *how* or *why-questions*, both at the top of the significance ranking of questions. Our proposal gets summaries with different contraction levels from graphs illustrating mechanisms and thus, providing an essential précis of a causal mechanism which is itself also essential in content.

Our approach for summarizing conceptual content from causal mechanisms is novel. It shares with the previous proposals the extraction of causal sentences, but moves away of them organizing the retrieved sentences in a causal graph or mechanism, showing causal dependences and timing causal influences in a graph.

This graph is used to generate in a new fashion way a summary that supports different degrees of contraction.

These aims are a small step in the approach of causal information retrieval, and a lot of challenges remain to be inquired if substantial causal questions, as *how* or *why* -questions intended to be answered and, so, consider them as a source for deeper summaries.

References

1. Spirtes P et al (2001) Causation, prediction and search. The MIT Press, Cambridge
2. Moffett J et al (1996) A model for a causal logic for requirements engineering. Requir Eng 1:27–46
3. Thagard P (1998) Explaining disease: correlations, causes and mechanisms. Mind Mach 8:61–78
4. Khoo CSG, Chan S, Niu Y (2000) Extracting causal knowledge from a medical database using graphical patterns. In: ACL Proceedings of the 38th annual meeting on association for computational linguistics, pp 336–343, Morristown, USA
5. Chang DS, Choi KS (2004) Causal relations extraction using cue phrase and lexical pair probabilities. IJCNLP, 2004, pp 61–70
6. Garcia D (1997) COATIS, and NLP system to locate expression of actions connected by causality links. In: Proceedings of the 10th european workshop in knowledge acquisition, modelling and management, EKAW'97, San Feliu de Guixols, Spain, 15–18 Oct 1997, pp 347–352
7. Hilton DJ (1990) Conversational processes and causal explanation. Psychol Bull 107(1):65–81

8. Vogt E et al (2003) The art of powerful questions: catalyzing insight, innovation and action. Whole Systems Associates, Mill Valley
9. Puente C, Sobrino A, Olivas JA, Merlo R (2010) Extraction, analysis and representation of imperfect conditional and causal sentences by means of a semi-automatic process. In: Proceedings IEEE international conference on fuzzy systems (FUZZ-IEEE 2010), Barcelona, Spain, pp 1423–1430
10. Varelas G, Voutsakis E, Raftopoulou P, Petrakis GME, Milios E (2005) Semantic similarity methods in WordNet and their application to information retrieval on the Web. WIDM'05, Bremen, Germany, 5 Nov 2005
11. Leacock C, Chodorow M (1998) Combininglocal context and WordNet similarity for word sense identification. In: Fellbaum 1998, pp 265–283
12. Wu Z, Palmer M (1994) Verb semantics and lexical selection. In: 32nd annual meeting of the association for computational linguistics, pp 133–138, Resnik
13. Puente C, Garrido E, Olivas JA, Seisdedos R (2013) Creating a natural language summary from a compressed causal graph. In: Proceedings of the Ifsa-Nafips'2013. Edmonton, Canada
14. Najork M (2007) Comparing the effectiveness of HITS and SALSA. In: CIKM'07 Proceedings of the sixteenth ACM conference on Conference on information and knowledge management. 6–8 Nov 2007, Lisboa, Portugal

Discovering the Dialog Rules by Means of a Soft Computing Approach

David Griol and José Manuel Molina

Abstract The current industrial development of commercial dialog systems deploys robust interfaces in strictly defined application domains. However, commercial systems have not yet adopted new perspectives for dialog management proposed in the academic settings, which would allow straightforward adaptation of these interfaces to various application domains. In this paper, we propose a new approach to bridge the gap between the academic and industrial perspectives in order to develop dialog systems using an academic paradigm while employing the industrial standards, which makes it possible to obtain new generation interfaces without the need for changing the already existing commercial infrastructures. Our proposal has been evaluated with a real dialog system providing railway information, which follows our proposed approach to manage the dialog by means of a set of fuzzy rules.

Keywords Spoken dialog systems · Dialog management · Soft computing · Evolving classifiers · Spoken human-machine interaction

1 Introduction

Spoken Dialog Systems (SDSs) are computer programs that receive speech as input and generate synthesized speech as output, engaging the user in a dialog that aims to be similar to that between humans [1, 2].

The design practices of conventional commercial dialog systems are currently well established in industry. In these practices, voice user interface (VUI) experts [3] handcraft a detailed dialog plan based on their knowledge about the specific task and the business rules (e.g., to verify the user's identity before providing

D. Griol (✉) · J.M. Molina
Computer Science Department, Carlos III University of Madrid,
Avda. de la Universidad, 30, 28911 Leganés, Spain
e-mail: david.griol@uc3m.es

J.M. Molina
e-mail: josemanuel.molina@uc3m.es

Á. Herrero et al. (eds.), *10th International Conference on Soft Computing Models in Industrial and Environmental Applications*, Advances in Intelligent Systems and Computing 368, DOI 10.1007/978-3-319-19719-7_32

certain information). In addition, designers commonly define the precise wording for the system prompts according to the dialog state and context, and also the expected types of user's utterances for each turn. As described in [4, 5], this approach is well-documented [6] and has been used to develop hundreds of successful commercial dialog systems.

This standard procedure to develop commercial dialog systems can be represented as a graph that describes the set of dialog states and tables containing the details of each state. Transitions between dialog states are determined by the user turns and the result of different system operations (e.g., the results of database queries).

The main objectives of this approach are usability and task completion, and the main challenge is to cope with the limitations of the front-end technology, as speech recognition errors are common. To solve this situation, especially when the domain model is quite simple and known by the users, commercial applications are usually implemented as directed dialogs, in which users are restricted and guided to provide specific pieces of information, thus restricting the possible user responses and minimizing the probability of speech recognition errors. This way, each interaction is designed to accept a restricted set of expected user reactions to the specific prompt played at that particular turn, providing the speech recognizer with an appropriately designed grammar with a small list of synonyms (thus, generic system prompts like "how can I help you today?" are seldom used).

This paradigm has facilitated the development of standards that are governing the speech industry, such as VoiceXML. This language allows creating dialog systems that select the next system response using a set of predefined rules for dialog management. Its major goals are to bring the advantages of web-based development and content delivery to interactive voice response applications, and to free the authors of such applications from low-level programming and resource management. It enables integration of voice services with data services using the familiar client-server paradigm.

A great effort is currently employed in commercial systems to design the described set of rules for dialog management and find empirical evidence of their appropriateness. This design is usually carried out in industry by hand-crafting dialog strategies tightly coupled to the application domain in order to optimize the behavior of the dialog system in that context, which is a very time-consuming process and has the disadvantage of lack of portability and adaptation to new contexts.

This has motivated the research community to find ways for automating dialog learning by using statistical models trained with real conversations. Statistical approaches can model the variability in user behaviors and allow exploring a wider range of strategies. Although the construction and parameterization of the model depends on expert knowledge of the task, the final objective is to develop dialog systems that have a more robust behavior, better portability, and are easier to adapt to different user profiles or tasks.

In this paper, we propose the application of soft computing techniques to automatically extract the set of rules for dialog management from a labeled dialog corpus. Our proposal models the dialog by means of a classification process, which considers the complete history until the current moment of the dialog as input. We model this classification process by means of fuzzy-rule-based evolving classifiers. This allows obtaining a set of fuzzy rules that can be directly employed to develop a VoiceXML-based dialog manager, then making possible to obtain new generation interfaces without the need for changing the already existing commercial infrastructures. Our proposal has been evaluated with a real dialog system providing railway information, which follows our proposed approach to manage the dialog by means of a set of fuzzy rules.

2 Our Proposal for Applying Soft Computing Techniques for Dialog Management

Figure 1 summarizes the five main tasks usually integrated in a spoken dialog system. Automatic Speech Recognition (ASR), Spoken Language Understanding (SLU), Dialog Management (DM), Natural Language Generation (NLG), and Text-To-Speech Synthesis (TTS). These tasks are typically implemented in different modules of the system's architecture. As can be observed, the dialog manager receives as input the semantic interpretation of a text string recognized by the ASR module. This information also includes confidence scores generated by the ASR and SLU modules [7].

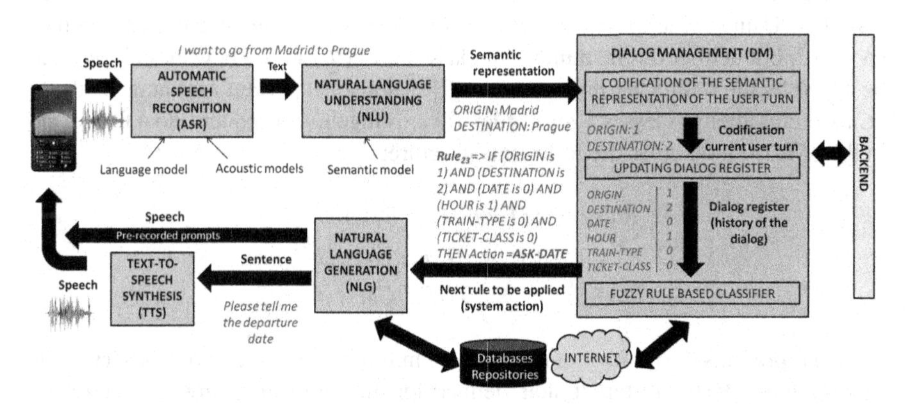

Fig. 1 Modular architecture of a spoken dialog system showing the proposed dialog management technique

Our proposal is focused on slot-filling dialog systems, for which dialog managers use a structure comprised of one slot per piece of information that the system can gather from the user. This data structure, which we call *Dialog Register* (*DR*), keeps the information provided by the user (e.g., slots) throughout the previous history of the dialog. The system can capture several data at once and the information can be provided in any order (more than one slot can be filled per dialog turn and in any order), thus supporting mixed-initiative dialogs in which the system asks the user a series of questions to gather information, and then consults an external knowledge source.

As described in Fig. 1, the dialog manager must consider the values for the slots provided by the user throughout the previous history of the dialog to select the next system action. For the dialog manager to take this decision, we have assumed that the exact values of the attributes are not significant. They are important for accessing databases and for constructing the output sentences of the system. However, the only information necessary to predict the next action by the system is the presence or absence of concepts and attributes. Therefore, the codification we use for each concept and attribute provided by the NLU module is in terms of three values, $\{0, 1, 2\}$, according to the following criteria:

- (0): The value for the slot has not been provided;
- (1): The value is known with a confidence score that is higher than a given threshold;
- (2): The value of the slot has a confidence score that is lower than the given threshold.

We propose to determine the rules that are implicit in a dialog corpus by means of the application of fuzzy-rule-based evolving classifiers. These classifiers are trained from a labeled corpus of training dialogs, which can be provided to the toolkit or be automatically generated by it using a specific module based on a recently developed automatic dialog simulation technique [8]. In this paper, we focus on the *eClass0* classifier. Using these classifiers, not only do the fuzzy rules not need to be prespecified, but neither do the number of classes for eClass (i.e., new class labels can be added by the online learning process) [9]. Another important advantage is that a *eClass0* classifier possesses a zero-order Takagi-Sugeno consequent, so a fuzzy rule in the *eClass0* model has the following structure:

$$\begin{aligned} Rule_i = IF(Feature_1 \text{ is } P_1)\, AND \ldots \\ \ldots AND\ (Feature_n \text{ is } P_n) \\ THEN\ Class = c_i \end{aligned} \tag{1}$$

where i represents the number of rule; n is the number of input features (observations corresponding to the different slots defined for the semantic representation of the user's utterances); the vector *Feature* stores the observed features, and the vector P stores the values of the features of one of the prototypes (coded in terms of three possible values, $\{0, 1, 2\}$) of the corresponding class $c_i \in \{\text{set of different classes}\}$. Each class is then associated to a specific system action (response).

The *eClass0* evolving classifier is trained by means of the set of steps that are described in [10]:

1. Classify each new sample in a group represented by a prototype (i.e., data sample that groups several samples which represent a specific system action). To do this, the sample is compared with all the prototypes previously created. This compar-

ison is done using the cosine distance and the smallest distance determines the closest similarity.
2. Calculate the potential of the new data sample to be a prototype. This value represents a function of the accumulated distance between a sample and all the other $k-1$ samples in the data space. Based on the potential of the new data sample to become a prototype, it could form a new prototype or replace an existing one.
3. Update all the prototypes considering the new data sample. All the existing prototypes are updated considering the new data sample. A new prototype is created if its value is higher than any other existing prototype. Existing prototypes could also be removed.

The following steps are carried out by the developed dialog managers after each user turn:

1. The values of the different slots provided by the SLU module for the current user turn is coded in terms of the previously described three possible values. Confidence scores also provided by this module are used to determine data reliability (e.g., in Fig. 1 the value of the associated confidence scores are used to code "Madrid" as a 1 and "Prague" as a 2).
2. The previous *Dialog Register* is updated with the new values for the slots determined in the previous step. In the example that Fig. 1 shows, the *Dialog Register* contains the slots *ORIGIN*, *DESTINATION*, *DATE*, *HOUR*, *TRAIN-TYPE*, and *TICKET-CLASS*. The previous *DR* containing just a 1 value for the *HOUR* slot is updated with the *ORIGIN* and *DESTINATION* values.
3. The classifier determines the fuzzy rule to be applied (i.e., next system action that will be translated into a sentence in natural language by the NLG module). The cosine distance is used to measure the similarity between the new sample to be classified and the rest of prototypes. For instance, the rule corresponding to the system action "Ask for the date" is selected in the dialog example included in Fig. 1.

3 Practical Application: A Spoken Dialog System Providing Railway Information

The WOz technique [11] allows the acquisition of a dialog corpus with real users without having a complete dialog system, for which the dialog corpus would be necessary. Human-wizard dialogs are usually preferred rather than human-human dialogs since people behave differently toward (what they perceive to be) machines and other people [12].

Three Spanish universities participated in the acquisition of a dialog corpus for a railway information system within the context of the DIHANA project [13]. Each university used a different WOz to carry out the acquisition. The three WOz have multiple information sources to determine the next system action: they heard the sentence pronounced by the user, received the output generated by the ASR module (sequence of words and confidence scores), the semantic interpretation generated by the SLU module the sequence of words recognized, and had a data structure that contains the complete history of the dialog.

In the acquisition, the WOz strategy was not constrained by a script. The three WOz were instructed with only a basic set of rules defined to acquire a corpus without excessive dispersion among them. These rules were recommended to be used by the WOz and are based on considering the confidence scores provided by the SLU module and the information in the current *Dialog Register*.

Two different situations for the dialog were considered. The dialog is in a *safe state* when all the data of the DR have a confidence score that is higher than the fixed threshold. The dialog is in a *uncertain state* when one or more data of the DR have a confidence that is lower than the threshold. A different set of recommended rules was defined for each state. The recommendations for a *safe state* were:

- To make an implicit confirmation and a query to the database if the user has already provided the objective of the dialog and, at least, the minimum necessary information (e.g. *I provide you with railway timetables from Madrid to Bilbao in first class*).
- To request the dialog objective or some of the required information.
- To select a mixed confirmation to give naturalness to the dialog. This selection is made on a variable number of safe turns instead of an implicit confirmation and query to the database. In a mixed confirmation, there are several items, and the confirmation only affects one of them (e.g. *You want railway timetables to Valencia. Do you want to leave from Madrid?*).

The recommendations for a *uncertain state* were:

- To make an explicit confirmation of the first uncertain item that appears in the DR (e.g. *Do you want to travel to Barcelona?*).
- To select a mixed confirmation to give naturalness to the dialog instead of a explicit confirmation. This is done on a variable number of uncertain turns of dialog instead of an explicit confirmation.

For the DIHANA task, we defined eight concepts and ten attributes. The eight concepts are divided into two groups:

1. *Task-dependent concepts*: they represent the concepts the user can ask for (*Hour*, *Price*, *Train-Type*, *Trip-Time*, and *Services*).
2. *Task-independent concepts*: they represent typical interactions in a dialog (*Affirmation*, *Negation*, and *Not-Understood*).

The attributes are: *Origin*, *Destination*, *Departure-Date*, *Arrival-Date*, *Class*, *Departure-Hour*, *Arrival-Hour*, *Train-Type*, *Order-Number*, and *Services*. Figure 2 shows an example of the semantic interpretation of an input sentence.

A total of 51 system responses were defined, which can be classified into confirmations of concepts and attributes, questions to require data from the user, answers obtained after a query to the database, and responses related to dialog formalities (e.g., opening, closing, acceptance, rejection, waiting, not-understood, etc.).

Input sentence:
[SPANISH] Sí, me gustaría conocer los horarios para mañana por la tarde desde Valencia.
[ENGLISH] *Yes, I would like to know the timetables for tomorrow evening leaving from Valencia.*

Semantic interpretation:
(*Affirmation*)
(*Hour*)
Origin: Valencia
Departure-Date: Tomorrow
Departure-Hour: Evening

Fig. 2 An example of the labeling of a user turn in the DIHANA corpus

A set of 900 dialogs was acquired in the DIHANA project. Although this corpus was acquired using a Wizard of Oz technique, real speech recognition and understanding modules were used. A corpus of 200 dialogs acquired for a previous project with a similar task was used to generate the language and acoustic models for the ASR module and to train a statistical model for the SLU module to carry out the acquisition. Some categories were incorporated to increase the coverage of the language model of the ASR module and for the SLU module. However, there were some situations in the acquisition of the DIHANA corpus in which these two modules failed. In these situations, the WOz worked considering only the speech output. The main characteristics of the acquired corpus are shown in Table 1.

Table 1 Main characteristics of the DIHANA corpus

Number of users	225
Number of dialogs per user	4
Number of user turns	6,280
Average number of user turns per dialog	7
Average number of words per user turn	7.7
Vocabulary	823
Duration of the recording (hours)	10.8

A total of 49 rules for the task were obtained learning a *eClass0* classifier using the described training corpus. Figure 3 shows the structure of these rules.

We propose three measures to evaluate the obtained set of rules, which are calculated by comparing the response automatically generated by applying this set for each input in the test partition with regard to the reference answer annotated in the corpus (the answer provided by the WOz). This way, the evaluation is carried out turn by turn. These three measures are:

$FRB - RailwayTask(eClass0)$:

IF $(Timetables$ is 1) AND $(Fares$ is 0) $AND \cdots AND$ $(Not - Understood$ is 0)
$THEN$ $Class =' Ask - Date'$

IF $(Timetables$ is 2) AND $(Fares$ is 0) $AND \cdots AND$ $(Not - Understood$ is 1)
$THEN$ $Class =' Confirm - Timetables'$

IF $(Timetables$ is 0) AND $(Fares$ is 1) $AND \cdots AND$ $(Not - Understood$ is 0)
$THEN$ $Class =' Provide - Fares'$

...

IF $(Timetables$ is 1) AND $(Fares$ is 2) $AND \cdots AND$ $(Not - Understood$ is 1)
$THEN$ $Class =' Close - Dialog'$

Fig. 3 Set of rules for the dialog manager obtained with *eClass0* for the railway task

- *%strategy*: the percentage of responses that exactly follow the strategy defined for the WOz to acquire the training corpus;
- *%coherent*: the percentage of responses that are coherent with the current state of the dialog although they do not follow the original strategy defined for the WOz.
- *%error*: the percentage of responses that would cause the failure of the dialog;

The measure *%strategy* is automatically calculated, evaluating whether the response selected follows the set of rules defined for the WOz. On the other hand, the measures *%coherent* and *%error* are manually evaluated by an expert in the task. The expert evaluates whether the answer selected using the set of rules allows the correct continuation of the dialog for the current situation or whether the answer causes the failure of the dialog (e.g., the dialog manager suddenly ends the interaction with the user, a query to the database is generated without the required information, etc.).

A 5-fold cross-validation process was used to carry out the evaluation of this manager. The corpus was randomly split into five subsets of 1,232 samples (20 % of the corpus). Our experiment consisted of five trials. Each trial used a different subset taken from the five subsets as the test set, and the remaining 80 % of the corpus was used as the training set. A validation subset (20 %) was extracted from each training set.

We have test the behavior of our proposal comparing it with different definitions of the classification function used to determine the next system response. In this work, we have used three approaches for the definition of the classification function: a multilayer perceptron (MLP), a multinomial naive Bayes classifier, and finite-state classifiers. We also defined three types of finite-state classifiers: bigram models, trigram models, and Morphic Generator Grammatical Inference (MGGI) models [14].

Table 2 shows the results obtained. As it can be observed, the Fuzzy-rule-based classifier provides satisfactory results in terms of the percentage of correct responses selected (*Matching* and *Coherence* measures) and responses that could cause the failure of the dialog (*Error* measure). With regard the rest of classifiers, the MLP classifier is the one providing the closest results to our proposal. The table also shows that among the finite-state model classifiers, the bigram and trigram classifiers are worse than the MGGI classifier, this is because they cannot capture long-term dependencies. The renaming function defined for the MGGI classifier seems to generate a model with too many states for the size of the training corpus, therefore, this classifier could be underestimated.

Table 2 Results of the evaluation of the different classification functions

Dialog manager	*Matching (%)*	*Coherence (%)*	*Error (%)*
Fuzzy-rule-based (FRB) classifier	76.7	89.2	5.6
MLP classifier	76.8	88.8	5.8
Multinomial classifier	63.4	76.7	10.6
Bigram classifier	28.8	37.3	42.2
Trigram classifier	31.7	42.1	44.1
MGGI classifier	46.6	67.2	24.8

4 Conclusions

The main objective of our work is to reduce the gap between academic and commercial spoken dialog systems by reducing the effort required to define optimal dialog strategies and implement the system. Our proposal combines the benefits of statistical methods for dialog management and VoiceXML. The former provide an efficient means to explore a wider range of dialog strategies, whereas the latter makes it possible to benefit from the advantages of using the different tools and platforms that are already available and simplify system development.

Our proposal employs evolving classifiers to automatically obtain a set of fuzzy rules that can be directly employed to develop a VoiceXML-based dialog manager, thus reducing the considerable effort and time that is required to manually define the dialog strategy. We have applied our proposed technique to develop a dialog manager for a system that provides railway information. The evaluation results show that the technique can predict coherent system answers in most of the cases.

For future work we are interested in applying our proposal to multi-domain tasks in order to measure the capability of our methodology to adapt efficiently to contexts that vary dynamically. We also want to combine our proposal to facilitate the interaction using also additional input and output modalities which are different to speech, both considering new features related to these additional modalities and combining our proposal with languages and standards defined for multimodal interaction and mobile devices. Finally, we want to develop a extraction model for the semantic representation which could better suited for the type of classifier integrated in our proposal.

Acknowledgments This work was supported in part by Projects MINECO TEC2012-37832-C02-01, CICYT TEC2011-28626-C02-02, CAM CONTEXTS (S2009/TIC-1485).

References

1. Pieraccini R (2012) The Voice in the Machine: Building Computers That Understand Speech. MIT Press, Massachusetts
2. McTear M, Callejas Z (2013) Voice Application Development for Android. Packt Publishing, Birmingham
3. Barnard E, Halberstadt A, Kotelly C, Phillips M (1999) A Consistent Approach To Designing Spoken-dialog Systems. In: Proceedings IEEE Workshop on Automatic Speech Recognition and Understanding (ASRU'99), pp 1173–1176. Keystone, Colorado, USA
4. Pieraccini R, Huerta J (2005) Where do we go from here? Research and commercial spoken dialog systems. In: Proceedings 6th SIGdial Workshop on Discourse and Dialog, pp 1–10. Lisbon, Portugal
5. Williams J (2009) The best of both worlds: Unifying conventional dialog systems and pomdps. In: Proceedings International Conference on Spoken Language Processing (InterSpeech'08), pp 1173–1176. Brisbane, Australia
6. Cohen M, Giangola J, Balough J (2004) Voice User Interface Design. Addison Wesley, Boston
7. García F, Hurtado L, Sanchis E, Segarra E (2003) The incorporation of confidence measures to language understanding. In: International Conference on Text Speech and Dialogue (TSD'03). LNCS series 2807, pp 165–172. Ceské Budejovice, Czech Republic
8. Griol D, Carbó J, Molina J (2013) An automatic dialog simulation technique to develop and evaluate interactive conversational agents. Appl Artif Intell 27(9):759–780
9. Angelov P, Zhou X (2008) Evolving fuzzy-rule-based classifiers from data streams. IEEE Trans Fuzzy Syst 16(6):1462–1475
10. Ordóñez F, Iglesias J, de Toledo P, Ledezma A, Sanchis A (2013) Online activity recognition using evolving classifiers. Expert Syst Appl 40(4):1248–1255
11. Fraser M, Gilbert G (1991) Simulating speech systems. Comput Speech Lang 5:81–99
12. Lane I, Ueno S, Kawahara T (2004) Cooperative dialogue planning with user and situation models via example-based training. In: Proceedings of Workshop on Man-Machine Symbiotic Systems, pp 2837–2840. Kyoto, Japan
13. Griol D, Hurtado L, Segarra E, Sanchis E (2008) A statistical approach to spoken dialog systems design and evaluation. Speech Commun 50(8–9):666–682
14. Segarra E, Hurtado L (1997) Construction of language models using morfic generator grammatical inference MGGI methodology. In: Proceedings Eurospeech'97, pp 2695–2698

Plant Identification: Two Dimensional-Based Vs. One Dimensional-Based Feature Extraction Methods

Tarek Gaber, Alaa Tharwat, Vaclav Snasel and Aboul Ella Hassanien

Abstract In this paper, a plant identification approach using 2D digital leaves images is proposed. The approach made use of two methods of features extraction (one-dimensional (1D) and two-dimensional (2D) techniques) and the Bagging classifier. For the 1D-based method, PCA and LDA techniques were applied, while 2D-PCA and 2D-LDA algorithms were used for the 2D-based method. To classify the extracted features in both methods, the Bagging classifier, with the decision tree as a weak learner, was used. The proposed approach, with its four feature extraction techniques, was tested using Flavia dataset which consists of 1907 colored leaves images. The experimental results showed that the accuracy and the performance of our approach, with the 2D-PCA and 2D-LDA, was much better than using the PCA and LDA. Furthermore, it was proven that the 2D-LDA-based method gave the best plant identification accuracy and increasing the weak learners of the Bagging classifier leaded to a better accuracy. Also, a comparison with the most related work showed that our approach achieved better accuracy under the same dataset and same experimental setup.

Keywords Plant identification · Principal component analysis · Linear discriminant analysis · Bagging classifier · Decision tree · Weak learners · 2D-LDA · 2D-PCA · Leaf image · Leaves images

---

Scientific Research Group in Egypt (SRGE) http://www.egyptscience.net.

T. Gaber (✉) · V. Snasel
Faculty of Computers and Informatics, Suez Canal University, Ismailia, Egypt

A. Tharwat
Faculty of Engineering, Suez Canal University, Ismailia, Egypt

V. Snasel
Faculty of Electrical Engineering and Computer Science, VSB-TU of Ostrava, It4innovations, Ostrava, Czech Republic

A.E. Hassanien
Faculty of Computers and Information, Cairo University, Giza, Egypt

Á. Herrero et al. (eds.), *10th International Conference on Soft Computing Models in Industrial and Environmental Applications*, Advances in Intelligent Systems and Computing 368, DOI 10.1007/978-3-319-19719-7_33

1 Introduction

Plants have a crucial role in the environment. Without plants, it is impossible for the earth's ecology to exist. There are various species of plants, which are subject to the danger of extinction. Therefore, there is a need for protecting plants and classifying them to different species. For this purpose, the identification techniques have become a hot area of research [1]. Manual plant identification requires extensive knowledge and it utilizes complex terminology in a way that even a professional botanist needs to spend much time in a field to achieve plant identification [2]. Such time could be significantly minimized if information technology has been employed to accomplish an automatic plants identification. Although, plant identification can be manually performed by experts, but this is a very time-consuming and inefficient process. Therefore, making the plant identification an automatic process will provide a great help to medicine, industry, and foodstuff production, as well as to biologists, chemists, and environmentalists.

This paper describes an approach addressing the plant identification problem by using features extracted from digital images of plant leaves. The leaf-based features are chosen instead of flowers, fruits, root, stem, etc., as most plants' leaves do not have seasonal nature of the flowers and the fruits and to avoid the inequality in the root and the stem characteristics. In addition, it is a low-cost and a convenient way to get leaf images dataset. Because of these reasons, there have been several studies on plant identification using color, shape, venation, and texture of leaf images.

The rest of the paper is organized as follows, Sect. 2 summarizes the related work of the plant identification based on machine learning. Section 3 highlights the feature extraction methods used in the design of our approach. Our proposed approach with the four feature extraction algorithms are presented in Sect. 4 while the experimental results are given is Sect. 5. The discussion of the obtained results and the conclusion of the paper are presented in Sects. 6 and 7, respectively.

2 Related Works

There are many studies about plant identification based on digital images. Satti et al. [3] classified plant leaves based on 2D images. They used Flavia image dataset and applied many preprocessing steps on the leaf images. Their approach achieved accuracy 85.9 % and 93.3 % using k-Nearest Neighbour (k-NN) and Artificial Neural Networks (ANN) classifiers, respectively. Arora et al. [4] applied the Speed up Robust Features (SURF) to extract the features from leaf images and then used the random forest classifier and tested their approach using Plant Leaves II dataset. In other research, Caglayan et al. [5] utilized color and shape features to classify 32 different kinds of plants. They used Support vector Machine (SVM), k-NN, Random Forest, and Naive Bayes classifiers and the best accuracy achieved reached to 96 %. In [6], Arun et al. transformed the leaf images into grayscale and applied boundary enhancement operations. They then used the PCA to extract features and then used SVM and

k-NN for classification. They used Flavia dataset and achieved the accuracy of 78 % to 81.3 % using k-NN classifier.

Also, Valliamma et al. [7] proposed identification approach for flower images dataset. They applied Preferential Image Segmentation (PIS) and other enhancement operations to the images. They then used the image-threshold to obtain some features and then used the probabilistic curve for classification. They used dataset of 500 flowers images. In [8], plant leaf images were converted to grayscale, the region of interest was segmented and the features were extracted. Probabilistic Neural Networks classifier was then used of Flavia dataset and classification rate was 92.5 %.

3 Feature Extraction Method

A feature extract method is a process in which some properties of an object is measured and transformed into numeric values. There are many types of features for an image, e.g. shape (area, length, and width), texture, and color. These features are used to describe images' contents. The shape and texture features could be extracted by *Scale Invariant Feature Transform* (SIFT) [9] or *Local Binary Patterns* (LBP) [10] whereas the color features can be extracted by Color Moments, Color Histogram, etc. [11].

There are many feature extraction methods, e.g. *Principal Component analysis* (PCA) [12], and *Linear Discriminant Analysis* (LDA) [13]. As given below, these methods can be classified into 1D and 2D feature extraction techniques.

One-Dimensional Feature Extraction Technique: The idea of this class is based on first converting each image ($I(w \times h)$) into one feature vector (x_i) (column or row), where w and h represent the width and height of the image (I), respectively. Secondly, all feature vectors are concatenated to form a feature matrix ($X = \{x_1, x_2, \dots, x_N\}$), where N represents the total number of images, and finally the features are extracted from this matrix. PCA and LDA are two examples of this class and they are briefly explained.

PCA: The PCA is a linear subspace and unsupervised method. It is used to find a linear transformation, U, which reduces, d-dimensional feature vectors to h-dimensional feature vectors (where $h < d$). PCA has many applications such as dimensionality reduction [14], face recognition [15, 16], ear recognition [17]. The idea of PCA is to find a column vector space or eigenvectors that construct a projection surface representing the direction of maximum variance of the given feature vectors. All eigenvectors are mutually orthogonal and called *principal components* which are measured and determined based on their corresponding eigenvalues of the covariance matrix [15].

LDA: It is a supervised dimensionality reduction and feature extraction method. It first finds the projection space maximizing the ratio of the between-class variance to the within-class variance and hence guaranteeing maximum class separability [18]. LDA is used as in many applications such as face recognition [19], palm-print [20], and ear recognition [21].

Two-Dimensional Feature Extraction Techniques: This class of feature extraction techniques deals with the images in the matrix representation. It does not require vector representation as in the 1D-based technique so there is no need for the step of transforming each image into one vector prior to feature extraction step. Examples of these techniques are the 2D-PCA and the 2D-LDA.

2D-PCA: The main idea of 2D-PCA is based on the same idea of 1D-PCA, but in the former, a 2D image must be first transformed into 1D vector for two reasons. Firstly, constructing covariance matrix can be computed directly using the original 2D images without a need to any transformation from 2D to 1D. Secondly, the size of the covariance matrix in 2D is much smaller than that of 1D. Thus, less time is required to calculate the eigenvectors in 2D-PCA than in 1D-PCA [16].

2D-LDA: This is very similar to the 1D-LDA algorithm, but in 2D-LDA, the image must be first transformed to a vector form. In addition, 2D-LDA is used mainly to (1) solve the problem of Small Sample Problem (SSS) occurring when the dimensions of feature vectors are higher than the number of the training samples, (2) the competitive accuracy, and (3) requiring less CPU time [13].

3.1 The Bagging Classifier

The Bagging classifier is an ensemble method creating its ensemble by training different classifiers on a random distribution of a training dataset. From this dataset, features of many images are drawn randomly with replacement to form a new training set. For each weak classifier of the Bagging ensemble, the new training set is used to build a Bagging model. Since the Bagging re-samples the training set with replacement, so some instances are chosen multiple times while the others are left out. The Bagging is effective with unstable learning algorithms such as neural networks and decision trees [22].

4 Proposed Approaches

Two different techniques are proposed to identify plant leaves based on their digital images. As illustrated in Fig. 1, these two techniques follow a similar methodology, but for feature extraction, one makes use of PCA and LDA, and another utilizes 2D-LDA and 2D-PCA. After extracting features, in both techniques, the Bagging classifier is used to identify the type of the unknown leave image.

In the training phase, the classifier's model is built using the Bagging classifier while in the testing phase this classifier model is used through the Bagging classifier as well to identify the unknown sample images. The detailed steps of our approach are given below.

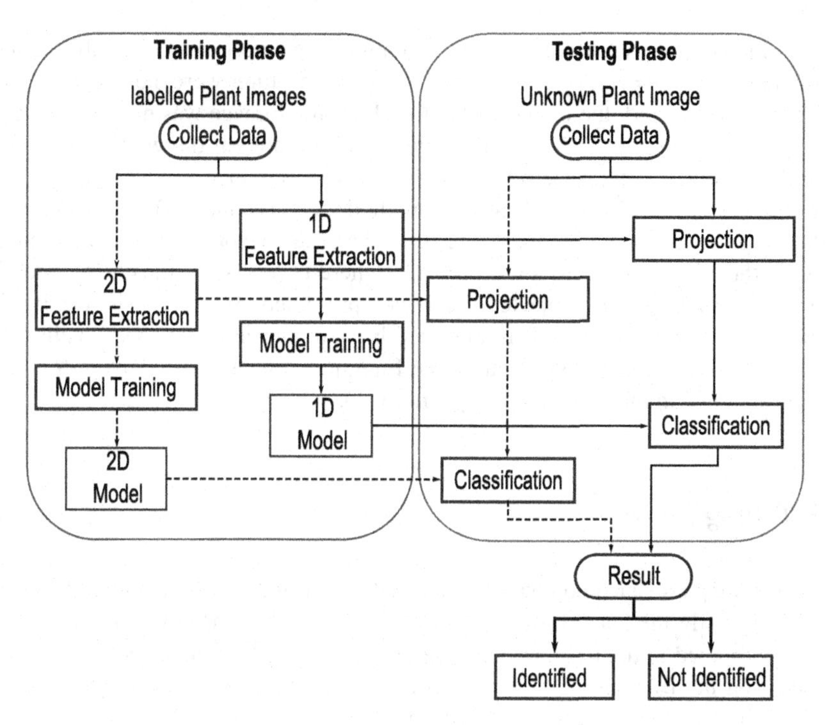

Fig. 1 Plant identification system using leaves' images

4.1 Training Phase

Algorithm 1 : Training Phase

1: Enter all the labelled leaves images ($I_{i=1}^{N}$), where N represents the total number of the training images.
2: **if** (One-Dimensional technique) **then**
3: Convert all images $I_i(w \times h), i = 1, \ldots, N$ into vectors $v_i(1 \times wh)$, w and h represent the width and the height of each training image.
4: Combine all feature vectors into one feature matrix ($A = [v_1, v_2, \ldots, v_N]$).
5: **else**
6: Deals with images in 2D form (i.e. matrix representation).
7: Combine all feature vectors into feature matrix ($A = [I_1, I_2, \ldots, I_N]$).
8: **end if**
9: Compute the projection surface (U) of the features matrix, (A).
10: Project A on the projection surface (U) to obtain the features D where $D = AU$.
11: Train the Bagging classifier using the extracted features, D.

In this phase, a number of leave images, $I_{i=1}^{N}$ were selected, where N represents the total number of training images. For each image, features were extracted using 1D-based and 2D-based methods. In the 1D-based method, each plant image was transformed into one vector v and then all training images vectors were combined into a feature matrix, $A = [v_1, v_2, \dots, v_N]$ from which the features were then extracted. In the 2D-based method, the image basis of each training image was not changed but represented as 2D matrix, $I(w \times h)$, where w and h represent the width and height of each training image, respectively. All 2D matrices were then combined into a feature matrix $A = [I_1, I_2, \dots, I_N]$. In both methods (i.e. 1D and 2D), the features of each image are extracted by projecting the image (i.e. vector or matrix representation) on the eigenspace as follows, $D = AU$, where U represents the eigenspace. In case of 1D methods, the projected image is represented by scalers whereas in 2D methods the projected image is represented by w-dimensional vectors. Finally, the extracted features, D, are used to train the Bagging classifier. The steps of training phase explained in more details in Algorithm (1).

4.2 Testing Phase

In the testing phase, an unknown leave image (T) is tested for its plant identification. To do so, firstly the leave features are extracted by projecting it on the projection space, computed in the training phase, i.e. $\xi = TU$. The computed vector ξ is classified using the Bagging classifier's model that has been also built in the training phase. Detailed steps are given in Algorithm (2)

Algorithm 2 : Testing Phase

1: Enter an unknown leave image (T).
2: **if** (One-Dimensional Technique) **then**
3: Convert this image $T(w \times h)$ into a vector form, $v_T(1 \times wh)$.
4: **else**
5: Deals with image in 2D form (i.e. matrix representation).
6: **end if**
7: Project the unknown 2D image on the projection surface to get ξ.
8: Match between ξ with D using the Bagging model built during the training phase to find the class label of the unknown image (T).

5 Experimental Results

To evaluate our proposed approach, we used the Flavia dataset consisting of 1907 colored leaves images with size (1600 $\times$ 1200) and collected from 32 different species. In this paper, a number of images were selected from 20 different species and for each plant, there are six images were used. The selected images are in different orientations, illumination, and quality.

The proposed approach with its four variants (i.e., using PCA, LDA, 2D-PCA, and 2D-LDA) were evaluated through two different scenarios investigating the

Table 1 Accuracy and CPU time of the proposed 4 variants with different number of training images and 10 weak learners of the Bagging classifier

No. of training images	Accuracy (in %)							CPU time (in secs)						
	1D		2D		Caglayan et al. [5]			1D		2D		Caglayan et al. [5]		
	PCA	LDA	PCA	LDA	Color	Shape	Color+ Shape	PCA	LDA	PCA	LDA	Color	Shape	Color+ Shape
1	42.5	52	50	52	47	25	50.5	0.008	0.013	0.006	0.009	0.004	0.004	0.006
2	50.6	60	58.8	58.8	57.7	32	57.7	0.009	0.014	0.006	0.01	0.004	0.005	0.011
3	56.8	65	65	65	56.7	35	56.8	0.011	0.016	0.007	0.012	0.005	0.005	0.014
4	63.6	77.5	80	80	62.5	41.5	64.5	0.014	0.021	0.007	0.014	0.006	0.007	0.029
5	68.9	75	85	90	75	50.5	76.7	0.023	0.037	0.007	0.016	0.007	0.007	0.034

Table 2 Accuracy and CPU time of the 2D-based method with different number of training images and weak learners of the Bagging classifier

Feature extraction method	No. of weak learners	Accuracy (in %)					CPU time (in secs)
		No. of training images					
		N=1	N=2	N=3	N=4	N=5	
2D-LDA	Ntrees=5	52	65.6	60	77.8	85	0.009
	Ntrees=10	52	58.8	65	80	90	0.014
	Ntrees=50	55	65.6	70	85	90	0.056
	Ntrees=100	60	68.8	71.7	87.5	95	0.142
	Ntrees=200	65	72.5	75	90	95	0.225
2D-PCA	Ntrees=5	47	56.6	56.7	77.8	85	0.007
	Ntrees=10	50	58.8	65	80	85	0.007
	Ntrees=50	54	60.4	70	80	92	0.025
	Ntrees=100	60	62.5	73.3	85	90	0.052
	Ntrees=200	65	67.5	76.7	87.5	95	0.185

performance and accuracy. In these scenarios, the Bagging classifier ensemble, with different number of trees (weak classifiers or ensemble size), was used to match the unknown image with the trained images.

In the first scenario, the accuracy of the 1D-based and 2D-based techniques were investigated through testing different number of training images (k, and $k = 1, 2, \ldots, 5$) of each plant type. The training samples were selected randomly from the database while the testing samples, i.e. the $(6 - k)$ images, were used during the testing phase. The accuracy and CPU time of this scenario are summarized in Table 1.

The second scenarios was designed based on the results of the first one in which the 2D-based methods has given better results than that of the 1D-based one. So, the aim of this scenario was to further understand the effect of changing the number of training data and to evaluate the accuracy and the performance stability over the standardize data. In this scenario, the 2D-PCA and 2D-LDA were used to extract the leave features. The Bagging classifier was then used in different experiments with changing its weak learners (i.e. 5, 10, 50, 100, and 200). In addition, the number of training images was also changed from 1 to 5. The results obtained from this scenario are summarized in Table 2.

6 Discussion

From the first scenario' results, shown in Table 1, the following remarks can be drawn. In terms of accuracy issues, firstly, the accuracy of all Four variants (i.e. PCA, LDA, 2D-PCA, and 2D-LDA) are improved when the number of training images is increased. Secondly, the accuracy of the 2D-based methods (i.e. 2D-PCA and 2D-LDA) is better than that of the 1D-based methods (i.e. 1D-PCA and 1D-LDA). Thirdly, the 2D-LDA-based method achieves the best accuracy and the 1D-PCA-based one accomplishes the worst accuracy.

In terms of the CPU performance, from Table 1 the following point can be also drawn. The 2D-PCA is the most efficient algorithm among all other methods and the 1D-LDA is the worst one. To explain this remark, the following example is given. If the image size of all training images is (100×100). Then, to compute the covariance matrix of 2D-PCA, it is required to multiply two matrices of (100×100). But when using the 1D-PCA, all training images are converted into one vector (1×10000), and the covariance matrix is computed by multiplying two matrices $(N \times 10000) \times (10000 \times N)$, where N represents the total number of training images. Thus, 2D-PCA takes CPU time much lower than 1D-PCA takes.

From Table 2, it can be noticed that (a) the accuracy of the proposed approach increases when the number of the trees of the Bagging classifiers increase. This is due to the fact that in order to reach to the global minimum of the error, the Bagging classifier needs many iterations (i.e. high number trees). However, this is achieved on the cost of taking more CPU time (see Table 2); (b) the 2D-LDA-based method achieves identification accuracy better than that of the 2D-PCA-based method, but this is also accomplished with more CPU time.

To further evaluate our proposed approach, we compared it with the approach suggested by Caglayan et al., [5]. We first implemented Caglayan's approach which makes use of the features of color, shape, and color along with shape features of plant leaves to identify plants. We then run it under the same dataset and with the same setup of our experiments (see Sect. 5). The results of Caglayan's approach comparing with proposed approach are shown in Table 1. From this table, it can be noticed that the proposed approach (with the 2D-based methods) achieved accuracy much better than Caglayan's one.

As a general remark, from Tables 1 and 2, it can be noticed that, the accuracy of the proposed approach with its variants is proportional to the number of training images and the best accuracy is achieved when the number of training images is six.

7 Conclusion

In this paper, an approach to identify plant through its 2D leaves images was proposed. The approach consists of two main phases: feature extraction and classification. For the feature extraction phase, four algorithms were investigated: i.e. PCA, LDA, 2D-PCA and 2D-LDA, while for the classification phase, the Bagging classifier, with decision tree as a weak learner, was employed. The four variants of the proposed approach were evaluated using Flavia dataset. The experimental results showed that when using the 2D-PCA and 2D-LDA, our approach achieves higher accuracy than when using the PCA and LDA. In addition, our approach achieved the highest accuracy when using the 2D-LDA as a feature extraction algorithm for the plant identification. Also, increasing the number of decision trees (i.e. the weak learners) for the Bagging classifier leaded to improving the identification accuracy. Moreover, the performance of our approach with the 2D-PCA and 2D-LDA was better than that of with the PCA and LDA. Furthermore, a comparative evaluation with the most related work showed that our approach achieved better accuracy under the same dataset. In the future work, the same leaves dataset will be tested against different types of feature extraction methods such as the Independent Component Analysis (ICA) and 2D-ICA techniques.

Acknowledgments This paper has been elaborated in the framework of the project "New Creative Teams in Priorities of Scientific Research", reg. no. CZ.1.07/2.3.00/30.0055, supported by Operational Programme Education for Competitiveness and co-financed by the European Social Fund and the state budget of the Czech Republic and supported by the IT4Innovations Centre of Excellence project, reg. no. (CZ.1.05/1.1.00/02.0070), funded by the European Regional Development Fund and the national budget of the Czech Republic via the Research and Development for Innovations Operational Programme.

References

1. Ma LH, Zhao ZQ, Wang J (2013) ApLeafis: An Android-Based Plant Leaf Identification System. In: Huang DS, Bevilacqua V, Figueroa JC, Premaratne P (eds) ICIC 2013, vol 7995., LNCSSpringer, Heidelberg, pp 106–111
2. Rademaker CA (2000) The classification of plants in the united states patent classification system. J World Patent Inf 22(4):301–307
3. Satti V, Satya A, Sharma S (2013) An automatic leaf recognition system for plant identification using machine vision technology. Int J Eng Sci Technol 4:0975–5462. ISSN
4. Arora A, Gupta A, Bagmar N, Mishra S, Bhattacharya A (2012) A plant identification system using shape and morphological features on segmented leaflets: team iitk, CLEF 2012. In: CLEF 2012 Evaluation Labs and Workshop, Online Working Notes, Rome, Italy, 17–20 Sept 2012
5. Caglayan A, Guclu O, Can AB (2013) A plant recognition approach using shape and color features in leaf images. In: Petrosino A (ed) ICIAP 2013, Part II, vol 8157., LNCSSpringer, Heidelberg, pp 161–170
6. Arun Priya C, Balasaravanan T, Thanamani AS (2012) An efficient leaf recognition algorithm for plant classification using support vector machine. In: International Conference on Pattern Recognition, Informatics and Medical Engineering (PRIME), IEEE, pp 428–432
7. Valliammal N, Geethalakshmi S (2011) Automatic recognition system using preferential image segmentation for leaf and flower images. Int J Comput Sci Eng 1(4):13–25
8. Uluturk C, Ugur A (2012) Recognition of leaves based on morphological features derived from two half-regions. In: International Symposium on Innovations in Intelligent Systems and Applications (INISTA), IEEE, pp 1–4
9. Tharwat A, Gaber T, Hassanien AE, Shahin MK, Refaat B (2015) SIFT-Based Arabic sign language recognition system. In: Abraham A, Krömer P, Snasel V (eds.) Afro-European Conference for Industrial Advancement, vol 334, pp 359–370. AISCSpringer, Heidelberg
10. Tharwat A, Gaber T, Hassanien AE, Hassanien HA, Tolba MF (2014) Cattle identification using muzzle print images based on texture features approach. In: Proceedings of the 5th International Conference on Innovations in Bio-Inspired Computing and Applications, IBICA, 23–25 June 2014, Springer, Ostrava, Czech, pp 217–227
11. Semary NA, Tharwat A, Elhariri E, Hassanien AE (2014) Fruit-based tomato grading system using features fusion and support vector machine. In: Proceedings of the 7th IEEE International

Conference Intelligent Systems IS'2014, vol 323, 24-26 Sept 2014, Springer, Warsaw, Poland, pp 401–410
12. Jolliffe I (2005) Principal component analysis. Wiley Online Library, New York
13. Ye J, Janardan R, Li Q (2004) Two-dimensional linear discriminant analysis. In: Neural Information Processing Systems, NIPS, 13–18 Dec 2004, Vancouver, British Columbia, Canada, pp 1569–1576
14. Moore B (1981) Principal component analysis in linear systems: controllability, observability, and model reduction. IEEE Trans Autom Control 26(1):17–32
15. Turk, MA, Pentland AP (1991) Face recognition using eigenfaces. In: Proceedings of IEEE Computer Society Conference on Computer Vision and Pattern Recognition (CVPR'91), IEEE, pp 586–591
16. Yang J, Zhang D, Frangi AF, Yang JY (2004) Two-dimensional pca: a new approach to appearance-based face representation and recognition. IEEE Trans Pattern Anal Mach Intell 26(1):131–137
17. Tharwat A, Ibrahim A, Ali H (2012) Personal identification using ear images based on fast and accurate principal component analysis. In: 8th International Conference on Informatics and Systems (INFOS), IEEE, pp 56–59
18. Welling M (2005) Fisher linear discriminant analysis. Department of Computer Science, University of Toronto, Toronto
19. Lu J, Plataniotis KN, Venetsanopoulos AN (2003) Face recognition using lda-based algorithms. IEEE Trans Neural Netw 14(1):195–200
20. Cui JR (2012) Multispectral palmprint recognition using image-based linear discriminant analysis. Int J Biometrics 4(2):106–115
21. Yuan L, Mu ZC (2007) Ear recognition based on 2d images. In: First IEEE International Conference on Biometrics: Theory, Applications, and Systems, (BTAS), IEEE, pp 1–5
22. Breiman L (1996) Bagging predictors. J Mach Learn 24(2):123–140

A First Approach in the Class Noise Filtering Approaches for Fuzzy Subgroup Discovery

C. J. Carmona and J. Luengo

Abstract The presence of noise in data is a common problem that produces several negative consequences, and is an unavoidable problem, which affects the data collection and data preparation processes in Data Mining applications, where errors commonly occur. The performance of the models built under such circumstances will heavily depend on the quality of the training data. Hence, problems containing noise are complex problems and accurate solutions are often difficult to achieve without using specialized techniques. A particular supervised learning field as subgroup discovery has overlooked the analysis of noise and its impact in the description obtained. In this paper, the noise impact in subgroup discovery is analyzed in a complete experimental study, using recent filtering techniques for several class noise levels. Specifically, the analysis is performed through the FuGePSD algorithm which is a state-of-the-art SD algorithm based on genetic programming and fuzzy logic.

Keywords Subgroup discovery · Class noise · Noise filters

1 Introduction

Real-world data is never perfect and often suffers from corruptions that may harm interpretations of the data, models built and decisions made. Noise can negatively affect the system performance in terms of accuracy, building time, size and interpretability of the model built [23] and it is specially relevant in supervised problems, where it alters the relationship between the informative features and the measure output. For this reason noise has been specially studied in classification and regression where noise hinders the knowledge extraction from the data and spoils the models

C.J. Carmona (✉) · J. Luengo
Dept. of Civil Engineering, University of Burgos, 09006 Burgos, Spain
e-mail: cjcarmona@ubu.es

J. Luengo
e-mail: jluengo@ubu.es

Á. Herrero et al. (eds.), *10th International Conference on Soft Computing Models in Industrial and Environmental Applications*, Advances in Intelligent Systems and Computing 368, DOI 10.1007/978-3-319-19719-7_34

when compared to models learned from clean data for the same domain, which represent the real implicit knowledge of the problem [10].

Several approaches have been studied in the literature to deal with noisy data and to obtain better models, traditionally in classification problems. *Robust learners* [2] are characterized for being less influenced by noise, but designing a robust learner is not trivial. C4.5 [18] is a typical example thanks to its pruning phase. Several works [19] claim that complete or partial noise correction made by *data polishing methods* improves test performance results in comparison with no preprocessing, but it is only feasible in small data sets due to a high computational cost. Finally, the most popular choice are *noise filters* [3, 13], as they act as a preprocessing step, identifying and eliminating the noisy instances from the training data.

However, the effect of noise and capabilities of descriptive algorithms in the presence of noise has been mostly ignored, and since this framework of data mining also relies on supervised examples, the negative effects of noise cannot be ignored. Subgroup discovery (SD) [12] is a descriptive data mining technique using supervised learning, i.e. it is a half-way between classification and description, where the knowledge is represented through rules. The analysis of quality measures in SD is a key factor in order to observe the correct operation of the SD algorithms. The values of these quality measures can be affected for the presence of noise in the data.

In this work we are interested in analysing the performance of SD learning in the presence of noise, and to study different approaches to deal with it. Since noise filtering is a popular preprocessing step that does not require any modification of the SD algorithms, we will use three recent filters for class noise: EF, CVCF and IPF. From 14 base supervised data sets, different amounts of class noise will be introduced, from 5 to 20 %, generating an increasingly noisy scenario to check the suitability of the filtering for SD. We will use a state-of-the-art SD technique, FuGePSD, which is the most recent evolutionary fuzzy system (EFS) for SD presented up to the moment and able to deal with low amounts of noise. The fuzzy confidence measure is used to evaluate the performance of FuGePSD, and the analysis is supported by the use of the Wilcoxon's Signed Rank non-parametric statistical test.

The rest of this contribution is organized as follows. Section 2 introduces the background concepts of subgroup discovery and filtering techniques used. Next, Sect. 3 describes the experimental framework and includes the experimental results and their analysis. Finally, Sect. 4 presents some concluding remarks.

2 Preliminaries

This section presents main concepts, algorithms and noise filters used in the paper. Section 2.1 presents main properties of SD and the FuGePSD algorithm, Sects. 2.2, 2.3 and 2.4 presents noise filters employed in the experimental study.

2.1 Subgroup Discovery

SD is a descriptive data mining technique based on supervised learning. The concept of SD was initially introduced by Kloesgen [14] and Wrobel [21]. The main purpose of SD is to seek and explore relationships between different properties or variables with respect to a target variable, and representations of the knowledge are performed through rules which consist of induced subgroup descriptions [9, 16]. Each rule R can be formally defined as:

$$R : Cond \rightarrow Target_{value}$$

where $Target_{value}$ is a value for the variable of interest (target variable) for the SD task (which also appears as *Class* in the literature), and *Cond* is commonly a conjunction of features (attribute-value pairs) which is able to describe an unusual statistical distribution with respect to the $Target_{value}$.

Despite the use of a target variable, SD is a descriptive induction task using supervised learning while classification is a predictive task. Main differences between SD and classification can be observed in [12].

The most important elements considered for an SD approach are: the target variable, the search strategy, the descriptive language of the subgroups and the quality measures used. Reviews about major properties, features, algorithms and real-world problems solved through the application of SD algorithms can be found in [4, 12].

Throughout the literature there are a wide number of SD algorithms based on exhaustive or stochastic strategies, among others. Recently, a new algorithm called FuGePSD has been presented [5]. This algorithm is an EFS [11] which are basically a fuzzy system augmented by a learning process based on evolutionary computation [8]. Fuzzy systems are usually considered in the form of fuzzy-rule based systems (FRBSs), which are composed of "IF-THEN" rules where both the antecedent and consequent can contain fuzzy logic statements and EAs are well known and widely used global search techniques with the ability to explore a large search space. In summary, the properties of this type of systems make them highly suitable for the development of SD approaches. In fact, the use of fuzzy rules, based on fuzzy logic [22], already allow to consider uncertainty, and also to represent the continuous variables in a manner which is close to human reasoning. In this way, interpretable fuzzy rules consider continuous variables as linguistic ones, where values are represented through fuzzy linguistic labels (*LLs*). The fuzzy set corresponding to each *LL* can be specified by the user or defined by means of uniform partitions if knowledge is not available.

Equation 1 represents a canonical fuzzy rule:

$$R : IF\ X_1 = (LL_1^2)\ AND\ X_3 = (LL_3^1)\ THEN\ Target_{value} \tag{1}$$

where:

- $X = \{X_m/m = 1, \ldots, n_v\}$ is a set of features used to describe the subgroups, and n_v is the number of descriptive features.

- $T = \{Target_{value}/j = 1, \ldots, n_{tv}\}$ is a set of values for the target variable, and n_{tv} is the number of values for the target variable.
- $LL_{n_v}^{l_{n_v}}$ is the LL number l_{n_v} of the variable n_v.

FuGePSD is based on genetic programming [15] with the ability to extract descriptive fuzzy rules for the SD task. It employs a tree with a variable-length structure to represent the individuals of the population together several mechanisms and specific operators in order to maximise the quality measures. A complete description of the algorithm FuGePSD can be found in [5].

2.2 Ensemble Filter

The *Ensemble Filter* (EF) [3] uses a set of learning algorithms to create classifiers in several subsets of the training data that serve as noise filters for the training set. The identification of potentially noisy instances is carried out by performing an Γ-FCV on the training data with μ classification algorithms, called filter algorithms. In the developed experimentation for this contribution we have utilized the three filter algorithms used by the authors in [3], which are C4.5, 1-NN and LDA [17]. The complete process carried out by EF is described below:

- Split the training data set D_T into Γ equal sized subsets.
- For each one of the μ filter algorithms:
  - For each of these Γ parts, the filter algorithm is trained on the other $\Gamma - 1$ parts. This results in Γ different classifiers.
  - These Γ resulting classifiers are then used to tag each instance in the excluded part as either correct or mislabeled, by comparing the training label with that assigned by the classifier.
- At the end of the above process, each instance in the training data has been tagged by each filter algorithm.
- Add to D_N the noisy instances identified in D_T using a consensus voting scheme, taking into account the correctness of the labels obtained in the previous step by the μ filter algorithms.
- Remove the noisy instances from the training set: $D_T \leftarrow D_T \setminus D_N$.

2.3 Cross-Validated Committees Filter

The *Cross-Validated Committees Filter* (CVCF) [20] uses ensemble methods in order to preprocess the training set to identify and remove mislabeled instances in classification data sets. CVCF is mainly based on performing an Γ-FCV to split the full training data and on building classifiers using decision trees in each training subset. The authors of CVCF place special emphasis on using ensembles of decision trees such as C4.5. The basic steps of CVCF are the following:

- Split the training data set D_T into Γ equal sized subsets.
- For each of these Γ parts, C4.5 (as suggested by the authors) is trained on the other $\Gamma - 1$ parts. This results in Γ different classifiers.
- These Γ resulting classifiers are then used to tag each instance in the training set D_T as either correct or mislabeled, by comparing the training label with that assigned by the classifier.
- Add to D_N the noisy instances identified in D_T using a voting scheme (the majority scheme in our experimentation), taking into account the correctness of the labels obtained in the previous step by the Γ classifier built.
- Remove the noisy instances from the training set: $D_T \leftarrow D_T \setminus D_N$.

2.4 Iterative-Partitioning Filter

The *Iterative-Partitioning Filter* (IPF) [13] is a preprocessing technique based on the *Partitioning Filter* [24]. It is employed to identify and eliminate mislabeled instances in large data sets. Most noise filters assume that data sets are relatively small and capable of being learned after only one time, but this is not always true and partitioning procedures may be necessary.

IPF removes noisy instances in multiple iterations until a stopping criterion is reached. The iterative process stops if, for a number of consecutive iterations s, the number of identified noisy instances in each of these iterations is less than a percentage p of the size of the original training data set. Initially, we have a set of noisy instances $D_N = \emptyset$ and a set of good data $D_G = \emptyset$. The basic steps of each iteration are:

- Split the training data set D_T into Γ equal sized subsets.
- For each of these Γ parts, C4.5 is trained on this part as recommended by the authors. This results in Γ different trees.
- These Γ resulting classifiers, are then used to tag each instance in the training set D_T as either correct or mislabeled, by comparing the training label with that assigned by the classifier.
- Add to D_N the noisy instances identified in D_T using majority voting, taking into account the correctness of the labels obtained in the previous step by the Γ classifier built.
- Remove the noisy instances and the good data from the training set: $D_T \leftarrow D_T \setminus \{D_N \cup D_G\}$.

At the end of the iterative process, the filtered data is formed by the remaining instances of D_T and the good data of D_G; that is, $D_T \cup D_G$.

3 Experimental Analysis

First, this section describes the processes to induce class noise into original base data sets and the methodology for the analysis of the results (Sect. 3.1). Next, the results and their analysis are presented in Sect. 3.2.

3.1 Experimental Framework and Analysis Methodology

The experimentation is based on 14 real-world classification problems from the KEEL-data set repository[1] [1] shown in Table 1, with their characteristics. Their initial amount of class noise is unknown and thus no assumptions about the noise level can be made. In order to control the amount of noise in each data set, different class noise levels $x\%$ are introduced with the *uniform class noise* scheme [19]. Following this noise introduction procedure, $x\%$ of the examples are corrupted by randomly replacing its current class label for any other possible one, drawn from a discrete uniform distribution.

Table 1 Base data sets used to introduce noise, including the number of attributes and their type (real, integer or nominal), the number of examples and the number of classes for each one

Data set	# Attributes (R/I/N)	# Examples	# Classes
Balance	4 (4/0/0)	625	3
Banana	2 (2/0/0)	5300	2
Cleveland	13 (13/0/0)	303	5
Ecoli	7 (7/0/0)	336	8
German	20 (0/7/13)	1000	2
Heart	13 (1/12/0)	270	2
Ionosphere	33 (32/1/0)	351	2
Iris	4 (4/0/0)	150	3
Led7digit	7 (7/0/0)	500	10
Newthyroid	5 (4/1/0)	215	3
Phoneme	5 (5/0/0)	5404	2
Pima	8 (8/0/0)	768	2
Vehicle	18 (0/18/0)	846	4
Wine	13 (13/0/0)	178	3

The accuracy estimation of the classifiers in a data set is obtained by means of 3 runs of a stratified 5-fold cross-validation. 5 partitions are used because, if each partition has a large number of examples, the noise effects will be more notable,

[1] http://www.keel.es/datasets.php.

facilitating their analysis. New class noise data sets will be created from the aforementioned forty base data sets, considering the noise levels ranging from $x = 0\,\%$ (base data sets) to $x = 20\,\%$, by increments of 5 %.

In order to check the behavior of the different methods when dealing with class noise, the results of the SD algorithm are evaluated by using a modified quality measure of confidence, the fuzzy confidence [6]:

$$FCnf(R_i) = \frac{\sum_{E^k \in E / E^k \in TargetValue_k} APC(E^k, R_i)}{\sum_{E^k \in E} APC(E^k, R_i)} \tag{2}$$

where, *APC* (Antecedent Part Compatibility) is the degree of compatibility between an example and the antecedent component of a fuzzy rule, i.e., the degree of membership for the example to the fuzzy subspace delimited by the antecedent part of the rule. An example E^k verifies the *APC* of a rule if

$$APC(E^k, R_i) = T(\mu_{LL_1^1}(e_1^k), \ldots, \mu_{LL_{n_v}^{l_{n_v}}}(e_{n_v}^k)) > 0 \tag{3}$$

The FCnf results' analysis are supported by a Wilcoxon's statistical test [7]. This is a non-parametric pairwise test that aims to detect significant differences between two sample medians. For each level of class noise, the three filters (CVCF, EF and IPF) and the no filtering approach will be compared using Wilcoxon's test.

3.2 Noise Filtering for Subgroup Discovery

We want to analyze whether using a filtering approach is beneficial for the performance of FuGePSD, as a recent and representative SD algorithm. For this reason, we compare the capabilities of FuGePSD without filtering, as being a fuzzy approach to SD should make it more robust against noise, over the increasingly noisy versions of the base data sets. We call this absence of filtering "No filter" in Table 2, where the results of FuGePSD after preprocessing the noisy data sets with CVCF, EF and IPF are also shown. We also indicate the base case, in which no class noise is introduced in the data set, in order to study whether the introduction of noise causes a significant decrement in FCnf as the amount of noise increases.

As can be seen from Table 2, there is a change of which approach is the best as the amount of noise increases. While for 5 % EF seems to be the best approach, being the best for 7 data sets, IPF and CVCF become better options than EF as the noise increases. It is also worthy to note that no filtering can be a reasonable choice for lower amounts of noise, as FuGePSD is able to deal with these small amounts of noise thanks to its fuzzy nature. However, as the noise is too much to be ignored, filtering is mandatory.

We must mention that for some data sets, introducing noise improves the FCnf value when no filtering, due to a beneficial change in the class labels, but it is not the

Table 2 Fuzzy confidence FCnf results for class noise for FuGePSD

	0 %	5 %				10 %				15 %				20 %			
	All	No filter	CVCF	EF	IPF	No filter	CVCF	EF	IPF	No filter	CVCF	EF	IPF	No filter	CVCF	EF	IPF
Balance	0.62	**0.68**	0.49	0.53	0.47	**0.69**	0.45	0.49	0.49	**0.63**	0.46	0.44	0.45	**0.64**	0.44	0.43	0.41
Banana	0.59	**0.59**	0.58	0.57	0.58	**0.59**	0.58	0.57	0.58	**0.58**	0.57	0.56	0.57	**0.57**	**0.57**	**0.57**	**0.57**
Cleveland	0.54	0.52	**0.53**	0.50	0.52	**0.51**	0.49	0.46	0.46	0.45	0.48	0.46	**0.49**	0.46	**0.52**	0.48	0.51
Ecoli	0.50	**0.46**	0.10	0.00	0.11	0.44	**0.51**	0.00	0.11	0.43	**0.52**	0.00	0.00	0.42	**0.49**	0.00	0.29
German	0.73	0.71	0.74	**0.75**	0.74	0.69	**0.74**	0.73	0.73	0.69	**0.72**	**0.72**	**0.72**	0.68	0.69	**0.70**	**0.70**
Heart	0.71	0.70	0.75	0.73	**0.76**	0.69	0.70	0.70	**0.72**	0.68	0.69	0.69	**0.71**	0.69	**0.76**	0.74	0.75
Ionosphere	0.62	0.64	0.90	**0.91**	**0.91**	0.46	0.90	0.90	**0.91**	0.50	0.88	0.87	**0.89**	0.60	0.83	0.83	**0.84**
Iris	0.76	0.74	**0.83**	**0.83**	**0.83**	0.67	**0.85**	**0.85**	**0.85**	0.66	0.79	**0.80**	0.79	0.68	0.74	0.74	**0.75**
Led7digit	0.47	0.46	0.60	**0.62**	**0.62**	0.44	**0.58**	**0.58**	**0.58**	0.40	0.54	0.57	**0.58**	0.40	0.54	0.54	**0.57**
Newthyroid	0.67	0.64	**0.84**	**0.84**	**0.84**	0.66	**0.84**	**0.84**	**0.84**	0.61	0.78	**0.79**	0.78	0.62	**0.86**	**0.86**	**0.86**
Phoneme	0.80	**0.79**	0.74	0.73	0.74	**0.75**	0.71	0.69	0.71	**0.76**	0.70	0.70	0.71	**0.75**	0.70	0.69	0.70
Pima	0.76	0.74	**0.78**	**0.78**	**0.78**	0.72	**0.77**	0.76	0.75	0.73	**0.74**	**0.74**	**0.74**	0.70	**0.75**	0.74	0.73
Vehicle	0.51	0.50	**0.64**	**0.64**	0.63	0.49	0.63	0.62	**0.64**	0.47	**0.60**	0.58	0.59	0.46	**0.58**	0.57	0.56
Wine	0.86	0.80	0.90	**0.92**	0.91	0.77	0.90	**0.92**	0.90	0.72	0.81	**0.85**	0.81	0.73	0.80	**0.81**	0.80

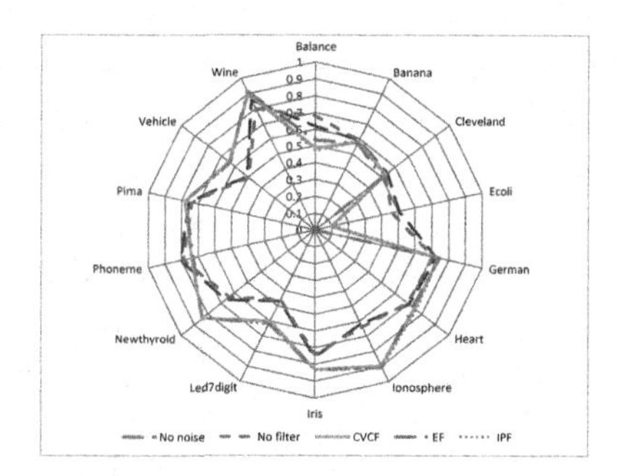

Fig. 1 FCnf of FuGePSD for 5 % class noise

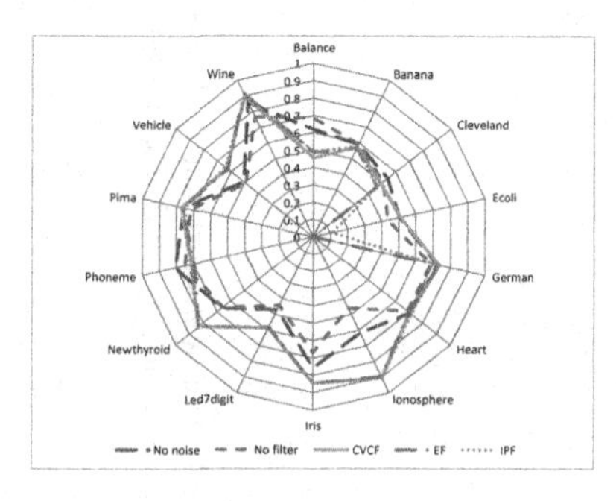

Fig. 2 FCnf of FuGePSD for 10 % class noise

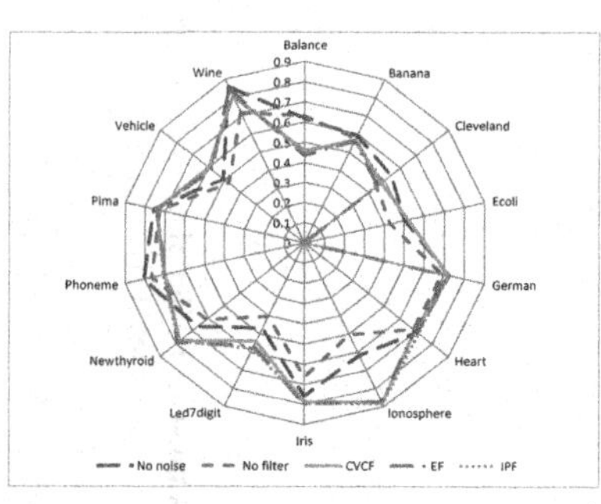

Fig. 3 FCnf of FuGePSD for 15 % class noise

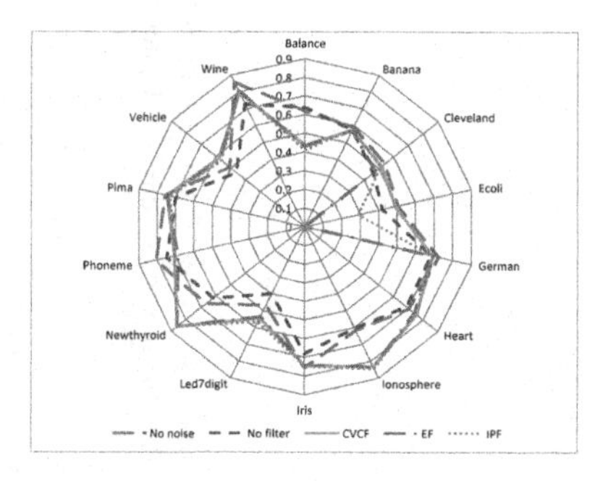

Fig. 4 FCnf of FuGePSD for 20 % class noise

Table 3 Summary of the Wilcoxon test for **class noise** of FuGePSD. • = the method in the row improves the method of the column. ∘ = the method in the column improves the method of the row. Upper diagonal of level significance $\alpha = 0.9$, Lower diagonal level of significance $\alpha = 0.95$

	5 %					10 %					15 %					20 %				
	No noise	No filter	CVCF	EF	IPF	No noise	No filter	CVCF	EF	IPF	No noise	No filter	CVCF	EF	IPF	No noise	No filter	CVCF	EF	
No noise	–	•				–	•				–	•				–	•			
No filter		–				∘	–	∘			∘	–	∘			∘	–	∘		
CVCF			–					–					–				•	–	•	
EF				–					–	∘				–	∘			∘	–	
IPF					–					–					–					–

common case as it only happens for the *balance*, *ecoli* and *phoneme* data sets. It is also important to note that the behavior of the filtering approaches vary depending on the data set. For better visualize this variability, we include the representation of the FCnf values in a radial depiction for each noise level. Figures 1, 2, 3 and 4 show the FCnf values for 5 10, 15 and 20 % of class noise respectively. From them it is easy to observe how *No filter* is similar to *No noise* with 5 % of noise, but becomes a bad option from 10 % onwards. We can also point out how some data sets benefit more from filtering, as *ionosphere*, *german*, *newthyroid*, *iris* and *heart*. For these data sets, the filtering even causes FuGePSD to work better than the *No noise* base case, as the filtering is able to clean the class boundaries and allowing FuGePSD to obtain a better description of the class clusters. For other data sets, as *balance* or *phoneme*, no filtering is appropriate for FuGePSD, causing a decrement in FCnf, but these are scarce cases compared to the total of data sets.

In Table 3 we include the summarized results of the Wilcoxon's Signed Rank test, in which we compare the different approaches for each noise level. Significative differences only appear with higher levels of noise. However, at all noise levels no filtering is worse than the 0 % noise level (*No noise*) but the filtering approaches do not show this difference. Thus, no filtering is significatively worse than the original, noise-induced free, case even at low amounts of noise. As a consequence, filtering in noisy frameworks for SD should be always taken into account. When the amount of noise is non-ignorable (from 10 % onwards), some filtering approaches are better suited for FuGePSD. As can be seen from Table 3, CVCF is specially indicated when the level of noise is high (15 and 20 %). As CVCF is never worse than the rest of filters, and it outperforms no filtering from 10 % onwards, it seems to be the best option for FuGePSD in this contribution and a promising choice for SD approaches.

4 Concluding Remarks

In this contribution, we have analyzed the influence of the class noise in the SD problem, that is a descriptive data mining technique using supervised learning. Specifically, the analysis has been performed with the FuGePSD algorithm which is the most recent EFS presented in the literature. It has an excellent behavior with data sets with continuous variables.

To do so, different class noise amounts (5, 10, 15 and 20 %) have been introduced in the original data sets. The study shows different situations with respect to the increment of noise. As FuGePSD is a fuzzy SD approach, is robust to uncertainty and noise to a certain extent, and thus no filtering can be a reasonable choice for lower amounts of noise. On the other hand, when the amount of noise increases, its treatment cannot be ignored and the application of specialized techniques as noise filters is beneficial. In the latter case, we remark that IPF and CVCF become better options when noise cannot be ignored. In addition, statistical tests support that CVCF seems to be the best option in this contribution and a promising choice for SD approaches.

This work opens future efforts in order to enlarge and observe the influence of the class and/or attribute noise in several SD proposals and so to obtain more comprehensible and complete analysis in this data mining field.

Acknowledgments Supported by the the Spanish Ministry of Economy and Competitiveness under projects TIN2012-33856 (FEDER Founds), the Spanish Ministry of Science and Technology under Projects TIN2011-28488 and TIN2010-15055, and also by the Regional Projects P10-TIC-6858 and P12-TIC-2958.

References

1. Alcalá-Fdez J, Fernández A, Luengo J, Derrac J, García S, Sánchez L, Herrera F (2011) KEEL data-mining software tool: Data set repository, integration of algorithms and experimental analysis framework. Journal of Multiple-Valued Logic and Soft Computing 17(2–3):255–287
2. Bonissone P, Cadenas JM, Carmen M (2010) Garrido, and R. Andrés Díaz-Valladares. A fuzzy random forest. International Journal of Approximate Reasoning 51(7):729–747
3. Brodley CE, Friedl MA (1999) Identifying Mislabeled Training Data. Journal of Artificial Intelligence Research 11:131–167
4. Carmona CJ, González P, del Jesus M, Herrera F (2014) Overview on evolutionary subgroup discovery: analysis of the suitability and potential of the search performed by evolutionary algorithms. WIREs Data Mining and Knowledge Discovery 4(2):87–103
5. Carmona CJ, Ruiz-Rodado V, del Jesus M, Weber A, Grootveld M, González P, Elizondo D (2015) A fuzzy genetic programming-based algorithm for subgroup discovery and the application to one problem of pathogenesis of acute sore throat conditions in humans. Information Sciences 298:180–197
6. del Jesus MJ, González P, Herrera F, Mesonero M (2007) Evolutionary Fuzzy Rule Induction Process for Subgroup Discovery: A case study in marketing. IEEE Transactions on Fuzzy Systems 15(4):578–592
7. Demšar J (2006) Statistical comparisons of classifiers over multiple data sets. Journal of Machine Learning Research 7:1–30
8. A. E. Eiben and J. E. Smith. Introduction to evolutionary computation. Springer, 2003
9. Gamberger D, Lavrac N (2002) Expert-Guided Subgroup Discovery: Methodology and Application. Journal Artificial Intelligence Research 17:501–527
10. García S, Luengo J, Herrera F (2015) Data Preprocessing in Data Mining. Springer Publishing Company, Incorporated
11. Herrera F (2008) Genetic fuzzy systems: taxomony, current research trends and prospects. Evolutionary Intelligence 1:27–46
12. Herrera F, Carmona CJ, González P, del Jesus MJ (2011) An overview on Subgroup Discovery: Foundations and Applications. Knowledge and Information Systems 29(3):495–525
13. Khoshgoftaar TM, Rebours P (2007) Improving software quality prediction by noise filtering techniques. Journal of Computer Science and Technology 22:387–396
14. W. Kloesgen. Explora: A Multipattern and Multistrategy Discovery Assistant. In Advances in Knowledge Discovery and Data Mining, pages 249–271. American Association for Artificial Intelligence, 1996
15. J. R. Koza. Genetic Programming: On the Programming of computers by Means of Natural Selection. MIT Press, 1992
16. Lavrac N, Cestnik B, Gamberger D, Flach PA (2004) Decision Support Through Subgroup Discovery: Three Case Studies and the Lessons Learned. Machine Learning 57(1–2):115–143
17. G. J. Mclachlan. Discriminant Analysis and Statistical Pattern Recognition (Wiley Series in Probability and Statistics). Wiley-Interscience, 2004

18. J. R. Quinlan. C4.5: programs for machine learning. Morgan Kaufmann Publishers, San Francisco, CA, USA, 1993
19. C.-M. Teng. Correcting Noisy Data. In Proceedings of the Sixteenth International Conference on Machine Learning, pages 239–248, San Francisco, CA, USA, 1999. Morgan Kaufmann Publishers
20. S. Verbaeten and A. V. Assche. Ensemble methods for noise elimination in classification problems. In Fourth International Workshop on Multiple Classifier Systems, pages 317–325. Springer, 2003
21. S. Wrobel. An Algorithm for Multi-relational Discovery of Subgroups. In Proceedings of the 1st European Symposium on Principles of Data Mining and Knowledge Discovery, volume 1263 of LNAI, pages 78–87. Springer, 1997
22. L. A. Zadeh. The concept of a linguistic variable and its applications to approximate reasoning. Parts I, II, III. Information Science, 8–9:199–249,301–357,43–80, 1975
23. Zhu X, Wu X (2004) Class Noise vs. Attribute Noise: A Quantitative Study. Artificial Intelligence Review 22:177–210
24. X. Zhu, X. Wu, and Q. Chen. Eliminating class noise in large datasets. In Proceeding of the Twentieth International Conference on Machine Learning, pages 920–927, 2003

SOCO15-SS04 - Soft Computing Methods in Manufacturing and Management Systems

Computation of Mechanical Properties of Parts Manufactured by Fused Deposition Modeling Using Finite Element Method

Filip Górski, Wiesław Kuczko, Radosław Wichniarek and Adam Hamrol

Abstract The paper focuses on a problem of evaluation of strength of products manufactured using Fused Deposition Modelling technology. The presented research was aimed at developing a method for computation of deformation and stresses in any product manufactured using FDM technology, with consideration of important, strength-affecting process parameters. Finite Element Method calculations were realized using digital CAD models of the products, obtained by simulation of FDM process in a virtual environment. These models represent the macrostructure of additively manufactured elements. Results of calculations for sample geometries were verified by experimental studies. The obtained results are mostly correct, changes in approach are needed to make the computations possible to be performed in shorter time and with better results.

Keywords Additive manufacturing · Finite element method · Fused deposition modelling

F. Górski (✉) · W. Kuczko · R. Wichniarek · A. Hamrol
Poznan University of Technology, Chair of Production Engineering and Management, Poznan, Poland
e-mail: filip.gorski@put.poznan.pl

W. Kuczko
e-mail: wieslaw.kuczko@put.poznan.pl

R. Wichniarek
e-mail: radoslaw.wichniarek@put.poznan.pl

A. Hamrol
e-mail: adam.hamrol@put.poznan.pl

Á. Herrero et al. (eds.), *10th International Conference on Soft Computing Models in Industrial and Environmental Applications*, Advances in Intelligent Systems and Computing 368, DOI 10.1007/978-3-319-19719-7_35

1 Introduction

Despite the fact that many companies, especially in the European Union and the USA, regularly make use of additive (layered) manufacturing technologies (also known as 3D printing) for prototyping of new products, it is still a group of technologies with capabilities unknown to most potential users and is rarely used in production of functional prototypes or final products. However, this state has been changing for a while now, thanks to many new, inexpensive solutions, recently brought to the market. One of the most widespread additive manufacturing technologies is Fused Deposition Modelling (FDM). The manufacturing in the FDM technology is carried out by extrusion of heated thermoplastic material (usually ABS or PLA) through the nozzle in the head, which moves in the XY plane. The material is deposed layer by layer – after one layer is finished, the table moves in Z direction and the process of material extrusion continues until the shape is finished. Information about geometry of consecutive layers is sent to the numerically controlled machine directly from the digital representation of the product, stored in the CAD model [1].

A final product manufactured using the Fused Deposition Modelling technology can be characterized some coefficients, which are influenced by many factors [2]. Unlike in most manufacturing technologies, values of parameters of the additive manufacturing process can be more significant than the properties of the part material – two different sets of process parameters applied to the same geometry can result in obtaining two products of entirely different properties. Each set of process parameters: orientation of the product in the working chamber, layer thickness and method of filling of the layer contour, will make the part structure look different, which will result in different values of coefficients such as strength, accuracy or surface quality.

Mutual relations between the FDM process parameters and properties of the obtained products are not fully discovered yet. Attempts at their experimental determination have been made [2, 4, 5], but obtaining full characteristics of these relations is still an open research problem. This paper describes an attempt at preparation of methodology of computation of the product strength, depending on orientation in the working chamber, by using a digital model of the product structure for Finite Element Method calculations.

2 Research Problem

The macrostructure of parts made by FDM technology consists of material threads deposed in alternate directions, creating layers bound together without material fusion (layer structure in the Fig. 1). It makes the manufactured elements behave in a specific way under any load. Even a simple load (single force or torque) or non-complex geometries will result in a complex stress state inside the element.

Fig. 1 View of a single layer of a sample manufactured by FDM technology – visible contour and filling threads (thread width approx. 0.5 mm, layer thickness 0.254 mm)

Mechanical properties of products manufactured using FDM technology are strictly related to parameters of the layer deposition process, particularly to the most important one – orientation in the working chamber. The base plane of the model table is a slicing plane – it defines how an element is divided into layers; consequently, different orientations will result in different layer arrangement. The product strength is the least beneficial in places where the material is joined – mostly at layer boundaries (the strength of bond between layers is several times lower than strength of the pure material [3, 4, 8]). In a single part, it results with anisotropic mechanical properties [6]. Manufacturing the same geometry with entirely different orientation will result in totally different values of, for example, tensile strength or impact strength – the differences can reach even a few hundred percent [3, 10–12]. The strength of any FDM part will always be lower than the strength of product of the same geometry, but with monolithic structure (e.g. obtained by injection molding) – this is related to volume errors in form of air gaps occurring inside the manufactured element [8].

A broad range of papers and dissertations devoted to determination of relations between various parameters of the FDM process and strength coefficients of obtained products in an experimental way can be found. The experiments described in these papers were usually carried out as typical strength tests (bending, impact, tensile and other tests, such as a compressive strength test [12, 14]), performed on samples of the same shape but with diversified internal structure (manufactured with variable process parameters and diversified materials, e.g. special purpose materials such as ULTEM [4]). Such studies are conducted by many research facilities all over the world [7, 9, 13]. As the relations are usually non-linear, such studies, however necessary, lead to conclusions applicable only to a narrow set of process conditions (specific material, machine, group of shapes, etc.).

Similarly as in the case of the laminar composites [15], building an analytical mathematical model allowing prediction of properties of the products, with consideration of various process parameters, is a very complex problem. So far, no universal theory has been built that would allow this for the products obtained by FDM technology. Existing mathematical models are based entirely on experimental data [12] or include only parameters of lower significance, such as type of internal filling pattern [7].

There are some research papers that present attempts at prediction of the mechanical properties of products manufactured using FDM technology with support of numerical calculation methods (namely Finite Element Method), taking the product orientation into consideration [16]. In the exemplary work [16], partial convergence of simulation results with the experimental test results was achieved, but in the case of some orientations, the error of calculation results is approximately 50 % (calculated deformation is underestimated). It was due to leaving the product macrostructure aside – the authors did not directly consider the layered structure. As such, this approach is valid only in certain situations.

Unsatisfactory results of previous attempts by other researchers [16, 17] and lack of any tools for FDM product strength prediction made the authors of this paper try to solve this problem. The authors' industrial experience shows that such a tool would be a valuable asset.

The presented work was aimed at development of a method of prediction (calculation) of the mechanical properties (by determination of stress and strain under any load) of any product manufactured by the FDM technology, taking into consideration the process parameters influencing the macrostructure. The process parameters influence was considered by using special, digital (CAD) models of products, obtained as a result of simulation of the layer deposition process in a virtual environment in the Finite Element Analysis calculations. Such models represent the structural features of the product manufactured using FDM technology. The computations performed using these models were verified by experimental strength tests (bending tests).

The studies presented in the paper are a part of a broader research, aimed at developing a semi-automatic algorithm for selection of the optimal values of parameters of the additive manufacturing processes, to meet the criteria defined by the product end user. The authors have already made an attempt at creating an expert algorithm for selection of the additive process for a certain task [18] and are aiming at expanding the methodology by using genetic algorithms for the process parameters optimization.

3 Research Plan

3.1 *Methodology of Building the Digital Thread Models and FEM Computations*

To realize the aim of the work, which is obtaining a method for strength calculations of products manufactured with FDM technology, a methodology of generating the CAD models resembling the real macrostructure of the products was prepared. These models are named digital thread models, as they consist of virtual threads of materials. The following requirements were set for the thread models to fulfill:

- fully accurate layer arrangement representation,
- the best possible representation of the internal layers filling,

- simplified representation of the thread (simplified shape – only straight lines),
- neglecting random, not repeatable manufacturing errors and repeatable manufacturing errors, which do not influence the strength in a significant way.

A basis for preparation of the digital thread model is a parametric, three-dimensional solid model of the part. The first stage is a preparation of tool paths in form of NC code, on the basis of the part geometry and selected process parameters, which are the part orientation, single layer thickness and method of filling the layer contour. This task is realized by the custom made software. It is used on top of the CATIA v5 CAD system and it allows division of a solid model into layers and it prepares trajectories of FDM head. The next stage is a simplified simulation of the layer deposition process in the virtual environment – for this purpose, a special Virtual Reality application was created. A result of the simulation is a cloud of points, which need to be connected into lines to re-create the lines along which the head moved during the material deposition. These points are saved to a text file containing their positions in a Cartesian coordinate system. The last stage is using the gathered data to build a parametric thread model in the CATIA v5 system.

Faithful modeling of the adhesive bonds between subsequent layers and threads would be a very labor-consuming task. Therefore, it was decided to use the simplified model (Fig. 2), for which the following assumptions have been made:

(1) The whole product is a single solid body – no adhesion is defined, the material between layers is joined together.
(2) There is no contact between threads of the internal filling (in reality, this contact is present and it strengthens the whole part) – there is a distance between threads.
(3) The thread shape is simplified – rounded edges are replaced with broken lines, to facilitate the discretization process.

The assumptions 1 and 2 balance each other, as 1 increases the product strength and 2 decreases it. The distance between threads of material of the internal filling is a crucial parameter here, along with the thread cross-section dimensions. Proper balancing required several trial-and-error attempts at finite element analysis, using various sets of parameters.

As a model situation for testing whether the finite element analysis is applicable in strength calculations of products manufactured with FDM technology, the ISO three-point bending test was selected (samples in form of rectangular beams of dimensions: 80 mm × 10 mm × 4 mm). This test was carried out on the real samples, to experimentally validate results of the calculations.

The digital thread models of samples for bending tests were prepared in three most commonly used orientations: the flat orientation (X and Y orientation equal to 0°), the side orientation (X = 90°, Y = 0°) and the vertical orientation (Y = 90°). For all samples, the same parameters were set while building the digital thread models and corrections to the thread models (aimed at obtaining greater convergence between experiments and calculations) were introduced simultaneously in all three samples in order to obtain a universal set of guidelines regarding the thread model preparation.

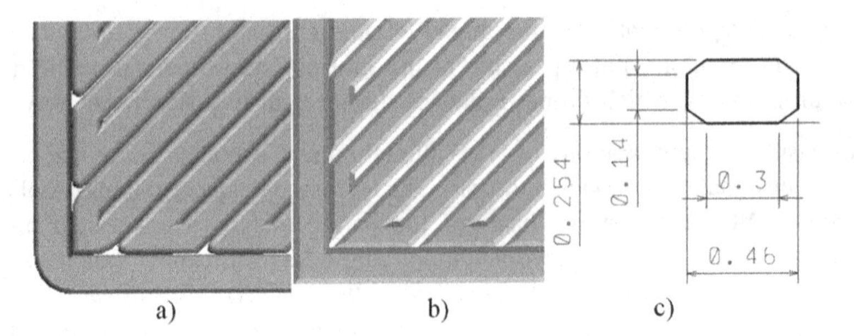

Fig. 2 Differences between a faithful (a) and a simplified (b) digital thread model, (c) dimensions of simplified cross-section of a single internal thread

COMSOL Multiphysics software was used for calculations. Properties of the ABS material used to manufacture samples for bending tests were determined during experimental studies conducted in the preliminary phase of research. A new model of material with the appropriate parameters (Young's modulus – 1.7 GPa, Poisson's ratio – 0.35, Density – 1050 kg/m^3) was created in the COMSOL environment:

To prepare a model of the static bending test, the following boundary conditions were used: supports (Prescribed Displacements) – two for each sample, with distance exactly the same as in the real bending test (64 mm), loads (Boundary Load) – applied bending force, values taken as averages from bending tests and symmetry – used in one case to reduce the amount of discrete elements.

Proper division of the thread model into finite elements is the most problematic stage of preparation of the model for calculations. The thread models were meshed using typical linear tetrahedral elements. Too dense mesh would take too long to calculate, so the task was to find the set of parameters that would give the smallest number of elements without any errors during the meshing process. This was done by trial and error. An example of generated mesh is shown in the Fig. 3.

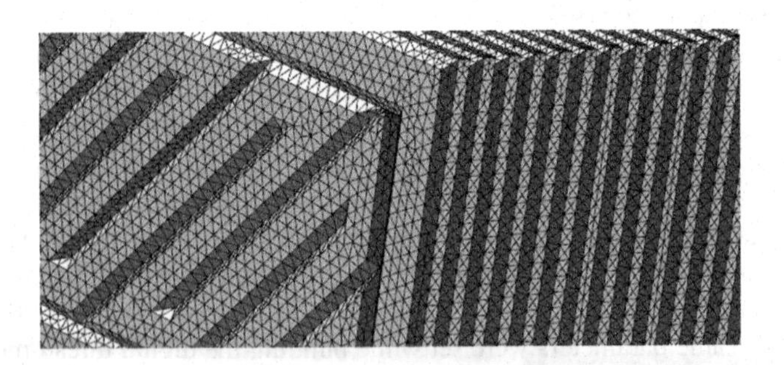

Fig. 3 Generated tetrahedral mesh for sample of vertical orientation

3.2 Methodology of the Bending Tests

The bending tests were conducted according to the ISO 178 standard on ABS samples manufactured using the Dimension BST 1200 machine. The test was performed as a three-point bending. Determination of the bending strength according to the ISO 178 standard is performed on the basis of bending the sample up to a conventional deflection, equal to 1.5 of the sample thickness (6 mm for used samples). If such a deflection is achieved during the test, the flexural stress at conventional deflection σ_{fC} is calculated. If a sample is destroyed before the conventional deflection can be achieved, the bending strength σ_{fM} is calculated instead, using exactly the same relation, but with different interpretation. In the described research, all the conditions described in the standard were maintained. Ten samples of each orientation were prepared for the studies, which gives a total number of 30 samples, divided in 3 series. A Zwick Roell Z020 machine was used for tests and the test speed was set at 2 mm/min.

4 Results and Discussion

4.1 Results of Bending Tests

Table 1 presents the results of the bending tests. An additional column was added with percentage of destroyed samples.

Table 1 Results of experimental bending tests

Sample	dL (mm)	F_{max} (N)	σ_{fM} or σ_{fC} (MPa)	% destroyed samples
Flat (no. 1)	6.0	58.5	35.1	0 %
Side (no. 2)	6.0	68.5	41.1	0 %
Vert. (no. 3)	2.7	30.2	18.1	100 %

The samples manufactured with vertical orientation were destroyed in a way unique for the products manufactured by the FDM technology – by layer separation. This effect is similar to the brittle fracture effect in rigid materials. In the performed studies, all the samples of types 1 and 2 (flat and side) achieved the conventional deflection of 6 mm, while all the samples of type 3 (vertical) were destroyed. This means the samples manufactured in vertical orientation need to be treated in a different way – calculated stress should be mostly above the strength of the ABS material, while in the samples of type 1 and 2, the stress should generally not exceed that strength, except for occasional peaks in weaker spots.

4.2 Results of FEM Computations

Results of computations of all the samples are presented in the Table 2. The table contains the following data:

- maximum force F_{max} – the average value of strength from the bending tests,
- deflection dL_e – mean deflection from the bending and dl_{calc} - computed deflection,
- comp. time – approx. time of calculations on the used DELL T7500 workstation,
- numerical results error – percentage deviation between calculated and experimentally obtained deflection, calculated as $(|dL_e\text{-}dL_{calc}|/dL_e)*100$ %,
- maximum von Mises stress – peak value of the internal stress calculated using the von Mises hypothesis.

Table 2 Summary of results of finite element analysis on digital thread models, compared with results of experiments

Sample	Parameter					
	F_{max} (N)	dL_e (mm)	dL_{calc} (mm)	Comp. time	Error	Max. stress (MPa)
Flat (no. 1)	58.5	6.0	6.47	~169 h	7.8 %	68
Side (no. 2)	68.5	6.0	5.62	~72 h	6.3 %	60
Vertical (no. 3)	30	2.7	2.79	~11.5 h	3.3 %	593

In the simulation no. 1, long time was a result of using of a non-default solver, as a default solver reported an error of lack of operating memory every time. Deviation from the 6 mm value achieved in the experiments was of negative character, which means that the predicted strength is lower than the actual strength, as opposed to the simulation no. 2. In the simulation no. 3, deviation from the 2.7 mm value achieved in the experiments was approx. 3 %, making it the sample with the lowest calculation error. Peak values of the stress were much higher than in the other samples (Fig. 4), so it can be stated that in the case of this sample, it would be possible to predict the sample destruction on the basis of the numerical simulation.

The assumption of correctness of the simulation results was verified statistically (using t-Student test), by a hypothesis that percentage deviation between calculated and experimentally determined part stiffness (F_{max}/dL ratio) is lower than 10 % every time.

Fig. 4 An exemplary distribution of stress after FEM calculations – simulation no. 3

5 Conclusions and Summary

The deflection most close to reality was achieved in the simulation no. 3. This is also a model, which did not require preparation of any auxiliary geometry in the course of the FEA procedure, which might have influenced the calculation results in the simulations 1 and 2. Deviation between experiments is visibly positive for sample 2 and negative for sample 1. At this stage, though, there can be no clear guidelines formulated regarding improvement of the geometry of digital thread models, other than reviewing the applied model simplifications (Fig. 2) and their parameters. Still, the obtained error levels are considerably low and acceptable at this level of the computation methodology development.

Predicted internal stress values are inconclusive. Simulation no. 3 would allow prediction of the destruction of the sample under load, as calculated stress values locally greatly exceed the tensile and bending strength of the ABS material. In the other samples, the stress locally equals the bending strength and exceeds it in just a minor part of the nodes, which can also be recognized as a generally good prediction. Still, it cannot be said that the obtained stresses are realistic, as the value of 593 MPa is far beyond destruction point of any thermoplastic material. This value only showed up in several nodes (per almost 9 million of generated tetrahedrons), so it could be omitted, but such an unrealistic value should not be present at all. Anyway, direct comparison of failure stress in any orientation to the yield strength is not meaningful, as the strength of the FDM parts is probably driven by fracture mechanics (due to their structure) rather than generalized yielding and plasticity, so change of approach would be needed to obtain more satisfying results regarding stresses.

Time of calculations of all the samples is unacceptably long, considering that a top-class workstation was used and the geometries were simple. At this stage of development, it would be difficult to conduct calculations on products with greater dimensions or complexity because of a very high amount of generated finite elements. In the used FEM software, it is difficult to optimize the number of generated elements, as density of the finite element mesh for the digital thread models prepared according to the developed methodology must be kept inside a strictly defined interval. Using values of parameters out of a certain range will result in errors during meshing or too dense mesh. Still, the error during the meshing might be because of the incapability of the computing ability, so it might not be impossible to perform the simulation in a reasonable time. The same conclusion applies for the total run time of simulation.

In general, results of the strength calculations in the COMSOL environment can be evaluated as satisfactory – implemented method of using digital models with macrostructure resembling the layered and threaded structure of elements manufactured by FDM technology is effective, universal (can be – in theory – applied to any geometry) and allows all process parameters to be taken into account. In comparison with other attempts, which can be found in the literature [16], where deviation reached 50 %, the developed method is a huge step forward. It was possible only by using the models with macrostructure resembling the structure of a real product.

The greatest flaw of the proposed calculation method are its high requirements regarding the computer performance. It is the most important aspect to improve if the method is going to be further developed for industrial applications. A set of dedicated computational software tools would probably need to be created, especially for the models described in the paper. The future steps in methodology development will be adjustment of the digital model generation parameters, further computations using various geometries and variable parameters other than orientation, in order to generalize and improve the obtained results.

The methodology – when fully expanded and optimized – will be also used as a core of a soft-computing based algorithm of optimization of FDM part orientation on the basis of the strength criterion. Algorithm of part optimization will be based on a genetic algorithm. It will be implemented as a dedicated software – it is still being developed. An initial population of individuals will be created partially on the basis of the most used orientations, partially by some indirect criteria and partially randomly. The genotype will be composed out of two angles – X and Y orientation values of the FDM-made part. For each individual, an adjustment function value will be calculated, using manufacturing time and orientation of critical areas of the part. After selection of the best individuals, the next generation will be created through genetic operators of crossover and mutation. After a certain number of generations, 3–5 best individuals will be conducted to full modeling and calculations, according to the FEA-based methodology described in the paper. The genetic algorithm and its results will be described in further works by the authors.

Acknowledgments The studies described in the paper were a part of a grant financed by the Polish National Science Centre (agreement no. UMO-2011/01/N/ST8/07603).

References

1. Chua CK, Leong KF, Lim CS (2010) Rapid prototyping: principles and applications. World Scientific Publishing Co. Pte. Ltd., Singapore, pp 25–35. doi: 10.1142/9789812778994_0002
2. Bellini A, Guceri S (2003) Mechanical characterization of parts fabricated using fused deposition modeling. Rapid Prototyping J 9:252–264. doi:10.1108/13552540310489631
3. Górski F, Kuczko W, Wichniarek R (2013) Influence of process parameters on dimensional accuracy of parts manufactured using Fused deposition modelling technology. Adv Sci Technol Res J 19(7):27–35. doi: 10.5604/20804075.1062340
4. Bagsik A, Schöppner V (2011) Mechanical properties of fused deposition modeling parts manufactured with ULTEM*9085". In: Proceedings of ANTEC 2011, Boston
5. Ahn SH, Montero M, Odell D et al (2001) Material characterization of fused deposition modeling (FDM) ABS by designed experiments. In: Proceedings of rapid prototyping and manufacturing conference, SME 2001
6. Ahn SH, Montero M, Odell D et al (2002) Anisotropic material properties of fused deposition modeling (FDM) ABS. Rapid Prototyping J 8(4):248–257. doi:10.1108/13552540210441166
7. Ahn SH, Baek C, Lee S et al (2003) Anisotropic tensile failure model of rapid prototyping parts—fused deposition modeling (FDM). Int J Modern Phys B 17(8–9). doi: 10.1142/S0217979203019241
8. Górski F, Wichniarek R, Andrzejewski J (2012) Influence of part orientation on strength of ABS models manufactured using fused deposition modeling technology. Polym Process 9:428–435
9. Giannatsis J, Sofos K, Canellidis V et al (2011) Investigating the influence of build parameters on the mechanical properties of FDM parts. In: Proceedings of the 5th international conference on advanced research in virtual and rapid prototyping, Leiria, Portugal, 28.09–1.10 2011, pp 525–529. doi: 10.1201/b11341-85
10. Sun Q, Rizvi GM, Bellehumeur CT et al (2008) Effect of processing conditions on the bonding quality of FDM polymer filaments. Rapid Prototyping J 14(2):72–80. doi: 10.1108/13552540810862028
11. Es-Said OS, Foyosa J, Noorania R et al (2000) Effect of layer orientation on mechanical properties of rapid prototyped samples. Mater Manuf Process 15. 10.1080/104269100 08912976
12. Sood AK, Ohdar RK, Mahapatra SS (2012) Experimental investigation and empirical modeling of FDM process for compressive strength improvement. J Adv Res 3(1):81–90. doi:10.1016/j.jare.2011.05.001
13. Pandey PM, Reddy NV, Dhande SG (2006) Part deposition orientation studies in layered manufacturing. J Mater Process Technol 185(1–3):125–131. doi:10.1016/j.jmatprotec.2006. 03.120
14. Lee CS, Kim SG, Kim HJ et al (2007) Measurement of anisotropic compressive strength of rapid prototyping parts. J Mater Process Technol 187–188:627–630. doi:10.1016/j.jmatprotec. 2006.11.095
15. Hyer MW (2009) Stress analysis of fiber-reinforced composite materials. DEStech Publications, Inc
16. Hambali RH, Celik HK, Smith PC et al (2010) Effect of build orientation on FDM parts: a case study for validation of deformation behaviour by FEA. IN: Proceedings of iDECON 2010—international conference on design and concurrent engineering, Universiti Teknikal Malaysia Melaka, Melaka 2010, pp 224–228
17. Gajdos I, Slota J, Spisak E (2008) Visualisation of FDM prototypes. In: International conference of additive technologies—iCAT 2008, 17-17.08.2008
18. Górski F, Kuczko W, Wichniarek R et al (2010) Choosing optimal rapid manufacturing process for thin-walled products using expert algorithm. J Ind Eng Manage 10/2010

New Method for Assessment of Raters Agreement Based on Fuzzy Similarity

Magdalena Diering, Krzysztof Dyczkowski and Adam Hamrol

Abstract In measurement systems one of the components affecting its variation is a human factor. Man - as a process operator - measures or rates the product. Since his decisions may have significant impact on the customer satisfaction, his reliability and usefulness for the measuring tasks should be evaluated. This article describes authors proposal for a new measurement system analysis methodology for unmeasurable features. In this method many features of the product are rated during one study, and the value of them can be expressed in a nominal or an ordinal (with imprecise data) measurement scale, and each of the features can be weighted (as less or more important from the customer point of view). The goal of the methodology is to gain information about raters' ability to define the value of each feature in relation to the specification and customer requirements. The authors propose to use a similarity measure for rates assignment of product features values. The method is based on the new fuzzy similarity coefficient *SC*, and the level of agreement of the final decisions of the raters is analyzed based on Gwet's AC_1 coefficient.

Keywords Measurement system analysis (MSA) · Level of agreement · Unmeasurable characteristic · Similarity coefficient · AC_1 coefficient · Fuzzy sets · Similarity measure · Cardinality of fuzzy sets

1 Introduction

Making a decision by operators about the compatibility of manufactured products with customer requirements is executing a primary function of industrial measurement systems for unmeasurable characteristics. Depending on the rank of the fea-

M. Diering (✉) · A. Hamrol
Faculty of Mechanical Engineering and Management, Poznań University of Technology, Pl. M. Sklodowskiej-Curie 5, 60-965 Poznań, Poland
e-mail: magdalena.diering@put.poznan.pl

K. Dyczkowski
Department of Imprecise Information Processing Methods, Faculty of Mathematics and Computer Science, Adam Mickiewicz University, Umultowska 87, 61-614 Poznań, Poland
e-mail: chris@amu.edu.pl

Á. Herrero et al. (eds.), *10th International Conference on Soft Computing Models in Industrial and Environmental Applications*, Advances in Intelligent Systems and Computing 368, DOI 10.1007/978-3-319-19719-7_36

ture, the decisions may be more or less important and have significant impact on the effects. Since each operator's decision pulls behind a particular benefit or loss to the organization, understanding the variation resulting from the interaction of the measurement system components and the evaluation of its usefulness are fundamental in solving the basic problems in the production process [1]. Hence, the industrial measurement systems should be analyzed in order to confirm their adequacy for the measuring tasks and reliability of the data obtained from measuring process.

In a study of measurement systems there are applicable statistical methods and mathematical models. In every case, they should be accordingly selected, i.e. with taking into account the specificity of the manufacturing process, and also the measurement techniques and technologies. Analysis of the measurement system should - in the authors opinion - give an answer to at least some of the questions:

- What is the level of the type I error of each rater (as the rater the authors define a man who measures or rates the product or its features, that is manufacturing process operator, quality controller or expert; type I error means that conforming part is consistent wrongly rejected)?
- What is the level of the type II error of each rater (type II error - nonconforming part wrongly accepted)?
- What is the level of "contradictory" rates/decisions or "ambiguous" rates/decisions for the each rater?
- What is the level of agreement of each individual rater with an expert (rater-expert reliability)?
- What is the level of agreement between operators (inter-rater reliability)?
- What is the internal agreement of each rater (taking into account the "accidental" rates which are called chance agreement) that is the repeatability of raters assessments (internal/within-rater reliability)?
- Which of the features turned out to be the most difficult to rate to?

On this basis, corrective, preventive and improving actions should be taken. Next, as a further step, the more important questions should be asked to quality managers [2]:

- How are we doing - are we getting better or worse?
- Did the changes we made improve measurement system reliability?
- What's not working as it should?
- Which raters (operators and/or experts) need some help?
- Where are the largest opportunities for improving measuring process?
- What are the bottlenecks or limiting factors in our measurement system?

2 Research Problem

In practice, in the case of measurement systems for unmeasurable characteristics, the best known are: the cross-tab method (based on Cohen's Kappa coefficient), and effectiveness study. Usually the companies operate methodology recommended by the AIAG group [3]. The main limitation of these methods is that they are presented as applicable to be used only for assessing decisions about the characteristics of the nominal scale and precise data. The level of agreement of the final decisions of operators is analyzed without analyzing the similarities of their assessments in relation to the each features of the product. In addition, cross-tables do not include the level of internal agreement of the each raters (repeatability), and effectiveness study ignores the so-called chance of agreement [4].

Meanwhile, in assessing unmeasurable characteristics of the product in practice various measuring scales are used (depending on the situation, the nature of the manufacturing process and customer acceptance criteria). Besides, in decision-making about the state of the product operators rates and assess many of its features, and raters information about their assessment can be soft (incomplete, imprecise or uncertain). For example, the raters opinion during visual inspection of the color of the produced items may be formulated (depending on the specifications of the product) as follow:

(a) The color is "correct" or "incorrect", that is assessment with a dichotomous nominal scale. In mathematical models one of the categories is assigned a value of 1 (1 - meets the requirements, conforming part), and second-value of 0 (0 - does not meet the requirements, nonconforming).
(b) The color is "very dark", "dark", "light", "very light", that is the assessment with an ordinal scale (in the case of the color most often it is assessed by comparison with the color of a master (standard) part). In mathematical models categories are graduated and linguistic categories are described by membership function values of fuzzy sets i.e. values from interval [0, 1], for example: "very dark" - 0.9, "dark" - 0.7, "light" - 0.4, and "very light" - 0.1.

The situation may be even more complicated if, for example, the color of the front side of the product have a greater importance to the client (if it affects the aesthetic qualities of the final product) than the color of the back side of the same element (which may be negligible upon receipt of goods). Importance (weights, priorities) of the different products features affect the way of evaluation by the operators. The weights of the features - according to the authors - should also be taken into account in assessing the reliability of the measurement system, and therefore in the mathematical model as well. Both the cross-tab method and the effectiveness study in many cases do not allow to reflect production conditions to evaluate measurement system in its natural environment. This may have a huge impact on the quality and the effect of the whole study. What is more, changed conditions of the study (in comparison to production conditions) can cause abnormal raters behavior during the study, and the wrong decisions of operators, wrong decisions of executive engineer (expert), discouraging raters to study and others. In summary, these methods do not cover the demand process engineers in the analysis and assessment of the reliability of the systems for unmeasurable features. However, statistical engineering and mathematical sciences provide many different models and approaches to assess the level of rater's agreement with reference value (in practice - with experts decision), but - because of their various limitations - only few of them have been successfully implemented for engineering applications. They are used in various analysis by analysts and statisticians, but are not in common use as part of measurement system study. Hence, the authors undertook research the aim of which is to develop not only models, but also methodologies for the analysis of industrial measurement systems for unmeasurable characteristics, taking into account the different production conditions. This article describes a proposal for a measurement system assessment methodology in which many features of the product are rated, and the value of these features can be expressed in nominal or ordinal scale, and each of the features can be weighted (as less or more important from the customer point of view). The goal of the authors is to "build" a useful tool for engineers. Thus, the methodology will be validated with practitioners and quality engineers.

3 Fuzzy Method for Measure Raters Agreement

3.1 A Characteristics of the Approach

For industrial measurement system analysis for unmeasurable features the authors propose an original methodology. The goal of the methodology is to take into account not only the final decision about the quality of measured product, but also to gain the information about raters ability to define the value of each feature according to the experts requirements. The authors propose to use a similarity measure for rates assignment of product features values. The new method is based on the similarity coefficient *SC* (which is described in details in Sect. 3.2 of the paper). The novel authors' methodology allows to rate many features of the product, and the value of these features can be expressed in nominal or ordinal scale (optional for each feature). The authors' model is applicable to the cases: multiple raters (at least one), multiple unmeasurable features (at least one) in the nominal scale or ordinal measurement scale (and it is possible to express the value of the feature using imprecise data), any number of parts (in practice it is accepted that it is at least 30 elements), multiple series of ratings (1–3). Thus, the study scheme (defined as the selection of the elements participating in the study) is flexible and better - in comparison with commonly used in the practice the cross-tab method - to reflect the conditions in which every day measurements are taken and thus, to obtain accurate information about the reliability of the measurement system used. Besides, the framework of the proposed methodology includes assessing the level of internal agreement of each raters, that is their ability to repeat the same ratings in the following series of assessments (because it would be questionable to analyze the level of agreement between raters if one may not agree with himself). As in previous studies that were not taken into account, in the context of engineering applications, this approach is novel and unique. The characteristics of the model proposed by the authors is described by comparison to other approaches (Table 1). In this comparison the most commonly are presented: cross-tab method, the effectiveness study (often used in parallel with

Table 1 Methods for measure raters agreement - comparison of chosen approaches

No.	Comparison criteria	Effectiveness study [3]	Cross-tab method [3, 6]	The authors method based on K_{win} coefficient [4, 5]	The new authors method based on *SC* coefficient
1	Number of raters	2–3	2–3	Any number (no limit)	Any number (no limit)
2	Number of series (trials) for each operator	2–3	2–3	1	2–3
3	Feature's measurement scale	Nominal dichotomous	Nominal dichotomous	Nominal	Nominal, Ordinal
4	Number of categories (possible values) of single feature assessment	2	2	Any number (no limit)	Any number (no limit)
5	Number of categories (possible values) of single decision (final decision about the product)	2 (identical to the categories of single feature assessment)	2 (identical to the categories of single feature assessment)	Any number (no limit, but identical to the categories of single feature assessment)	Any number (no limit; may be different than the categories of single feature assessment)
6	Number of features in the study	1 (assessment of a single feature or decision about the final product)	1 (assessment of a single feature or decision about the final product)	1 (assessment of a single feature or decision about the final product)	Any number (possible to assess the many features of the product and the decisions about the final product)
7	Nature of the data	Precise (crisp)	Precise (crisp)	Precise (crisp), but taking into account the lack of data (incomplete/missing data is possible)	Precise (crisp) (for nominal scale) and/or imprecise (soft) (for ordinal scale)
8	Acceptance criteria for the system	% Effectiveness, Miss Rate, False Alarm Rate	$Kappa \in [-1, 1]$ (in practice $\in [0, 1]$)	$K_{win} \in [0, 1]$	$SC \in [0, 1]$ and $K_{win} \in [0, 1]$
9	Taking into account the importance of assessed features (the ability to give weights, priorities)	Not applicable	Not applicable	Not applicable	Yes
10	Taking into account the level of internal agreement of the raters	Yes (but without taking into account chance-agreement)	Not applicable (in practice it is recommended to provide cross-tab method parallel with the effectiveness study)	Yes	Yes

the cross tables), another author's method based on the basic Gwet'a agreement coefficient AC_1 (details of which are discussed in the previous authors work [5]) and the new authors method with similarity coefficient SC. Besides the comparison of the number of raters, or the number of trials (series) of the assessment, the table includes several other criteria, among other: the number of assessed features of the product, a measurement scale adopted for each feature individually, importance of each features (weights, priorities), the level of raters internal agreement and others.

3.2 The Model

In our model we use fuzzy sets concept to represent ratings. Fuzzy sets theory was introduced by Zadeh in 1965 [7] and since then, has been widely used in many fields of research. There are a lot of literature on theory and application of fuzzy sets (eg. [8–10]). More information on cardinality and similarity of fuzzy sets used in this paper one can find in [11–13].

Formally as an input data for the problem analysis is as follows:

- set of raters: $R = \{r_1, \ldots, r_l\}$, where l is a number of raters
- set of rated products (or parts): $P = \{p_1, \ldots, p_k\}$, where k is a number of rated products
- set of rated features (or attributes) $A = \{a_1, \ldots, a_m\}$, where m is a number of rated products features
- set of series of ratings $S = \{s_1, \ldots, s_n\}$, where n is a number of series of ratings
- set of features weights (which represents importance of each feature): $W = \{w_1, \ldots, w_m\}$

Each rater from set R makes an assessment of each product on each attribute in n series. The expert does the same in one series i.e. rates all features from set A for all products from set P.

The rating of k-th product p_k done by expert is represented by fuzzy set $\mathcal{E}_{pk}$:

$$\mathcal{E}_{pk} = oe_1/a_1 + oe_2/a_2 + \ldots + oe_m/a_m,$$

where $oe_i \in [0, 1]$ is rating of expert for i-th attribute.

The rating of k-th product p_k done by r-th rater (taking into account n series of ratings) is represented by fuzzy set $\mathcal{R}_{r,pk}$:

$$\mathcal{R}_{r,pk} = \sum_{a_i \in A \times s_j \in S} or_{ij}/(a_i, s_j),$$

where $or_{ij} \in [0, 1]$ is a rating of r-th rater for i-th attribute in j-th series.

A similarity Sim_p between expert and rth rater of rating of p_k-th product is defined by formula:

$$Sim_p(\mathcal{E}_{p_k}, \mathcal{R}_{r,p_k}) = 1 - \sum_{j=1}^{n} \frac{\sum_{i=1}^{m}(w_i|oe_i - or_{ij}|)}{n\sum_{i=1}^{m} w_i}. \quad (1)$$

We compute the similarity Sim_p for each product and receive as a result fuzzy set C_r which represents agreement between expert and r-th rater on all products:

$$C_r = Sim_p(\mathcal{E}_1, \mathcal{R}_{r,1})/p_1 + Sim_p(\mathcal{E}_2, \mathcal{R}_{r,2})/p_2 + \ldots + Sim_p(\mathcal{E}_k, \mathcal{R}_{r,k})/p_k.$$

Fuzzy set C_r represents information how much expert agreed with given rater on every product.

As a measure of agreement between expert and given rater we take cardinality $card(C_r)$ of fuzzy set C_r with weighting function f_t:

$$card_t(C_r) = \sum_{i=1}^{k} f_t(C_r(p_i)). \quad (2)$$

For more information on theory of fuzzy sets cardinalities and construction of weighting functions see [10, 11, 13].

The role of weighting function $f_t : [0, 1] \rightarrow [0, 1]$ is to decide with given threshold point $t \in [0, 1]$ when to treat agreement level as sufficient to be counted. The authors propose to use the following weighting function:

$$f_t(x) = \begin{cases} x & \text{if } x \geq t, \\ 0 & \text{otherwise.} \end{cases} \quad (3)$$

We can interpret $card(C_r)$ as a number of products rated similar between expert and given rater (i.e. cardinality of fuzzy set of expert and raters similarities).

Finally we get rating agreement coefficient SC_r for each rater:

$$SC_r = \frac{card(C_r)}{k}. \quad (4)$$

Iteratively we compute rating agreement coefficients for all raters.

Based on Table 1 and on the basis of formulas 1, 2, 3, 4, it can be concluded that the model proposed by the authors allows - as the only from the well-known in the literature approaches - to analyze the ratings of the many features (attributes) of the product in one study. The model takes into account the weight of each characteristics and its nature (the model allows to measure the level of agreement with imprecise data, also includes a measuring scale individually adopted for each feature).

3.3 The Methodology

For the model described in Sects. 3.1 and 3.2, a novel authors' methodology is proposed (in order to engineering applications):

1. To undertake a study at least 30 parts (products, subjects) should be prepared (in the literature can be found different recommendations, usually 30–50 parts). About 50 % of the subjects should be out of the specification and at least 50 % of all subjects should be taken from the border of the specification (from the "grey area"), i.e. should be difficult (hard) to clearly assess. In the study should be parts which - in experts opinion - can be easy to rate (in analogy to measurable characteristics - parts from the middle of tolerance), and those that will be difficult in the assessment (parts close to the limits of acceptance). Analysis of the results of those difficult points to opportunities for improvement and the analysis of these easy - for necessary of corrective and remedial actions in the area of measurement system.
2. Prior to the study, make sure that the executive expert has his within-rater reliability on the accepted level (to make sure that experts decisions are reliable source of information). Expert rates are treated as reference values and master to the operators ratings. Meanwhile, in practice it turns out that the expert also can make a mistake (and sometimes he does!) in the evaluation of the products. Experts internal-rater reliability can be estimated, for example, based on the coefficient AC_1, but with the note that instead of number of raters there is a number of series of the same operator ratings [4, 5].
3. As with all known methods of measurement systems analysis, including the methods proposed by the authors, each subject should be marked and easy to identify by executive (but not by the raters!) - conditions of independence results should be assured.
4. The study should be carried out in the place in which the measurement system is used in practice, that is with the standard (daily) environment and measurement conditions.
5. During the study, the executive person should write operators results in the data collection sheet (Table 2). Operators rate each characteristic of each object using nominal or ordinal measurement scales, and then make a decision about the product, classifying it into one of at least two categories.

Table 2 A worksheet (with sample data) for measurement system study (for unmeasurable characteristics), according to the model described in section 3.2 (r_i - rater, e - expert, a_i - feature, w_i - fetures weight, s_i - series, p_i - part)

	r_1									r_2									r_3									e		
	a_1			a_2			a_3			a_1			a_2			a_3			a_1			a_2			a_3			w_1	w_2	w_3
	w_1		1	w_2		1	w_3		0.5	w_1		1	w_2		1	w_3		0.5	w_1		1	w_2		1	w_3		0.5	1	1	0.5
	s_1	s_2	s_3	s_1	s_2	s_3	s_1	s_2	s_3	s_1	s_2	s_3	s_1	s_2	s_3	s_1	s_2	s_3	s_1	s_2	s_3	s_1	s_2	s_3	s_1	s_2	s_3	a_1	a_2	a_3
p_1	0.4	0.5	0.4	0.4	0.3	0.4	0.0	0.0	0.0	0.6	0.7	0.7	0.3	0.3	0.4	1.0	1.0	0.0	0.7	0.7	0.7	0.4	0.4	0.4	0.0	0.0	1.0	0.7	0.4	1.0
p_2	0.6	0.6	0.6	0.4	0.5	0.4	0.0	0.0	0.0	0.6	0.6	0.6	0.4	0.2	0.4	0.0	0.0	0.0	0.5	0.6	0.5	0.4	0.4	0.4	0.0	0.0	0.0	0.6	0.4	0.0
p_3	0.6	0.7	0.6	0.5	0.5	0.5	0.0	0.0	0.0	0.8	0.8	0.8	0.6	0.5	0.7	0.0	1.0	1.0	0.8	0.7	0.7	0.6	0.6	0.6	1.0	1.0	1.0	0.8	0.5	1.0
p_4	0.7	0.8	0.8	0.5	0.5	0.5	0.0	0.0	0.0	0.6	0.6	0.7	0.4	0.3	0.3	0.0	0.0	0.0	0.7	0.7	0.7	0.5	0.5	0.5	0.0	0.0	0.0	0.7	0.5	0.0

Table 3 Level of raters agreement - acceptance criteria

Benchmark scale	Level of agreement
$0 \leq SC_r < 0.50$	Poor
$0.50 \leq SC_r \leq 0.80$	Good
$0.80 < SC_r \leq 1$	Very good

6. The completed worksheet should be computed according to the model described in Sect. 3.2.
7. Evaluation of the level of raters agreement should be based on the acceptance criteria. Benchmark scale is shown in Table 3.
8. After analyzing the similarities in all parts, the measurement system reliability study for the raters decisions about the product can be performed. Here, the applicable is a method proposed in [5], based on the coefficient AC_1 [4].

If evaluation of the raters ability to assign the proper value to the characteristics (compared to the customer's specification or expert) is at acceptable level (SC is greater than 0.5), then the assessment of their level of decision agreement with the expert may be studied. For this purpose, the value of K_{win} (cf. [5]) is applicable:

$$K_{win} = \frac{\sum_{i=1}^{l} AC_1 K_w}{\sum_{i=1}^{l} K_w}, \tag{5}$$

where:

- K_{win} - aggregated measure of the level of internal raters agreement (within-rater reliability) and the level of agreement between the raters (inter-rater reliability),
- AC_1 - Gwet's AC_1 coefficient (cf. [4]),
- K_w - Kappa type coefficient, which measures level of internal agreement of each rater (within-rater reliability); to calculate K_w the AC_1 is used, where the number of raters replaces the number of repeated series of ratings by the same rater.

Interpretation of results should allow to answer questions posed in Sect. 2.

3.4 An Example

To show the procedure of computation as an input we take sample data form Table 2. In the real case, in accordance to the methodology, the number of elements would be much greater. Computations will be done in three steps:

1. In first step we compute similarity Sim_p resulting fuzzy sets of similarities for each rater:

$$C_1 = 0.68/p_1 + 0.99/p_2 + 0.73/p_3 + 0.97/p_4,$$

$$C_2 = 0.68/p_1 + 0.97/p_2 + 0.89/p_3 + 0.91/p_4,$$

$$C_3 = 0.87/p_1 + 0.97/p_2 + 0.93/p_3 + 1/p_4.$$

2. Next we compute cardinalities of fuzzy sets C_r with threshold $t = 0.7$:

$$card_t(C_1) = 2.69,\ card_t(C_2) = 3.67,\ card_t(C_3) = 3.77.$$

3. Finally we get similarity coefficient SC_r for all raters:

$$SC_1 = 0.67,\ SC_2 = 0.92,\ SC_3 = 0.94.$$

The level of agreement in assessing each attribute (feature) of the product for each of the raters with the expert is at the acceptable level. It means that raters knowledge (know-how) about features and the product (about its variation) is reliable source of information. Thus, the next step of the measurement system analysis may be assessment of the level of agreement of raters decision with the expert. An example of this computation one can find in [5].

4 Conclusions

Proposed novel model is the basis for the authors to build an expert assessments supporting system based on fuzzy models. In summary, the result of the authors work is presented in the article model which is an original, flexible solution that can be used in the evaluation of any industrial measurement system for unmeasurable characteristics (for the characteristics of the nominal or ordinal measuring scales and imprecise data). In order to engineering applications, for the described model the authors' methodology is proposed.

References

1. Hamrol A (2008) Quality management with examples (in Polish). Polish Scientific Publishers PWN
2. Kaydos W (1998) Operational performance measurement: increasing total productivity. CRC Press, Boca Raton
3. AIAG Work Group (2010) Measurement systems analysis, 4th edn. Reference manual, AIAG work group, Daimler Chrysler Corporation, Ford Motor Company, General Motors Corporation
4. Gwet KL (2014) Handbook of inter-rater reliability: the definitive guide to measuring the extent of agreement among raters. Advanced Analytics, LLC
5. Diering M, Dyczkowski K, Hamrol A (2015) Estimating the level of assessments agreement in visual inspection—the problems in determining kappa coefficients (in Polish). In: Innovation in management and production engineering

6. Cohen J (1960) A coefficient of agreement for nominal scale. Educ Psychol Meas 20(1):37–46
7. Zadeh LA (1965) Fuzzy sets. Inf Control 8(3):338–353
8. Klir G, Yuan B (1995) Fuzzy sets and fuzzy logic: theory and applications. Prentice Hall, New Jersey
9. Dubois D, Prade H (2000) Fundamentals of fuzzy sets, vol 7. Kluwer, New York
10. Wygralak M (2013) Intelligent counting under information imprecision: applications to intelligent systems and decision support. Springer, Berlin
11. Dyczkowski K (2007) A less cumulative algorithm of mining linguistic browsing patterns in the world wide web. In: EUSFLAT conference, pp 129–135
12. Stachowiak A, Żywica P, Dyczkowski K, Wójtowicz A (2015) An interval-valued fuzzy classifier based on an uncertainty-aware similarity measure. In: Intelligent systems' 2014, pp 741–751. Springer, Berlin
13. Wygralak M (2003) Cardinalities of fuzzy sets. Springer, Berlin

The Role of Artificial Neural Network Models in Ensuring the Stability of Systems

Anna Burduk

Abstract In order to ensure smooth functioning of a production system, the stability of its processes must be guaranteed, while on the other hand it must be possible to make quick decisions encumbered with the lowest possible risk. Innovations concerning products or processes are a necessary condition to remain on the market, but they always carry the risk of losing the stability. The risk results from the uncertainty associated with making decisions as to the future, as well as from the fact that the implementation of innovations is one of the factors that disturb the current manner of the company's operation.

Keywords Production system · Stability of production · Models of production system

1 Introduction

The concept of stability is derived from the systems theory. Most definitions found in the literature refer to the concept of the state of balance and define the stability of a system as its ability to return to the state of balance after the disturbances that caused the instability have ceased. The stability of a control system is its most important feature that characterizes the ability to accomplish the tasks, for which it has been built [2, 7].

If the value of the parameter $P(t_i)$, which characterizes the production system at the time t_i, is within the predetermined interval $P_1 \leq P(t_i) \leq P_2$, this will indicate a correct course of the process. Otherwise, corrective measures should be taken. Corrective measures usually consist in changing the values of control variables (inputs to the system) in such a way, so that the values of the parameters

A. Burduk (✉)
Wrocław University of Technology, 27 Wybrzeże Wyspiańskiego St, 50-370 Wrocław, Poland
e-mail: Anna.Burduk@pwr.wroc.pl

Á. Herrero et al. (eds.), *10th International Conference on Soft Computing Models in Industrial and Environmental Applications*, Advances in Intelligent Systems and Computing 368, DOI 10.1007/978-3-319-19719-7_37

characterizing the controlled variables (outputs from the system) return to the process course standards established at the planning stage [2]. Production plans and parameters characterizing them usually constitute the standards. A correct decision will cause that the system will return to the steady state.

It can be said that a production system is in the steady state, if values of the parameters defining it are within the ranges specified in the planning function and recorded in a standard, i.e. a production plan, as schematically shown in Fig. 1.

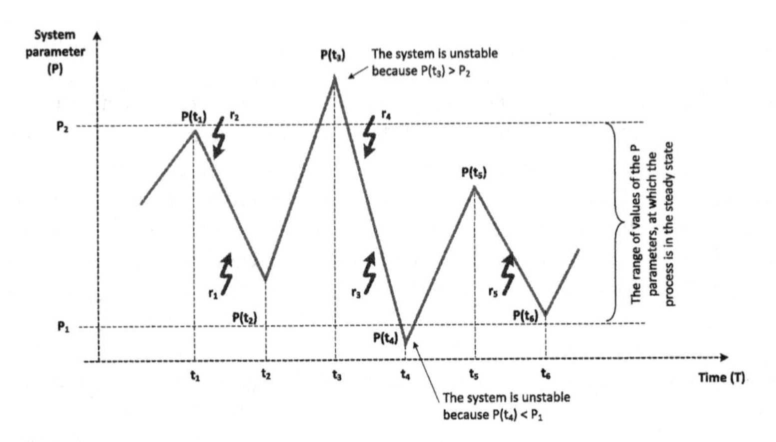

Fig. 1 The variability of the system parameter $P(t_i)$ is caused by the impact of disturbing factors (r_i)

In order to ensure the stability of production systems, on the one hand an appropriate control is needed, while on the other hand it is necessary to analyse, evaluate and eliminate the random factors causing the disturbances (risk factors). Control in the context of production systems means making decisions based on the information or data coming from the controlled system. The impact of single- or multi-criteria decisions on the production system can be verified very well on a model of the production system, which contains important system elements and their parameters as well as the relationships between them.

2 Application of Artificial Neural Networks (ANN) in Control of a Production System

The primary objective of modelling the dynamics of a production process is to identify the temporal variability of its physical quantities or states [1, 4–6]. To this end, a time series, i.e. an ordered sequence of values of a certain variable over time, should be determined. A time series may have a form of the vector $[y(t_1), y(t_2), \ldots, y(t_N)]$. Due to the fact that process parameters may differ in individual phases of the process, the time series vector can take the form of a vector defined in N-dimensional space. Individual components of this vector will be the states of the production process stages in the past, which in turn can be regarded as points in a multi-dimensional output space. Thus the task of analysing the temporal variability of the production process can be reduced to searching N-dimensional space for a certain trajectory, on which the analysed output variable of the process "moves". Thus, a given quantity in the form of a time series is determined in order to predict its value in future moments [2]. Unidirectional multilayer networks (without feedback) are used in over 80 % of all applications of neural networks [8].

Artificial neural networks are usually used to solve problems associated with the approximation, interpolation, prediction, classification, recognition and control [8]. Image recognition, which also includes classification, grouping and processing, accounts for approx. 70 % of all industrial applications. In the management and operation of production systems, artificial neural networks are more and more often used for [8]:

- control of production processes, robots,
- analysis of manufacturing problems,
- diagnostics of electronic systems of machines,
- selection of personnel and input materials,
- optimization of the business activity, waste disposal, robot movements,
- planning overhauls of machines,
- forecasting.

The main advantages of artificial neural networks in the context of their use for production system modelling may include [1, 3, 8]:

- parallel and distributed information processing, which allows performing analyses in a reasonably short time, for different areas of the production system,
- possibility of modelling phenomena, systems and processes only on the basis of historical or measurement data, without the need to build a mathematical model, which in the case of production systems may be very complicated,
- high efficiency in analysing large amounts of data, which offers an opportunity to analyse processes with considerable disturbances, dynamically variable processes, fuzzy data, and even chaotic processes,
- they do not require any programming, i.e. the role of a programmer is reduced to designing a network structure adequate to a given type of problem and to directing the training process in a competent manner,
- ability to model any non-linear complex dependences occurring in a production system.

Weaknesses of artificial neural networks in the context of their use for modelling production systems may include the multi-step reasoning, i.e. a case, where conclusions are drawn based on previous results.

3 Application of the ANN Model to Ensure the Stability of the Copper Ore Mining Process

The purpose of building the ANN model is to ensure such control of the copper ore (i.e. output) loading and haulage process, so that the process remains stable in the assumed period. The control will consist in an adequate selection of parameter values for process inputs in order to obtain an output value (i.e. the production volume) consistent with the predetermined production plan. The process will remain stable, if the production volume is consistent with the volume set in the production plan.

3.1 Characteristics of the Production System

The study was conducted in one of mining companies located in Lower Silesia, Poland. The mine, for which the neural network was built, covers an area of 158 km^2 underground, while mining operations take place at a depth from 610 to 850 m. One of the mining divisions of the mine (G-1 division) was selected for the analyses.

The purpose of the mining system is to provide copper ore to an ore enrichment plant. The exploitation process is the main process in the entire mining system. The process itself takes place in many places at the same time, i.e. in multiple mining fields in their different parts. The place where the mining operations are conducted is called a mine face.

One of the mining process stages (i.e. the loading and haulage process) was selected for the further analysis and building an ANN model. This process is one of the most important processes in the mining process. Its purpose is to move the ore from the mine face to the grate, where the ore is crushed and discharged onto a belt conveyor. Main stages of this process are shown in Fig. 2.

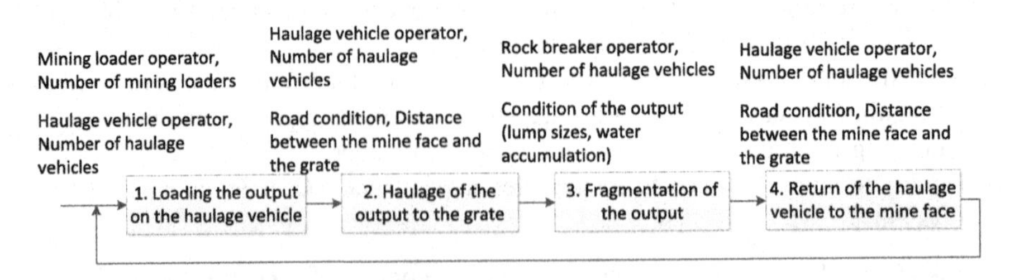

Fig. 2 Main stages of the loading and haulage process

1. Loading the output onto a haulage vehicle. This operation is performed with the use of a mining loader in the mine face and consists in loading a haulage vehicle with copper ore.
2. Haulage of ore to the grate. The haulage vehicle, after having been loaded, goes to the discharge point (grate). Haulage roads may have a different inclination angle, while their condition may vary depending on the type of rocks forming the road. If the rocks are fairly soft, the road becomes muddy over time and ruts are formed in it. As time goes on, the transport on such a road becomes more and more difficult and therefore longer.
3. Unloading the output on the grate. The grate is located over the belt conveyor and is a discharge point, where the transport of the copper ore to the ore enrichment plant is started. Its role consists in adequate fragmentation of the output and retaining all impurities, such as metal anchors, props and other elements that could damage the belt conveyor.
4. The haulage vehicle, after having been unloaded on the grate, goes back to the mine face to be loaded again.

In addition to failures of mining machinery, the elements disturbing the loading and haulage process include also variable conditions in the environment. They cause that the time of transporting the output and the time of return of the haulage vehicle from the grate to the mine face are very variable.

3.2 Method of Building the Neural Network

In order to consider the mining process to be stable, it should deliver the established amount of copper ore to ore enrichment plants. Since the haulage vehicle has a constant and limited capacity (20 t), the amount of the ore getting to the processing plants depends on the number of haulage vehicles unloaded on the grate. The shorter the time of the loading and haulage process, the more haulage vehicles can be unloaded on the grate during a work shift. The time of the loading and haulage process depends on many factors.

During observation and measurements of operation times in the loading and haulage process, it has been found that the following factors have the greatest impact on the duration of the process and at the same time on the number of runs to be made by haulage vehicle operators:

- Length of the haulage road from the mine face to the grate.
- Distance travelled by a mining loader when loading a haulage vehicle. This distance may be from 3 to 30 m.
- Condition of haulage roads, which is affected by the type of rocks on the floor, the inclination angle, and the degree of water accumulation. Heavy road transport causes, that ruts are formed and a layer of mud reaches even 80 cm.

These parameters as well at the grate number were selected as independent variables, when building the ANN model, as presented in Fig. 3.

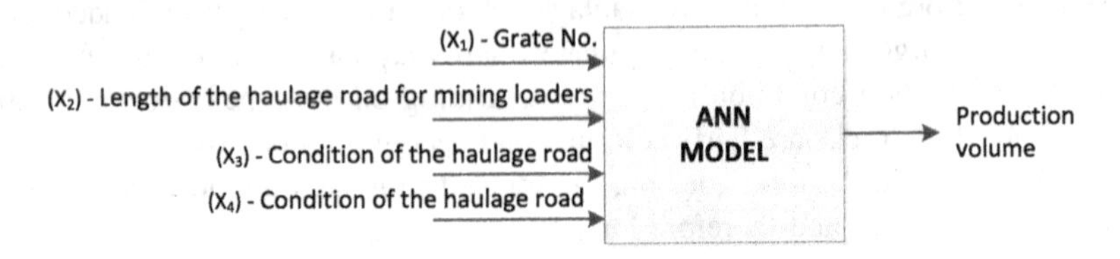

Fig. 3 Independent variables and the dependent variable used to build the ANN model

In order to predict the amount of the excavated ore under the loading and haulage process in the G1 division, at the assumed input values, a unidirectional neural network (perceptron) was built. The results of observations and measurements of times in the loading and haulage process were used as the learning data set. In total, 211 measurements were made during 20 days, in 3 work shifts. The measurements were made by shift foremen with the use of forms prepared especially for this purpose. Four values have been introduced to describe the road condition:

1. mostly even and dry road,
2. mostly even and wet road,
3. mostly rough and dry road,
4. mostly rough and wet road.

The experiment was performed in the SAS Enterprise Miner 6.2 environment. The first step was to investigate the correlation between independent variables and the dependent variable. The results containing the correlation value are shown in Table 1.

Table 1 Values of the correlation between variables

Independent attribute (variable)	Correlation value
Grate No.	−0.06787
Length of the haulage road for mining loaders	0.01009
Length of the haulage road for haulage vehicles	−0.32767
Road condition	−0.07535

The obtained results indicate that it is pointless to use a linear regression method (absolute values of the correlation are below 0.5) for the analysed problem. Therefore it is justified to use neural networks which form non-linear regression models.

As a part of further experiments, a model of a multilayer perceptron network was built, for which the number of neurons in the hidden layer was changed. In order to confirm the results of the correlation analysis, a neural network based on the generalized linear model was also built. The model was built in the SAS Enterprise Miner program ver. 6.2.

For the neural network models built, a series of experiments for a different number of independent variables was conducted. The purpose of these experiments was to establish for which combination of independent variables the neural network will determine the value of the output in the best way. When building the models, different numbers of independent variables were considered. Their selection was dictated by previous experiments, i.e. it depended on the absolute value of the correlation (see: Table 1). In the experiment No. 1, all input attributes are used, while in the experiment No. 2 the attribute "Length of the haulage road for mining loaders" (the lowest absolute value of the correlation) was discarded. In the experiment No. 3 the attribute "Grate number" (the next lowest absolute value of the correlation) was discarded in addition. The results are presented in Table 2, where the values obtained represent the network selection criterion, i.e. the mean square error. These results concern the analysis of the input data set, which was also used for network training process.

Table 2 Results of the experiments conducted with the use of a neural network

Neural network model	Mean squared error		
	Experiment No. 1	Experiment No. 2	Experiment No. 3
MPN – NN = 3	1228.59	1643.71	2375.39
MPN – NN = 16	1072.43	1369.98	1851.50
MPN – NN = 32	427.08	866.69	1033.93
MPN – NN = 48	**327.15**	764.22	1019.25
MPN – NN = 64	348.80	772.59	999.05
GLM	2440.74	2450.18	2537.86

Where: MPN - a multilayer perceptron network, NN - number of neurons in the hidden layer, GLM - generalized linear model

The analysis of the results confirms that linear models are not suitable for solving this problem. In the case of each experiment, the worst results (with the highest mean square error) were obtained for a neural network built according to the generalized linear model. The best results were obtained for a multilayer perceptron network with 48 neurons under the experiment No. 1. This neural network model was used for further experiments.

3.3 *Determination of the Stability of the Loading and Haulage Process*

The selected neural network model was used to determine the stability of the process of ore mining in the G1 division. For this purpose, test data were prepared and the "score" node of the SAS Enterprise Miner 6.2 environment was used.

The test data contain various variants of changes in input attributes (independent variables). For such data, the selected neural network model predicts the values of the output, which are interpreted in the context of the stability of the mining process. Sample test data along with the predicted production volume are presented in Tables 3, 4 and 5. The planned output volume was set at 330 t. For the needs of the study, an assumption was made that the ore production is stable, if the absolute value of the variation in the production does not exceed 20 tonnes. This corresponds to unloading two haulage vehicles per shift. Table 3 shows the production volume predicted by the ANN model depending on the length of the haulage road.

Table 3 Predicted production volume for the grate No. 4 at variable lengths of the haulage roads

Network inputs				Network outputs
Gate No.	Road for mining loaders (m)	Road for haulage vehicles (m)	Road condition (m)	Anticipated output (t)
4	50	700	4	339
4	50	750	4	340
4	50	800	4	334
4	50	850	4	293
4	50	900	4	257

The data included in Table 3, presented in the context of the process stability, are shown additionally in Fig. 4.

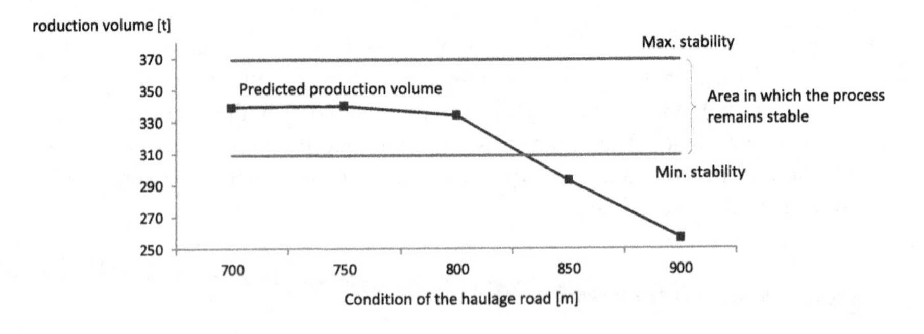

Fig. 4 Predicted production volume for the grate No. 4 at variable lengths of the haulage roads

As it results from Table 3 and Fig. 4, the process becomes out of balance, if the haulage road is extended to 850 m. At this length it is not possible to execute the assumed production plan. This is an indication for the decision-maker that the values of input variables should be changed, for example by improving the condition of the roads. Table 4 presents the production volume predicted by the ANN model depending on the condition of the haulage roads.

Table 4 Anticipated output for the grate No. 3 at variable road conditions

Grate No.	Road for mining loaders (m)	Road for haulage vehicles (m)	Road condition	Anticipated output (t)
3	60	1000	1	348
3	60	1000	2	277
3	60	1000	3	230
3	60	1000	4	183

The data presented in Table 4. are illustrated also in Fig. 5 in the context of the stability of the production process.

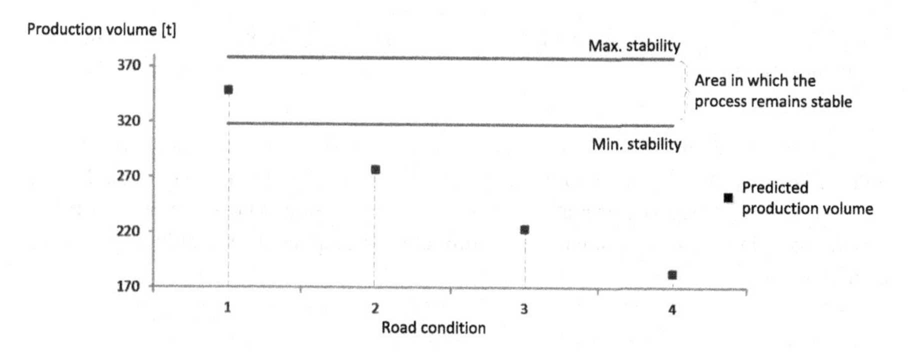

Fig. 5 Predicted production volume for the grate No. 3 at a variable condition of haulage roads

As it results from Table 4 and Fig. 5, for the assumed lengths of the haulage road and the distance travelled by a mining loader when loading a haulage vehicle, the process will remain stable only if the haulage road is even and dry. In other cases it is not possible to execute the assumed production plan without changing the values of other input parameters.

Table 5 shows the production volume predicted by the ANN model, depending on the length of the haulage road travelled by a mining loader when loading a haulage vehicle.

Figure 6 illustrates the data from Table 5 in the context of the stability of the loading and haulage process, i.e. the planned production volume of 330 t $\pm$ 20 t, depending on the length of the road travelled by a mining loader.

Table 5 Predicted production volume for the grate No. 4 for a variable length of the haulage road for the a backhoe loader used to load haulage vehicles

Grate No.	Road for mining loaders (m)	Road for haulage vehicles (m)	Road condition	Anticipated output (t)
4	30	1200	3	323
4	40	1200	3	302
4	50	1200	3	196
4	60	1200	3	178

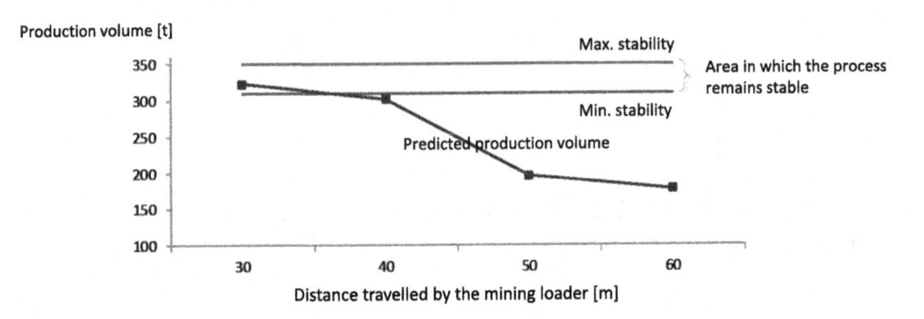

Fig. 6 Predicted production volume for the grate No. 4 at variable length of the haulage road for the mining loader

As it results from Table 5 and Fig. 6, the process will remain stable for the set values, only if the distance to be travelled by a mining loader when loading a haulage vehicle is no longer than 30 m. Otherwise, in order to execute the assumed production plan, values of other input parameters (independent variables) should be changed.

4 Conclusion

Production system modelling allows ensuring the stability of a production system through: understanding and assessing the impact of the decisions made on the production system and its various functional areas, designing or reorganizing the production system in a manner that does not disturb its current and future operation, controlling the production system by selecting such parameters of system inputs, so that to obtain the designed values of the output parameters, as well as identifying, assessing and eliminating the impact of the factors that disturb correct functioning of the production system.

References

1. Azadegan A, Probic L, Ghazinoory S, Samouei P (2011) Fuzzy logic in manufacturing: a review of literature and a specialized application. Int J Prod Econ 132(2)
2. Bubnicki Z (2005) Modern control theory. Springer, Berlin
3. Calvo-Rolle JL, Corchado E (2014) A bio-inspired knowledge system for improving combined cycle plant control tunin. Neurocomputing 126:95–105
4. Krenczyk D, Dobrzanska-Danikiewicz A (2005) The deadlock protection method used in the production systems. J Mater Process Technol 164:1388–1394
5. Krenczyk D, Skołud B (2011) Production preparation and order verification systems integration using method based on data transformation and data mapping. Lecture notes in computer science, vol. 6697, pp. 297–404
6. Roux O, Jamali M, Kadi D, Chatelet E (2008) Development of simulation and Optimization platform to analyse maintenance policies performance for manufacturing systems. Int J Comput Integr Manuf 21: 407–414
7. Vladimirs J, Vitalijs J (2012) Modelling the behaviour of stability of production systems of economics. Econ Bus 22
8. Wieczorek T (2008) Neuronowe modele procesów technologicznych, Monograph. Publishing House of the Silesian University of Technology, Gliwice

Coordination in the Supply Chain

Pawel Pawlewski

Abstract Coordination has become a significant factor of the integration of various parts of the organization as well as various organizations of the supply chain. This seems to be a key factor in the success of logistics management. It is also an element allowing for a common list of tasks to be accomplished as well as common objectives to be achieved in the selected system (micro- or meta-). The objective of the publication is the theoretical discourse on the topic of coordination in the supply chains. General premises regarding the topic have been presented in the first part of the publication. The starting point of the theoretical argumentation is the presentation of the meaning of the term supply chain (part 2) as well as coordination (part 3). This has allowed for the types of coordination, which can occur in supply chain systems to presented in the subsequent parts. The text is concluded with a summary. The presented text is of a review nature.

Keywords Supply chain · Coordination theory · Coordination techniques

1 Introduction

Contemporarily, the coordination of supply activities is one of the main challenges in business and created supply chains. The supply chains are multi-actor systems. As a whole, they struggle with the lack of inner rationality, unverified information, uncertainty or lack of knowledge. Far too often, the contemporary supply chains are a sequence of loosely related actions of inner kind (within the supply chain) or occurring outside the company, yet within the frames of the system. Supply chain (SC) structure defines the way various organizations within the supply chain are

P. Pawlewski (✉)
Poznan University of Technology, Ul.Strzelecka 11, 60-965 Poznań, Poland
e-mail: pawel.pawlewski@put.poznan.pl

Á. Herrero et al. (eds.), *10th International Conference on Soft Computing Models in Industrial and Environmental Applications*, Advances in Intelligent Systems and Computing 368, DOI 10.1007/978-3-319-19719-7_38

arranged and related to each other. The SC structure falls into four main types [1]: Convergent: each node in the chain has at least one successor and several predecessors. Divergent: each node has at least one predecessor and several successors. Conjoined: which is a combination of each convergent chain and one divergent chain. Network: which cannot be classified as convergent, divergent or conjoined, and is more complex than the three previous types [2]. SC management is defined as "the systemic, strategic coordination of the traditional business functions and the tactics across these business functions within a particular company and across businesses within the supply chain, for the purposes of improving the long term performance of the individual companies and the supply chain as a whole" [3]. This leads to a gap in the coordination. The barriers in the coordination of companies' actions (e.g. organizational or related with information flow) restrain the control centralization of goods flow in a supply chain. In numerous cases it is neither possible or desirable [4]. That is why, such forms of organizations, which are based on decentralized activities performed by autonomic units according to their own principles and objectives, are capable of maximizing profits and shortening the operation times in the whole supply chain. The use of coordination mechanisms, known from the coordination theory, may contribute to the increased efficiency of supply chains, which are systems consisting of separate and individual enterprises [5]. Therefore, coordination as a manifestation of logistic management should play a crucial role in distributed systems, such as contemporary supply chains.

The coordinating actions are fundamental, those that (1) stimulate the supply chain through the creation of a SC growth concept, (2) regulate the supply chain by redistributing the possessed resources (3) integrate the supply chain by linking resources, monitoring and an assessment of the actions [6]. These problems are very complex so to solve them the new computational techniques based on soft computing models are necessary [7].

The article presents results of the authors' investigations in the field of SC coordination. Its main purpose is to identify the coordination techniques in the supply chain structures and to prove the need for creating reference models for coordination activities. The research highlights of the performed works are as follows: presentation of the meaning of the term SC as well as coordination, identification of coordination methods, description of the performed investigations, identification of the need for a reference model for coordination. The article is divided into 5 sections. Section 2 refers to the coordination theory; it presents the theoretical discourse on the topic of coordination in the supply chains. The types of coordination in a supply chain are described in Sect. 3. Section 4 defines ten models of coordination techniques and Sect. 5 focuses on the presentation of results of experimental research performed by the authors. Finally, Sect. 5 provides conclusions and suggestion for the further stages of the project.

2 Coordination Theory

A supply chain is something in between an individual enterprise (microstructure) and global economy (macrostructure). It is understood as a metastructure [8], which is characterized by a dynamic and holarchical structure of cooperating holons (enterprises). The larger the chain is, the less coherent the created system becomes. Therefore, in such a metastructure the relations become more or less durable. The chain consists of links that constantly participate in it (the so-called core of a SC) and links changing dynamically, depending on the implemented task (the so-called satellite links). When the cooperation is over, they are separated from the core of the supply chain and the cooperation is abandoned [9, 10].

On a regular basis, supply chains are made of numerous participants (links), such as subcontractors, contractors, manufacturers, distribution centres etc. The structure is considered as centralized when there is only one decision-maker who tries to optimize the whole metasystem [11].

However, even in a decentralized system, each unit will have their own individual (local) objectives. They are often contrary to the objectives of other participants of the metasystem (e.g. manufacturers prefer to produce large quantities of goods to decrease the production costs, yet it means increasing the amount of stock in distribution centres as well as growth in storage costs). Each company in a SC attempts to locally optimize its system. Therefore, decentralized SC are dominant in the market. The centralized metasystem leads to global optimization, while the decentralized one leads to local optimizations [12]. To achieve the effect of global optimization in a decentralized SC, the contradictory objectives of particular participants should be adjusted with use of coordination mechanisms.

Coordination is present in various systems, e.g. social, open, dynamic, distributed, analytical, biological, etc. It is based on numerous research fields, i.e. organizational theory, management sciences, psychology, computer science and game theory. Owing to its interdisciplinary character and the lack of commonly used name, the terms coordination theory or coordination science were postulated for use [13]. There are a lot of definitions of the term of coordination (Table 1). Their variety indicates the different viewpoints and interpretations of the problem as well as difficulties in describing the term.

Taking the above definitions into account, we can state that coordination is a cooperation of subjects, which (1) is directed at attaining mutual objectives, (2) consists of systemizing, ordering and coordinating processes and various system components, (3) proceeds in a set time period, (4) affects the behaviour of cooperating subjects.

The coordinated actions are arranged in a way that there are no interruptions. They are harmonized, adapted and adjusted [5].

There are two key notions of the coordination theory. The first states that the dependencies and mechanisms of coordination are of a general nature. It means that they can be found in various systems and organizations, e.g. coordination in a workgroup consisting of various specialists. Here, the idea of coordination is to

Table 1 The chosen definitions

No.	Researcher, year	Coordination is defined as
1	Kotarbiński (1946)	Reconciliation
2	Malone (1987)	Pattern of decision-making and communication between participants, who perform activities to achieve objectives
3	Malone, Crowston (1994)	Process of managing dependencies between activities
4	Hewitt (1994)	Planning, monitoring and adjusting internal and interorganizational logistics processes
5	Gupta, Weerawat (2006)	Action, when the rationally acting decision-maker in a supply chain takes decisions which are effective in the whole supply chain
6	Kaipia (2007)	Activity which can take place within cross-functional coordination or between organizations (interorganizational coordination)

assign tasks according to specific skills and, therefore, restrain (the number of) specialists working on a particular task. The coordination in a more complex system, such as a SC, is based on similar principles. Specialist tasks are assigned to enterprises which can successfully deal with the set task. The applied pattern is the same.

The other notion of coordination theory indicates that the same problem may be dealt with various coordination mechanisms. Thus, alternative processes can occur for various coordination mechanisms. The mechanism used to manage the dependencies between the task (order) and the enterprise that implements the task, is a good example here: (1) a customer can choose a contractor according to the first found – first hired rule, (2) a customer can a choose a contactor according to a recommendation (of another enterprise), (3) a customer can choose a contactor based on a tender, (4) a customer can choose a contractor from its database of subcontractors. This notion of coordination indicates that it is possible to create alternative processes by using alternative mechanisms of coordination.

The investigations on the coordination theory are focused on coordinating activities inside the organization as well as activities between organizations, i.e. in a supply chain [14, 15]. These research works, along with many others, emphasize the importance of coordinating activities for the implementation of the set objectives related to reducing costs, enhancing customers' satisfaction, increasing flexibility, etc. All activities and their coordination are aimed at passing the product or service to an end customer. Therefore, coordination refers to integrating activities of various parts of an organization (Fig. 1b) as well as various organizations in a SC.

Figure 1b shows an idea of an enterprise and five specific activities which can be performed without internal relations. It is more probable, however, that they are interrelated and bond to the need for coordination. In this case, an enterprise can be perceived as a bundle of activities [16]. The enterprise deals with performing and coordinating tasks (e.g. using resources in the best possible way).

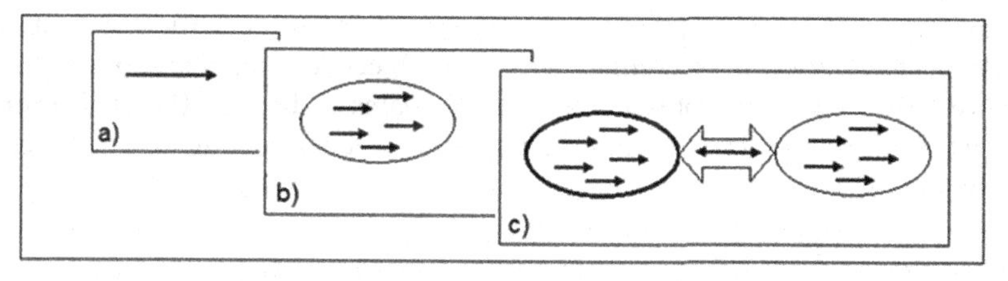

Fig. 1 Coordination (a) An individual activity shown as an arrow, (b) Enterprise using five individual activities, (c) Activity links between two enterprises [17]

If the concept of coordination is expanded to the cooperation of two companies, the need for coordination at a higher level of the business relations between them appears. Between enterprises there is a coordination shown as activity links [17] (Fig. 1c) leading to interrelated actions, which are synchronized and adjusted. The actions undertaken by two (or more) enterprises in a business relation become more or less related, depending on the development of the relation.

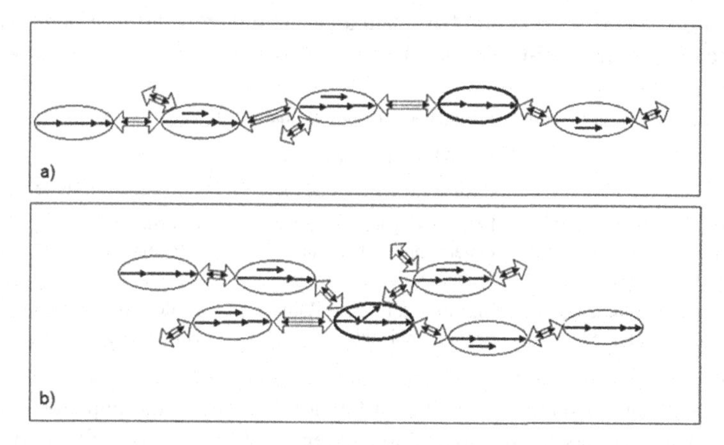

Fig. 2 Coordination through activity links (a) Chain of activities, (b) Fragment of a complex activity structure [17]

An activity chain is made of activities related in various business relationships within a supply chain (Fig. 2). The concept of activity chains means defining the sequence of activities precisely. Each enterprise in a supply chain is interested in coordinating activities and creating the best possible conditions for the integration of the members of the chain.

Various chains of supplies related with one another (through activity links) constitute a complex activity structure (network). Each activity link is a fragment of a larger entity. However, not all the internal activities of an enterprise must be related with a given activity chain (Fig. 2b). Enterprises have a lot of suppliers and customers and, therefore, the activities must be coordinated by activity links according to varied customers' needs. It means that each enterprise in a network coordinates its activities according to its own subjective comprehension. Taking the above into account, we present two supply chains in Fig. 2. The change of one activity link results in a change in the whole model and the relation between enterprises [17].

3 Coordination in the Supply Chain

Coordination is based on three components: group effort, synergy of actions and general objective. It helps to unite efforts and attempts of particular subjects in the whole system. The coordination should guarantee the effectiveness of the group effort by means of uniting actions of various groups within the process, by sustaining the relations between the general goal and the individual or the group types of activities [18].

There are three types of coordination. They can be identified based on two dimensions: correlation and the subject of coordination (Fig. 3). The correlation of coordination can be divided into two main areas: complementarity of processes and coherency of understanding (Table 2).

Table 2 Classification of coordination types in supply chain [19]

		Mutuality on coordination	
		Complementarity of processes	Coherency of understanding
Focus of coordination	Operational linkages	Logistics synchronization Object: product/service, process	Information sharing object: information
	Organizational linkages	Incentive alignment Object: benefits and risk	Collective learning object: Knowledge and capability

The use of the presented types of coordination guarantees the implementation of the mutual set of tasks as well as reciprocal benefits [20]. The benefits involve e.g. (1) expeditious reaction to the competitors' actions and customers' demands, (2) reduction of the bullwhip effect, (3) integration of three streams occurring in the supply chain (material, information and financial flows).

Coordination of actions, understood as the logistics synchronization, refers to mediation/negotiation conducted by the participants of a SC. Its aim is to adjust the varied range of products/services to customers' individual needs. It concerns, for instance, stock management, product availability, implementation of transport between partners; all these factors interrupt the efficiency of the chain. Logistics synchronization makes it also possible to solve conflicts related with the relocation of goods.

The coordination of actions, in the form of information exchange, guarantees precise and up-to-date information for all participants of a SC. It is obvious that not all participants of the metastructure have the same supply of information (information asymmetry). The seller of a product can have more precise information about the future demand for the product than its producer, yet in a limited scale. The information sharing enhances the transparency of participants of the chain and it is a factor that connects them all.

The incentive alignment makes it possible to coordinate the proper way of decision-making and, as a result, award or punish for the undertaken decisions or actions. It should be emphasised that incentive alignment affects cooperation of partners. Especially its more modern forms are worth considering as they are not based on the local cost and benefit analyses and they enable the analysis of global efficiency.

Coordination based on collective learning or collaborative learning makes it possible to sustain the whole system, i.e. the supply chain. The investigations in the field of ecology confirms this idea, as they show that survival of an organism depends on its learning rate. The organism dies if its learning rate is smaller than the dynamics of changes [21]. An important advantage of collective and collaborative learning is the active thought exchange, which arouses the interest of participants and supports critical thinking (convergent thinking). The idea of coordination based on collective thinking is to solve problems and analyse existing phenomena in order to come to specific conclusions.

4 Supply Chain Coordination Techniques

As we mentioned before, coordination means cooperation of subjects directed at attaining mutual set objectives. It consist of systemizing, ordering and agreeing on processes and various components of the system. Coordination proceeds in the set time and affects the behaviour of cooperating subjects. Owing to the varied types of cooperation, it is advisable to develop reference models of coordination. A reference model of coordination is treated as a pattern that the real-life coordination may refer to and, therefore, it can undertake appropriate activities. Actually, it is a point of reference for the actions related with analysis, rationalization and optimization of both the coordination in supply chains and the supply chains themselves. The reference model, understood as an organization and process pattern, uses physical objects that build up a logistics system and a SC in the micro-, meso- and macroscale. Such a pattern must include information about processes that are necessary for their implementation, as well as human commitment and factions that must be performed. The first thing to build such patterns, i.e. reference models, has been to identify the techniques which can be used for the coordination of a SC.

In structures such as supply chains or networks, the following coordination techniques can be used [22–24]:

- Negotiating – it is the best known coordination technique. It is considered to be the basis for all the other presented techniques, as each of them contains some elements of negotiations, which are a communication process.

- Coordination through building organizational structures – it is the simplest coordination technique. It assumes creating a certain structure (in reference to a SC it is a configuration). This structure defines the scope of appointed activities and works, as well as rights and duties resulting from that. It also controls the mutual interrelations between enterprises. Therefore, the role played is the system can be explicitly specified.
- Coordination through mutual information standards – it creates such conditions for information sharing between enterprises so that procedures can be simplified and the possible IT solutions can be applied.
- Contracting – it is a technique used in case of well defined (in terms of structure: hierarchical) activities. The activities are assigned as a result of contracting and subjected to decomposition. The technique requires a high volume of communication between enterprises. However, it lacks solutions that can be used in conflict situations, especially when contactors become/are antagonistic actors.
- Harmonizing – it is a technique that refers to rules and rights of organic nature.
- Open coordination – it creates a system of mutually related organizations using the existing links between them. It is based on five principles: (1) subsidiarity (balance between common objectives of a system and individual aims), (2) convergence (of the results of actions), (3) management through objectives (using quantity and/or quality indices), (4) observability (mutual comparing and revealing the best practices), (5) comprehensiveness (coordination of activities).
- Animation – it uses initiation (animation) of other system elements for the activity. The technique uses the synergy effect: the sum of knowledge, information and competence of all system elements considerably excesses the knowledge, information and competence of a single element, even if it performs managerial functions.
- Pre-delivery inspection – it is based on evaluating the activity and discovering creditable and imitable phenomena.
- Information resonance – it refers to the technique of conveying and enhancing information.
- Multi-actor planning – the technique assumes that it is possible to build up an overall activity plan for all the stakeholders in the system, taking into account objectives and possibilities of all the involved.

5 Experimental Investigations on Coordination

The researchers started investigating the coordination of actions in a supply chain, as they wanted to learn to what degree the presented models are used in practice. Fifty economically unrelated enterprises took part in the investigation. Based on the size of a company, they were divided into four groups: microenterprises that employ fewer than 10 people (8 companies took part in the survey), small-sized enterprises, which employ 10–49 employers (15 surveyed), medium-sized enterprises employing 50–249 people (13 surveyed) and large enterprises that employ over 250 people (14 surveyed). The division into groups was made only based on the number of employees. The investigated enterprises operate in various markets: local (14 companies), domestic (21) and global (16).

While indicating the coordination techniques used in enterprises, the respondents could choose more than one answer. Most often they indicated the use of three coordination techniques (23 %), the least frequent answer given by respondents indicated use of six or seven techniques (4 % respectively); 16 % of the interviewed enterprises use as many as eight out of ten coordination techniques.

The results of the survey show that, in case of large enterprises (Fig. 3), harmonizing is the most commonly used technique (64 % answers). The technique is related with theories and rules of organic nature. The second most popular technique is coordination through building organizational structures (50 % answers). This technique is used in case of well defined actions undergoing decomposition. The actions can be determined as a result of mapping business processes and ascribed to companies cooperating either permanently or within specific orders. In case of medium-sized companies, the negotiating technique, understood as the mutual agreement, turned out t be the most popular one (62 % answers). This technique is preferred also by small-sized and microenterprises (73 % and 75 % respectively). Negotiating is the most commonly used coordination technique. It is considered to be the basis for all the other presented techniques and that is probably the major factor that makes this technique so popular.

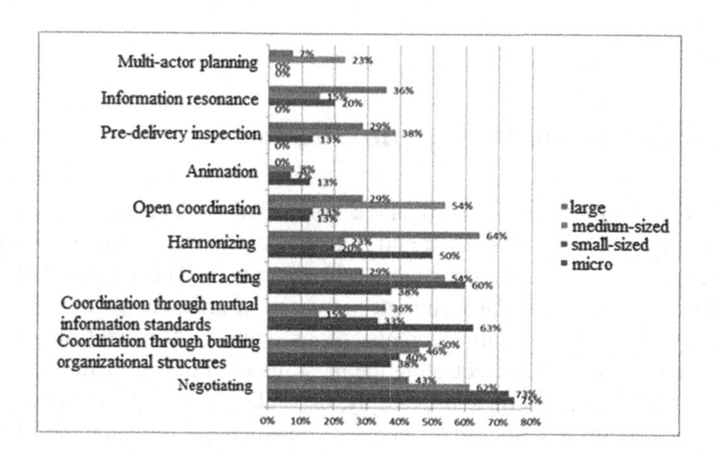

Fig. 3 Coordination technique and the size of enterprise [24]

Regardless of the market in which enterprises operate, the most commonly technique is negotiating. In case of enterprises operating in a global scale, we can observe a vast variety of the applied coordination techniques (Fig. 4). The negotiating technique is dominant (50 % answers), yet coordination through building organizational structures and harmonizing are only slightly less popular (44 %

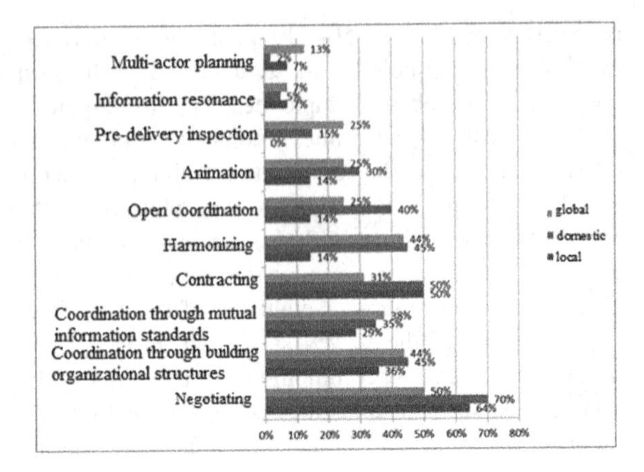

Fig. 4 Coordination technique and functioning in the market [24]

answers respectively). The least commonly used technique in the investigated group is multi-actor planning (13 % answers).

Nevertheless, it is much more commonly used than in case of enterprises operating in the local and domestic market. This technique assumes that it possible to build up a complete and comprehensive plan of activities for all stakeholders operating within the system. We can consider it as an attempt to formalize the whole supply chain, by taking into consideration the objectives of all enterprises operating in it and their potential. In case of medium-sized and microenterprises, it is contracting that is the second most popular technique (70 and 64 % answers respectively).

6 Conclusions and Further Investigations

The article present results of investigations conducted by the authors in the field of coordination of SC. We described and defined coordination techniques and presented results of investigations carried out in enterprises. The studies explored the degree to which the particular techniques are used in practice, taking into account the size of an enterprise and the range of its activity. The next stages of our investigations are as follows: selecting the most important coordination techniques: defining the selection criteria and, possibly, expanding the scope of study, creating reference models for the selected techniques, constructing simulation models for the developed reference models, in a way that a simulation model can be quickly adjusted to a given investigated enterprise, developing the methodology for evaluating the investigated coordination technique in a real enterprise compliant with the reference model and improvement method.

Acknowledgments The presented research works are carried out under the LOGOS project (Model of coordination of virtual supply chains meeting the requirements of corporate social responsibility) under the grant agreement number PBS1/B9/17/2013.

References

1. Beamon BM, Chen VCP (2001) Performance analysis of conjoined supply chains. Int J Prod Res 39:3195–3218
2. Sitek P, Wikarek J (2013) A hybrid approach to modeling and optimization for supply chain management with multimodal transport. In: IEEE conference: 18th international conference on methods and models in automation and robotics (MMAR), pp 777–782
3. Awasthi A, Grzybowska K, Chauhan S, Goyal SK (2014) Investigating organizational characteristics for sustainable supply chain planning under fuzziness. In: Kahraman, C, Öztayşi, B (eds) Supply chain management under fuzziness. Studies in fuzziness and soft computing, vol 313. Springer, Berlin
4. Lee HL, Billington C (1993) Material management in decentralized supply chains. Oper. Res. 41(5):835–847
5. Grzybowska K (2011) Coordination in supply chain—logistics management—theoretical approach [in Polish]. In: Witkowski, J, Bąkowska-Morawska U (eds) Strategies and logistics in service sector. Wydawnictwo Uniwersytetu Ekonomicznego we Wrocławiu, pp 259–268
6. Grzybowska K, Kovács G (2014) Logistics process modelling in supply chain—algorithm of coordination in the supply chain—contracting. In: International joint conference SOCO'14-CISIS'14-ICEUTE'14. Advances in intelligent systems and computing, vol 299, pp 311–320
7. Sitek P, Wikarek J (2014) Hybrid solution framework for Supply Chain Problems. Distribiuted computing and artificial intelligence (DCAI 2014). Book Series: Advances in intelligent systems and computing, vol 290, pp 11–18
8. Grzybowska K (2010) Coherent supply chain—efficiency growth of metastructures in crisis time [in Polish]. "Success depends on Changes", Research Works UE Nr 128, pp 319–326
9. Grzybowska K (2010) Creating trust in the supply chain. In: Grzybowska K (ed) New insights into supply chain. Publishing House of Poznan University of Technology, pp 9–21
10. Gajdzik B, Grzybowska K (2012) Example models of building trust in supply chains of metallurgical enterprises. Metalurgija 51(4):555–558
11. Simchi-Levi D, Kaminsky P, Simchi-Levi E (2000) Designing and managing the supply chain: concepts, strategies, and case studies. McGraw-Hill, Singapore
12. Toktas-Palut P, Ülengin F (2011) Coordination in a two-stage capacitated supply chain with multiple suppliers. Eur J Oper Res 212:43–53
13. Malone TW, Crowston K (1994) The interdisciplinary study of coordination. ACM Comput Surv 26(1)
14. de Souza R, Zice S, Chaoyang L (2000) Supply chain dynamics and optimization. Integr Manuf Syst 11(5):348–364
15. Natarajan, HP (2004) Optimization models to support negotiation and coordination in supply chains, PhD Dissertation
16. Harland CM, Lamming RC, Zheng J, Johnsen TE (2001) A taxonomy of supply networks. J Supply Chain Manage 37(4):21–27
17. Bankvall L (2008) Activity coordination from a firm perspective—towards a framework. In: Proceedings IMP-conference in Uppsala, Sweden
18. Håkansson H, Snehota I (eds) (1995) Developing relationships in business networks. Routledge, London
19. Simatupang TM, Wright AC, Sridharan R (2002) The knowledge of coordination for supply chain integration. Bus Process Manage J 8(3):289–308
20. Xue X, Li X, Shen Q, Wang Y (2005) An agent-based framework for supply chain coordination in construction. Autom Constr 14:413–430
21. Pedler M (1997) Action learning in practice. Gower Publishing Ltd., Hampshire
22. Nwana H, Lee L, Jennings N (1996) Co-ordination in software agent systems. British Telecom Tech J 14(4):79–89

23. Mintzberg H (1983) Structure in fives: designing effective organizations. Prentice Hall, New York
24. Grzybowska K (2012) Coordinations techniques in the supply chain [in Polish]. Wareh Manage Logist 12:50–52

ERP, APS and Simulation Systems Integration to Support Production Planning and Scheduling

Damian Krenczyk and Mieczyslaw Jagodzinski

Abstract The paper presents a method of the integration of ERP and advanced planning and scheduling (APS) systems extended with automatic generators of simulation models. The approach allows the use of simulation and visualization for rapid verification of production plans. Both integration module and model generator use data exchange and data transformation methods. The concept of data-driven modeling also allows to verify the obtained solution in terms of quality of production flow with full visualization of the processes occurring in the system. A practical example of simulation verification of a solution achieved in the APS system, using the method of automatic generation of simulation models has been shown. During the verification phase of the implemented methodology, the IFS Application - ERP system, Production Order Verification System (SWZ) for multi-assortment, concurrent production planning and Enterprise Dynamics simulation system integration have been used.

Keywords ERP · Production planning · Simulation · Visualization · Data mapping · Data transformation · Model generation

1 Introduction

Today's market conditions force manufacturers to seek more and more efficient methods of production planning. Fierce competition enforces reduction of production costs which is possible, inter alia, through the exploitation of production resources remaining at the disposal of the company as much as possible. Degree of

---

D. Krenczyk (✉)
Faculty of Mechanical Engineering, Silesian University of Technology, Gliwice, Poland
e-mail: damian.krenczyk@polsl.pl

M. Jagodzinski
Institute of Automatic Control, Silesian University of Technology, Gliwice, Poland
e-mail: mieczyslaw.jagodzinski@polsl.pl

Á. Herrero et al. (eds.), *10th International Conference on Soft Computing Models in Industrial and Environmental Applications*, Advances in Intelligent Systems and Computing 368, DOI 10.1007/978-3-319-19719-7_39

resource utilization is dependent on the efficiency of determining of the production plans. Fundamental problem associated with production planning is scheduling of production orders. Scheduling problems are often too complex to get the optimum solution for the finite time period - in many cases they belong to the NP-hard problems - often because planners have to settle for sub-optimal or sufficient solutions [1–3].

In modern companies integrated information systems supporting management processes are used extensively at different levels (strategic to operational), including the area of production planning. However, the computer systems are usually applied in a heterogeneous information system environment [4, 5]. Most often offered solutions are the systems belonging to Enterprise Systems (ES) class, from Manufacturing Resource Planning (MPR) to Enterprise Resource Planning (ERP) and complementary solutions - Customer Relationship Management (CRM), Product Data Management Systems (PDM) or Manufacturing Execution System (MES). Unfortunately, the modules offered within this class of systems, including the most common tools to support production scheduling, are limited to the basic techniques of determining the schedules - forward scheduling, backward scheduling, the sequence selectable by the operator or based on simple dispatching rules.

IT solutions offering more advanced methods of determining schedules are usually created by scientific entities and are usually implemented as a customized software for a specific customer or offered as an independent specialized IT tool. These solutions, due to the high complexity of the scheduling problem, are based on artificial intelligence methods, heuristic methods, approximate methods, constraint propagation methods and others belonging to the soft computing methods. Soft computing techniques have developed significantly in last ten years. Many soft computing methods have recently come to forefront for solving unlimited number of complex real-world problems particularly related to optimization research area [6–10]. Application of production scheduling problem-solving approximate methods (soft computing) and obtained thereby results due to the complexity of the problem, do not take into account all the constraints of the production system. One of the problems is to ensure the correct functioning of the production system (without any deadlocks and starvations) during the implementation of the control procedures in the control system. The use of the advanced discrete-event simulation and visualization systems could be very helpful in this area. Despite the multiple possibilities offered by the use of simulation and visualization systems, there are many problems that make simulation tools not commonly used in enterprises. The main factors include substantial labour-consumption and time-consumption, associated with collecting and analyzing data, construction of the production system simulation model and significant costs involved.

The solution to these problems may be to use the method of automatic generation of simulation models for discrete-event simulation systems, which could be also an efficient tool for the integration of planning and simulation systems. The increased interest in the methods of automatic generation of simulation models is observed since the 80's, when Mathewson [11] defined the generator as an interactive software tool that translates the logic of a model described in a relatively general symbolism into the code of a simulation language. The concept of automatic generation of simulation models is most often implemented by preparing dedicated modules that generate the code whose individual lines contain instructions in the internal programming languages of computer simulation systems. The functions of the code create the components of the model, corresponding to the real resources of the production system with a fixed course of production routes (static data), and information concerning the production flow on the resources (dynamic data). An example of such solution, automatic simulation model generator, is proposed by Son and Wysk [12]. The generator has been implemented using the Arena simulation system language and the models are built on the basis of data on a static model of resources and planning data on the production flow. Similar concepts are proposed in [13–15].

Another group of methods represent approaches using dedicated modules that are part of the production planning support system (production control), where the simulation model is rigidly saved with the data about the production resources (their operating parameters, efficiency, distribution, etc.) or/and auxiliary resources. Parameterization of the simulation model is carried out using data on orders planned to be executed in the production system [16–19]. The approach proposed in this paper constitutes a development of the above-mentioned concept of the ability to automatically generate simulation models that could be implemented in advanced commercial systems simulation with 3D visualization of processes flow in the system. Presented simulation model generator is also responsible for planning/scheduling and simulation systems integrating module is used as a tool for verification of generated production plans.

This paper presents a practical solution using the concepts of integration of ERP systems with an advanced scheduling system combined with simulation and visualization computer system (Fig. 1). Integration of simulation system (supporting the decision making process about the possibility of implementation of the

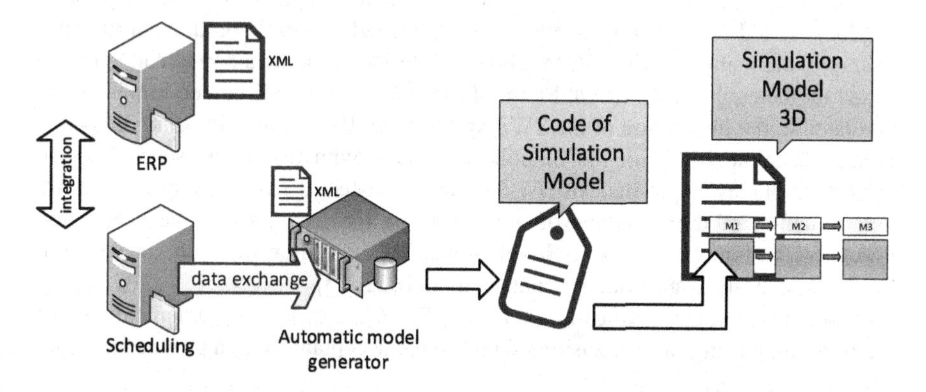

Fig. 1 The concept of systems integration

solution obtained from the planning system - the process of verification) is implemented based on the method of automatic simulation model generation based on data obtained from other systems - data-driven simulation [20, 21]. Section 2 presents ERP and production planning/scheduling systems description and the way they have been integrated with the use of data mapping and data transformation methods. Section 3 presents a universal method of automatic generation of simulation models code allowing the integration of systems described in Sect. 2 with advanced simulation system. Section 4 presents an example of a practical realization of this concept. Finally, Sect. 5 concludes the research with a discussion of results and directions for future work.

2 ERP and Scheduling Systems Integration

Practical implementation of the concept of integration of planning and simulation systems was carried out on the basis of the commercial IFS Application ERP class system and SWZ planning and order verification system. SWZ system has been developed in the framework of the research work at the Silesian University of Technology [22–25].

IFS Application system is a complete ERP class system solution and is an integrated product, supporting the main areas of business management: Manufacturing, Service and Asset, Projects and Supply Chain. IFS Manufacturing supports planning, control, execution and analysis in many types of manufacturing (Repetitive, Batch Manufacturing, Make to Stock, Configure to order, Make to order, Engineer to order), from mass production of high-volume products to mass customization and several different production models in all manufacturing process phases. IFS Manufacturing handles various EDI standards, including EDIFACT and ANSI X12 as well as XML messages. IFS Component Architecture is an open architecture, and the integration process with other application is provided through IFS Web Services or the IFS Connect integration broker. Product lifecycle management using IFS Applications gives support for total product configuration management. Product-related information is effectively managed for each product phase. For example, the manufacturing BOM (Bill of Material) could be automatically created based on the engineering BOM. Phase of the research work discussed in this paper, is related to the integration of the SWZ system with IFS Application, and intends to complement the IFS with the module associated with the orders scheduling on system resources. Input data of SWZ system transferred from IFS system allows determination of the sequence of operations on the resources of the production system. After transferring it into Simulation and visualization software, the data will become the basis to generate simulation model [22, 23].

Production Orders Verification System – SWZ is a computer implementation of methods supporting rapid decision taking on the acceptability of a production order for multi-assortment repetitive production systems. It necessitates the abandonment of determining the set of all possible solutions for the determination of a subset of feasible solutions. SWZ is based on the rapid decision-making support methodology on the acceptability of a production order using constraints satisfaction techniques and is tantamount to testing a sequence under arbitrarily selected conditions. The fulfillment of all conditions guarantees the possibility of order execution. Lack of solution provides information about the necessary abandonment of specified conditions of the order, or having to meet the needs associated with an increase in available capacity, storage space, etc. Sufficient conditions, include: system balance condition, buffer capacity condition and due time realization possibility condition – processes included in the production order will be executed within the due time required by the customer [25–27].

The process of integration is accomplished through the use of methods of data ex-change and data transformation using neutral data models. The exchange of data between IFS and SWZ systems, due to the versatility and convenience in use, is achieved using the Extensible Markup Language XML [24, 26]. For production system and production order models the document structure definition is developed using XML Schema. XML Schema contains the definition of the XML document structure, describing the manufacturing systems production resources, i.e.: machines, inter-operational buffers, input and output buffers for all products in production system, and data on production processes, i.e.: technological routes, data on the setup times, cycle times and data on individual sequences of manufacturing operations execution for all production processes. For the purpose of transformation process automation Extensible Stylesheet Language Transformations XSLT was used [24, 26].

Implementation of the conversion process has been performed using the standalone XSLT processor as components of SWZ module software. Calculations on the data are realized with the use of XML Path Language.

Data exchange module was implemented into SWZ system. Implementation of the data transformation (integration) module in SWZ system consists of the following steps: loading the XML file, containing information on the planned production, from the IFS, validation of the loaded file based on XML Schema, transformation of the file using XSLT (data mapping and calculations using XPatch), generating an XML file for SWZ and validation of the loaded file based on XML Schema (Fig. 2).

3 Simulation Model Generator

Implementation of the method for automatic code generation of simulation models of production systems together with data about production flow for simulation and visualization uses data mapping and data transformation techniques as well. Therefore definitions of the XML document structure, containing data related to the planning/scheduling system were also developed. It was decided that the definition

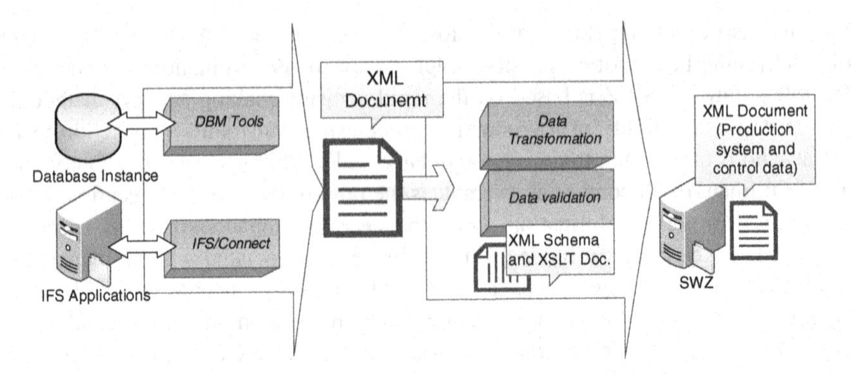

Fig. 2 IFS and SWZ integration

of the structure of the XML document will be performed with the use of XML Schema which enables to define restrictions on the mapped data and create new definitions, structure, or combine information from the different schemes (integrated systems). This gives the ability to map and transform the data obtained from both systems, containing the information required to prepare a simulation model.

XML Schema developed for the method for automatic generation of simulation models contains the definition of the XML document structure, describing the manufacturing systems production resources, i.e.: machines, inter-operational, input and output buffers for all products in system, and data on production processes, i.e.: technological routes, setup times data, cycle times and data on individual sequences of manufacturing operations execution for all production processes [24–26].

For the purpose of automation of the process of data transformation into the input file for the simulation software, Extensible Stylesheet Language Transformations called XSLT, allowing to transform an XML document into another XML document, web page, a text document or different type of file was used (Fig. 3).

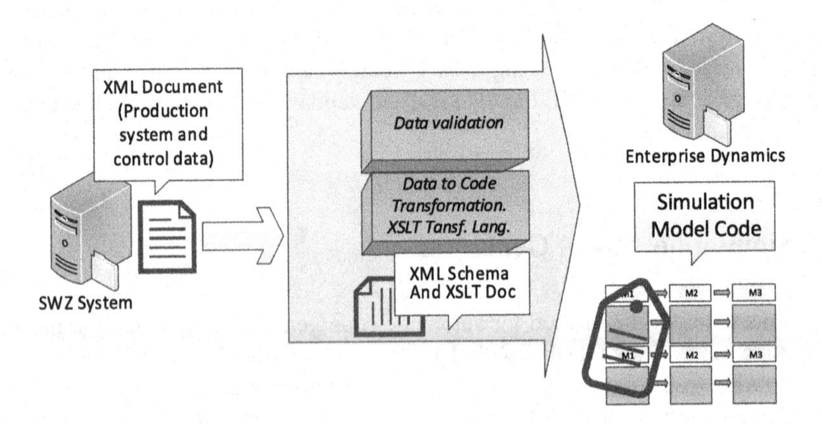

Fig. 3 XML document to 4DScript Code transformation

Based on the data from SWZ stored in the XSLT document, XSLT processor automatically transforms data to 4Dscript code input file for the commercial Enterprise Dynamics (ED) simulation system. Generated 4DScript file contains the code, responsible for creating resources that make up the manufacturing system, i.e.: production resources, buffers, warehouses, elements that generate products and information resources, i.e.: tables containing data about the setup and cycle times of operations on production resources, tables containing the schedules of resources, data on processes routes as a function of carrying out the connection between appropriate atoms in the model, the scene parameters and scripts for the implementation of process simulation and visualization by the data contained in information resources.

The script code, generating ED system internal language is a sequential process and has to be performed in the order of necessity, of defined individual layers of the model. It is, for example, impossible to generate parameterization code of a production operation cycle times on the production resources without generating the code that creates such resources in the amount resulting from the input data.

4 Practical Example

Practical verification of the proposed systems integration method based on data mapping and data transformation is implemented in the transformation module of SWZ system (which generates an input file for simulation system with a description of the model).

The data describing the production system, consisting of four resources M_1, M_2, M_3, M_4, M_5 has been imported from IFS into SWZ system (Fig. 4). An example of the production flow is shown in Fig. 5.

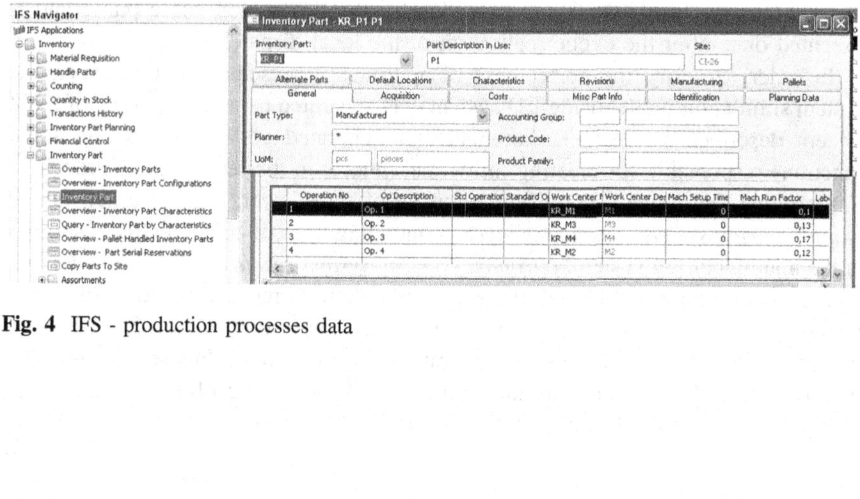

Fig. 4 IFS - production processes data

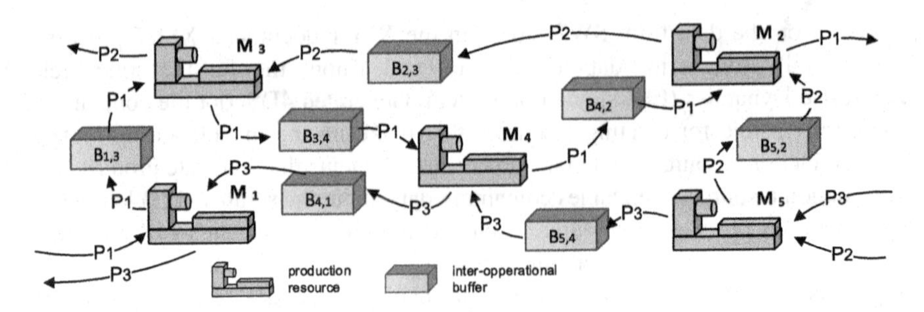

Fig. 5 System of multiassortment production

The production routes and times are recorded in the processes matrix:

$$MP_1 = \begin{bmatrix} 1 & 3 & 4 & 2 \\ 6 & 8 & 10 & 7 \\ 0 & 0 & 0 & 0 \end{bmatrix}, \; MP_2 = \begin{bmatrix} 5 & 2 & 3 \\ 4 & 5 & 6 \\ 0 & 0 & 0 \end{bmatrix}, \; MP_3 = \begin{bmatrix} 5 & 4 & 1 \\ 3 & 7 & 8 \\ 0 & 0 & 0 \end{bmatrix}, \qquad (1)$$

the first row of the matrix corresponds to the resources over which the route of the process goes; in the second line cycle times on proper resources are given; the third line contains setup times. The dispatching meta-rules containing the sequence of operations executed in the system in starting-up phase, steady state and cease phases have been generated:

$$\mathrm{R1} = \{(1,1,1);(1,3);(3,3)\}, \quad \mathrm{R2} = \{(2);(1,2);(2,1,1,1)\}, \quad \mathrm{R3} = \{(1,1);(1,2);(1,2,2)\},$$
$$\mathrm{R4} = \{(1,3);(1,3);(1,1,3)\}, \quad \mathrm{R5} = \{(2,2,3,3);(2,3);()\}.$$

The first part of the meta-rule is the starting-up rule that is executed once and assures the production flow synchronisation into the expected system cycle. The second part of the meta-rule is cyclically executed and guarantees the steady state of the system. The third part is the procedure of the production cease phase and is executed once after the cyclic realisation of the local dispatching rules.

In order to verify the correctness of approaches presented in this paper, the system simulation model using ED system was generated using the module of SWZ system described in Sect. 3. Automatically generated code contains the stored functions responsible for creating individual objects in the resulting 3D model for ED system. The generated objects compose of three layers of the target model: *manufacturing systems resources layer* - containing lines of code creating resources in the manufacturing system, *parameters of the system resources layer* - containing lines of code that parameterize the previously defined resources and *information resources layer* - containing code creating information resources, i.e.: data tables on the execution times of production operations on resources. Figure 6 presents a fragment of code that contains the information forming the objects of manufacturing systems resources layer.

```
1
Sets(CreateAtom(AtomByName([Server], Library), Model, [M2])),
setloc(18,3,0),
SetChannels (3,3,s),
 SetExprAtt(1,[czasy(Value(StripString(Name(First(c)),[P])),Value(StripString(Name(c),[M])))],s),
SetExprAtt(2,[Value(StripString(Name(First(c)),[P]))],s),
SetExprAtt(18,[openic(lrrkz(Value(StripString(Name(c),[M]))),+(mod(input(c),lrrkz(Value(StripString(Name(c),[M])),299)),1)),c)],
OnReset(s):=[Do(Inherit,CloseAllIc(c),InStrategy)],
```

Fig. 6 Generated 4DScript Code

Model was parameterized by data on the production system and the production flow together with the dispatching rules containing the sequence of operations executed on the systems resources. Prepared model is shown in Fig. 7.

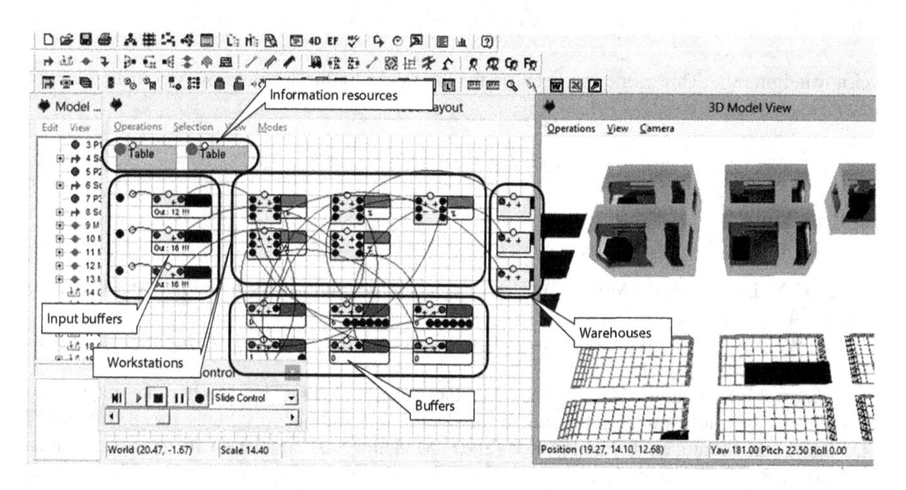

Fig. 7 Simulation model (ED system)

The results of experiments were reports containing information about the coefficient of production resources utilization, time of production cycle and lead times for each product. For the steady state cycle time was Tc = 17 time units. The coefficient of the system resources utilisation was 75 % and lead time was 1732 (for a batch size of 100). Visualization of production flow also allowed to verify the correct functioning of the production system (to detect any deadlocks and starvations), to verify sufficient capacity of inter-operational buffers and to synchronize production flow to the bottleneck (critical resource). Generated simulation model file can be downloaded from imms.home.pl/rapidsim/soco2015.mod.

5 Summary

In the paper concept of planning, scheduling and simulation production system integration has been presented. Due to the fact that simulation models are created using automatic code generation methods, it is possible to make extensive use of simulation systems to support decision-making process of acceptance of the production orders, production schedules and rapid verification of production plans. Presented method can be successfully used with most computer simulation systems, which are characterized by openness, i.e., have the ability to exchange data or to create new functionalities with the use of internal scripting programming languages, that also allow integration with other systems or electronic data interchange (EDI) systems. The direction of further works is to extend the presented method for further functionalities related to transport subsystem and the possibility of interactive entering data on the resources location on the production floor.

Acknowledgments This work has been partly supported by the Institute of Automatic Control under Grant BK/265/RAU1/2014.

References

1. Lee C-Y, Lei L, Pinedo M (1997) Current trends in deterministic scheduling. Ann Oper Res 70:1–41
2. Tan W, Khoshnevis B (2000) Integration of process planning and scheduling—a review. J Intell Manuf 11(1):51–63
3. Kalinowski K, Grabowik C, Kempa W, Paprocka I (2014) The graph representation of multi-variant and complex processes for production scheduling. Adv Mater Res 837:422–427
4. Salmela A, Montonen J, Jarvenpaa P (2007) Modeling and simulation for customer driven manufacturing system design and operations planning. In: 2007 winter simulation conference, pp 1853–1862
5. Drake GR, Smith JS, Peters BA (1995) Simulation as a planning and scheduling tool for flexible manufacturing systems. In: Proceedings of the WSC, pp 805–812
6. Corchado E, Sedano J, Curiel L, Villar JR (2012) Optimizing the operating conditions in a high precision industrial process using soft computing techniques. Expert Syst. 29:276–299
7. Sedano J, Berzosa A, Villar JR, Corchado E, De La Cal E (2011) Optimising operational costs using soft Computing techniques. Integr Comput-Aided Eng 18(4):313–325
8. Woźniak M, Graña M, Corchado E (2014) A survey of multiple classifier systems as hybrid systems. Inf Fusion 16:3–17
9. Ahmed F, Deb K (2013) Multi-objective optimal path planning using elitist non-dominated sorting genetic algorithms. Soft Comput 17(7):1283–1299
10. Burnwal S, Deb S (2013) Scheduling optimization of flexible manufacturing system using cuckoo search-based approach. Int J Adv Manuf Technol 64(5–8):951–959
11. Mathewson SC (1984) The application of program generator software and its extensions to discrete event simulation modeling. IIE Trans 16(1):3–18
12. Son YJ, Wysk RA (2001) Automatic simulation model generation for simulation-based, real-time shop floor control. Comput Ind 45(3):291–308
13. Bengtsson N et al (2009) Input data management methodology for discrete event simulation. In: Proceedings of the 2009 winter simulation conference (WSC), pp 1335–1344

14. Pidd M (1992) Guidelines for the design of data driven generic simulators for specific domains. Simulation 59(4):237–243
15. Heilala J et al (2010) Developing simulation-based decision support systems for customer-driven manufacturing operation planning. In: Proceedings of the 2010 winter simulation conference, pp 3363–3375
16. Lee YT, Luo Y (2005) Data exchange for machine shop simulation. In: Proceedings of the winter simulation conference, pp 1446–1452
17. Mertins K, Rabe M, Gocev P (2008) Integration of factory planning and ERP/MES systems: adaptive simulation models. In: Koch T (ed) Lean business systems and beyond, vol 257. Springer, Boston, pp 185–193
18. Burduk A (2012) Assessment of risk in a production system with the use of the FMEA analysis and linguistic variables. Lecture notes in computer science, vol 7209, pp 250–258
19. Chlebus E, Burduk A, Kowalski A (2011) Concept of a data exchange agent system for automatic construction of simulation models of manufacturing processes. In: Corchado E, Kurzyński M, Woźniak M (eds) HAIS 2011, vol 6679., LNCSSpringer, Heidelberg, pp 381–388
20. Lee S, Son Y-J, Wysk RA (2007) Simulation-based planning and control: from shop floor to top floor. J Manuf Syst 26(2):85–98
21. Sihn W (2003) Simulation-based configuration, animation and simulation of manufacturing systems. Progress in virtual manufacturing systems, pp 215–218
22. Rojek I, Jagodzinski M (2012) Hybrid artificial intelligence system in constraint based scheduling of integrated manufacturing ERP systems. Lecture notes in computer science, vol 7209, pp 229–240
23. Jagodzinski M (2010) IFS applications solutions for the agile enterprise. Appl Comput Sci 6 (1):54–63
24. Krenczyk D, Zemczak M (2014) Practical example of the integration of planning and simulation systems using the RapidSim software. Adv Mater Res 1036:1662–8985
25. Krenczyk D, Skolud B (2014) Transient states of cyclic production planning and control. Appl Mech Mater 657:1662–7482
26. Krenczyk D (2014) Automatic generation method of simulation model for production planning and simulation systems integration. Adv Mater Res 1036:1662–8985
27. Krenczyk D, Skołud B (2011) Production preparation and order verification systems integration using method based on data transformation and data mapping. In: Corchado E, Kurzyński M, Woźniak M (eds) HAIS 2011. Lecture notes in artificial intelligence, Series: Lecture notes in computer science, vol 6697. Springer, Heidelberg, pp 297–404

Multi-agent System to Support Decision-Making Process in Ecodesign

Ewa Dostatni, Jacek Diakun, Damian Grajewski, Radosław Wichniarek and Anna Karwasz

Abstract In the paper the application of multi-agent system to support decision-making process in ecodesign is presented. The ecodesign term is highlited either as the design problem or from the point of view of regulations. The structure of agent-system supporting the designer during the design process is showed. The basis of special kind of product model, that is the extension of standard 3D product model, called recycling-oriented product model (RmW), are described. The example results of analysis, based on real household appliance model, are presented.

Keywords Agent technology · Ecodesign · Product modeling · Product design

1 Introduction

The process of designing new products requires a number of criteria to be considered. The designers must be able to predict the final results of the project, so that when the finished product is introduced on the market, it is approved by consumers and brings the expected benefits. The basic problem related to design is that

E. Dostatni · J. Diakun (✉) · D. Grajewski · R. Wichniarek · A. Karwasz
Faculty of Mechanical Engineering and Management, University of Technology, Poznań, Poland
e-mail: jacek.diakun@put.poznan.pl

E. Dostatni
e-mail: ewa.dostatni@put.poznan.pl

D. Grajewski
e-mail: damian.grajewski@put.poznan.pl

R. Wichniarek
e-mail: radoslaw.wichniarek@put.poznan.pl

A. Karwasz
e-mail: anna.karwasz@put.poznan.pl

Á. Herrero et al. (eds.), *10th International Conference on Soft Computing Models in Industrial and Environmental Applications*, Advances in Intelligent Systems and Computing 368, DOI 10.1007/978-3-319-19719-7_40

designers have to use current information to predict future situations, which may not happen if the predictions turn out to be incorrect. The crucial aspect of decision-making in design is accurate prediction of market developments and expectations of future users. In their decisions, the designers are guided by logical factors, shaped by their knowledge and skills, and by psychological factors, which include their character, emotional and environmental influences [1]. To make the right decisions the designer should above all be able to make use of all the available information and see the analogy in phenomena or problems that are different or seemingly similar to each other. This will allow the designer to use the information from previous projects, leading to a significant reduction of project time, helping to avoid any mistakes made in other projects, etc. [2, 3]. The design process requires a large number of decisions to be made. The most important ones involve the choice of design solution options, which have a major impact on the quality of the design. Ecodesign is a new approach to product design, which assumes that products delivered on the market should have the minimum impact on the environment throughout their life cycle, and should be easy to recycle at end-of-life. This approach to design is also enforced by the legislation of individual countries and by the European Union [4–6]. It generates the need to extend the data necessary for decision-making in the design process. The designer must consider not only the criteria of product functionality, its durability, quality, and cost, but also the issues related to product disassembly (time and cost), environmental impact of all of its elements, marking of the materials, compatibility of materials, and types of joints to be used. It is therefore necessary to develop IT tools to aid the designer by providing data necessary to make informed decisions. The system, which not only provides the data but also supports the very decision-making process is the ideal solution. Such systems are based on multi-agent systems, which allow for the collection of data from dispersed sources, make decisions and which may be implanted in distributed structures of design offices.

2 The Use of Multi-agent Systems in a Distributed Environment

The terms “decentralized artificial intelligence” and “agent” emerged in literature in the early 80s, but dynamic development of IT solutions based on this paradigm ensued in the mid 90s of the 20th century. The interpretations of the agent’s role are diverse [7–10] but undoubtedly have the following common features: observation, autonomy, mobility, communication and intelligence [11].

The features described above directed their applications in various areas. Nowadays, agent technology also has a broad application in production preparation and determines the new paradigm in computer-aided engineering [12].

To perform the task it was assigned, each of the programmed agents must have a certain level of intelligence. Specific features of the agents are used in communication technologies [13–16], data acquisition, aiding operations in distributed environments [17], simulation [18] and decision-supporting systems [19–21]. Agents were also implemented for modeling and simulation of production systems [22, 23] and for integration of activities in specific manufacturing areas [24–26].

This broad scope of agent technology implementation motivated the authors to search for a new and innovative field of implementation towards supporting the activities in recycling-oriented eco design. The agents follow the designers' activities, analyze the structure of the product being designed and conduct an analysis of the product, taking into account recycling issues. The agents suggest how to improve the product design in order to make it more environmentally friendly [11].

Soft computing includes i.e. artificial neural networks, fuzzy logic, genetic algorithms, chaos theory and decentralized artificial intelligence (especially agent technology). In some areas it is necessary to use soft computing because of better performance effects [27–29]. Soft computing enables acquisition and analysis of distributed data for i.e. systems supporting negotiations. The agent ontology definition is necessary for this purpose, in order to communicate based on common language [30].

3 Functionality of the Agent System Supporting Ecodesign

The agent technology is used in the system to support decision-making in the design of recycling-oriented products. The multi-agent system is to conduct a recycling-oriented assessment of a product designed in a CAD 3D environment, based on data from a product model (RmW - recycling product model) made in a CAD 3D system. The system makes it possible to control and monitor recycling parameters of products on an ongoing basis in subsequent versions of the design, and suggest potential remodeling which may improve the parameters. It supports designers working in a distributed environment. Each designer who participates in the work on the project may work in a different node of the distributed network. He may use an IT tool (the agent system), which analyzes his work and based on the observations it conducts a recycling-oriented assessment of the product. Moreover, the system offers suggestions of construction modifications to improve the recyclability of the product [31].

The system cooperates with an existing CAD 3D extension, exporting the CAD 3D design to two XML files – a material database and a structural description of the product. The import is done automatically without the participation of the designer. Programmed agents read the information stored in the XML files and conduct the analysis. In the system the software agents observe and acquire information, exchange it, and operate in a defined, common virtual environment. The key features of agents used in the system are:

- autonomy - they make decisions and act only based on the interaction with the environment
- locality - no agent has (and does not need to have) complete information about the state of the system
- decentralization - there is no agent that directly controls the operation of the entire system.

Agents communicate with each other and observe information sent by other agents. In the multi-agent system the agent environment is defined by a distributed message exchange service, implemented as a central server. It provides agents with three basic operations:

- sending a message to the environment
- receiving a message (that matches the given template)
- reading a message (that matches the given template) – but without deleting it from the environment.

Message exchange between agents in the system was developed based on table architecture and tuple spaces [32, 33]. Every message sent in the system has a specific type and associated data. The basic template of a message is the type, although they may be more detailed. The tasks of the system are conducted by independent software agents, and each of them is in charge of a separate phase of the environmental product assessment. The agents coordinate their work using a common, central message server and the knowledge base.

The main task of the agents is to assist the designer in the selection of material, taking into account its ecological properties and compatibility with other materials, and in the selection of the type and number of joints. Based on a series of analyses it is to advise which element of the designed product has the greatest adverse impact on the environment.

4 The Structure of the Multi-agent System

Figure 1 shows the general architecture of the agent system. The agent system is responsible for maintaining the agent environment and for the access to the knowledge base agent, which allows the system to retain the information on the history of suggestion evaluation and on the history of project versions. The knowledge is used to provide suggestions to improve the recycling rate of the designed product. One agent server handles all the agents of the system. The agent server does not require any configuration, it only provides the possibility to observe the recorded knowledge base.

During the recycling-oriented analysis of the product the agent server temporarily stores messages from the agent environment. The messages are hidden and are not available to the user. It also saves the successive versions of the analyzed designs together with previously suggested modifications and complete results of

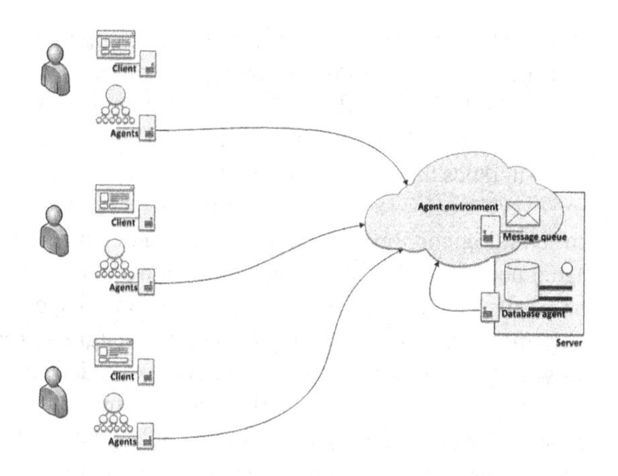

Fig. 1 General architecture of the agent system supporting ecodesign (Source: [31])

the recycling-oriented analysis. The data are necessary to demonstrate differences between the successive versions of the design, and they give the user the possibility to monitor the changes in recycling rates. They constitute the knowledge resources necessary to offer suggestions to the designer.

5 Tasks of the Agents

There are two types of agents in the system. The first type is generally referred to as agents, and the other one as microagents. Agents are: the interface agent, conversion agent, the structural agents (which consist of microagents), design modifications agent, tips agent, material agent, report agent, and the integration agent. The microagents are: order agent, update agent, joint control agent, weight update agent, agent counting the joints, agent counting the materials, and the measurement agent. The functions of all the agents are described below.

The interface agent is responsible for monitoring the work of the product designer. When the agent discovers changes in the CATIA design, it sends the design to be re-analyzed. When the analysis is complete, its results are displayed through the interface. The changes in the design are discovered based on an intermediate XML file, created whenever a CATIA file is saved.

The **conversion agent** converts the intermediate XML file from a common data interchange format to an internal format used by the rest of the system. The introduction of the conversion agent provides a greater flexibility in choosing a common data format and relieves the programmers using CATIA. The internal format is additionally described using XML Schema, which allows you to create programs that support it in any object-oriented programming language.

The **structural agents** (microagents) are a group of agents which move along the tree of the designed product and update its recycling parameters. Since the majority the system functions are defined as actions on a tree, it is easy to extend the system by new features and it is much easier to implement them. The increase in system performance is an additional benefit. The order agent guarantees that the product tree is analyzed in a strictly defined order(post-order) – it ensures that the structural agents visiting the given tree vertex had visited its successors before. The update agent updates the product tree based on intermediate results that have been generated for every tree vertex by the various structural agents. It is noteworthy that the structural agents operate parallelly and independently of each other. If a structural agent needs intermediate results of other agents, the system guarantees that they will be recalled in the correct order. The joint control agent analyzes the design coherence in terms of its joints. If some joints in the designed product were not defined and the recycling-oriented product assessment proceeded, its results could be warped. The remaining agents would complete the missing information with their own assumptions and suppositions, which in turn could mislead the designer. The weight update agent updates the information about the weight of each element of the product. If it discovers that changes were made to the designed product (e.g. an element was added or removed), it updates the calculation of weight for each element and for the entire product. The accuracy of the data is particularly important in the calculation of the recycling rate. The agent counting the joints prepares the statistics of joints used within each element and its subtree. The agent counting the materials prepares the statistics of materials used within each element and its subtree. The measurement agents calculate the measurements of the product. The agents are: the recycling rate agent, joint variations agent, and material variations agent.

Design modifications agent compares the current and previous version of the design and prepares a list of changes that were made.

Tips agent makes an algorithmic analysis of the design and creates a list of possible positive changes.

The tips agent qualifies tips as suggestions based on the history of approved hints and it checks which tips have been used by the designer in the new version of the product.

Design history agent stores the information about designs.

Suggestion history agent stores the information about tips used in all projects.

Material agent stores the material databases for each project and offers information about their compatibility.

Report agent creates reports based on the results of the analysis.

Integration agent collects the results provided by all the agents and sends them to the interface agent.

The **knowledge base agent** saves all the knowledge resources necessary to offer suggestions and tips.

As previously mentioned, the agents communicate with each other, send information and perform the assigned tasks. They support the designer in making decisions to improve the recycling properties of the designed product and they carry out the recycling-oriented product assessment. In the paper soft computing is used for acquisition of distributed data in design environment and for agent negotiations in order to impart suggestions concerning ecodesign. The ontology for agent system was also developed [34].

6 Recycling-Oriented Design with the Use of the Agent System

Modeling in CAD systems with the use of the agent system allows for a comprehensive assessment of the design solutions and improvement of their recyclability. During design, the following aspects are assessed: compatibility of materials used in product prototypes, types of joints and recyclability indicators (recycling rate, among others). The IT tool used in the recycling-oriented analysis has a modular structure. The first module is used to operate the RmW - the recycling-oriented product model in the CAD 3D environment. The other one operates the agent system (product analysis, calculation of the evaluation measurements, collection and use of information on completed projects).

The recycling-oriented assessment of each product modeled in a CAD 3D system takes place according to the same procedures:

- developing digital 3D models,
- defining geometric constraints for the designed assembly,
- developing the RmW,
- recycling-oriented product analysis.

The recycling-oriented product assessment, conducted by the system automatically already in the design phase, requires that a recycling-oriented product model (RmW) be created.

Usually a designer working on a product in a CAD system models the geometry and develops its kinematics and wiring diagrams. The designer may also perform the durability calculations of the designed product, e.g. using the finite element method (FEM). However, standard CAD 3D systems lack tools used specifically to design recycling-related properties. The existing IT solutions – both autonomous software and the specific modules (relatively recent) available in some CAD 3D systems – are mainly used to conduct a general environmental analysis (e.g. according to the LCA approach), or they are material databases of various types. The recycling-oriented product assessment requires that additional elements be introduced to the CAD system: data on the recyclability properties of the designed product. The data, together with a standard geometric 3D model, are called the recycling-oriented product model (RmW). The model consists of the following elements [31]:

- extended product structure,
- extended material attributes,
- disassembly attributes (data on the disassembly process),
- product categorization.

The product structure in RwM is based on the typical structure, created by the constructor in the CAD 3D system, and it is an extension where additional data are added. Due to the significance of the joints used in the product and their impact on the recyclability, and due to the lack of a clear representation of this construction feature in a standard 3D model, additional data were implemented, clearly and in detail describing the joints. Four basic assumptions concerning the representation of joints in the product were made:

- assumption 1 – the representation of the broadest possible group of joints used in products should be considered in the model
- assumption 2 – product components are divided into three groups: connecting elements (fasteners), combined elements and connecting-combined elements
- assumption 3 – the joints in the 3D model will be represented in a method other than geometric constraints
- assumption 4 – the model of joints will be coded in the 3D model using only typical mechanisms of 3D modeling.

The inclusion of the recycling-related issues into the tools that support the designer's work requires as well that the list of standard material attributes be expanded. This list has been expanded to include: harmfulness of material, material disposal cost, graphic designation of the material according to the standards and other legal acts on the requirements for materials used in the product.

Another element to be taken into account in the tools supporting the work of the designer is the combination of materials in the so-called groups (in which different materials are grouped together) and determining the so-called compatibility of each group of materials. Compatibility is understood here as the recyclability of a piece of the product (an entire assembly or several connected elements) as one element. Determining the compatibility of groups of materials, we indirectly determine the compatibility of two materials.

In the proposed solution we assumed that recycling of the product will involve a non-destructive disassembly. Thus, the basic (elementary) disassembly operation will be to dismantle the joints in the product. Therefore, on a given level of product structure the components are considered disassembled when all joints have been dismantled. The data characterizing the disassembly include: dismantling a joint during disassembly, the tools used in the dismantling operation, and the time needed to dismantle the joint. Dismantling a joint is a logical attribute yes/no (subject to dismantling/not subject to dismantling). In the existing regulatory framework some products (such are domestic appliances) were divided into groups and certain minimum requirements were set for their recycling rates. Assigning the product to a given group is the final element of the RmW, and it allows for a direct comparison of the product assessment values resulting from the legal regulations with the current values at a given stage of design or analysis. As a result, already at this stage we may verify whether the designed product meets the legal requirements laid down for this group, which determines whether the product may be introduced on the market. Figure 2a shows a sample view of a CAD interface where RmW is implemented.

The data described above, implemented to RmW, are the basis for the analysis conducted by the agent system. The agent system supporting green, recycling-oriented design features the following functions [31]:

- analysis of product structure, including the logical correctness of the defined joints between elements – the design is the process in which data are supplemented in stages, in certain situations; when the stored data do not allow for a clear description of the product, further analysis could yield meaningless results; the aim of this function is to detect these moments and block further analysis,
- automatic calculation and updating the weight of components
- detection of changes made to the product design by the user – product design analysis is started when modifications are detected; the user will have access to current information
- calculating statistics and recyclability indicators of the entire product and its individual elements
- detecting changes in the recycling rate and recyclability indicator – thanks to the feature the user can see which changes in the product structure caused the increase, reduction or exceeding of acceptable levels and rate of recycling
- detecting and identifying elements, inseparable and incompatible elements
- detecting and identifying elements that have the greatest negative impact on the recycling rate
- creating a numerical summary of the used materials and types of joints
- detecting the use of hazardous materials and warning the user
- suggestions and tips on changing materials or joints to improve recyclability parameters of the designed product
- creating reports.

One of the basic tasks of the system is to offer tips and suggestions to improve the product's recyclability parameters. The suggestions include:

1. changing the material
2. changing the joint from inseparable to separable (thus allowing for the disassembly).

The information that the designer approves type (1) suggestions is saved in the suggestion base (e.g. material A was changed to material B). As a result, when material A may be changed to some compatible materials, the system will arrange the suggestions according to the frequency of their actual application. Similarly with type (2) suggestions: the system saves information e.g. that joint A was changed to joint B.

Based on the value of the total recycling rate (CWR) and on the classification of the product elements, the system suggests the user to introduce changes, which will improve the value of the rate. The system suggests changing the elements, which cannot be disassembled and which are connected to incompatible materials. That may be done in two ways:

- by changing the joint to a separable and marked as subject to disassembly
- by changing the material from which the element is made to a material compatible with other materials in the inseparable joint.

The designer may choose from many types of joints and compatible materials. Suggestions offered by the system may help in making the decision. The suggestion is based on the tip, which results from the static analysis of the design, and on the data collected by the system – the history of approved suggestions. The data are used to further detail the tips found. For example, a tip may be to change the material of which element E is made: from material M to one of the compatible materials: X, Y, or Z. The historical data show that designers changed material M to material X 10 times, to material Y – 5 times, and the change to material Z has not been recorded yet. In this case, the priority suggestion will to be to change material M in element E to material X, the secondary suggestion will be to change the material to Y, and the change to material Z will remain a tip. All tips and suggestions are stored in the knowledge base of the agent system. Figure 2b shows a sample view of a agent system interface, which demonstrates the tips offered by the system during product design.

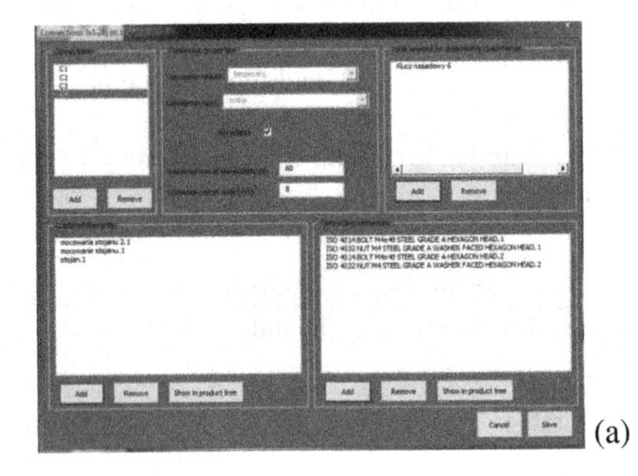
(a)

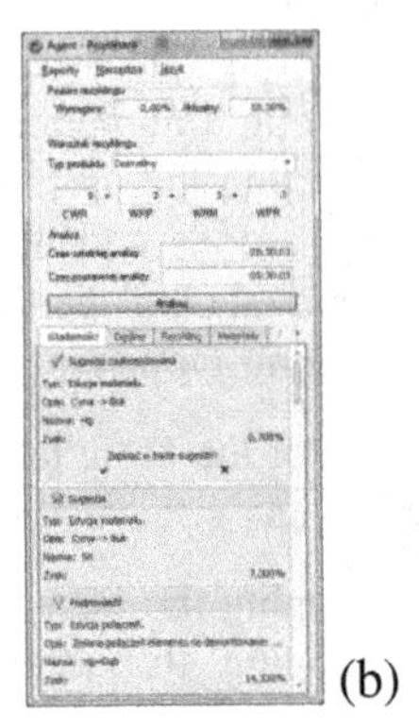
(b)

Fig. 2 Selected user interfaces view at work: a) RmW interface view, b) agent system interface – after the analysis (Source: [31])

7 Conclusions

Agents in the multiagent system have been developed on the basis of our own method of recycling-oriented product assessment, based on the total recycling rate – TRR [31]. It is the sum of three sub-indices. The sub-indices are minimants: the lower their value, the easier the recycling. It was assumed that based on the value of the total recycling rate, the classification of product elements, and knowledge gathered during previous design works, the agent system suggests design modifications to improve the recycling-oriented assessment of the designed product.

The method takes into account the cost of recycling. Two basic factors have been considered: the first is the cost of disassembly, and the second is the cost of disposal and recycling of materials used in the product. The use of the agent technology to support the design of environmentally friendly products enabled the automation of the decision-making process in design and the use of the system in design offices with distributed structure.

The presented approach to modeling (recycling-oriented product model - RmW) describes the designed product more precisely than standard way in CAD 3D modeling systems available nowadays on the market. It gives the possibility to derive another assessment measures that proposed in the paper from the RmW. Because of the possibility of RmW implementation in another CAD 3D modeling system, the design and analysis could be conducted in different than Catia design environment. The system described in the paper may be also used by the recycling-industry companies for assessment of possible ways of product dismantling in order to choose the most profitable disassembly and recycling option.

References

1. Bąbiński C (1964) Engineering in industrial design (in Polish). WNT, Warsaw
2. Ellis DO (1962) Systems Philosophy. Prentice-Hall, Englewood Cliffs N.J
3. Weiss Z., Dostatni E. Decisions made in the product design process which involve recycling, In: Knosala R (ed) conference proceedings computer integrated management (in Polish), Zakopane
4. Directive 2002/95/EC of the European Parliament and of the Council of 27 January 2003 on the restriction of the use of certain hazardous substances in electrical and electronic equipment (RoHS)
5. Directive 2005/32/EC Ecodesign for energy-using products (EuP)
6. ISO/TR 14062 (2001) Environmental Management—Integrating environmental Aspects into product design and development
7. Brenner W, Zarnekow R, Wittig H (1998) Intelligente Softwareagenten Grundlagen und Anwendungen. Springer, Berlin
8. Caglayan AK, Harrison CG (1998) Intelligente Software-Agenten, Carl Hanser Vrlag München Wien (1998)
9. Rao AS, Georgeff MP (1991) Modeling rational agents within a BDI architecture, In: Guru R (ed) Proceedings of the international conference on principles of knowledge representation and reasoning (KR-91). Morgan Kaufmann, San Mateao, pp 473–484
10. Wooldridge MJ (1999) Intelligent agents. MIT Press, Cambridge
11. Dostatni E, Diakun J, Hamrol A, Mazur W (2013) Application of agent technology for recycling-oriented product assessment. Ind Manag Data Syst 113(6):817–839
12. Van Dyke PH (1996) What can agents do in industry, and why? In: Proceedings of PAAM, London, No. April, pp 205–223
13. Bigus JP, Bigus J (1998) Constructing Intelligent agents with Java. A Programmer's guide to smart applications. Wiley Computer Publishing, New York
14. Weiss G (1999) Multi-agent systems: a modern approach to distributed artificial intelligence. MIT Press, Cambridge
15. Amruta M, Sheetal V, Mukhopadhyay D (2014) Agent based negotiation using cloud—An approach in e-commerce, In: Satapathy SC, Avadhani PS, Udgata SK (eds) 48th Annual convention of computer society of India (CSI), ICT and critical infrastructure: proceedings of ohe 48th Annual Convention Of Computer Society Of India, vol I Book Series: Advances in intelligent systems and computing, vol 248, pp 489–496

16. Aziz ASA, Hanafi SE, Hassanien AE (2014) Multi-agent artificial immune system for network intrusion detection and classification. In: DeLaPuerta JG, Ferreira IG, Bringas PG, Klett F, Abraham A, DeCarvalho ACPLF, Herrero A, Baruque B, Quintian H, Corchado E (2014) International joint conference SOCO'14-CISIS'14-ICEUTE'14, Book Series: Advances in intelligent systems and computing, vol 299, pp 145–154
17. Dent V, Harris S, Hall W, Martinez K (2001) Agent technology concepts in a heterogeneous distributed searching environment. VINE: J Inf Knowl Manag Syst 31(2):55–63
18. Aziza R, Borgi A, Zgaya H (2014 A multi-agent simulation: the case of physical activity and childhood obesity. In: Omatu S, Bersini H, Corchado JM (eds) 11th international symposium on distributed computing and artificial intelligence (DCAI), Book Series: Advances in intelligent systems and computing, vol 290, pp 359–367
19. Frayret J-M, D'Amours S, Rousseau A, Harvey S, Gaudreault J (2007) Agent-based supply chain planning in the forest products industry. Int J Flex Manuf Syst 19(4):358–391
20. Nilsson F, Darley V (2006) On complex adaptive systems and agent-based modeling for improving decision-making in manufacturing and logistics settings. Int J Oper Prod Manag 26 (12):1351–1373
21. Hilletofth P, Lättilä L (2012) Agent based decision support in the supply chain context. Ind Manag Data Syst 112(8):1217–1235
22. Jennings N, Wooldridge MJ (1998) Agent technology: foundations, applications, and market. Springer, New York
23. Shen W, Norrie DH, Barthe JP (2000) Multi-agent systems for concurrent intelligent design and manufacturing. CRC Press, London
24. Lu T, Yih Y (2001) An agent-based production control framework for multiple-line collaborative manufacturing. Int J Prod Res 39(10):2155–2176
25. Yu R, Iung B, Panetto H (2003) A multi-agent-based e-maintenance system with case-based reasoning decision support. Eng Appl Artif Intell 16:321–333
26. Park YJ, Choi HR, Kim HS (2003) Automated negotiation for order transactions of injection mold manufacturer. In: Proceedings of ICEC2003, vol 5. Pittsburgh, PA pp 488–497
27. Chen P, Zhang J (2008) Research on applications of soft computing agents. In: International symposium on intelligent information technology application workshops IITAW '08, Shanghai
28. Chen P, Zhang J, Castillo O, Melin P (2004) Hybrid intelligent system using fuzzy logic, neural networks and genetic algorithms. Nonlinear Stud 11(1):1–3
29. Jazayeriy H, Azmi-Murad M, Sulaiman N, Udzir NI (2011) A review on soft computing techniques in automated negotiation. Sci Res Essays 6(24):5100–5106
30. Khosla R (2003) Multi-layered distributed agent ontology for soft computing systems, in: knowledge-based intelligent information and engineering systems. Lecture notes in computer science, vol 2773. Springer, Berlin, pp 445–452
31. Dostatni E, Diakun J, Karwasz A, Grajewski D, Wichniarek R (2014) Ecodesign of products in CAD 3D environment with the use of agent technology (in Polish). Publishing House of Poznań University of Technology, Poznań
32. Freeman E, Hupfer S, Arnold K (1999) JavaSpaces principles, patterns, and practice. Addison Wesley, London
33. Costa P, Mottola L, Murphy AL, Picco GP (2006) Teenylime: Transiently shared tuple space middleware for wireless sensor networks, In: MidSens '06 Proceedings of the 1st international workshop on middleware for sensor networks, pp 43–48, Melbourne
34. Grajewski D, Dostatni E, Jankowiak B (2013) Knowledge acquisition method and its effective application in the process of ecodesign. Acad J Manuf Eng 11(1):44–53

Preparatory Stages of the Production Scheduling of Complex and Multivariant Products Structures

Krzysztof Kalinowski and Marcin Zemczak

Abstract The paper presents the way of creating a scheduling model for manufacturing systems where complex and multivariant structures of products are considered. Described method of scheduling tasks modelling, based on graphs, includes stages of building graphs of operations planning sequence of orders (taking into account their scheduling strategies and alternatives of realisation), input sequencing of the set of production orders and also creation of the aggregated graph of operations planning of the set of orders. The developed model is the basis for the implementation in the APS and simulation type of scheduling software, supporting production planning and scheduling. Because of NP-complete character of considered scheduling tasks, it is a potential application area of soft computing graph searching techniques.

Keywords Production scheduling · Scheduling preparatory stages · Modelling · Graphs

1 Introduction

The complexity of real production systems and realized tasks requires a considerable amount of calculations in planning and control processes, what can be noticed especially in scheduling. Those problems are mainly related to: multiassortment,

K. Kalinowski (✉)
Faculty of Mechanical Engineering, Institute of Engineering Processes Automation and Integrated Manufacturing Systems, Silesian University of Technology, Konarskiego 18A, 44-100 Gliwice, Poland
e-mail: krzysztof.kalinowski@polsl.pl

M. Zemczak
Faculty of Mechanical Engineering and Computer Science, Department of Production Engineering, University of Bielsko-Biala, Willowa 2, 43-309 Bielsko-Biala, Poland
e-mail: mzemczak@ath.bielsko.pl

Á. Herrero et al. (eds.), *10th International Conference on Soft Computing Models in Industrial and Environmental Applications*, Advances in Intelligent Systems and Computing 368, DOI 10.1007/978-3-319-19719-7_41

non-repeatable production flow, complex structures of manufacturing processes or the need for planning of both main and additional resources, with varying degrees of specialization. Scheduling in real manufacturing systems, in the vast majority belongs to the class of NP hard (NP-complete) problems so there are probably no possibilities for finding optimal solution. In such situations it is reasonable to use methods from the soft computing domain.

The problem of preparing complex model for the purpose of determining production schedules is considered in this paper. Proposed model supports hierarchical structures of manufacturing processes with alternatives of their realisation (different routes) that are generally used in ERP-class systems. A production order for a complex product is defined as a common order for both the final product and its components and parts. In such structures there is a combination of machining and assembling operations. The similar problem as a two stage assembly scheduling problem is discussed in [1, 5, 12, 13, 17]. They proved its NP-hard complexity even for two machines, and proposed different heuristics for finding a solution, for single criterions like makespan, maximum lateness, etc.

The multivariant process is understood as a process, which can be realized in different variants in the same system, depending on the current situation in the production environment, available resources (sites, tools and instrumentation), adopted manufacturing time and cost, etc. The problem of scheduling in presence of possibility of use of different production routes is considered in [3, 8, 14, 16]. They use advanced heuristics for solving problems of choosing the best routes in the manufacturing system. The variants may concern the whole manufacturing process, a subset of operations, a single operation, treatment, movement, etc. - each level of the adopted structure of the manufacturing process. Creating independent variants of the whole manufacturing process may result from the possibility of using different technologies, in which all or most of the operations are separable, implemented using different technical means. So, the choice of process variant is made rather at the planning stage and after the start of the production it is not possible to change it. Variants of the subset of operations are dependent at the disposal means of production. Depending on the level of specialization (or universality) of these means and realized production volume there is a possibility of concentration of operations and treatments or their decomposition. Accordingly, the number of operations in particular variants of the process may vary. Variants of individual operations refer to the use of the resources with the various working parameters that affect the value of the operation time, setup time and/or the cost of the operation. The processes with a multivariant structure allow adjusting the production route to the current, existing conditions, both at the planning stage, when optimizing the schedule as well as at the execution phase, when planned realization requires some changes. Analysis of the possibility of changing a variant of the process should be carried out in each decision situation in which the model of production system, the set of orders or the objective function are changed [7].

2 The Preparatory Scheduling Stages

The process of creating a solution for the scheduling process, according to the adopted model, taking into account the complexity of the configuration of resources, processes and orders was divided into three main stages. Depending on the implementation, some stages can be carried out before a schedule construction, i.e. the insertion operation to schedule, and other during its creation. In the presented method, the preparatory stages of schedule generation lead to creation of aggregated graph of operations planning sequence of the set of orders, describing the order of insertion operations to the schedule [6]. The following stages are selected:

- creation of the set of graphs of operations planning sequence, for each planned order ($Gkpo_j$); implementation of the scheduling strategy: scheduling subprocesses and operations of orders,
- input sequencing of production orders – considering the given relations between the orders, setting priorities,
- creation of the aggregated graph of operations planning sequence of the set of orders – the structure including all precedence constraints between operations, which enables searching a solutions (creating a schedule) by selected method.

2.1 Creation of a Graph of Operations Planning Sequence of an Order

The sequence of operations for scheduling of a single production order is mainly determined by the processes structure and the scheduling strategy. Scheduling strategy is usually specified in the input data, according to the accepted model of treating of the customer order. Generally, two basic strategies for scheduling orders: forward and backward are distinguished. There are also other, combined strategies, e.g. bidirectional that simultaneously use two above mentioned basic strategies [2, 10, 11, 15]. Scheduling strategies have a major impact on the sequence of the location of operations in the schedule, but they are not always sufficient to uniquely ordering of all operations. Methods of scheduling according to the forward and backward strategies for sample order Z_1 consisting of four operations and the problem of choosing variants of operations – the order route are shown in Fig. 1.

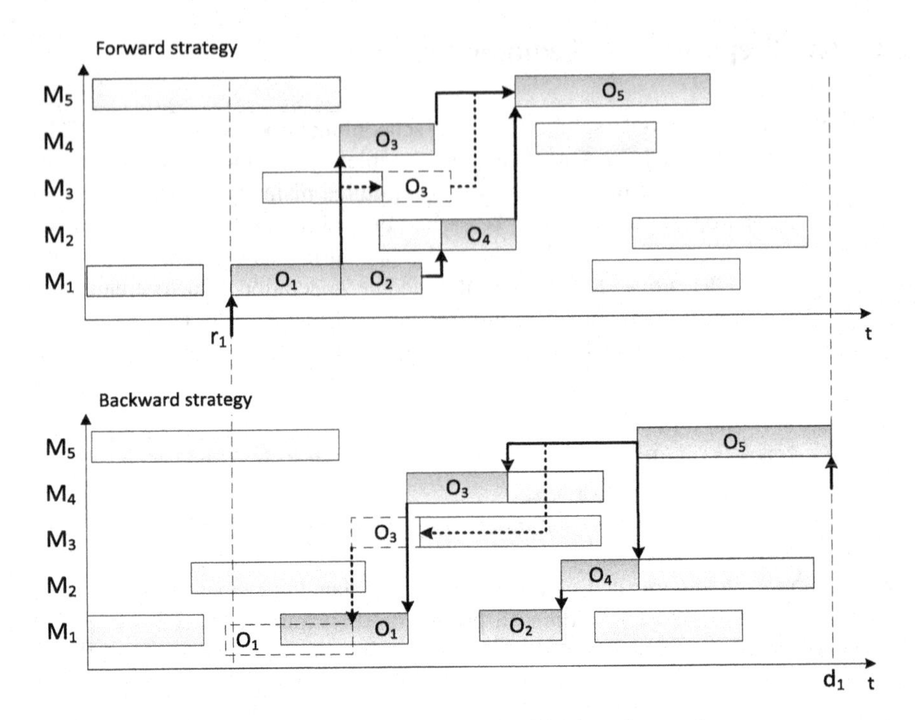

Fig. 1 Example of scheduling according to forward and backward strategies

The ready time of order execution was determined by r_1, the due date by d_1. In the case of the forward strategy operations are inserted into the schedule in order from first to last, and the beginning time of an operation is scheduled at the earliest possible date – release date of the order (r_1), finishing of suborder(s) or completion of the previous operation. In the case of backward scheduling operations are planned from the last operation to the first, so that the order has been completed as close as possible to the due date d_1. In the case of parallel operations in the sense of technology it is possible to establish any order between them by given scheduling algorithm.

On the basis of the adopted scheduling strategy and the product structure the graph of operations planning sequence of order is created. In Fig. 2 two graphs representing the same, described above order, according to forward and backward strategies are presented. The example structure of the product consists of three processes, P_1 and P_2, should be realised first (e.g. parts processing), and then P_3 (assembly of final product). The sequence of operations planning of order Z_1, according to the forward strategy, takes place from the first to the last operation, therefore, operations of processes P_1 and P_2 are scheduled first, and then followed by operations of P_3. The sequence of operations planning of order Z_2, according to the backward strategy, takes place from the last to the first operation, so operations are scheduled first of process P_3 and then of processes P_1 and P_2.

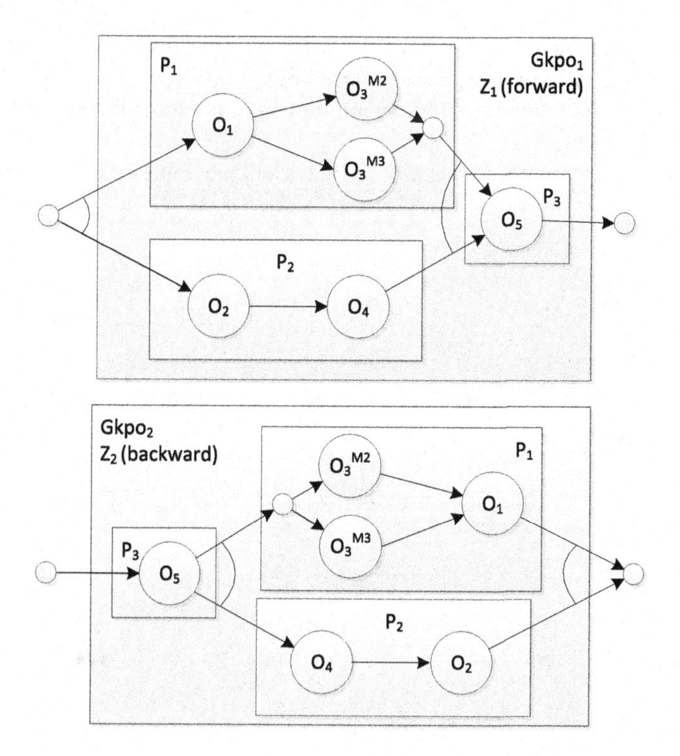

Fig. 2 Graphs of operations planning sequence for the forward and backward strategies

2.2 Input Sequencing in the Set of Production Orders

The aim of this stage is to determine the sequence of production orders consideration. The assumption, according to which any sequence of order is acceptable, causes that the number of all possible sequences in a set of production orders is equal to the number of permutations of this set. Finding the optimal sequence in that case requires a complete review of all the solutions, which in case of the larger number of orders can be too time-consuming. The space of feasible solutions can be reduced by setting precedence constraints between orders – e.g. giving the value of priority index to orders that enforces the appropriate succession of their examination. In this situation, the role of scheduling algorithms is determining the sequence of subsets of orders with the same priority value, which can significantly reduce the number of possible solutions. For models of the production system without a parallel resources and tasks with simple structures of processes, decision-making problems in the construction schedule is limited to the scheduling operation phase.

In Fig. 3 an example of relations between the three orders: Z_1, Z_2, Z_3 are shown. At the stage of input scheduling there are three types of precedence relations between orders:

- orders are equivalent - do not have assigned value of priority (degree of urgency of the order), or the values are the same (Fig. 3a),
- orders are partially arranged - there are groups of orders with the same value of priority (Fig. 3b),
- orders are fully arranged - each order has a unique value of priority (Fig. 3c).

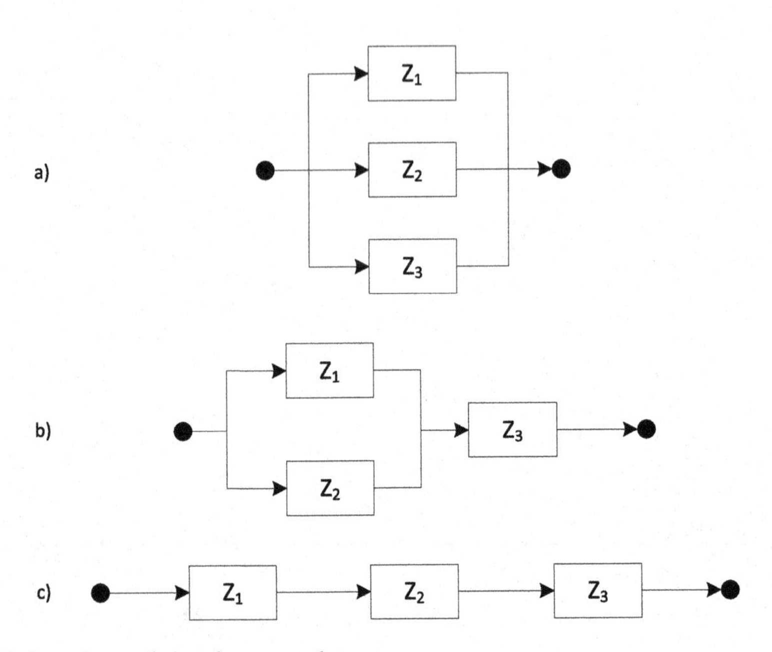

Fig. 3 Precedence relations between orders

The values of priorities can be assigned to orders arbitrarily by planners and/or dispatchers of production system. They rank the orders according to their preferences, using criteria based on experience and intuition. Despite the identified characteristics of different scheduling tasks, in which particular priority rules give much better results than others, we cannot guarantee that a particular rule will give the effective sequence of orders in the similar, considered case. Therefore, at the stage of orders sequencing it is recommended creating and comparing a larger number of different sequences, using various priority rules. If the sequence of orders is not strictly defined in the input data the set of different possible sequences can be prepared by applied sequencing algorithms. They create rankings of orders, based on the defined performance measures (e.g. total production time or due dates) using different techniques, from simple priority rules up to advanced metaheuristics.

2.3 Aggregation of Graphs of Operations Planning Sequence in the Set of Orders

Relations between orders determine a way of creating an aggregated graph of operations planning sequence in the set of orders. Figure 4 shows an example of aggregated graph for the operations of three orders Z_1, Z_2 and Z_3 according to the scheme shown in Fig. 3b.

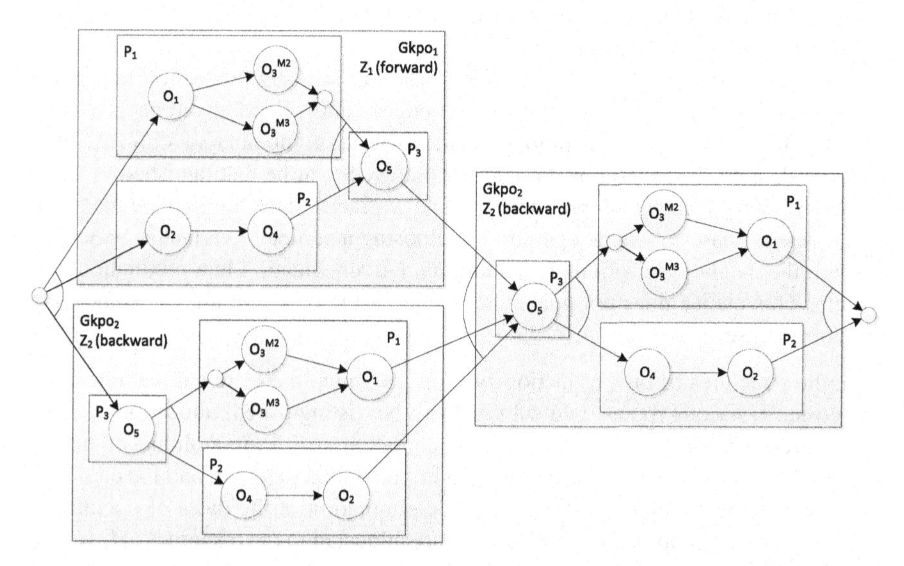

Fig. 4 The aggregated graph of operations planning sequence of the set of orders

All orders relate to the same product described by the structure that consists of three subprocesses. Order Z_1 is planned according to the forward strategy and uses prepared according to this strategy graph $Gkpo_1$. Orders Z_2 and Z_3 are planned by backward strategy, so they use graph $Gkpo_2$ prepared for planning according to backward strategy. A complete aggregated graph of operations planning sequence of the set of orders includes all operations of all orders considered in given scheduling decision problem. If there is a multivariant process in the structure of a product, represented by an AND/OR graph, the aggregated graph is also AND/OR type.

3 Searching the Aggregated Graph

In the aggregated graph of operations planning sequence of the set of orders all planning precedence constraints are included. This graph is the basis for the development of scheduling algorithms relevant to the adopted model. Based on this graph creating a feasible solution requires to determine proceeding in the following:

- selection of a general method for searching the graph – in order to search the space various algorithms, based on the depth-first search or breadth-first search schemes may be used. The searching algorithm can also use both schemes at the same time, representing the so-called mixed scheme. There are also many different algorithms for implementing these strategies, extended by additional procedures for choosing the order of search nodes of the graph, the most promising directions, e.g. a best-first search, beam search, etc. These strategies differ in the way of passing through nodes, but a common feature of them is a single execution of the action associated with a node, regardless of the number of passes through the node.
- selection method of determining the order of visiting nodes of the graph processes and operations on the parallel edges. This selection can be supported by different schedule generation schemes (SGS). A number of different SGS were defined. In general, serial and parallel SGS can be distinguished.
- selection of variants of production flow: variants of technology, variants of processes phases or single operations. Choosing a particular variant is associated with the local optimization, depending on a given situation in a production system. This choice does not guarantee obtaining the best solution (schedule) in a global sense.

In the structures of the production systems, in which there are parallel resources and possible various routes, two subtasks can be distinguished: allocation of tasks to resources and scheduling. Their implementation can be monolithic or hierarchical. In the monolithic approach they are implemented together, and the choice of resource is done simultaneously with the determination of the dates of operations. In the hierarchical approach, all operations are allocated to the resources at first, and then their starting times are calculated.

4 Summary

The paper presents a modelling method of production scheduling tasks that include complex structures and alternatives of routes. In particular, preparatory stages that lead to creation of the aggregated graph of operations planning sequence of the set of orders are specified. In the process of building a schedule in that conditions, which is treated as a multi-step decision-making process, a number of subproblems affecting on the value of the final solution is solved. The major decision problems that should be considered when designing algorithms for searching the developed graph are described.

Determination of solutions in that multistage process can be accomplished using many different strategies and algorithms, but the presence of multiple optimizations of subtasks may lead to application of a "greedy" type of algorithms. Greedy algorithm at any stage of the proceedings makes most promising choice of partial solutions. Searching local optimality of solutions may concern e.g. the optimal path in the graph of operations planning of orders and problems of operations, processes and parallel orders handling. The basic problem in the optimisation of the various stages of the construction schedule is that the suboptimality of decisions does not guarantee the global optimality of the created solutions. The accuracy of decisions taken in sequence, however, can be verified only after the completion of the schedule.

References

1. Allahverdi A, Al-Anzi FS (2006) A PSO and a Tabu search heuristics for the assembly scheduling problem of the two-stage distributed database application. Comput Oper Res 33:1056–1080
2. Artigues Ch, Lopez P, Ayache PD (2005) Schedule generation schemes for the job-shop problem with sequence-dependent setup times: dominance properties and computational analysis. Ann Oper Res 138: 21–52
3. Brandimarte P (1993) Routing and scheduling in a flexible job shop by tabu search. Ann Oper Res 41: 157–183
4. Dzitkowski T, Dymarek A (2014) Active reduction of identified machine drive system vibrations in the form of multi-stage gear units. Mechanika 20(2):183–189
5. Hariri AMA, Potts CN (1997) A branch and bound algorithm for the two-stage assembly scheduling problem. Eur J Oper Res 103(3):547–556
6. Kalinowski K, Grabowik C, Kempa W, Paprocka I (2014) The graph representation of multivariant and complex processes for production scheduling. Adv Mater Res 837:422–427
7. Kalinowski K (2011) Decision making stages in production scheduling of complex products. J Mach Eng 11(1–2): 68–77
8. Kalinowski K (2009) Scheduling of production orders with assembly operations and alternatives. In: Proceedings of the 19th international conference on flexible automation and intelligent manufacturing, FAIM 2009, Middlesbrough, p 85
9. Kaliszewski I (2006) Soft computing for complex multiple criteria decision making. In: International series in operations research & management science, vol 85
10. Kim J, Ellis R (2010) Comparing schedule generation schemes in resource-constrained project scheduling using elitist genetic algorithm. J Constr Eng Manage 136: 160–169
11. Klein R (2000) Bidirectional planning: improving priority rule-based heuristics for scheduling resource-constrained projects. Eur J Oper Res 127: 619–638
12. Navaei J, Fatemi Ghomi SMT, Jolai F, Shiraqai ME, Hidaji H (2013) Two-stage flow-shop scheduling problem with non-identical second stage assembly machines. Int J Adv Manuf Technol 69(9–12) 2215–2226
13. Potts CN, Sevast'janov SV, Strusevich VA, Van Wassenhove LN, Zwaneveld CM The two-stage assembly scheduling problem: complexity and approximation. Oper Res 43(2): 346–355
14. Rossi A, Dini G (2010) Flexible job-shop scheduling with routing flexibility and separable setup times using ant colony optimisation method. Expert Syst Appl 37(1):678–687
15. Scholl A, Klein R (1997) SALOME: a bidirectional branch-and-bound procedure for assembly line balancing. INFORMS J Comput Fall 9: 319–334
16. Yazdani M, Amiri M, Zandieh M (2010) Flexible job-shop scheduling with parallel variable neighborhood search algorithm. Expert Syst Appl 37(1):678–687
17. Zemczak M, Kalinowski K (2014) Computer aided order sequencing in a mixed-model assembly system. Zarządzanie Przedsiębiorstwem 2014(2): 42–48
18. Zemczak M, Krenczyk D (1036) A new procedure of production orders sequencing in mixed-model production systems. Adv Mater Res 864–868:2014
19. Zolkiewski S Testing composite materials connected in bolt joints. J Vibroeng 13(4): 817–822

Author Index

A
Akaichi, Jalel, 15
Alemany, J., 27
Alonso, Pedro, 229
Álvarez, David, 101
Álvarez, María Luz Alonso, 251
Aranda-Corral, Gonzalo A., 79
Arroyo, Ángel, 117
Auñón, Pablo García, 317
Ayllón, David, 261

B
Banković, Zorana, 153
Boaventura Cunha, J., 307
Botti, V., 27, 191
Bouchet, Agustina, 229
Burduk, Anna, 427

C
Cabanes, Itziar, 285
Calvo-Rolle, José Luis, 273
Camara, Monica, 241
Carmona, C.J., 387
Carrascosa, C., 203
Castaño, O., 341
Casteleiro-Roca, José Luis, 273
Celorrio, Luis, 143
Chira, Camelia, 241
Cho, Sung-Bae, 215
Cocaña-Fernández, Alberto, 261
Contraş, Diana, 165
Corchado, Emilio, 65, 117, 273

D
de la Cal, E.A., 49
de la Cal, Enrique, 39
de Moura Oliveira, P.B., 307
del Val, E., 27, 191
Diakun, Jacek, 463
Díaz, Irene, 229
Díaz, Joaquín Borrego, 79
Díaz-Villar, P., 133
Diering, Magdalena, 415
Dolezel, Petr, 91
Dostatni, Ewa, 463
Dyczkowski, Krzysztof, 415

E
Enrique Sierra, J., 329
Estévez, Julián, 297

F
Fernández, Ramón, 101
Florido, E., 341
Freire, Hélio, 307

G
Gaber, Tarek, 375
García, Jesus Manuel de la Cruz, 317
Garrido, E., 353
Gil-Pita, Roberto, 261
González, Jerónimo, 251
González, Silvia, 251
González, Víctor M., 39
Górski, Filip, 403
Grajewski, Damian, 463
Graña, Manuel, 297
Griol, David, 3, 365

H
Hamrol, Adam, 403, 415
Hassanien, Aboul Ella, 375
Herrero, Álvaro, 49, 65, 117
Huang, Lan, 111

I
Irigoyen, Eloy, 285

Á. Herrero et al. (eds.), *10th International Conference on Soft Computing Models in Industrial and Environmental Applications*, Advances in Intelligent Systems and Computing 368, DOI 10.1007/978-3-319-19719-7

J
Jagodzinski, Mieczyslaw, 451
Jiménez-Fernández, S., 133
Julian, V., 203

K
Kalinowski, Krzysztof, 475
Karwasz, Anna, 463
Krenczyk, Damian, 451
Kuczko, Wiesław, 403

L
Larrea, Mikel, 285
Limam, Hela, 15
López-García, Pedro, 153
Lopez-Guede, Jose Manuel, 297
Luengo, J., 387

M
Mallol-Poyato, R., 133
Manzanedo, Miguel A., 65
Mariska, Martin, 91
Martínez-Álvarez, F., 341
Martínez, C., 191
Matei, Oliviu, 165
Menéndez, Manuel, 39
Molina, José Manuel, 3, 365
Montes, Susana, 229
Muñoz-Velasco, Emilio, 179

O
Olivas, J.A., 353

P
Páez, Juan Galán, 79
Pawlewski, Pawel, 439
Peñas, Matilde Santos, 317
Pérez, Juan Albino Méndez, 273
Piñón-Pazos, Andrés José, 273
Prieto, Carlos, 241
Puente, C., 353

Q
Quirós, Pelayo, 229

P
Ranilla, José, 261
Rincon, J.A., 203

S
Sáiz-Bárcena, Lourdes, 65
Salcedo-Sanz, S., 133
Sánchez, Lidia, 101
Sánchez, Luciano, 261
Santos, Matilde, 329
Sedano, Javier, 39, 49, 241, 251
Snasel, Vaclav, 375
Sobrino, A., 353
Solteiro Pires, E.J., 307

T
Terán, Joaquín, 251
Tharwat, Alaa, 375
Tricio, Verónica, 117
Troncoso, A., 341

V
Vălean, Honoriu, 165
Valero, S., 27
Vergara, P.M., 49
Villar, José Ramón, 39, 49, 241, 251

W
Wałęga, Przemysław, 179
Wang, Guishen, 111
Wichniarek, Radosław, 403, 463

Y
Yu, Jae-Min, 215

Z
Zemczak, Marcin, 475
Zubizarreta, Asier, 285

Printed in the United States
By Bookmasters